How Humans Evolved

How Humans Evolved

Second Edition

Robert Boyd
Joan B. Silk

UNIVERSITY OF CALIFORNIA, LOS ANGELES

W. W. Norton & Company
New York London

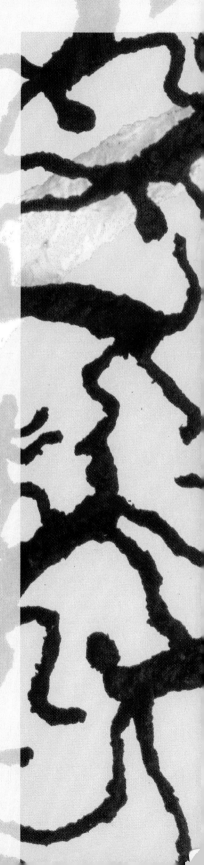

Copyright © 2000, 1997 by W. W. Norton & Company, Inc.

Printed in the United States of America

The text of this book is composed in Bembo with the display set in Phaistos
Composition by UG / GGS Information Services, Inc.
Manufacturing by Courier—Kendallville
Editor: John Byram
Copy editor: Kathryn Talalay
Project editor: Lynne Cannon-Menges
Managing editor: Jane Carter
Manufacturing director: Roy Tedoff
Book design by Maura Rosenthal
Cover illustration: Courtesy of the Getty Conservation Institute, Los Angeles, California, © 1999 The
J. Paul Getty Trust

Library of Congress Cataloging-in-Publication Data

Boyd, Robert, Ph. D.
  How humans evolved / Robert Boyd, Joan B. Silk. — 2nd ed.
    p.   cm.
  Includes bibliographical references and index.
  ISBN 0-393-97477-4 (pbk.)
  1. Human evolution.   I. Silk, Joan B.   II. Title.
GN281.B66   2000   99-28071
599.93′8—dc21

W. W. Norton & Company, Inc., 500 Fifth Avenue, New York, N.Y. 10110
                  www.wwnorton.com

W. W. Norton & Company Ltd., 10 Coptic Street, London WC1A 1PU

1 2 3 4 5 6 7 8 9 0

For Sam and Ruby

Contents

NATALITY?

Preface

When we wrote the first edition of *How Humans Evolved*, we had two major goals. First, we wanted to create a text that emphasized the processes that have shaped human evolution. It would be a text that integrated recent theoretical and empirical work in evolutionary theory, population genetics, and behavioral ecology with information about the hominid fossil record. Second, it would be a book that seriously considered the role of evolution in shaping contemporary human behavior. Although there is little consensus (and considerable controversy) about evolutionary approaches to human behavior, we wanted to produce a book that confronted these issues openly and clearly. The positive response to the textbook indicates that many of our colleagues shared this vision.

In the second edition of *How Humans Evolved* our objectives remain much the same. However, in the years that have elapsed since the preparation of the first edition, there have been many important new developments in human evolutionary studies. New fossils have been discovered that change our understanding of the history of the human lineage. Emerging evidence about primate cognitive abilities has made us question certain assumptions about the evolution of mind. New theory and data have been applied to understanding the evolution of additional stages in the human life cycle. These developments, and many others, have prompted us to rewrite Chapters 9 and 13 entirely, add new boxes to several chapters, and make substantive changes to a number of chapters in Parts II, III, and IV. New readings from the literature have been selected for nearly half of the chapters, and the list of references for further reading has been updated. The second edition has also given us an opportunity (hopefully) to eliminate all of the errors that escaped our detection and the more competent proofreading by our editor, reviewers, and an exceptional copyeditor.

In the second edition, we have retained features that users of the book have found useful. These include the "key idea" statements, which are set in italics and marked in the margin with a small icon, and the discussion questions at the end of each chapter. The key idea statements are meant to provide a concise statement of the main idea in the material that follows. We recommend that students use the key ideas like yellow highlighters—to help keep track of important concepts and facts and to structure their review of the material. Important and unfamiliar terms are defined in the text when they are first used, and are also defined in the glossary. The discussion questions that appear at the end of each chapter are meant to help students synthesize material presented in the text. Some questions are meant to help students review factual material, and others are intended to help students understand the processes or theoretical principles described in the chapter. Some questions are open-ended and designed to encourage students to think about what the ideas mean. Students have found these questions useful in mastering the material and preparing for exams.

Each chapter of *How Humans Evolved* includes one short reading that is relevant to the material in the chapter. These readings are drawn from a diverse set of sources, ranging from technical articles in scientific journals to more popular literature. These readings are meant to broaden the range of material included in the textbook, allowing us to explore certain topics in more detail, and providing readers with a more direct picture of how research is done. The list of references for further reading at the end of each chapter provides a starting point for students who want to delve more deeply into the material covered in that chapter.

ANCILLARY MATERIALS

For instructors, there is an Instructor's Manual prepared by Elizabeth Erhart of Southwest Texas State University, which includes a detailed outline of each chapter, answers to all of the discussion questions, and a test bank. The test questions are also available on diskette in MS-DOS and Macintosh formats, and will be provided free of charge upon adoption of this textbook. The publisher has also created full-color transparencies of the figures included in the book for instructor use.

ACKNOWLEDGMENTS

The new edition benefited greatly from constructive, critical readings by a number of our colleagues, users of the book, and nonusers. We are especially grateful to Tom Plummer, who again provided valuable help with Part III, and to Patricia Wright for extremely helpful comments on the whole of Part II. Beverly Strassman and Daniel Povinelli reviewed new material based on their work. We thank the following reviewers of the book: Andrew Irvine and Douglas Crews for Chapters 1 through 3; Mark Stoneking and Darrell La Lone for Chapters 2 through 4; James Paterson, Thad Bartlett, Marilyn Norconk, and Patricia Wright for Chapters 5 through 7; Margaret Clarke, Sharon Gursky, and Horst Steklis for Chapters 7 through 9; Mark Griffin, Clark Larsen, and Ann Palkovich for Chapters 13 through 15; Joan Stevenson for Chapters 16 through 18; and Lynette Leidy and Rebecca Storey for Chapters 17 through 19. Although we are certain that we have not satisfied all those who read and commented on parts of the book, we found all of the comments to be very helpful as we revised the text.

Richard Klein provided us with many exceptional new drawings of fossils that appear in Part III, an act of generosity that we deeply appreciate. We also give special thanks to Neville Agnew and the Getty Conservation Institute, which granted us permission to use the image of the Laetoli footprints on the cover of the book and allowed us to publish several photographs of the conservation project.

We remain grateful to our colleagues who provided invaluable help with the first edition of the book by reviewing drafts, providing access to unpublished material, and supplying us with information: Part I—Mark Ridley, Alan Rogers, Scott Carroll; Part II—Dorothy Cheney, Robin Dunbar, Lynn Fairbanks, Sandy Harcourt, Joseph Manson, Susan Perry, Robert Seyfarth, Phyllis Lee, and John Mitani; Part III—Leslie Aiello, Glenn Conroy, Richard Klein, Henry McHenry, Steve Pinker, Tom Plummer, Tab Rasmussen, Alan Walker, and Tim White; and Part IV—Monique Borgerhoff Mulder, Martin Daly, Kristin Hawkes, Nancy Levine, Jeff Long, Jocelyn Peccei, Frank Sulloway, Don Symons, and Margo Wilson. We also appreciate the detailed comments provided by several reviewers who read the first edition at the request of our publishers, particularly Barry Bogin, Richard Klein, Eric Smith, Craig Stanford, and several anonymous reviewers.

Many users of the book commented on the quality of the illustrations. For this we must thank the many friends and colleagues who allowed us to use their photographs: Bob Bailey, Nick Blurton Jones, Sue Boinski, Monique Borgerhoff Mulder, Scott Carroll, Marina Cords, Robert Gibson, Peter Grant, Kim Hill, Kevin Hunt, Lynne Isbell, Charles Janson, Nancy Levine, Carlão Limeira, Joe Manson, Bill McGrew, John Mitani, Claudio Nogueira, Susan Perry, Craig Stanford, Karen Strier, Alan Walker, Katherine West, and John Yellen. The National Museums of Kenya kindly allowed us to reprint a number of photographs. In addition, we thank Richard Byrne who provided a new photograph for the second edition.

We also acknowledge the thousands of students and dozens of teaching assistants at UCLA who have used various versions of this material over the years. Student evaluations of the original lecture notes, the first draft of the text, and the first edition have been helpful as we revised and rewrote various sections. Our teaching assistants have helped us to identify many parts of the text that needed to be clarified, corrected, or reconsidered.

We thank all the people at Norton who helped us produce this book, particularly our excellent editor, John Byram. We thank Paul Fyfe for help in coordinating the new readings and obtaining illustrations; Kathryn Talalay for copyediting expertise; and Elizabeth Erhart for preparing the revised Instructor's Manual.

Finally, we would like to thank our family and friends for their contributions to both the first and second editions of this book. They tolerated our preoccupation with the project from its inception and celebrated the delivery of the first edition. We are grateful for their continued support.

P R O L O G U E

Why Study Human Evolution?

Origin of man now proved—Metaphysics must flourish—He who understand baboon would do more toward metaphysics than Locke.

—CHARLES DARWIN, *M Notebook,* August 1838

In 1838, Charles Darwin discovered the principle of evolution by natural selection and revolutionized our understanding of the living world. Darwin was 28 years old, and it was just two years since he had returned from a five-year voyage around the world as a naturalist on the HMS *Beagle* (Figure 1). Darwin's observations and experiences during the journey had convinced him that biological species change through time and that new species arise by the transformation of existing ones, and he was avidly searching for an explanation of how these processes worked. In late September of the same year, Darwin read Thomas Malthus's *Essay on Population,* in which Malthus (Figure 2) argued that human populations invariably grow until they are limited by starvation, poverty, and death. Darwin realized that Malthus's logic also applied to the natural world, and this intuition inspired the conception of his theory of evolution by natural selection. In the intervening century and a half, Darwin's theory has been augmented by discoveries in genetics and

Figure 1

When this portrait of Charles Darwin was painted, he was about 30 years old. He had just returned from his voyage on the HMS *Beagle* and was still busy organizing his notes, drawings, and vast collections of plants and animals.

Figure 2

Thomas Malthus was the author of *Essay on Population,* a book Charles Darwin read in 1838 and that profoundly influenced the development of his theory of evolution by natural selection.

amplified by studies of the evolution of many types of organisms. It is now the foundation of our understanding of life on earth.

This book is about human evolution, and we will spend a lot of time explaining how natural selection and other evolutionary processes have shaped the human species. Before we begin, it is important to consider why you should care about this topic. Many of you will be working through this book as a requirement for an undergraduate class in biological anthropology, and will read the book in order to earn a good grade. As instructors of a class like this ourselves, we approve of this motive. However, there is a much better reason to care about the processes that have shaped human evolution: understanding how humans evolved is the key to understanding why people look and behave the way they do.

The profound implications of evolution for our understanding of humankind were apparent to Darwin from the beginning. We know this today because he kept notebooks in which he recorded his private thoughts about various topics. The quotation that begins this prologue is from the *M Notebook,* begun in July 1838, in which he jotted down his ideas about humans, psychology, and the philosophy of science. In the 19th century, metaphysics involved the study of the human mind. Thus Darwin was saying that, since he believed humans evolved from a creature something like a baboon, it followed that an understanding of the mind of a baboon would contribute more to an understanding of the human mind than would all of the works of the great English philosopher, John Locke.

Darwin's reasoning was simple. Every species on this planet has arisen through the same evolutionary processes. These processes determine why organisms are the way they are by shaping their morphology, physiology, and behavior. The traits that characterize the human species are the result of the same evolutionary processes that created all other species. If we understand these processes, and the conditions under which the human species evolved, then we will have the basis for a scientific understanding of human nature. Trying to comprehend the human mind without an understanding of human evolution is, as Darwin wrote in another notebook that October, "like puzzling at astronomy without mechanics." By this, Darwin meant that his theory of evolution could play the same role in biology and psychology that Isaac Newton's laws of motion had played in astronomy. For thousands of years, stargazers, priests, philosophers, and mathematicians had struggled to understand the motions of the planets without success. Then, in the late 1600s, Newton discovered the laws of mechanics, and showed how all of the

Figure 3

Sir Isaac Newton discovered the laws of celestial mechanics, a body of theory that resolved age-old mysteries about the movements of the planets.

intricacies in the dance of the planets could be explained by the action of a few simple processes (Figure 3).

In the same way, understanding the processes of evolution enables us to account for the stunning sophistication of organic design and the diversity of life, and to understand why people are the way they are. As a consequence, understanding how natural selection and other evolutionary processes shaped the human species is relevant to all of the academic disciplines that are concerned with human beings. This is a vast intellectual domain that includes medicine, psychology, the social sciences, and even the humanities. Beyond academia, understanding our own evolutionary history can help us answer many questions that confront us in everyday life. Some of these questions are relatively trivial. Why do we sweat when hot or nervous? Why do we crave salt, sugar, and fat, even though large amounts of these substances cause disease (Figure 4)? Why are we better marathon runners than mountain climbers? Other questions are more profound. Why do only women nurse their babies? Why do we grow old and eventually die? Why do people look so different around the world? As you shall see, evolutionary theory provides answers or insights about all of these questions. Aging, which eventually leads to death, is an evolved characteristic of humans and most other creatures. Understanding how natural selection shapes the life histories of organisms tells us why we are mortal, why our life span is about 70 years, and why other species live shorter lives. In an age of horrific ethnic conflicts and growing respect for multi-

Figure 4

A strong appetite for sugar, fat, and salt may have been adaptive for our ancestors who had little access to sweet, fatty, and salty foods. We have inherited these appetites and have easy access to these foods. As a consequence, many of us suffer from obesity, high blood pressure, diabetes, and heart disease.

cultural diversity, we are constantly reminded of the variation within the human species. Evolutionary analyses tell us that genetic differences between human groups are relatively minor, and that our notions of race and ethnicity are culturally constructed categories, not biological realities.

All of these questions deal with the evolution of the human body. However, understanding evolution is also an important part of our understanding of human behavior and the human mind. The claim that understanding evolution will help us to understand contemporary human behavior is much more controversial than the claim that it will help us to understand how human bodies work. But it should not be. The human brain is an evolved organ of great complexity, just like the endocrine system, the nervous system, and all of the other components of the human body that regulate our behavior. Understanding evolution helps us to understand our minds and behavior because evolutionary processes forged the brain that controls human behavior, just as they forged the brain of the chimpanzee and the salamander.

One of the great debates in Western thought centers on the essence of human nature. One view is that people are basically honest, generous, and cooperative creatures who are corrupted by an immoral economic and social order. The opposing view is that we are fundamentally amoral, egocentric beings whose antisocial impulses are held in check by social pressures. This question turns up everywhere. Some people believe that children are little barbarians who are civilized only through sustained parental effort, while others think that children are gentle beings that are socialized into competitiveness and violence by exposure to negative influences like toy guns and violent TV programs (Figure 5). The same dichotomy underpins much political and economic thought. Economists believe that people are rational and selfish, while other social scientists, particularly anthropologists and sociologists, question and sometimes reject this assumption. We can raise an endless list of interesting questions about human nature. Does the fact that in most societies women rear children and men make war mean that men and women differ in their innate predispositions? Why do men typically find younger women attractive? Why do some people neglect and abuse their children, while others adopt and lovingly raise children who are not their own?

Understanding human evolution does not reveal the answers to all of these questions, or even provide a complete answer to any one of them. However, as

Figure 5

One of the great debates in Western thought focuses on the essential elements of human nature. Are people basically moral beings, corrupted by society? Or fundamentally amoral creatures socialized by cultural conventions, social strictures, and religious beliefs?

we shall see, it can provide useful insights about all of them. An evolutionary approach does not imply that behavior is "genetically determined," or that learning and culture are unimportant. In fact, we will argue that learning and culture play crucial roles in human behavior. Behavioral differences among peoples living in different times and places mainly result from flexible adjustments to different social and environmental conditions. Understanding evolution is useful precisely because it helps us to understand why humans respond in different ways to different conditions.

Overview of the Book

Humans are the product of organic evolution. By this we mean that there is an unbroken chain of descent that connects every living human being to a bipedal, apelike creature that walked through tall grasses of the African savanna 3 million years ago (mya), to a monkeylike animal that clambered through the canopy of great tropical forests covering much of the world 35 mya, and finally to a small, egg-laying, insect-eating mammal that scurried about at night during the age of the dinosaurs, 100 mya. To understand what we are now, you have to understand how this transformation took place. We tell this story in four parts.

Part One: How Evolution Works

More than a century of hard work has given us a good understanding of how evolution works. The transformation of apes into humans involved the assembly of many new, complex adaptations. For example, in order for early humans to walk upright on two legs, there had to be coordinated changes in many parts of their bodies, including their feet, legs, pelvis, backbone, and inner ear. Understanding how natural selection gives rise to such complex structures, and why the genetic system plays a crucial role in this process, are essential for understanding how new species arise. An understanding of these processes also allows us to reconstruct the history of life from the characteristics of contemporary organisms.

Part Two: Primate Behavior and Ecology

In the second part of the book, we consider how evolution has shaped the behavior of nonhuman primates. This helps us to understand human evolution in two ways. First, humans are members of the primate order, and we are more similar to other primates, particularly the great apes, than we are to wolves, raccoons, or other mammals. Studying how primate morphology and behavior are affected by ecological conditions helps us to determine what our ancestors might have been like and how they may have been transformed by natural selection. Second, we study primates because they are an extremely diverse order and are particularly variable in their social behavior. Some are solitary, others live in monogamous pairs, and some live in large groups that contain many adult females and males. Data derived from studies of these species help us to understand how social behavior is molded by natural selection. We can then use these insights to interpret the hominid fossil record and the behavior of contemporary people (Figure 6).

Figure 6

We will draw on information about the behavior of living primates, like this chimpanzee, to understand how behavior is molded by evolutionary processes, to interpret the hominid fossil record, and to draw insights about the behavior of contemporary humans.

Part Three: The History of the Human Lineage

General theoretical principles are not sufficient to understand the history of any lineage, including our own. The transformation of a shrewlike creature into the human species involved many small steps, and each step was affected by specific environmental and biological circumstances. To understand human evolution, we have to reconstruct the actual history of the human lineage and the environmental context in which these events occurred. Much of this history is chronicled in the fossil record. These bits of mineralized bone, painstakingly collected and reassembled by paleontologists, document the sequence of organisms that link early mammals to modern humans. Complementary work by geologists, biologists, and archaeologists allows us to reconstruct the environments in which the human lineage evolved (Figure 7).

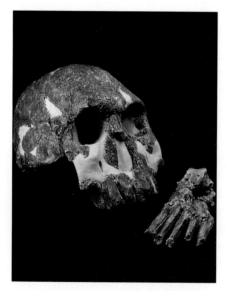

Figure 7

Fossils painstakingly excavated from many sites in Africa, Europe, and Asia provide us with a record of our history as a species. This creature, *Homo habilis*, lived in East Africa about 2 million years ago, walked upright on two legs, made and used stone tools, and had a brain substantially bigger than the brains of contemporary apes. This is the oldest known member of our own genus.

Figure 8

A Bushman pauses to gather berries.

Part Four: Evolution and Modern Humans

Finally, we turn our attention to modern humans and ask why we are the way we are. Why is the human species so variable? How do we acquire our behavior? Why do women undergo menopause? How has evolution shaped human psychology and behavior? How do we choose our mates? Why do we care for some children and neglect others? We will explain how an understanding of evolutionary theory and a knowledge of human evolutionary history provide a basis for addressing questions like these (Figure 8).

The history of the human lineage is a great story, but it is not a simple one. The relevant knowledge is drawn from many disciplines in the natural sciences, such as physics, chemistry, biology, and geology, and from the social sciences, mainly anthropology, psychology, and economics. Learning this material is an ambitious task, but it offers a very satisfying reward. The better you understand the processes that have shaped human evolution and the historical events that took place in the human lineage, the better you will understand how we came to be and why we are the way we are.

PART ONE

How Evolution Works

Adaptation by Natural Selection

Explaining Adaptation before Darwin

Animals and plants are adapted to their conditions in subtle and marvelous ways.

Even the casual observer can see that organisms are well suited to their circumstances. For example, fish are clearly designed for life under the water, and certain flowers are designed to be pollinated by particular species of insects. More careful study reveals that organisms are more than just suited to their environments—they are complex machines, made up of many exquisitely constructed components, or **adaptations**, that interact to help the organism survive and reproduce.

The human eye provides a good example of an adaptation. Eyes are amazingly useful: they allow us to move confidently through the environment, to locate critical resources like food and mates, and to avoid dangers, like predators and cliffs. Eyes are extremely complex structures made up of many interdependent parts (Figure 1.1). Light enters the eye through a transparent opening, then passes

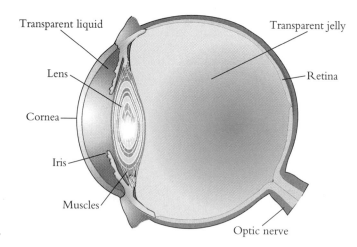

Figure 1.1

A cross section of the human eye.

through a diaphragm called the iris, which regulates the amount of light entering the eye and allows the eye to function in a wide range of lighting conditions. The light then passes through a lens that projects a focused image on the retina on the back surface of the eye. Several different kinds of light-sensitive cells then convert the image into nervous impulses that encode information about spatial patterns of color and intensity. These cells are more sensitive to light than the best photographic film. The detailed construction of each of these parts of the eye makes sense in terms of the eye's function—seeing. If we probed into any of these parts, we would see that they too are made of complicated, interacting components whose structure is understandable in terms of their function.

Moreover, differences between human eyes and the eyes of other animals make sense in terms of the types of problems each creature faces. Consider, for example, the eyes of fish and humans (Figure 1.2). The lens in the eyes of humans and other terrestrial mammals is much like a camera lens; it is shaped like a squashed football and has the same index of refraction (a measure of light-bending capacity) through-

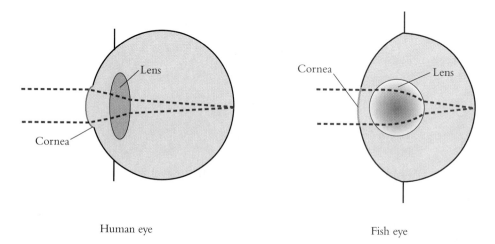

Human eye   Fish eye

Figure 1.2

Like other terrestrial mammals, human eyes have more than one light-bending element. A ray of light entering the eye (dashed lines) is bent first as it moves from the air to the cornea, and then bent again as it enters and leaves the lens. In contrast, fish eyes have a single lens that bends the light throughout its volume. As a result, fish eyes have a short focal length and high light-gathering power.

out. In contrast, the lens in fish eyes is a sphere located at the center of the curvature of the retina, and the index of refraction of the lens increases smoothly from the surface of the lens to the center. It turns out that this kind of lens, called a spherical gradient lens, provides a sharp image over a full 180° visual field, a very short focal length, and high light-gathering power—all desirable properties. Terrestrial creatures like us cannot use this design because light is bent when it passes from the air through the cornea, the transparent cover of the pupil, and this fact constrains the design of the remaining lens elements. In contrast, light is not bent when it passes from water through the cornea of aquatic animals, and the design of their eyes takes advantage of this fact.

Before Darwin there was no scientific explanation for the fact that organisms are well adapted to their circumstances.

As many 19th-century thinkers were keenly aware, complex adaptations like the eye demand a different kind of explanation than other natural objects do. This is not simply because adaptations are complex, since many other complicated objects exist in nature. Adaptations require a special kind of explanation because they are complex in a particular, highly improbable way. For example, the Grand Canyon, with its maze of delicate towers intricately painted in shades of pink and gold, is byzantine in its complexity (Figure 1.3). However, given a different geological history, the Grand Canyon might be quite different—different towers, in different

Figure 1.3

The Grand Canyon is an impressive geological feature, but much less remarkable in its complexity than the eye.

hues—yet we would still recognize it as a canyon. The particular arrangement of painted towers of the Grand Canyon is improbable, but the existence of a spectacular canyon with a complex array of colorful cliffs in the dry sandstone country of the American Southwest is not unexpected at all, and, in fact, wind and water produced several other canyons in this region. In contrast, any substantial changes in the structure of the eye would prevent the eye from functioning, and then we would no longer recognize it as an eye. If the cornea was opaque, or the lens was on the wrong side of the retina, then the eye would not transmit visual images to the brain. It is highly improbable that natural processes would randomly bring together bits of matter having the detailed structure of the eye because only an infinitesimal fraction of all arrangements of matter would be recognizable as a functioning eye.

In Darwin's day, most people were not troubled by this problem because they believed that adaptations were the result of divine creation. In fact, the theologian William Paley used a discussion of the human eye to argue for the existence of God in his book, *Natural Theology,* published in 1802. Paley argued that the eye is clearly *designed* for seeing, and where there is design in the natural world there certainly must be a heavenly designer.

While most scientists were satisfied with this reasoning, a few, including Charles Darwin, sought other explanations.

Darwin's Theory of Adaptation

Charles Darwin was expected to become a doctor or clergyman, but instead revolutionized science.

Charles Darwin was born into a well-to-do, intellectual, and politically liberal family in England. Like many prosperous men of his time, Darwin's father wanted his son to become a doctor. But after failing at the prestigious medical school at the University of Edinburgh, Charles went on to Cambridge University, resigned to becoming a country parson. He was, for the most part, an undistinguished student, and was much more interested in tramping through the fields around Cambridge in search of beetles than in studying Greek and mathematics. After graduation, one of Darwin's biology professors, William Henslow, provided him a chance to pursue his passion for natural history as a naturalist on the HMS *Beagle*.

The *Beagle* was a Royal Navy vessel whose charter was to spend two to three years mapping the coast of South America, and then to return to London, perhaps by circling the globe (Figure 1.4). Darwin's father forbade him to go, preferring that he get serious about his career in the clergy, but Darwin's uncle (and future father-in-law) Josiah Wedgwood intervened. The voyage turned out to be the turning point in Darwin's life. His work during the voyage established his reputation as a skilled naturalist. His observations of living and fossil animals ultimately convinced him that plants and animals sometimes change slowly through time, and that such evolutionary change was the key to understanding how new species came into existence. This view was rejected by most scientists of the time and was considered heretical by the general public.

Figure 1.4
The HMS *Beagle* in Beagle Channel on the southern coast of Tierra del Fuego.

Darwin's Postulates

Darwin's theory of adaptation follows from three postulates: the struggle for existence, variation in fitness, and the inheritance of variation.

In 1838, shortly after the *Beagle* returned to London, Darwin formulated a simple mechanistic explanation for *how* species change through time. His theory follows from three postulates:

1. The ability of a population to expand is infinite, but the ability of any environment to support populations is always finite.
2. Organisms within populations vary, and this variation affects the ability of individuals to survive and reproduce.
3. The variations are transmitted from parents to offspring.

Darwin's first postulate means that populations grow until they are checked by the dwindling supply of resources in the environment. Darwin referred to the resulting competition for resources as "the struggle for existence." For example, animals require food to grow and reproduce. When food is plentiful, animal populations grow until their numbers exceed the local food supply. Since resources are always finite, it follows that not all individuals in a population will be able to survive and reproduce. According to the second postulate, some individuals will possess traits that enable them to survive and reproduce more successfully (producing more offspring) than others in the same environment. The third postulate holds that if the advantageous traits are inherited by offspring, then these traits will become more common in succeeding generations. Thus, traits that confer advantages in survival and reproduction are retained in the population, and traits that are disadvantageous disappear. When Darwin coined the term **natural selection** for this process, he was making a deliberate analogy to the artificial selection practiced by animal and plant breeders of his day. A much more apt term would be "evolution by variation and selective retention."

Figure 1.5

(a) The islands of the Galápagos, which are located off the coast of Ecuador, house a variety of unique species of plants and animals. (b) Cactus finches from Charles Darwin, *The Zoology of the Voyage of H.M.S. Beagle.*

(a)   (b)

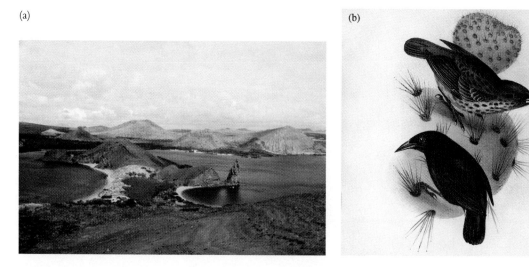

An Example of Adaptation by Natural Selection

Contemporary observations of Darwin's finches provide a particularly good example of how natural selection produces adaptations.

In his autobiography, first published in 1887, Darwin claimed that the curious pattern of adaptations he observed among the several species of finches that live on the Galápagos Islands off the coast of Ecuador—now referred to as Darwin's finches—was crucial in the development of his ideas about evolution (Figure 1.5). Recently discovered documents suggest that Darwin was actually quite confused about the Galápagos finches during his visit, and they played little role in his discovery of natural selection. Nonetheless, Darwin's finches hold a special place in the minds of most biologists.

Peter and Rosemary Grant, biologists at Princeton University, conducted a landmark study of the ecology and evolution of one particular species of Darwin's finches on one of the Galápagos Islands. The study is remarkable because the Grants were able to directly document how Darwin's three postulates lead to evolutionary change. The island, Daphne Major, is home to the medium ground finch *(Geospiza fortis),* a small bird that subsists mainly by eating seeds (Figure 1.6). The Grants and

Figure 1.6

The medium ground finch, *Geospiza fortis,* uses its beak to crack open seeds. (Photograph courtesy of Peter Grant.)

(a)

(b)

Figure 1.7

(a) This is how Daphne Major looked in March 1976, after a year of good rains. (b) This is how Daphne Major looked in March 1977, after a year of very little rain.

their colleagues caught, measured, weighed, and banded nearly every finch on the island each year, some 1500 birds in all. They also kept track of critical features of the birds' environment, such as the distribution of seeds of various sizes, and observed the birds' behavior. A few years into the Grants' study, a severe drought struck Daphne Major (Figure 1.7). During the drought, plants produced many fewer seeds, and the finches soon depleted the stock of small, soft, easily processed seeds, leaving only large, hard, difficult-to-process seeds (Figure 1.8). The bands on the birds' legs enabled the Grants to track the fates of individual birds during the drought, and the regular measurements that they had made of the birds allowed them to compare the traits of birds that survived the drought with the traits of those that perished. They also made detailed records of the environmental conditions, which allowed them to determine how the drought affected the finch's habitat. It was this vast body of data that enabled the Grants to document the action of natural selection among the finches of Daphne Major.

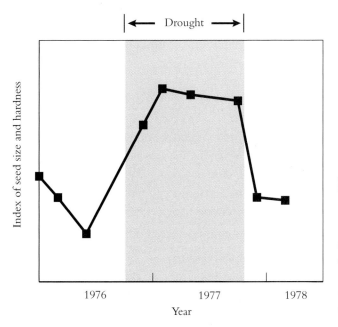

Figure 1.8

During the two-year drought, the size and hardness of seeds available on Daphne Major increased because birds consumed all of the desirable small, soft seeds, leaving mainly larger and harder seeds. Each point represents an index of seed size and hardness at a given time.

Studying Evolution in Darwin's Finches

Half past seven on Daphne Major. Peter and Rosemary Grant sit themselves down on stones, a few steps from their traps. Peter opens a yellow notebook with waterproof pages. "Okay," he says. "Today is the twenty-fifth."

It is the twenty-fifth of January, 1991. There are four hundred finches on the island at this moment, and the Grants know every one of the birds on sight, the way shepherds can tell every sheep in their flocks. In other years there have been more than a thousand finches on Daphne Major, and Peter and Rosemary could still recognize each one. The flock was down to three hundred once. The number is falling toward that now. The birds have gotten less than a fifth of an inch of rain in the last forty-four months: in 1,320 days, 5 millimeters of rain.

The Grants, and the Grants' young daughters, and a long line of assistants, keep coming back to this desert island like sentries on a watch. They have been observing Daphne Major for almost two decades, or about twenty generations of finches. By now Peter and Rosemary Grant know many of the birds' family trees by heart—again like shepherds, or like Bible scholars, who know that Abraham begat Isaac, and Isaac begat Jacob; and Abraham also begat Jokshan, who begat Dedan, who begat Asshurim, Letushim, and Leummim.

In each generation there are always a few birds, just one or two in a hundred, that keep away from the Grants and refuse to be caught. This morning Rosemary, after a week of watching and plotting, has just captured two of the wariest, most difficult finches on the island. She caught them both in the space of a single minute, high on the island's north rim, next to a fallen cactus pad, in black box traps baited with green bananas. "How about that," she cried, when the traps' doors clicked shut. And when Peter strode through the cactus trees and across the lava rubble to join her, Rosemary lifted up her first prize, fluttering in a blue pouch. "I deserve a bottle of wine for this!"

Now the Grants are sitting beside the traps at the edge of a cliff, 100 meters above the Pacific Ocean. Except for the honking and whistling of two masked boobies, courting on a rock nearby, the scene is quiet. The ocean is more than pacific; it is flat as a pond. The morning's weather is what Charles Darwin described in his diary when he first saw the Galápagos archipelago, "a steady, gentle breeze of wind & gloomy sky."

From the upper rim of Daphne Major, on clearer mornings than this one, Rosemary and Peter can see the island of Santiago, where Darwin camped for nine days. They can also see the island of Isabela, where Darwin spent one day. They can make out more than a dozen other islands and black lava ruins that Darwin never had a chance to visit, including an islet known officially as Sin Nombre (that is, Nameless) and another black speck called Eden.

"If I have seen further," Isaac Newton once wrote, with celebrated modesty, "it is by standing upon the shoulders of Giants." The dark volcanoes of the Galápagos are Darwin's shoulders. These islands meant more to him than any other stop in his five-year voyage around the world. "Origin of all my views," he called them once—the origin of the *Origin of Species*. The Grants are doing what Darwin could not do, going back to the Galápagos year after year; and the Grants are seeing there what Darwin did not imagine could be seen at all.

Rosemary unlatches their tool kit, a tackle box. From it, Peter extracts a pair of jeweler's spectacles, a plastic mask with bulging lenses, which make him look like Robinson Crusoe from Mars. "Okay, Famous Bird," Peter says. "*Ow!* Famous Bird has decided to bite the hand that feeds him." He grasps the finch with one hand, and its head sticks out observantly from his fist. The bird is about the size of a sparrow, and jet-black, with a black beak and shiny dark eyes.

Rosemary hands Peter a pair of calipers. "Now, here we go," Peter says. "Wing length, 72 millimeters."

Rosemary jots the number in the yellow notebook.

"Tarsus length, 21.5." (The tarsus is the bird's leg.)

Rosemary writes it down.

"Beak length, 14.9 millimeters," Peter recites. "Beak depth, 8.8. Beak width, 8 millimeters."

"Black Five plumage." The Grants rate the birds' plumage from zero, which is brown, to five, totally black. Black Five means a mature male.

"Beak black." Normally these birds' beaks are pale, the color of horn. A black beak means the bird is ready to mate.

Peter dangles the bird in a little weighing cup. "Weight, 22.2 grams."

"This bird has lived a long time," he muses. "Thirteen years." There are only three others of its generation still alive on the island, and none older. "But I don't think there's a single one of his offspring flying around. Not *one* has made it to the breeding season." The bird has been a father many times, and never once a grandfather.

Peter puts a gray ring and a brown ring on the bird's left ankle. He puts a light green ring over a metal one on its right ankle. Bands like these, and an ingenious color code, help the Grant team to keep track of their flocks from dawn to dusk, from the cliffs at the base of the island to this guano-painted rubble at the rim.

Peter holds the bird in his fist one more time and inspects its beak in profile. In rushing up to join Rosemary at the rim, he has forgotten his camera. Otherwise he would photograph the bird just so, from a distance of 27 centimeters. That is the Grants' standard mug shot for one of Darwin's finches.

Source: From pp. 3–6 in *The Beak of the Finch*, by Jonathan Weiner. Copyright © 1994 by Jonathan Weiner. Reprinted by permission of Alfred A. Knopf, Inc.

The Grants' data show how the processes identified in Darwin's postulates lead to adaptation.

The events on Daphne Major embodied all three of Darwin's postulates. First, the supply of food on the island was not sufficient to feed the entire population, and many finches did not survive the drought. From the beginning of the drought in 1976 until the rains came nearly two years later, the population of medium ground finches on Daphne Major declined from 1200 birds to only 180.

Second, beak depth (the top-to-bottom dimension of the beak) varied among the birds on the island, and this variation affected the birds' survival. The Grants and their colleagues had observed before the drought began that birds with deeper beaks were able to process large, hard seeds more easily than birds with shallower beaks were. Deep-beaked birds usually concentrated on large seeds while shallow-beaked birds normally focused their efforts on small seeds. The open bars of the upper histogram in Figure 1.9 show what the distribution of beak sizes in the population would have been like before the drought. The height of each open bar in Figure 1.9b gives the number of birds with beaks in a given range of depths, for example, 8.8 to 9.0 mm or 9.0 to 9.2 mm. During the drought, the relative abundance of small seeds decreased, forcing shallow-beaked birds to shift to larger and harder seeds. Shallow-beaked birds were then at a distinct disadvantage because it was harder for them to crack these seeds. The distribution of individuals within the population changed during the drought because finches with deeper beaks were more likely to survive than finches with shallow beaks. The shaded bars of the histogram in Figure 1.9b show what the distribution of beak depths would have been like among the survivors. Because many died, fewer birds remain in each category among the survivors. However, mortality was quite specific. The proportion of shallow-beaked birds that died greatly exceeded the proportion of

Figure 1.9

A schematic diagram of how directional selection increased mean beak depth among medium ground finches on Daphne Major. (a) The probability of survival for birds of different bill depths is plotted. Birds with shallow beaks are less likely to survive than birds with deep beaks. (b) The height of each bar represents the numbers of birds whose beak depths fall within each of the intervals plotted on the *x*-axis with beak depth increasing to the right. The histogram with open bars shows the distribution of beak depths before the drought began. The histogram with shaded bars shows the distribution of beak depths after a year of drought. Notice that the number of birds in each category has decreased. Since birds with deep beaks were less likely to die than birds with shallow beaks, the peak of the distribution has shifted to the right, indicating that the mean beak depth has increased.

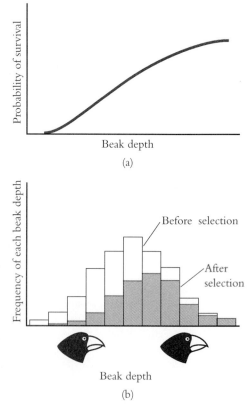

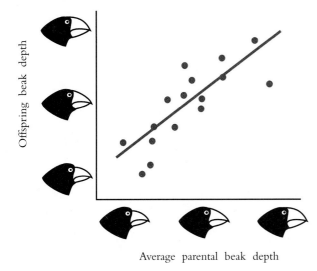

Figure 1.10

Parents with deeper-than-average beaks tend to have offspring with deeper-than-average beaks. Each point represents one offspring. Offspring beak depth is plotted on the vertical axis (deeper beaks farther up the axis), and the average of the two parents' beak depth is plotted on the horizontal axis (deeper beaks farther to the right). Each point represents one bird.

deep-beaked birds that died. As a result, the lower histogram shows a shift to the right, which means that the average beak depth in the population increased. Thus, the average beak depth among the survivors of the drought is greater than the average beak depth in the same population before the drought.

Third, parents and offspring had similar beak depths. The Grants discovered this because they captured and banded nestlings and recorded the identity of the nestlings' parents. When the nestlings became adults, the Grants recaptured and measured them. When compiling these data, the Grants discovered that parents with deep beaks on average produced offspring with deep beaks (Figure 1.10). Since parents were drawn from the pool of individuals who survived the drought, their beaks were on average deeper than those of the original residents of the island, and since offspring resemble their parents, the average beak depth of the survivors' offspring was greater than the average beak depth before the drought. This means that through natural selection, the average **morphology** (an organism's size, shape, and composition) of the bird population changed so that birds became better adapted to their environment. This process, operating over approximately two years, led to a 4% increase in the mean beak depth in this population (Figure 1.11).

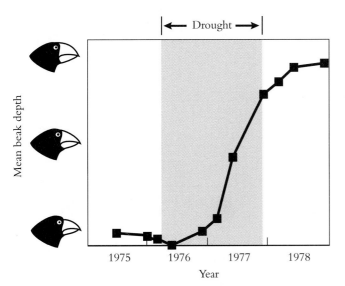

Figure 1.11

The average beak depth in the population of medium ground finches on Daphne Major increased during the drought of 1975–1978. Each point plots an index of average beak depth of the population in a particular year. Deeper beaks are plotted higher on the y axis.

xSelection preserves the status quo when the most common type is the best adapted.

So far, we have seen how natural selection led to adaptation as the population of finches on Daphne Major evolved in response to changes in their environment. Will this process continue forever? If it did, eventually all the finches would have deep enough beaks to efficiently process the largest seeds available. However, large beaks have disadvantages as well as benefits. The Grants showed, for instance, that birds with large beaks are less likely to survive the juvenile period than birds with small beaks, probably because they require more food to grow (Figure 1.12). Evolutionary theory predicts that, over time, selection will increase the average beak depth in the population until the costs of larger-than-average beak size exceed the benefits. At this point, finches with the average beak size in the population will be the most likely to survive and reproduce, and finches with deeper or shallower beaks than the new average will be at a disadvantage. When this is true, beak size does not change, and we say that an **equilibrium** exists in the population with regard to beak size. The process that produces this equilibrium state is called **stabilizing selection**. Notice that even though the average characteristics of the beak in the population will not change in this situation, selection is still going on. Selection is required to change a population, and selection is also required to keep a population the same.

It might seem that beak depth would also remain unchanged if this trait had no effect on survival (or put another way, if there were no selection favoring one type of beak over another). Then, all types of birds would be equally likely to survive from one generation to the next, and beak depth would remain constant. This logic would be valid if selection were the only process affecting beak size. However, real populations are also affected by other processes that cause **traits**, or **characters**, to change in unpredictable ways. We will discuss these processes further in Chapter 3. The point to remember here is that populations do not remain static over the long run unless selection is operating.

Figure 1.12

When birds with the most common beak depth are most likely to survive and reproduce, natural selection keeps the mean beak depth constant. (a) Birds with deep and shallow beaks are both less likely to survive than birds with average beaks. Birds with shallow beaks cannot process large, hard seeds, and birds with deep beaks are less likely to survive to adulthood. In (b) the open bars represent the distribution of beak depths before selection, and the shaded bars represent the distribution after selection. As in Figure 1.7, notice that there are fewer birds in the population after selection. However, since birds with average beaks are most likely to survive, the peak of the distribution of beak depths is not shifted and mean beak depth remains unchanged.

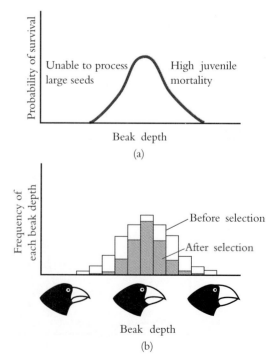

Species are populations of varied individuals that may or may not change through time.

As the Grants' work on Daphne Major makes clear, a species is not a fixed type or entity. Species change in their general characteristics from generation to generation according to the postulates Darwin described. Before Darwin, however, people thought of species as fixed, unchanging categories much the same way that we think of geometrical figures. A finch could no more change its properties than a triangle. If a triangle acquires another side, it is not a modified triangle, but rather a rectangle. In much the same way, to biologists before Darwin, a changed finch is not a finch at all. Ernst Mayr, the distinguished evolutionary biologist, calls this pre-Darwinian view of immutable species, essentialism. According to Darwin's theory, a **species** is a dynamic *population* of individuals. If the characteristics of a particular species appear static over a long period of time, it is because this type of individual is consistently favored by stabilizing selection. Both stasis (staying the same) and change result from natural selection, and both require explanation in terms of natural selection. Stasis is not the natural state of species.

Individual Selection

Adaptation results from the competition among individuals, not between entire populations or species.

It is important to notice that selection produces adaptations that benefit *individuals*. Such adaptation may or may not benefit the population or species. In the case of simple morphological characters such as beak depth, selection probably does allow the population of finches to compete more effectively with other populations of seed predators. However, this need not be the case. Selection often leads to changes in behavior or morphology that increase the reproductive success of individuals but decrease the average reproductive success of the group, population, and species.

The fact that almost all organisms produce many more offspring than are necessary to maintain the species provides an example of the conflict between individual and group interests. A female monkey may, on average, produce 10 offspring during her lifetime (Figure 1.13). In a stable population, perhaps only two of these offspring will survive and reproduce. From the point of view of the species, the other eight are a waste of resources. They compete with other members of their species for food, water, and sleeping sites. The demands of a growing population can lead to serious overexploitation of the environment, and the species as a whole might be more likely to survive if all females produced fewer offspring. However, this does not happen because natural selection among individuals favors females who produce many offspring.

To see why selection on individuals will lead to this result, let's consider a simple, hypothetical case. Suppose the females of a particular species of monkey are maximizing individual reproductive success when they produce 10 offspring. Females who produce more than or less than 10 offspring will tend to leave fewer descendants in the next generation. Further suppose that the species' likelihood of becoming extinct would be lowest if females produced only two offspring apiece. Now suppose that there are two kinds of females. Most of the population is composed of low-fecundity females who each produce just two offspring, but there

Figure 1.13

A female blue monkey holds her infant. (Photograph courtesy of Marina Cords.)

are a few high-fecundity females who each produce 10 offspring. (**Fecundity** is the term demographers use for the ability to produce offspring.) High-fecundity females have high-fecundity daughters, and low-fecundity females have low-fecundity daughters. The proportion of high-fecundity females will increase in the next generation because such females produce more offspring than do low-fecundity females. Over time, the proportion of high-fecundity females in the population will increase rapidly. As fecundity increases, the population will grow rapidly and may exhaust available resources. This will, in turn, increase the chance that the species becomes extinct. However, this fact is irrelevant to the evolution of fecundity before the extinction, because natural selection results from competition among individuals, not competition among species.

The idea that natural selection operates at the level of the individual is a key element in understanding adaptation. In discussing the evolution of social behavior in Chapter 8, we will encounter several additional examples of situations in which selection increases individual success but decreases the competitive ability of the population.

The Evolution of Complex Adaptations

The example of the evolution of beak depth in the medium ground finch illustrates how natural selection can cause adaptive change to occur rapidly in a population. Deeper beaks enabled the birds to survive better, and deeper beaks soon came to predominate in the population. Beak depth is a fairly simple character, lacking the intricate complexity of an eye. However, as we will see, the accumulation of small variations by natural selection can also give rise to complex adaptations.

Why Small Variations Are Important

There are two categories of variation: continuous and discontinuous.

It was known in Darwin's day that most variation is continuous. An example of **continuous variation** is the distribution of heights in people. Humans grade smoothly from one extreme to the other (short to tall) with all the intermediate types (in this case, heights) represented. However, Darwin's contemporaries also knew about **discontinuous variation**, in which a number of distinct types exist with no intermediates. In humans, height is also subject to discontinuous variation. For example, there is a genetic condition called **achondroplasia**, which causes affected individuals to be much shorter than other people, have proportionately shorter arms and legs, and bear a variety of other distinctive features. Discontinuous variants are usually quite rare in nature. Nonetheless, many of Darwin's contemporaries, who were convinced of the reality of evolution, believed that new species arise as discontinuous variants.

Discontinuous variation is not important for the evolution of complex adaptations because complex adaptations are extremely unlikely to arise in a single jump.

Unlike most of his contemporaries, Darwin thought that discontinuous variation did not play an important role in evolution. A hypothetical example, described by Oxford University biologist Richard Dawkins in his book *The Blind Watchmaker,* illustrates Darwin's reasoning. Dawkins recalls an old story in which an imaginary collection of monkeys sit at typewriters happily typing away. Lacking the ability to read or write, they strike keys at random. Given enough time, the story goes, the monkeys will reproduce all the great works of Shakespeare. Dawkins points out that this is not likely to happen in the lifetime of the universe, let alone the lifetime of one of the monkey typists. To illustrate why it would take so long, Dawkins presents these illiterate monkeys with a much simpler problem, reproducing a single line from *Hamlet,* "Methinks it is like a weasel." To make the problem even simpler for the monkeys, Dawkins ignores the difference between uppercase and lowercase letters and omits all punctuation except spaces. There are 28 characters (including spaces) in the phrase. Since there are 26 characters in the alphabet and Dawkins is keeping track of spaces, each time a monkey types a character, there is only a 1-in-27 chance that she will type the right character. There is also only a 1-in-27 chance that the second character will be correct. Again, there is a 1-in-27 chance that the third character will be right, and so on up to the 28th character. Thus the chance that a monkey will type the correct sequence at random is 1/27 times itself 28 times, or

$$\underbrace{\frac{1}{27} \times \frac{1}{27} \times \frac{1}{27} \times \cdots \times \frac{1}{27}}_{28 \text{ times}} \approx 10^{-40}$$

This is a *very* small number. To get a feeling for how small a chance there is of the monkeys typing the sentence correctly, suppose a very fast computer (faster than any currently in existence) could generate 100 billion (10^{11}) characters per second and ran for the lifetime of the earth, about 4 billion years, or 10^{17} seconds. Then, the chance of randomly typing the line "Methinks it is like a weasel" even once during the whole of the earth's history would be about 1 in a trillion! Typing the whole play is obviously astronomically less likely, and although *Hamlet* is a very complicated thing, it is much less complicated than a human eye. There's no chance that a structure like the human eye would arise by chance in a single trial. If it did, it would be as the astrophysicist Sir Fred Hoyle said, "like a hurricane blowing through a junk yard and chancing to assemble a Boeing 747."

Complex adaptations can arise through the accumulation of small random variations by natural selection.

Darwin argued that continuous variation is essential for the evolution of complex adaptations. Once again, Richard Dawkins provides an example that makes clear Darwin's reasoning. Again imagine a room full of monkeys and typewriters, but now the rules of the game are different. The monkeys type the first 28 characters at random, and then during the next round they attempt to copy the same initial string of letters and spaces. Most of the sentences are just copies of the previous string, but because monkeys sometimes make mistakes, some strings have small variations, usually in only a single letter. During each trial, the monkey trainer selects the string that most resembles Shakespeare's phrase "Methinks it is like a weasel" as the string to be copied by all the monkeys in the next trial. This process is repeated until the monkeys come up with the correct string. Calculating the exact number of trials required to generate the correct sequence of characters is quite difficult, but it is easy to simulate the process on a computer. Here's what happened when Dawkins performed the simulation. The initial random string is

WDLMNLT DTJBKWIRZREZLMQCO P

After one trial Dawkins got

WDLMNLT DTJBSWIRZREZLMQCO P

After 10 trials:

MDLDMNLS ITJISWHRZREZ MECS P

After 20 trials:

MELDINLS IT ISWPRKE Z WECSEL

After 30 trials:

METHINGS IT ISWLIKE B WECSEL

After 40 trials:

METHINKS IT IS LIKE I WEASEL

The exact phrase was reached after 43 trials. Dawkins reports that it took his 1985-vintage Macintosh only 11 seconds to complete this task.

Selection can give rise to great complexity starting with small random variations because it is a *cumulative* process. As the typing monkeys show us, it is spectacularly unlikely that a single random combination of keystrokes will produce the correct sentence. However, there is a much greater chance that some of the many *small* random changes will be advantageous. The combination of reproduction and se-lection allows the typing monkeys to accumulate these small changes until the desired sentence is reached.

Why Intermediate Steps Are Favored by Selection

The evolution of complex adaptations requires that all of the intermediate steps be favored by selection.

There is a potent objection to the example of the typing monkeys. Natural selection, acting over time, can lead to complex adaptations, but it can do so only if each small change along the way is itself adaptive. While it is easy to assume that this is true in a hypothetical example of character strings, many people have argued that it is unlikely for every one of the changes necessary to assemble a complex organ like the eye to be adaptive. An eye is only useful, it is claimed, once all parts of the complexity are assembled; until then, it is worse than no eye at all. After all, what good is 5% of an eye?

Darwin's answer, based on the many adaptations for seeing or sensing light that exist in the natural world, was that 5% of an eye *is* often better than no eye at all. It is quite possible to imagine that a very large number of small changes—each favored by selection—led cumulatively to the wonderful complexity of the eye. Living mollusks, which display a broad range of light-sensitive organs, provide examples of many of the likely stages in this process (Figure 1.14):

1. Many invertebrates have a simple light-sensitive spot (Figure 1.14a). Photore-ceptors of this kind have evolved many times from ordinary epidermal (surface) cells—usually ciliated cells whose biochemical machinery is light-sensitive. Those individuals whose cells are more sensitive to light are favored when information about changes in light intensity are useful. For example, a drop in light intensity may often be an indicator of a predator in the vicinity.

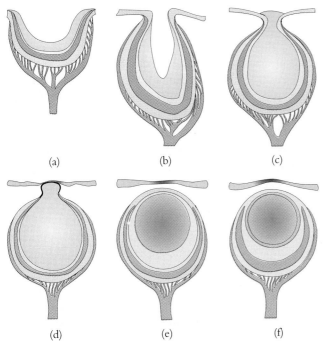

(a)   (b)   (c)

(d)   (e)   (f)

Figure 1.14

Living gastropod mollusks illustrate all of the intermediate steps between a simple eye cup and a camera-type eye: (a) The eye pit of a limpet, *Patella* sp.; (b) the eye cup of Beyrich's split shell, *Pleurotomaria beyrichi*; (c) the pinhole eye of a California abalone, *Haliotis* sp.; (d) the closed eye of a turban shell, *Turbo creniferus*; (e) the lens eyes of the spiny dye murex, *Murex brandaris*; (f) the lens eyes of the Atlantic dog whelk, *Nucella lapis*. (Lens is shaded in e and f.)

2. By locating the light-sensitive cells in a depression, the organism will get some additional information about the direction of the change in light intensity (Figure 1.14b). The surface of organisms is variable, and those individuals whose photoreceptors are in depressions will be favored by selection in environments where such information is useful. For example, mobile organisms may need better information about what is happening in front of them than do immobile ones.

3. Through a series of small steps, the depression could get deeper, and each step could be favored by selection because better directional information would be available (Figure 1.14c).

4. If the depression got deep enough, it could form images on the light-sensitive tissue, much the way pinhole cameras form images on photographic film (Figure 1.14d). In settings in which detailed images are useful, selection could then favor the elaboration of the neural machinery necessary to interpret the image.

5. The next step is the formation of a transparent cover (Figure 1.14e). This might be favored because it protects the interior of the eye from parasites and mechanical damage.

6. A lens could evolve either through gradual modification of the transparent cover, or through the modification of internal structures within the eye (Figure 1.14f).

Notice that evolution produces adaptations like a tinkerer, not an engineer. New organisms are created by small modifications of existing ones, not by starting with a clean slate. Clearly there will be many beneficial adaptations that will not arise because they are blocked at some step along the way when a particular variation is not favored by selection. Darwin's theory explains how complex adaptations can arise through natural processes, but it does not predict that every, or even most, adaptations that could occur, have occurred, or will occur. This is not the best of all possible worlds; it is just one of many possible worlds.

Sometimes unrelated species have independently evolved the same complex adaptation. This suggests that the evolution of complex adaptations by natural selection is not a matter of mere chance.

The fact that natural selection constructs complex adaptations like a tinkerer might lead you to think that the assembly of complex adaptations is a chancy business. If even a single step was not favored by selection, the adaptation could not arise. Such reasoning suggests that complex adaptations are mere coincidence. While chance does play a very important role in evolution, the power of cumulative natural selection should not be underestimated. The best evidence that selection is a powerful process for generating complex adaptations comes from a phenomenon called **convergence**, the evolution of similar adaptations in unrelated groups of animals.

The similarity between the marsupial faunas of Australia and South America and the placental faunas of the rest of the world provides a good example of convergence. In most of the world, the mammalian fauna is dominated by **placental mammals**, which nourish their young in the uterus during long pregnancies. However, both Australia and South America became separated from an ancestral supercontinent, known as Pangaea, long before placental mammals evolved. In Australia and South America, **marsupials** (nonplacental mammals, like kangaroos, that rear their young in external pouches) came to dominate the

Figure 1.15
The marsupial wolf that lived in Tasmania until early in the 20th century (drawn from a photograph of one of the last living animals). Similarities with placental wolves of North America and Eurasia illustrate the power of natural selection to create complex adaptations. Their last common ancestor was probably a small insectivorous creature like the shrew.

mammalian fauna, filling all available mammalian niches. Some of these marsupial mammals were quite similar to the placental mammals on the other continents. For example, there was a marsupial wolf in Australia that looked very much like placental wolves of Eurasia, even sharing subtle features of their feet and teeth (Figure 1.15). These marsupial wolves became extinct in the 1930s. Similarly, in South America, a marsupial saber-toothed cat independently evolved many of the same adaptations as the placental saber-toothed cat that stalked North America 10,000 years ago. These similarities are more impressive when you consider that the last common ancestor of marsupial and placental mammals was a small, nocturnal, insectivorous creature, something like a shrew, that lived about 120 mya (million years ago). Thus, selection transformed a shrew step by small step, each step favored by selection, into a saber-toothed cat—and it did it twice. This cannot be coincidence.

The evolution of eyes provides another good example of convergence. Remember that the spherical gradient lens is a good lens design for aquatic organisms because it has good light-gathering ability and provides a sharp image over the full 180° visual field. Complex eyes with lenses have evolved independently eight different times in distantly related aquatic organisms: once in fish, once in cephalopod mollusks like squid, several times among gastropod mollusks like Atlantic dog welk, once in annelid worms, and once in crustaceans (Figure 1.16). These are very diverse creatures whose last common ancestor was a simple creature that did not have a complex eye. Nonetheless, in every case they have evolved very similar spherical gradient lenses. Moreover, no other lens design is found in aquatic animals. Despite the seeming chanciness of assembling complex adaptations, natural selection has achieved the optimal design in every case.

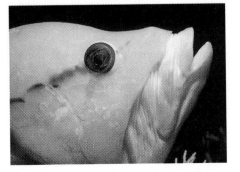

(a)

(b)

Figure 1.16
Complex eyes with lenses have evolved independently in a number of different kinds of aquatic animals, including this slingjaw wrasse (a) and squid (b).

Rates of Evolutionary Change

Natural selection can cause evolutionary change that is much more rapid than we commonly observe in the fossil record.

In Darwin's day, the idea that natural selection could change a chimpanzee into a human, much less that it might do so in just a few million years, was unthinkable. Though people are generally more accepting of evolution today, many still think of evolution by natural selection as a glacially slow process that requires millions of years to accomplish noticeable change. Such people often doubt that there has been enough time for selection to accomplish the evolutionary changes observed in the fossil record. And yet, as we will see in subsequent chapters, most scientists now believe that humans evolved from an apelike creature in only 5 million to 10 million years. In fact, some of the rates of selective change observed in contemporary populations are far faster than necessary for such a transition. The puzzle is not whether there has been enough time for natural selection to produce the adaptations that we observe. The real puzzle is why the change observed in the fossil record was so slow.

The Grants' observation of the evolution of beak morphology in Darwin's finches provides one example of rapid evolutionary change. The medium ground finch of Daphne Major is one of 14 species of finches that live in the Galápagos. Evidence suggests that all 14 are descended from a single South American species that migrated to the newly emerged islands about half a million years ago (Figure 1.17). This doesn't seem like a very long time. Is it possible that natural selection created 14 species in only half a million years?

To start to answer the question, let's calculate how long it would take for the medium ground finch to come to resemble its closest relative, the large ground finch *(Geospiza magnirostris)*, in beak size and weight (Figure 1.18). The large ground finch is 75% heavier than the medium ground finch, and its beak is about 20% deeper. Remember that beak size increased about 4% in two years during the 1977 drought. The Grants' data indicate that body size also increased by a similar amount. At this rate, Peter Grant calculated that it would take between 30 and 46 years for selection to increase the medium ground finch population's beak size and body weight to match that of the large ground finch. But these changes occurred in response to an extraordinary environmental crisis. The data suggest that selection doesn't generally push consistently in one direction. Instead, in the Galápagos evolutionary change seems to go in fits and starts, moving traits one way and then another. So, let's suppose that a net change in beak size like the one that occurred during 1977 occurs only once in every century. Then it would take about 2000 years to transform the medium ground finch into the large ground finch, still a very rapid process.

Similar rates of evolutionary change have been observed elsewhere when species invade new habitats. For example, about 100,000 years ago a population of elk (called red deer in Great Britain) colonized and then became isolated on the island of Jersey, off the French coast, presumably by rising sea levels. By the time the island was reconnected with the mainland about 6000 years later, the red deer had shrunk to about the size of a large dog. University of Michigan paleontologist Philip Gingerich compiled data on the rate of evolutionary change in 104 cases in

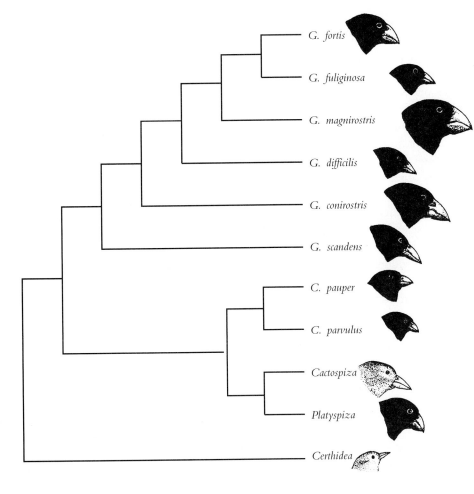

Figure 1.17

The relationship among various species of Darwin's finches can be traced by analyzing their protein polymorphisms. Species that are closely linked in the phylogenetic tree are more similar to one another genetically than to other species, because they share a more recent common ancestor. The tree does not include 3 of the 14 species of Darwin's finches.

which species invaded new habitats. These rates ranged from a low of zero (that is, no change) to a high of about 22% per year, with an average of about 0.1% per year.

The changes the Grants observed in the medium ground finch are relatively simple: birds and their beaks just got bigger. More complex changes usually take

Figure 1.18

The large ground finch *(Geospiza magnirostris)* has a beak that is nearly 20% deeper than the beak of its close relative, the medium ground finch *(G. fortis)*. At the rate of evolution observed during the drought of 1977, Peter Grant calculated that selection could transform the medium ground finch into the large ground finch in less than 46 years. (Photograph courtesy of Peter and Rosemary Grant.)

longer to evolve, but two kinds of evidence suggest that selection can produce big changes in remarkably short periods of time.

One line of evidence comes from artificial selection. Humans have performed selection on domesticated plants and animals for thousands of years, and for most of this period this selection was not deliberate. For example, modern cattle are descended from the auroch, an extinct creature something like a large antelope. People presumably kept the docile aurochs that were easy to manage and ate the rest. By inadvertently selecting for docility, a life-threatening trait in the wild, Stone Age farmers transformed the fierce auroch into the placid Hereford. There are many other familiar examples. All domesticated dogs, for instance, are believed to be descendants of wolves. Scientists are not sure when dogs were domesticated, but 10,000 years ago is a good guess, which means that in a few thousand generations, selection changed wolves into Pekinese, beagles, greyhounds, and St. Bernards. In reality, most of these breeds were created fairly recently, and most were the products of directed breeding. Darwin's favorite example of artificial selection was the domestication of pigeons. In the 19th century, pigeon breeding was a popular hobby, especially among working people who competed to produce showy birds (Figure 1.19). They created a menagerie of wildly different forms, all

(a)

(b)

(c)

(d)

Figure 1.19

In Darwin's day, pigeon fanciers created many new breeds of pigeons, including (a) pouters, (b) fantails, and (c) carriers, from (d) the common rock pigeon.

descended from the rather plain-looking rock pigeon. Darwin pointed out that these breeds are so different that if they were discovered in nature, biologists would surely classify them as members of different species. Yet they were produced by artificial selection within a few hundred years.

A second line of evidence comes from theoretical studies of the evolution of complex characters. Dan-E Nilsson and Susanne Pleger of Lund University, in Sweden, have built a mathematical model of the evolution of the eye in an aquatic organism. They start with a population of organisms, each with a simple eyespot, a flat patch of light-sensitive tissue sandwiched between a transparent protective layer and a layer of dark pigment. They then consider the effect of every possible small (1%) deformation of the shape of the eyespot on the resolving power of the eye. They determine which 1% change has the greatest positive effect on the eye's resolving power, and then repeat the process again and again, deforming the new structure by 1% in every possible way at each step. The results are shown in Figure 1.20. After 538 changes of 1%, a simple concave eye cup evolves; after 1033 changes of 1%, crude pinhole eyes emerge; after 1225 changes of 1%, an eye with an elliptical lens is created; and after 1829 steps the process finally comes to a halt because no small changes increase resolving power. The end result is an eye with a spherical gradient lens just like those in fish and other aquatic organisms. As Nilsson and Pleger point out, 1829 changes of 1% add up to a substantial amount of change. For instance, 1829 changes of 1% would lengthen a 10-cm (4-in.) human finger to a length of 8000 km (5500 miles)—about the distance from Los Angeles to New York and back. Nonetheless, making very conservative assumptions about the strength of selection, Nilsson and Pleger calculate that this would take only about 364,000 generations. For organisms with short generations, the complete structure of the eye can evolve from a simple eyespot in less than a million years, a brief moment in evolutionary time.

By comparison, most changes observed in the fossil record are much slower. Human brain size has roughly doubled in the last 2 million years—a rate of change of 0.00005% per year. This is 10,000 times slower than the rate of change that the Grants observed in the Galápagos. Moreover, such slow rates of change typify what can be observed from the fossil record. As we will see, however, the fossil

Stage

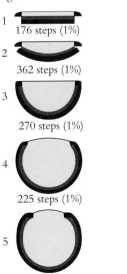

1
176 steps (1%)

2
362 steps (1%)

3
270 steps (1%)

4
225 steps (1%)

5
192 steps (1%)

6
308 steps (1%)

7
296 steps (1%)

8

Figure 1.20

A computer simulation of the evolution of the eye generates this sequence of forms. In the first step, the eye is a flat patch of light-sensitive cells that lies between a transparent protective layer and a layer of dark pigment. Eventually the eyespot deepens, the lens curves outward, and the radius lengthens. This process involves about 1800 changes of 1%.

record is incomplete. It is quite likely that some evolutionary changes in the past were rapid, but the sparseness of the fossil record prevents us from detecting them.

Darwin's Difficulties Explaining Variation

Darwin's *The Origin of Species* was a bestseller during his day, but his proposal that new species and other major evolutionary changes arise by the accumulation of small variations through natural selection was not widely embraced. Most educated people accepted the idea that new species arise through the transformation of existing species, and many scientists accepted the idea that natural selection is the most important cause of organic change (although by the turn of the century even this consensus broke down, particularly in the United States). But only a minority endorsed Darwin's view that major changes occur through the accumulation of small variations.

Darwin couldn't convince his contemporaries that evolution occurred through the accumulation of small variations because he couldn't explain how variation is maintained.

Darwin's critics raised a telling objection to his theory: the actions of blending inheritance (see below) and selection would both inevitably deplete variation in populations and make it impossible for natural selection to continue. These were potent objections that Darwin was unable to resolve in his lifetime because he and his contemporaries did not yet understand the mechanics of inheritance.

Everyone could readily observe that many of the characteristics of offspring are an average of the characteristics of their parents. Most people, including Darwin, believed this phenomena to be caused by the action of **blending inheritance**, a model of inheritance that assumes the mother and father each contribute a hereditary substance that mixes, or "blends," to determine the characteristics of the offspring. Shortly after publication of *The Origin of Species,* a Scottish engineer named Fleeming Jenkin published a paper in which he clearly showed that with blending inheritance there could be little or no variation available for selection to act on. The following example shows why Jenkin's argument was so compelling. Suppose a population of one species of Darwin's finches displays two forms, tall and short. Further suppose that a biologist controls mating so that every mating is between a tall individual and a short individual. Then, with blending inheritance all of the offspring will be the same intermediate height, and their offspring will be the same height as they are. All of the variation for height in the population will disappear in a single generation. With random mating, the same thing will occur, though it will take longer. If inheritance were purely a matter of blending parental traits, then Jenkin would have been right about its effect on variation. However, as we will see in Chapter 3, genetics accounts for the fact that offspring are intermediate between their parents and does not assume any kind of blending.

Another problem arose because selection works by removing variants from populations. For example, if finches with small beaks are more likely to die than finches with larger beaks over the course of many generations, eventually all that will be left are birds with large beaks. There will be no variation for beak size, and Darwin's second postulate holds that without variation there can be no evolution by natural selection. For example, suppose the environment changes so that individuals with small beaks are less likely to die than those with large beaks are. The average beak size in the population will not decrease because there are no small-

(a)

(b)

Figure 1.21

(a) The wolf is the ancestor of all domestic dogs, including (b) this poodle and St. Bernard. These transformations were accomplished in several thousand generations of artificial selection.

beaked individuals. Natural selection destroys the variation required to create adaptations.

Even worse, as Jenkin also pointed out, there was no explanation of how a population might evolve beyond its original range of variation. The cumulative evolution of complex adaptations requires that populations move far outside their original range of variation. Selection can cull away some characters from a population, but how can it lead to new types not present in the original population? This apparent contradiction was a serious impediment to explaining the logic of evolution. How could elephants, moles, bats, and whales all descend from an ancient shrewlike insectivore unless there is some mechanism for creating new variants not present at the beginning? For that matter, how could all the different breeds of dogs have descended from their one common ancestor, the wolf (Figure 1.21)?

Remember that Darwin and his contemporaries knew there were two kinds of variation: continuous and discontinuous. Because Darwin believed complex adaptations could arise only through the accumulation of small variations, he thought discontinuous variants were unimportant. However, many biologists thought the discontinuous variants, called "sports" by 19th-century animal breeders, were the key to evolution because they solved the problem of the blending effect. The following hypothetical example illustrates why. Suppose that a population of green birds has entered a new environment in which red birds would be better adapted. Some of Darwin's critics believed that any new variant that was slightly more red would have only a small advantage and would be rapidly swamped by blending. In contrast, an all-red bird would have a large enough selective advantage to overcome the effects of blending, and could increase its frequency in the population.

Darwin's letters show that these criticisms worried him greatly, and although he tried a variety of counterarguments, he never found a satisfactory one. The solution to these problems required an understanding of genetics, which was not available for another half century. As we will see, it was not until well into the 20th century that geneticists came to understand how variation is maintained, and Darwin's theory of evolution was generally accepted.

Further Reading

Browne, J. 1996. *Charles Darwin: Voyaging, A Biography*. Princeton University Press, Princeton, N.J.

Dennett, D. 1995. *Darwin's Dangerous Idea*. Touchstone, New York.

Dawkins, R. 1996. *The Blind Watchmaker*. W.W. Norton, New York.

Ridley, M. 1996. *Evolution*. 2d ed. Blackwell Scientific, Oxford.

Weiner, J. 1994. *The Beak of the Finch*. Alfred A. Knopf, New York.

Study Questions

1. It is sometimes observed that offspring do not resemble their parents for some character, even though the character varies in the population. Suppose this were the case for beak depth in the medium ground finch.
   (a) What would the plot of offspring beak depth against parental beak depth look like?
   (b) Plot the mean depth in the population among (i) adults before a drought, (ii) adults after a year of drought, and (iii) offspring.
2. Many species of animals are cannibalistic. This practice certainly reduces the ability of the species to survive. Is it possible that cannibalism could arise by natural selection? If so, with what adaptive advantage?
3. Some insects mimic dung. Ever since Darwin, biologists have explained this as a form of camouflage: selection favors individuals who most resemble dung because they are less likely to be eaten. Recently Harvard paleontologist Stephen Jay Gould has objected to this explanation. He argues that while selection could perfect such mimicry once it evolved, it could not cause the resemblance to arise in the first place. "Can there be any edge," Gould asks, "to looking 5% like a turd?" (quoted on p. 81 in R. Dawkins, 1996; *The Blind Watchmaker*, W.W. Norton). Can you think of a reason why looking 5% like a turd would be better than not looking at all like a turd?
4. In the late 1800s an American biologist named Hermon Bumpus collected a large number of sparrows that had been killed in a severe ice storm. He found that birds whose wings were about average length were rare among the dead birds. What kind of selection is this? What effect would this episode of selection have on the mean wing length in the population?

CHAPTER 2

Genetics

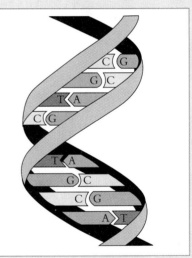

Mendelian Genetics

Although none of the main participants in the 19th-century debate about evolution knew it, the key experiments necessary to understand how genetic inheritance really worked had already been performed by an obscure Silesian monk, Gregor Mendel, living in what is now Slovakia (Figure 2.1). The son of peasant farmers, Mendel was recognized by his teachers as an extremely bright student, and he enrolled in the University of Vienna to study the natural sciences. While he was there, Mendel received a first-class education from some of the scientific luminaries of Europe. Unfortunately, Mendel had an extremely nervous disposition—every time he was faced with an examination he became physically ill, taking months to recover. As a result, he was forced to leave the university, after which he joined a monastery in the city of Brno, more or less because he needed a job. Once there, he continued to study inheritance, an interest he had developed in Vienna.

Figure 2.1

Gregor Mendel, about 1884, fifteen years after abandoning his botanical experiments.

Mendel discovered how inheritance worked through careful experiments with plants.

Between 1856 and 1863, using the common edible garden pea plant (Figure 2.2), Mendel isolated a number of traits with only two forms, or **variants**. For example, one of the traits he studied was the color of peas. This trait had two variants, yellow and green. He also studied pea texture, a trait that also had two variants, wrinkled and smooth. Mendel cultivated populations of plants in which these traits bred true, meaning that the traits did not change from one generation to the next. For example, **crosses** (matings) between plants that bore green peas always produced offspring with green peas, and crosses between plants that bore yellow peas consistently produced offspring with yellow peas. Mendel performed a large number of crosses between these different kinds of true-breeding peas. For example, a series of crosses between green and yellow variants yielded offspring that all bore yellow peas, matching only one of the parent plants (Figure 2.3). Mendel's next step was to perform crosses among the offspring of these crosses.

Figure 2.2

Mendel's genetics experiments were conducted on the common garden pea.

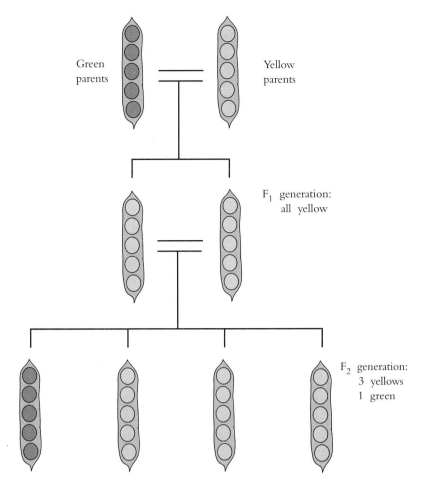

Green
parents

Yellow
parents

F_1 generation:
all yellow

F_2 generation:
3 yellows
1 green

Figure 2.3

In one of Mendel's
experiments, cross-
ing true-breeding
lines of green and
yellow peas led to
all yellow offspring.
Crossing these F_1 in-
dividuals led to an
F_2 generation with a
3:1 phenotypic ratio
of yellow to green
individuals.

Before going further, we need to establish a way to keep track of the results of the matings. Geneticists refer to the original founding population as the **F_0 generation**, the offspring of the original founders as the **F_1 generation**, and so on. In this case, the original true-breeding plants constitute the F_0 generation and the plants created by crossing true-breeding parents constitute the F_1 generation. The offspring of the F_1 generation will be the **F_2 generation**.

When members of the F_1 generation (all of whom bore yellow peas) were crossed, some of the offspring produced yellow seeds and some produced green seeds. Unlike most of the people who had experimented with plant crosses before, Mendel did a large number of these kinds of crosses and kept careful count of the numbers of each kind of individual that resulted. These data showed that in the F_2 generation there were three individuals with yellow seeds for every one with green seeds.

Mendel was able to formulate two principles that accounted for his experimental results.

Mendel derived two insightful conclusions from his experimental results:

1. The observed characteristics of organisms are determined jointly by two particles, one inherited from the mother and one from the father. The American geneticist T.H. Morgan later named these particles **genes**.
2. Each of these two particles or genes is equally likely to be transmitted when **gametes** (eggs and sperm) are formed. Modern scientists call this **independent assortment**.

These two principles account for the pattern of results in Mendel's breeding experiments, and as we shall see, they are the key to understanding how variation is preserved.

Cell Division and the Role of Chromosomes in Inheritance

Nobody paid any attention to Mendel's results for almost 40 years.

Mendel thought his findings were important, so he published them in 1866 and sent a copy of the paper to Karl Nägeli, a very prominent botanist. Nägeli was studying inheritance and should have understood the importance of Mendel's experiments. Instead, Nägeli dismissed Mendel's work, perhaps because it contradicted his own results or because Mendel was an obscure monk. Soon after this, Mendel was elected abbot of his monastery and was forced to give up his experiments. His ideas did not resurface until the turn of the 20th century, when several botanists independently replicated Mendel's experiments and rediscovered the laws of inheritance. In 1896, the Dutch botanist Hugo de Vries unknowingly repeated Mendel's experiments with poppies. Instead of publishing his results immediately, he cautiously waited until he had replicated his results with more than 30 plant species. Then in 1900, just as de Vries was ready to send off a manuscript describing his experiments, a colleague sent him a copy of Mendel's paper. Poor de Vries, his hot new results were already 30 years old! About the same time, two other European botanists, Carl Correns and Erich von Tschermak, also duplicated Mendel's breeding experiments, derived similar conclusions, and again discovered that they had all been done before. Correns and von Tschermak graciously acknowledged Mendel's primacy in discovering the laws of inheritance, but de Vries was less magnanimous. He did not cite Mendel in his treatise on plant genetics, and refused to sign a petition advocating the construction of a memorial in Brno commemorating Mendel's achievements.

When Mendel's results were rediscovered, they were widely accepted because scientists now understood the role of chromosomes in the formation of gametes.

By the time Mendel's experiments were rediscovered in 1900, it was well known that virtually all living organisms are built out of cells. Moreover, careful embryological work had shown that all the cells in complex organisms arise from a single cell through the process of cell division. Between the time of Mendel's initial discovery of the nature of inheritance and its rediscovery at the turn of the century, a crucial feature of cellular anatomy was discovered—the **chromosome**. Chromosomes are small, linear bodies contained in every cell and replicated during cell division (Figure 2.4). Moreover, scientists had also learned that chromosomes were replicated in a special kind of cell division that creates gametes. As we will see in subsequent sections, this research provides a simple material explanation for Mendel's results. Our current model of cell division, which was developed in small steps by a number of different scientists, is summarized in the following sections.

Mitosis and Meiosis

Ordinary cell division, called mitosis, creates two copies of the chromosomes present in the nucleus.

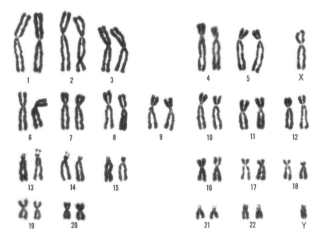

Figure 2.4

There are 23 pairs of human chromosomes that appear in the cell nucleus when cells divide [including a pair of sex chromosomes, for a male (as shown here), x and y]. Different chromosomes can be distinguished by their shape and by the banding patterns created by dyes that stain the chromosomes.

When plants and animals grow, their cells divide. Every cell contains within it a body called the **nucleus** (Figure 2.5); when cells divide, their nuclei also divide. This process of ordinary cell division is called **mitosis**. As mitosis begins, a cloud of material begins to form in the nucleus, and gradually this cloud condenses into a number of linear chromosomes. The chromosomes can be distinguished under the microscope by their shape and by the way that they stain. (Stains are dyes added to cells in the laboratory that allow researchers to distinguish different parts of a cell.) Different organisms have different numbers of chromosomes, but in **diploid** organisms, chromosomes come in **homologous pairs** (pairs whose members have similar shapes and staining patterns). All primates are diploid, but other organisms have a variety of arrangements. Diploid organisms also vary in the number of chromosome pairs their cells have. The fruit fly *Drosophila* has four pairs of chromosomes, humans have 23, and some organisms have many more.

Mitosis has two features that suggest that the chromosomes play an important role in determining the properties of organisms. First, mitosis involves duplication of the original set of chromosomes, so that each new daughter cell has an exact copy of the chromosomes present in its parent. This means that, as an organism

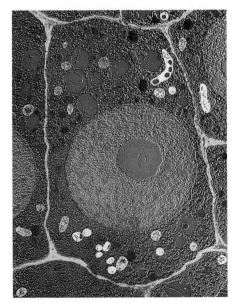

Figure 2.5

The cells of all plants and animals contain a body called the nucleus. The nucleus contains the chromosomes.

Diploid parent cell   Diploid daughter cells   Diploid parent cell   Haploid gametes

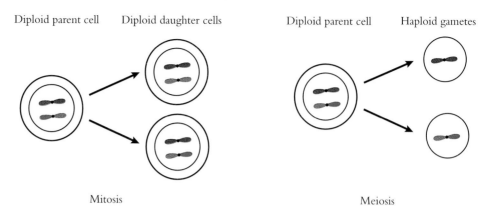

Mitosis   Meiosis

Figure 2.6

In diploid cells there are *n* pairs of homologous chromosomes; *n* varies widely among species, but here *n* = *1*. Members of homologous pairs may differ in the alleles they carry at specific sites along the chromosome. Mitosis duplicates the chromosomes, while meiosis creates gametes that carry only one member of each homologous pair of chromosomes.

grows and develops through a sequence of mitotic divisions, every cell will have the same chromosomes that were present when the egg and sperm united. Second, the material that makes up the chromosome is present even when cells are not dividing. Cells spend little of their time dividing; most of the time they are in a "resting" period, doing what they are supposed to do as liver cells, muscle cells, bone cells, and so on. During the resting period, chromosomes are not visible. However, the material that makes up the chromosomes is always present in the cell.

In meiosis, the special cell-division process that produces gametes, only half of the chromosomes are transmitted from the parent cell to the gamete.

The sequence of events that occur during mitosis is quite different from the sequence of events that occur during **meiosis**, the special form of cell division leading to the production of gametes. The key feature of meiosis is that each gamete contains only *one* copy of each chromosome, while cells that undergo mitosis contain a *pair* of homologous chromosomes. Cells that contain only one copy of each chromosome are said to be **haploid** (Figure 2.6). When a new individual is conceived, a haploid sperm from the father unites with a haploid egg from the mother to produce a diploid **zygote**. The zygote is a single cell that then divides mitotically over and over again to produce the millions and millions of cells that make up an individual's body.

Chromosomes and Mendel's Experimental Results

Mendel's two principles can be deduced from the assumption that genes are carried on chromosomes.

In 1902, less than two years after the rediscovery of Mendel's findings, Walter Sutton, a young graduate student at Columbia University, made the connection between chromosomes and the properties of inheritance discovered by Mendel's principles. Recall that the first of Mendel's two principles states that an organism's observed characteristics are determined by particles acquired from each of the parents. This fits with the idea that genes reside on chromosomes because individuals inherit one copy of each chromosome from each parent. The idea that observed characteristics are determined by genes from both parents is consistent with the observation that mitosis transmits a copy of both chromosomes to every daughter cell, so every cell contains copies of both the maternal and paternal chromosomes. Mendel's second principle states that genes segregate independently. The observation that meiosis involves the creation of gametes with only one of the two possible chromosomes from each homologous pair is consistent with two notions: (1) that one gene is inherited from each parent, and (2) that each of these genes is equally likely to be transmitted to gametes. Not everyone agreed with Sutton, but over the next 15 years, T. H. Morgan and his colleagues at Columbia performed many experiments that proved Sutton right.

Different varieties of a particular gene are called alleles. Individuals with two copies of the same allele are homozygous, while individuals with different alleles are heterozygous.

To see more clearly the connection between chromosomes and the results of Mendel's experiments with the peas, we need to introduce some new terms. The word *gene* is used to refer to the material particles carried on chromosomes. Later you will learn that genes are made of a molecule called DNA. **Alleles** are varieties of genes with the same effect on the organism. Individuals with two copies of the same allele are **homozygous** for that allele and are called homozygotes. When individuals carry copies of two different alleles, they are said to be **heterozygous** for those alleles and are called heterozygotes.

Consider the case where all the yellow individuals in the parental generation carry two genes for yellow pea color, one on each chromosome. We will use the symbol *A* for this allele. Thus, these plants are homozygous *(AA)* for yellow pea color. All of the individual plants with green peas are homozygous for a different allele, which we denote *a*; thus, green-pea plants are *aa*. As we will see, this is the only pattern that is consistent with Mendel's model. What happens if we cross two yellow-pea plants with each other, or two green-pea plants with each other? Since they are homozygous, all of the gametes produced by the yellow parents will carry the *A* allele. This means that all of the offspring produced by the crossing of two *AA* parents will also be homozygous for that *A* allele, and therefore will also produce yellow peas. Similarly, all of the gametes produced by parents that are homozygous for the *a* allele will carry the *a* allele; when the gametes of two *aa* parents unite, they will produce only *aa* individuals, with green peas. Thus, we can explain why each type breeds true.

A cross between a homozygous dominant parent and a homozygous recessive parent produces all heterozygotes in the F_1 generation.

Next, let's consider the offspring of a mating between a true-breeding green parent and a true-breeding yellow parent (Figure 2.7). The green parent produces only *a* gametes, and the yellow parent produces only *A* gametes. Thus, every one of their offspring inherits an *a* gamete from one parent and an *A* gamete from the other parent. According to Mendel's model, all the individuals in the F_1 generation will be *Aa*. Since Mendel discovered that all offspring of such crosses have yellow peas, it must be that heterozygotes bear yellow peas. To describe these effects,

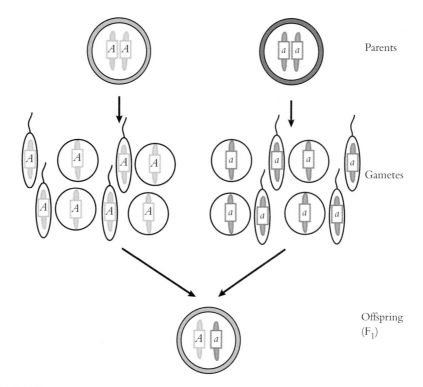

Parents

Gametes

Offspring
(F₁)

Figure 2.7

In Mendel's experiments, crosses of two true-breeding lines produced offspring that all
had the same yellow color phenotype. All of the gametes produced by homozygous *AA*
parents carry the *A* allele. Similarly, all of the gametes produced by homozygous *aa* parents carry the *a* allele. All of the zygotes from an *AA* × *aa* mating get an *A* from one
parent and an *a* from the other parent. Thus all F₁ offspring are *Aa*, and since *A* is dominant, they all produce yellow seeds.

geneticists use the following four terms: **Genotype** refers to the particular combination of genes or alleles that an individual carries, while **phenotype** refers to
the observable characteristics of the organism, such as the color of the peas in
Mendel's experiments. The *A* allele is **dominant** because individuals with only
one copy of the dominant allele have the same phenotype, yellow peas, as individuals with two copies of that allele. And the *a* allele is **recessive** because it has
no effect on phenotype in heterozygotes. As you can see in Table 2.1, *AA* and *Aa*
individuals have the same phenotype but different genotypes. Note that knowing

Table 2.1   The relationship between
genotype and phenotype in Mendel's
experiment on pea color.

GENOTYPE	PHENOTYPE
AA	Yellow
Aa	Yellow
aa	Green

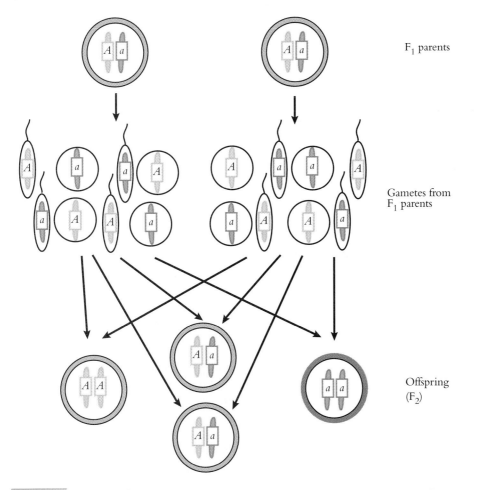

Figure 2.8

Crosses of the F_1 heterozygotes yielded a 3:1 ratio of phenotypes in Mendel's experiments. All the parents are heterozygous, or Aa. This means that half of the gametes produced by each parent carry the A allele, and the other half produce the a allele. Thus, $\frac{1}{4}$ of the zygotes will be AA, $\frac{1}{2}$ will be Aa, and the remaining $\frac{1}{4}$ will be aa. Since A is dominant, $\frac{3}{4}$ of the offspring will produce yellow seeds.

an individual's observable characteristics, or phenotype, does not necessarily tell you its genetic composition, or genotype.

A cross between heterozygous parents produces a predictable mixture of all three genotypes.

Now consider the second stage of Mendel's experiment, crossing members of the F_1 generation with each other to create an F_2 generation (Figure 2.8). We have seen that every individual in the F_1 generation is heterozygous, *Aa*. This means that half of their gametes contain a chromosome with an *A* allele and half with an *a* allele. (Remember, meiosis produces haploid gametes.) On average, if we draw pairs of gametes at random from the gametes produced by the F_1 generation, $\frac{1}{4}$ of the individuals will be *AA*, $\frac{1}{2}$ *Aa*, and $\frac{1}{4}$ *aa*.

To see why there is a 1:2:1 ratio in the F_2 generation, it is helpful to construct an event tree, like the one shown in Figure 2.9. First, let's pick the paternal gamete. Every male in the F_1 generation is heterozygous—he has one chromosome with

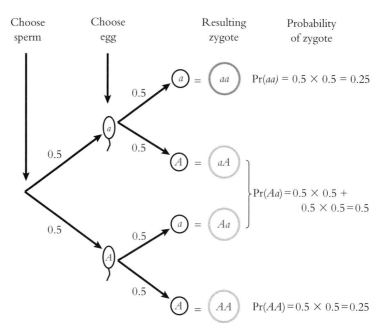

Choose sperm	Choose egg	Resulting zygote	Probability of zygote

Figure 2.9

This event tree shows why there is a 1:2:1 genotypic ratio among offspring in the F_2 generation. Imagine forming a zygote by first choosing a sperm and then an egg. The numbers along each branch give the probability that a choice at the previous node ends on that branch. The circles at the end of each branch are the resulting zygote, and the color of the ring denotes the phenotype, yellow or green peas. The first node represents the choice of a sperm. There is a 50% chance of selecting a sperm carrying the A allele, and a 50% chance of selecting a sperm carrying the a allele. Now choose an egg. Once again, there is a 50% chance of selecting an egg with each allele. Thus, the probability of getting both an A sperm and an A egg is 0.5 × 0.5 = 0.25. Similarly, the chance of getting an a egg and an a sperm is 0.5 × 0.5 = 0.25. The same kind of calculation shows that there is a 25% chance of getting an a sperm and an A egg, and a 25% chance of getting an A sperm and an a egg. Thus, the probability of getting an Aa zygote is 50%.

an A allele, and one with an a allele. Thus, there is a probability of $\frac{1}{2}$ of getting a sperm that carries an A and a probability of $\frac{1}{2}$ of getting a sperm that carries an a. Suppose you select an A by chance. Now pick the maternal gamete. Once again you have a $\frac{1}{2}$ chance of getting an A and a $\frac{1}{2}$ chance of getting an a. The probability of getting two As, one from the father and one from the mother, is

$$\Pr(AA) = \Pr(A \text{ from dad}) \times \Pr(A \text{ from mom}) = \tfrac{1}{2} \times \tfrac{1}{2} = \tfrac{1}{4}$$

If you repeated this process a large number of times—first picking male gametes and then picking female gametes—roughly $\frac{1}{4}$ of the F_2 individuals produced would be AA. Similar reasoning shows that $\frac{1}{4}$ will be aa. One-half will be Aa because there are two ways of combining the A and a alleles: the A could come from the mother and the a from the father, or vice versa. Since both AA and Aa individuals have yellow peas, there will be three yellow individuals for every green one among the offspring of the F_1 parents. Another way to visualize this result is to construct a diagram called a **Punnett square** like the one shown in Figure 2.10.

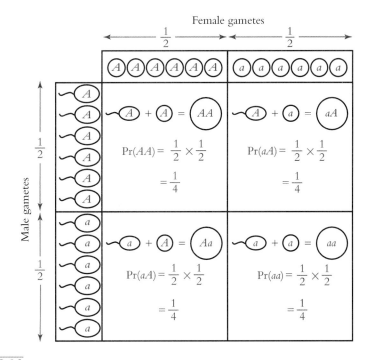

Figure 2.10

This diagram, called a Punnett square, provides another way to see why there is a 1:2:1 genotypic ratio among offspring in the F_2 generation. The horizontal axis is divided in half to reflect the equal proportion of A and a eggs. The vertical axis is divided according to the proportion of sperm carrying each allele, and again it is divided in half. The areas of squares formed by the intersection of the vertical and horizontal dividing lines give the proportion of zygotes that result from each of the four possible fertilization events: AA = 0.25 (one of the four squares), Aa = 0.25 + 0.25 = 0.50 (two of the four squares), and aa = 0.25 (one of the four squares). Thus, zygotes will have genotypes in the ratio 1:2:1.

Linkage and Recombination

Mendel also performed experiments involving two traits he believed showed that separate characters segregate independently.

Mendel also performed experiments that involved two characters. For example, he crossed individuals that bred true for two different traits—pea color and seed texture. He crossed plants with smooth yellow seeds and plants with wrinkled green seeds. All of the F_1 individuals were smooth-yellow, but the F_2 individuals occurred in the following ratio:

9 smooth-yellow : 3 smooth-green : 3 wrinkled-yellow : 1 wrinkled-green

This experiment is important because it demonstrates that sexual reproduction shuffles genes that affect different traits and thereby produces new combinations of traits, a phenomenon called **recombination**. This process is extremely important for maintaining variation in natural populations. We will return to this issue in Chapter 3.

To understand Mendel's experiment, recall that each of the parental characters (pea color and seed texture) breeds true, which means that each is homozygous for a different allele controlling seed color: the yellow line is *AA,* and the green line is *aa*. Since seed texture is also a true-breeding character, the parental lines must also be homozygous for genes affecting seed texture. Let's suppose that the parents with smooth seeds are *BB* and the parents with wrinkled seeds are *bb*. Thus, in the parental generation there are only two genotypes, *AABB* and *aabb*. By the F_2 generation, sexual reproduction has generated all 16 possible genotypes and two new phenotypes—smooth-green and wrinkled-yellow. The details are given in Box 2.1. The 9:3:3:1 ratio of phenotypes tells us that the genes determining seed color and the genes determining seed texture each segregate independently, as Mendel's second law predicts. Thus, knowing that a gamete has the *A* allele tells us nothing about whether it will have *B* or *b*—it is equally likely to carry either of these alleles. Mendel's experiments convinced him that all traits segregate independently. However, today we know that independent segregation occurs only when the traits measured are controlled by genes that reside on different chromosomes.

Genes are arranged on chromosomes like beads on a string.

It turns out that the genes for a particular character occur at a particular site on a particular chromosome. Such a site is called a **locus**, and several sites are called **loci**. Loci are arranged on the chromosomes in a line, like beads on a string. A particular locus may hold any of several alleles. The gene for seed color is always in the same position on a particular chromosome, whether it codes for green or for yellow seeds. Keep in mind that since organisms have more than one pair of chromosomes, the locus for seed color may be located on one chromosome and the locus for seed texture located on another chromosome. All of the genes carried on all the chromosomes are referred to as the **genome**.

Traits may not segregate independently if they are affected by genes on the same chromosome.

Mendel's conclusion that traits segregate independently is only true if the loci that affect the traits are on different chromosomes because links between loci on the same chromosome alter the patterns of segregation. When loci for different traits occur on the same chromosome, they are said to be **linked**; loci on different chromosomes are said to be **unlinked**. You might think that genes at two loci on the same chromosome would always segregate together as if they were a single gene. If this were true, then a gamete receiving a particular chromosome from one of its parents would get all of the same genes that occurred on that chromosome in its parent. There would be no recombination. However, chromosomes frequently get tangled and break as they are replicated during meiosis. Thus, chromosomes are not always preserved intact, and genes on one chromosome are sometimes shifted from one member of a homologous pair to the other (Figure 2.11). We call this process **crossing over**. Linkage reduces the rate of recombination, but does not eliminate it altogether. The rate at which recombination generates novel combinations of genes at two loci on the same chromosome depends on the likelihood of a crossing-over event occurring. If two loci are located close together on the chromosome, crossing over will be rare, and the rate of recombination will be low. If the two loci are located far apart, then crossing over will be common and the rate of recombination will approach the rate for genes on different chromosomes. This process is discussed more fully in Box 2.1.

BOX 2.1

More on Recombination

The first step to a deeper understanding of recombination is to see the connections between Mendel's two-trait experiment, independent segregation, and chromosomes. Mendel crossed a smooth-yellow *(AABB)* parent and a wrinkled-green *(aabb)* parent to produce members of an F_1 generation. Smooth-yellow parents produce only *AB* gametes, and wrinkled-green parents produce only *ab* gametes; all members of the F_1 generation are therefore *AaBb*. If we assume that the genes for seed color and seed texture enter gametes independently, then each of the four possible types—*AB, Ab, aB,* and *ab*—will be represented by $\frac{1}{4}$ of the gametes produced by members of the F_1 generation (Figure 2.11).

With this information we can construct a Punnett square predicting the proportions of each genotype in the F_2

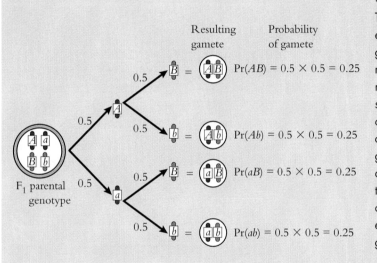

Figure 2.11

This event tree shows why an F_1 parent is equally likely to produce all four possible gametes when genes for two traits are carried on different chromosomes. The first node represents the choice of the chromosome carrying the gene for seed color (*A* or *a*), and the second node represents the choice of the chromosome that carries the gene for seed texture (*B* or *b*). The number along each branch gives the probability that the choice at the previous node ends on that branch. The circles at the end of each branch represent the resulting gametes.

generation (Figure 2.12). Once again, we divide the vertical and horizontal axes in proportion to the frequency of each type of gamete, in this case dividing each axis into four equal parts. The areas of the rectangles formed by the intersection of the vertical and horizontal dividing lines give the proportion of zygotes that result from each of the 16 possible fertilization events. Since the area of each cell in this matrix is the same, the phenotypic ratios of zygotes that result from each of the 16 possible fertilization events can be determined in this example by simply counting the number of squares that contain each phenotype. There are nine smooth-yellow squares, three wrinkled-yellow squares, three smooth-green squares, and one wrinkled-green square. Thus, the 9:3:3:1 ratio of phenotypes that Mendel observed was consistent with the assumption that genes on different chromosomes segregate independently. If the genes controlling these traits did not segregate independently, the ratio of phenotypes would be different, as we will see in the following paragraph.

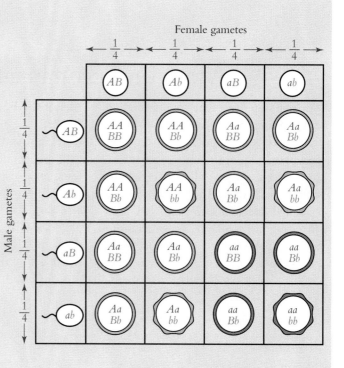

Female gametes

Figure 2.12

This Punnett square shows why there is a 9:3:3:1 phenotypic ratio among offspring of the F$_2$ generation when the genes for the two traits, seed color and seed texture, are carried on different chromosomes. The rings in each square show the color (green and yellow) and seed texture (wrinkled or smooth) of the phenotype associated with each genotype.

During meiosis chromosomes frequently become damaged, break, and recombine. This process, called crossing over, creates chromosomes with combinations of genes not present in the parent (see text). In the example in the previous paragraphs, remember that all members of the F$_1$ generation were *AaBb*. Now suppose that the locus controlling seed color and the locus controlling seed texture are carried on the same chromosome. Further assume that when the chromosomes are duplicated during meiosis, a fraction *r* of the time there is crossing over (Figure 2.13), and a fraction 1 − *r* of the time there is no

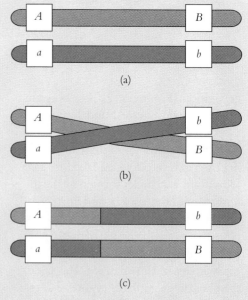

Figure 2.13

Crossing over during meiosis sometimes leads to recombination and produces novel combinations of traits. Suppose that the *A* allele leads to yellow seeds and the *a* allele to green seeds, while the *B* allele leads to smooth seeds and the *b* allele to wrinkled seeds. (a) Here, an individual carries one *AB* chromosome and one *ab* chromosome. (b) During meiosis, the chromosomes are damaged and crossing over occurs. (c) Now, the *A* allele is paired with *b* and the *a* allele is paired with *B*.

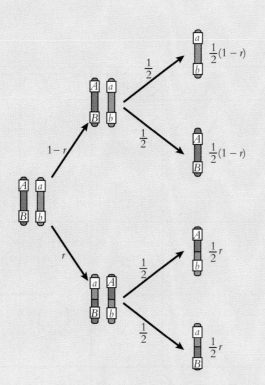

Figure 2.14

This event tree shows how to calculate the fraction of each type of gamete that will be produced when genes are carried on the same chromosome. The first node represents whether or not crossing over takes place. There is a probability r that crossing over takes place and produces novel trait combinations, and a probability $1 - r$ that there is no crossing over. At the second node, chromosomes are randomly assigned to gametes. The value given above each branch represents the probability of reaching that branch from the previous node. The likelihood of forming each type of genotype is the product of the probabilities along each pathway.

crossing over. Next, chromosomes segregate independently into gametes. The types of chromosomes present in the parental generation, Ab and ab, each occur in a fraction $(1 - r)/2$ of the gametes, while the novel, recombinant types Ab and aB each occur in a fraction $r/2$ of the gametes (Figure 2.14).

Now we can use a Punnett square to calculate the frequency of each of the 16 possible genotypes in the F_2 generation (Figure 2.15). As before, we divide the vertical and horizontal axes in propor-

tion to the relative frequency of each type of gamete, and the area of the rectangles formed by the intersection of these grid lines gives the frequency of each genotype. If recombination rates are low, most members of the F_2 generation will be of three genotypes— $AABB$, $AaBb$, and $aabb$—just as if there were only two alleles, AB and ab. However, if the recombination rate is higher, more of the novel recombinant genotypes will be produced.

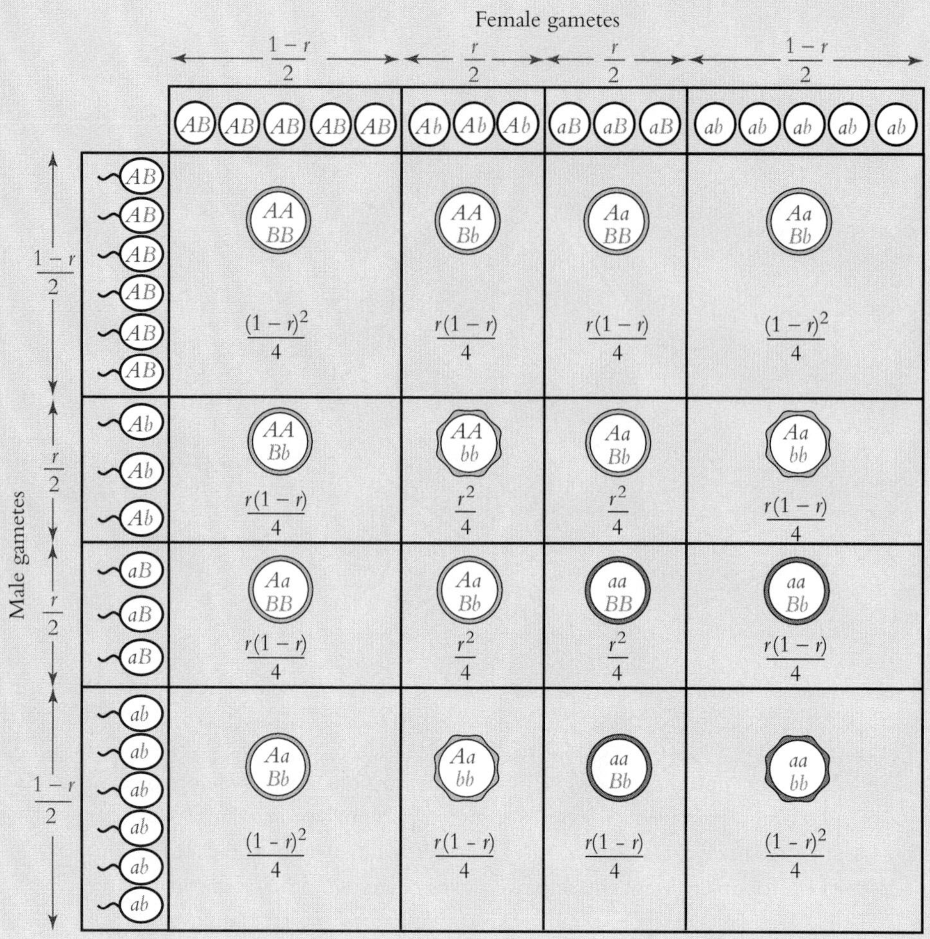

Figure 2.15

This Punnett square shows how to calculate the frequency of each phenotype among off-spring in the F₂ generation if the genes for seed color and seed texture are carried on the same chromosome. The horizontal axis is divided according to the proportion of each type of egg, *AB*, *Ab*, *aB* and *ab*. When genes are carried on the same chromosome, the frequency of each type of gamete is calculated as shown in Figure 2.14. The vertical axis is divided according to the proportion of each type of sperm, and the horizontal axis is divided to the proportion of each type of egg. The areas of the rectangles formed by the intersection of the vertical and horizontal dividing lines give the proportions of zygotes that result from each of the 16 possible fertilization events.

Molecular Genetics

Genes are segments of a long molecule called DNA, which is contained in chromosomes.

In the first half of the 1900s, geneticists made substantial progress in describing the cellular events that took place during meiosis and mitosis, and in understanding the chemistry of reproduction. For instance, by the middle of this century it was known that chromosomes contain two structurally complex molecules—protein and **deoxyribonucleic acid**, or **DNA**. It had also been determined that the particle of heredity postulated by Mendel was DNA, not protein, though exactly how DNA might contain and convey the information essential to life was still a mystery. But in the early 1950s, two young biologists at Cambridge University, Francis Crick and James Watson, made a discovery that revolutionized biology: they deduced the structure of DNA. Watson and Crick's elucidation of the structure of DNA was the wellspring of a great flood of research that continues to provide a deep and powerful understanding of how life works at the molecular level. We now know how DNA stores information and how this information controls the chemistry of life, and this knowledge explains why heredity leads to the patterns Mendel described in pea plants, and why there are sometimes new variations.

Understanding the chemical nature of the gene is critical to the study of human evolution: (1) molecular genetics links biology to chemistry and physics, and (2) molecular methods help us to reconstruct the evolutionary history of the human lineage.

Modern molecular genetics, the product of Watson and Crick's discovery, is a field of great intellectual excitement producing many new discoveries of practical value in medicine and agriculture. However, progress in molecular genetics has not yet fundamentally changed our understanding of how morphology and behavior evolve. Morphological and behavioral traits are usually affected by many genes that interact in ways that are for the most part still too complex to understand at the molecular level. There are several important exceptions to this generalization. For example, in recent years there has been rapid progress in our understanding of the genes that control the development of body size and shape, and the genetic basis of certain forms of behavior. Nonetheless, an understanding of molecular genetics is crucially important in understanding the evolution of the human phenotype for two reasons:

1. Molecular genetics links biology to chemistry and physics. One of the grandest goals of science is to provide a single, consistent explanatory framework for the way the world works. We want to place evolution in this grand scheme of scientific explanation. It is important to be able to explain not only how new species of plants and animals arise, but also the evolution of a wide range of phenomena—from the origin of stars and galaxies to the rise of complex societies. Modern molecular biology is of profound importance because it explains how life and evolution work at the level of physics and chemistry, and it thereby connects physical and geochemical evolution to Darwinian processes.
2. Molecular genetics provides important data for reconstructing evolutionary history. In recent years, molecular geneticists have used information about variation in DNA sequences to reconstruct the history of particular lineages. To see how this is done, you have to understand a little about the molecular biology of the gene.

Genes Are DNA

DNA is unusually well suited to be the chemical basis of inheritance.

The discovery of the structure of DNA was fundamental to genetics because the structure itself implied how inheritance must work. Each chromosome contains a single DNA molecule roughly two meters long that is folded up to fit in the nucleus. DNA molecules consist of two long strands, and each strand has a "back-bone" of alternating sequences of phosphate and sugar molecules. Attached to each sugar is one of four molecules, collectively called **bases: adenine, guanine, cytosine,** or **thymine**. The two strands of DNA are held together by very weak chemical bonds, called hydrogen bonds, which connect some of the bases on different strands. Thymine bonds only with adenine, and guanine bonds only with cytosine (Figure 2.16).

The repeating four-base structure of DNA allows the molecule to assume a vast number of distinct forms. Furthermore, a staggering number of DNA molecules exist in nature that are equally stable chemically. DNA is not the only complex molecule with many alternative forms, but other molecules have some forms that

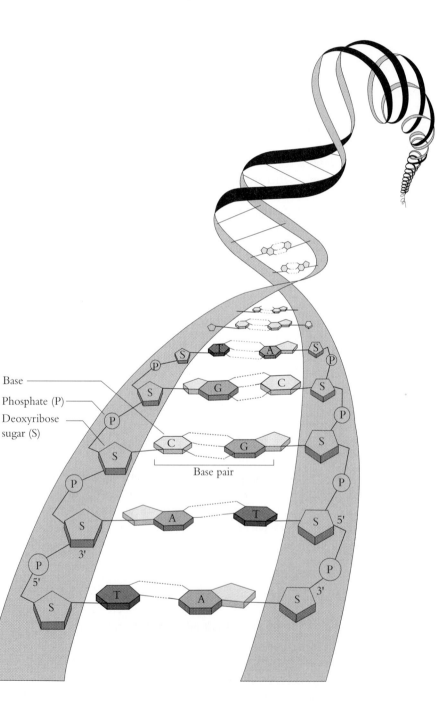

Figure 2.16

The chemical structure of DNA consists of two long backbones made of alternating sugar and phosphate molecules. One of four bases (adenine, guanine, cytosine, or thymine) is attached to each of the sugars. The two strands are connected to each other by hydrogen bonds (dotted lines) between certain pairs of bases. Thymine bonds only to adenine, and guanine bonds only to cytosine.

Base

Phosphate (P)

Deoxyribose sugar (S)

Base pair

are less stable than others. Such molecules would be unsuitable for carrying information because the messages would degrade (become garbled) as the molecules changed toward a more stable form. What makes DNA unusual is that all of its nearly infinite number of forms are equally stable.

Each DNA configuration is exactly like a message written in an alphabet with letters that stand for each of the four bases (T for thyamine, A for adenine, G for guanine, and C for cytosine). Thus,

TCGGTAGTAGTTACGG

is one message, while

ATCCGGATGCAATCCA

is another message. Since the DNA in a single chromosome is millions of bases long, there is room for a nearly infinite variety of messages.

In addition to preserving a message faithfully, hereditary material must be replicable. Without the ability to make copies of itself, the genetic message that directs the activities of living cells could not be spread to offspring, and natural selection would be impossible. DNA is replicated within cells by a highly efficient cellular machinery: it first unzips the two strands, and then, with the help of other specialized molecular machinery, it adds complementary bases to each of the strands until two identical sugar and phosphate backbones are built (Figure 2.17).

The Chemical Basis of Life

By acting as catalysts, proteins called enzymes are important in helping the genes of DNA build cells.

It's no exaggeration to say that one organism has the form and structure that it does, is the way it is, and not like something else, because of the very large number of particular chemical compounds that make up its cells and organs. But if organisms differ chemically, then how do different organisms use the same raw materials and produce the sets of chemicals that make them unique? What process makes the intestinal cells of a blue whale different from the cells that line the gut of a cockroach?

The answer is that the hereditary material—the DNA in genes—determines what the raw materials are transformed into when used to build cells (Figure 2.18). To understand how genes build cells, we need to understand the chemical process called **catalysis**. Consider what a cell might do to a **glucose** molecule, a simple sugar composed of carbon, oxygen, and hydrogen atoms arranged in a particular way. Suppose we put some glucose in a sterile environment containing oxygen at room temperature. What would happen? The answer depends on how long you would be willing to wait. In the short run not much would happen, but after an extremely long time the carbon and hydrogen would combine with oxygen, and the glucose would be converted to carbon dioxide (CO_2) and water (H_2O). This change might take a very, very long time, but it *would* eventually happen. Now, suppose we raised the temperature to 1000°F. In this case, sugar would be very rapidly converted to CO_2 and H_2O, quite a bit of heat would be released, and perhaps some light would be produced. In other words, the glucose would catch fire or even explode. What is happening here?

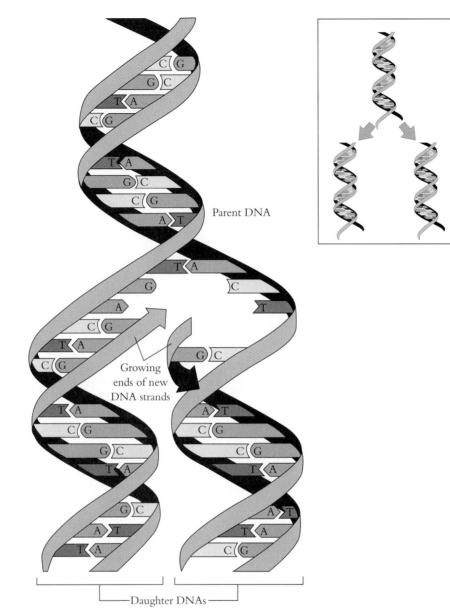

Parent DNA

Growing
ends of new
DNA strands

Daughter DNAs

Figure 2.17

When DNA is replicated, the two strands are separated and two daughter DNA strands are formed.

At room temperature the reaction that forms carbon dioxide and water from glucose and oxygen proceeds slowly because there is an intermediate state, a particular conformation of the molecules, that must be reached before the reaction occurs. Achieving the intermediate state requires a threshold amount of energy

Figure 2.18

Life in a nutshell: the cells of organisms are composed of a large number of very complicated chemicals. The cells of different organisms vary *because* they are made up of different chemicals. The world outside the organism is made (for the most part) of much smaller, simpler chemical compounds. DNA determines the structure of organisms—their phenotypes—by controlling which complicated chemicals get made from the simple raw materials available in the environment.

The Environment

Small number
of simple
chemical compounds

The Organism

Large number of
complex chemical
compounds

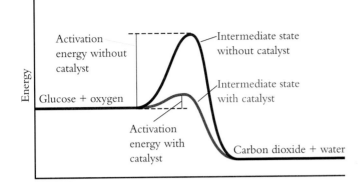

Figure 2.19

Glucose and oxygen react to form carbon dioxide and water. For glucose and oxygen to react, they must first achieve a higher energy state, represented by the mound in the middle. Once they achieve this state, they rapidly react to form CO_2 and H_2O. The rate at which the reaction occurs is determined by the rate at which glucose and oxygen combine to reach the intermediate state. Because molecules move more rapidly at higher temperatures, more molecules achieve the necessary activation energy, and the reaction occurs more rapidly. Organic catalysts increase rates of reaction by changing the way that the molecules interact, thereby reducing the activation energy necessary to achieve the intermediate state.

called the activation energy. At room temperature, only a tiny fraction of the molecules have enough energy (move fast enough) to reach the intermediate state. At higher temperatures, more of the molecules move much faster, so more molecules achieve the activation energy, and the reaction proceeds much more rapidly. Sometimes the addition of another material reduces the activation energy, enabling the reaction to proceed at a rapid rate at lower temperatures, even at room temperature (Figure 2.19). Such materials are called **catalysts**, and in biological systems many proteins play this role. Proteins that act as catalysts are called **enzymes**.

Genes regulate the chemical processes of living cells by determining which enzymes are present to catalyze cellular reactions.

The best way to understand how enzymes determine the characteristics of organisms is to think of an organism's biochemical machinery as a branching tree. Glucose serves as a food source for many cells, meaning that it provides energy and a source of raw materials for the construction of cellular structures. Glucose might initially undergo any one of an extremely large number of slow-moving reactions. The presence of particular catalytic enzymes will determine which reactions occur rapidly enough to alter the chemistry of the cell. For example, some enzymes lead to the metabolism of glucose and the release of its stored energy. Thus, enzymes act as switches to determine what chemicals will be present in the cell (Figure 2.20). At the end of the first branch there is another node representing all of the reactions that could involve the product(s) of the first branch. Here, one or more enzymes will determine what happens next.

This picture has been greatly simplified. Real organisms take in many different kinds of compounds; and each compound is involved in a complicated tangle of branches that biochemists call pathways. Real **biochemical pathways** are very

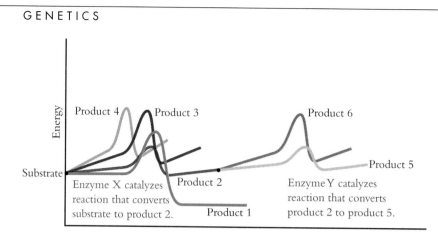

Figure 2.20

Enzymes control the chemical composition of cells by catalyzing some chemical reactions but not others. In this hypothetical example, the molecules that provide the initial raw material (called a substrate) of the pathway could undergo four different reactions yielding different molecules, labeled products 1 through 4. However, since enzyme X is present, the reaction that yields product 2 proceeds much more rapidly than the other reactions (note that it has the lowest activation energy), and all of the substrate is converted to that product. Product 2 could then undergo two different reactions yielding products 5 and 6. When enzyme Y is present, it lowers the activation energy, thus causing the reaction yielding product 5 to proceed much more rapidly, and only product 5 is produced. In this way, enzymes link products and reactants together into pathways that satisfy particular chemical objectives, such as the extraction of energy from glucose.

complex, as you can see in Figure 2.21, which shows the pathway for the conversion of glucose to energy in animal cells. One set of enzymes causes glucose to be shunted to a pathway that yields energy. A different set of enzymes causes the glucose to be shunted to a pathway that binds glucose molecules together to form glycogen, a starch that functions as energy storage. The presence of a third set would lead instead to the synthesis of cellulose, a complex molecule that provides the structural material in plants. Enzymes play roles in virtually all cellular processes—from the replication of DNA and division of cells to the contraction and movement of muscles.

DNA Codes for Protein

Once the structure of DNA was known, a new puzzle emerged: how does DNA determine which enzymes will be synthesized? It took more than a decade to discover the amazing answer. To understand this, it is first necessary to know about the structure of enzymes.

The sequence of amino acids in proteins determines each protein's enzymatic properties.

Like all **proteins**, enzymes are constructed of amino acids. There are 20 different amino acid molecules. All **amino acids** have the same chemical backbone, but they differ in the chemical composition of the side chain connected to this backbone (Figure 2.22). The sequence of amino acid side chains, called the **primary structure** of the protein, is what makes one protein different from others. You can think of a protein as a very long railroad train in which there are 20 different kinds of cars, each representing a different amino acid. The primary structure is a list of the types of cars in the order that they occur.

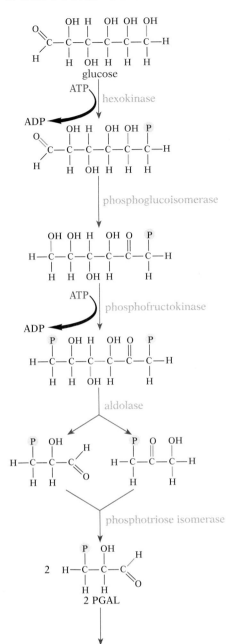

Figure 2.21

This is the glycolysis pathway, a biochemical pathway found in animals that converts glucose to energy that can be used to power cellular processes. Each step involves a chemical reaction controlled by a different enzyme.

When proteins are actually doing their catalytic business, they are folded in complex ways. The three-dimensional shape of the folded protein, called the **tertiary structure**, is crucial to its catalytic function. The way the protein folds depends on the sequence of amino acid molecules that make up its primary sequence. This means that the function of enzymes depends on the sequence of the amino acids that make them up. (Proteins also have secondary structure, and sometimes also quaternary structure, but to keep things simple, we will ignore them here.)

These ideas are illustrated in Figure 2.23, which shows the folded shape of a part of a **hemoglobin** molecule, a protein that transports oxygen from the lungs to the tissues via red blood cells. As you can see, the protein folds into a roughly spherical glob, and oxygen is bound to the protein near the center of the glob. **Sickle-cell anemia**, a condition common among people in West Africa and

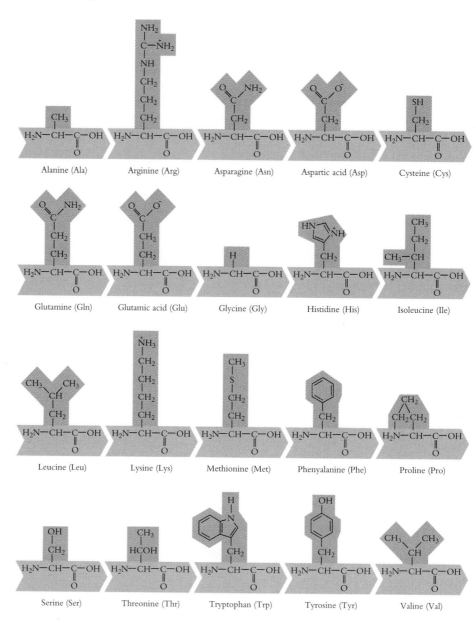

Figure 2.22

Amino acids all share the same chemical backbone, here colored blue. They differ according to the chemical structure of the side group attached to the backbone, here colored pink.

Alanine (Ala) Arginine (Arg) Asparagine (Asn) Aspartic acid (Asp) Cysteine (Cys)

Glutamine (Gln) Glutamic acid (Glu) Glycine (Gly) Histidine (His) Isoleucine (Ile)

Leucine (Leu) Lysine (Lys) Methionine (Met) Phenylalanine (Phe) Proline (Pro)

Serine (Ser) Threonine (Thr) Tryptophan (Trp) Tyrosine (Tyr) Valine (Val)

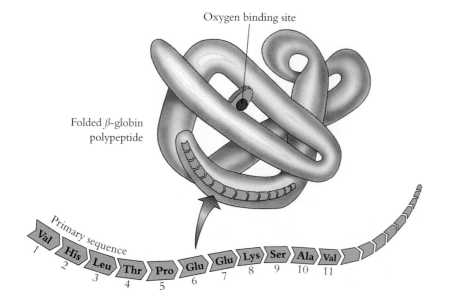

Oxygen binding site

Folded β-globin polypeptide

Primary sequence

Val His Leu Thr Pro Glu Glu Lys Ser Ala Val
1   2   3   4   5   6   7   8   9   10 11

Figure 2.23

The primary and tertiary structure of hemoglobin, a protein that transports oxygen on red blood cells. The primary structure is the sequence of amino acids making up the protein. The tertiary structure is the way the protein folds into three dimensions.

among African Americans, is caused by a single change in the primary sequence of amino acids in the hemoglobin molecule. Glutamic acid is the sixth amino acid in normal hemoglobin molecules, but in people afflicted with sickle-cell anemia, valine is substituted for glutamic acid. This single substitution changes the way that the molecule folds, and reduces its ability to bind oxygen.

DNA specifies the primary structure of protein.

Now we return to our original question: how does the information contained in DNA—its sequence of bases—determine the structure of proteins? Remember that DNA encodes messages in a four-letter alphabet. Researchers determined that these letters are combined into three-letter "words" called **codons**, each of which specifies a particular amino acid. Since there are four bases, there are 64 possible three-letter combinations for codons (four possibilities for the first base times four for the second times four for the third, or $4 \times 4 \times 4 = 64$). Sixty-one of these codons are used to code for the 20 amino acids that make up proteins. For example, the codons GCT, GCC, GCA, and GCG all code for alanine; GAT and GAC code for asparagine; and so on. The remaining three codons are "punctuation marks" that mean either "start, this is the beginning of the protein" or "stop, this is the end of the protein." Thus, if you can identify the base pairs, it is a simple matter to determine what proteins are encoded on the DNA.

You might be wondering why several different codons code for the same amino acid. This redundancy serves an important function. Since there are a number of processes that can damage DNA and cause one base to be substituted for another, redundancy decreases the chance that a random change will alter the primary sequence of the protein produced. Proteins represent complex adaptations, and so we would expect most changes to be deleterious (harmful). But since the code is redundant, many substitutions have no effect on the message of a particular stretch of DNA. The importance of this redundancy is underscored by the fact that the most common amino acids are the ones with the greatest number of codon variants.

Before DNA is translated into proteins, its message is first transcribed into messenger RNA.

DNA can be thought of as a set of instructions for building proteins, but the real work of synthesizing proteins is performed by other molecules. The first step in the translation of DNA into protein occurs when a facsimile of one of the strands of DNA, which will serve as a messenger or chemical intermediary, is made usually in the cell's nucleus. This copy is **ribonucleic acid**, or **RNA**. RNA is similar to DNA except that it has a slightly different chemical backbone, and the base **uracil** (denoted U) is substituted for thymine. RNA comes in several forms, all of which aid in protein synthesis. The form of RNA used in this first step is **messenger RNA (mRNA)**. mRNA then leaves the nucleus and migrates to the **cytoplasm**, the part of the cell outside the nucleus.

Each kind of amino acid is attached to a transfer RNA molecule that bears the anti-codon for that amino acid.

Meanwhile, another essential part of protein synthesis is going on in the cytoplasm. Amino acid molecules are bound to a different kind of RNA, called **transfer RNA (tRNA)**. These tRNA molecules have a triplet of bases, called an **anticodon,** at a particular site (Figure 2.24). Different tRNA molecules have

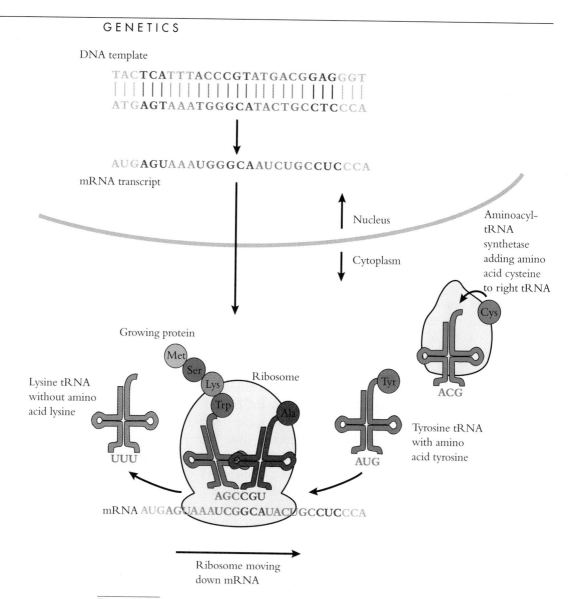

DNA template

TAC**TCATTTACCCGTATGACGGAG**GGT

ATG**AGTAAATGGGCATACTGCCTC**CCA

AUG**AGUAAAUGGGCAAUCUGCCUC**CCA

mRNA transcript

Nucleus

Cytoplasm

Aminoacyl-
tRNA
synthetase
adding amino
acid cysteine
to right tRNA

Cys

ACG

Growing protein

Met

Ser

Lys

Trp   Ala

Ribosome

Tyr

Lysine tRNA
without amino
acid lysine

UUU

AUG

Tyrosine tRNA
with amino
acid tyrosine

AGCCGU

mRNA AUGAGUAAAUCGGCAUACUGCCUCCCA

Ribosome moving
down mRNA

Figure 2.24

The information encoded in DNA determines the structure of proteins in the following
way. Inside the nucleus, an mRNA copy is made of the original DNA template. The
mRNA is coded in three-base codons. Here each codon is given a different color. For
example, the sequence AUG codes for the start of a protein and the amino acid methionine, the sequence AGU codes for serine, and the sequence AAA codes for lysine. The
mRNA then migrates to the cytoplasm. In the cytoplasm, special enzymes called aminoacyl-tRNA synthetases locate a specific kind of tRNA and attach the amino acid whose
mRNA codon will bind to the anticodon on the tRNA. For example, the mRNA codon for
cysteine is UGG and the appropriate anticodon is ACC because U binds to A and C
binds to G. In this diagram, matching mRNA codons and tRNA anticodons are given the
same color. The initiation of the process of protein assembly is complicated and involves
specialized enzymes. Once the process is started, each codon of the mRNA binds to the
ribosome. Then the matching tRNA is bound to the mRNA, the amino acid is transferred
to the growing protein, the tRNA is released, the ribosome shifts to the next codon, and
the process is repeated.

different anticodon sequences and also differ in other ways. Each type of tRNA is bound to the amino acid whose codon binds to the anticodon on the tRNA. For example, one of the codons for the amino acid alanine is the base sequence GCU, and GCU binds only to the anticodon CGA. Thus, the tRNA with the anticodon CGA binds only to the amino acid alanine. Because several different codons correspond to each amino acid, there are more types of tRNA than there are kinds of amino acids.

The tRNAs are bound to the appropriate amino acid by a type of enzyme with the tongue-twisting name **aminoacyl-tRNA synthetase**. For each type of tRNA there is a distinctive aminoacyl-tRNA synthetase that recognizes the tRNA and binds it to the appropriate amino acid. Scientists do not understand exactly how the enzymes accomplish this crucial task; in some cases the anticodon sequence seems to be important, while in other cases the enzyme seems to recognize other distinctive features of the appropriate tRNA.

A cellular organelle called the ribosome then synthesizes a particular protein by reading the mRNA copy of the gene.

The next step in the process involves the **ribosomes**. Ribosomes are small cellular **organelles** composed of protein and nucleic acid. Organelles are bounded cellular components that perform a particular function, analogous to the way organs like the liver perform a function for the body as a whole. The mRNA first binds to ribosomes at a binding site and then moves through the binding site one codon at a time. As each codon of mRNA enters the binding site, a tRNA with a complementary anticodon is drawn from the complex soup of chemicals inside the cell, and bound to the mRNA. The amino acid bound to the other end of the tRNA is then unattached from the tRNA and added to one end of the growing protein chain. The process repeats for each codon, continuing until the end of the mRNA molecule passes through the ribosome. Voilá, a new protein is ready for action.

Not All DNA Codes for Proteins

In eukaryotes the DNA that codes for proteins is interrupted by noncoding sequences called introns.

So far, our description of protein synthesis applies to almost all organisms. However, most of this information was learned studying *Escherichia coli,* a kind of bacteria common in the human gut. Like other bacteria, *E. coli* belongs to a group of organisms called the **prokaryotes** because it does not have chromosomes or a cell nucleus. In prokaryotes the DNA sequence that codes for a particular protein is uninterrupted. A stretch of DNA is copied to RNA and then translated into a protein. For many years, biologists thought that the same would prove to be true of **eukaryotes** (organisms like plants, birds, and humans that have chromosomes and a cell nucleus).

Beginning in the 1970s, new recombinant DNA technology allowed molecular geneticists to study eukaryotes. These studies revealed that in eukaryotes the segment of DNA that codes for a protein is almost always interrupted by at least one (and sometimes many) noncoding sequences, called **introns**. (The coding sequences are called **exons**.) Protein synthesis in eukaryotes includes one additional step not mentioned in our discussion so far: after the entire DNA sequence is copied to make an mRNA molecule, and while the mRNA is still in the nucleus,

Deciphering DNA

We can get an idea of what the physical map [the complete DNA sequence of the genome] looks like with a geographical analogy. Imagine the journey along the whole of your own DNA as being equivalent to one along the whole length of Britain, from Land's End to John o'Groat's via London. This is about a thousand miles altogether (which means that its American equal would be roughly equivalent to a trip from Palm Beach to New York up the Eastern Seaboard). To fit in all the DNA letters into a road map on this scale, there have to be fifty DNA bases per inch, or about three million per mile. The journey passes through twenty-three counties of different sizes. These administrative divisions, conveniently enough, are the same in number as the twenty-three chromosomes into which human DNA is packaged. The amount of DNA completely sequenced so far has covered a distance equivalent to a walk between Land's End and Falmouth, a few miles away—in the American sense, not much more than the length of the island of Manhattan. There is a long way to go yet.

The scenery for most of the trip is very tedious. Like much of modern Britain—or the northeastern states—it seems to be totally unproductive. About a third of the whole distance is covered by repeats of the same message. Fifty miles, more or less, is filled with words of five, six or more letters, repeated endlessly next to each other. Many are palindromes. They read the same backward as forward, like the obituary of Ferdinand de Lesseps—"A man, a plan, a canal: Panama!" Some of these "tandem repeats" are scattered in blocks all over the genome. The position and length of each block varies from person to person. The famous "genetic fingerprints," the unique inherited signature used in forensic work, depend on variation in the number and position of tandem repeats. To make a fingerprint an enzyme which cuts one particular repeated section is used. This slices the DNA into dozens of fragments. Other repeated sequences involve just the two letters C and A, multiplied thousands of times. Yet more of the genome is given over to occasional long and complicated messages which seem to say nothing.

It is dangerous to dismiss all this DNA as useless because we cannot understand what it says. The Chinese term "Shi" can—apparently—have seventy-three different meanings depending on how it is pronounced. It is possible to construct a sentence such as "The master is fond of licking lion spittle" just by using "Shi" again and again. This would seem like meaningless repetition to someone, like most of us, who cannot understand Chinese.

Much of the inherited landscape is littered with the corpses of abandoned genes, sometimes the same one again and again. The DNA sequences of these "pseudogenes" look rather like that of their working relatives, but they are riddled with decay and no longer produce anything. At some time in their history a crucial part of the machinery was damaged. Since then they have been gently rusting. Surprisingly enough, the same pseudogenes may turn up at several points along the journey.

After many miles of dull and repetitive DNA terrain, we begin to see places where something is being made. These are the working genes. There are some surprises in their structure, too. A functioning gene can be recognized by the order of the letters in the DNA alphabet, which start to read in words of three letters written in the genetic code, implying that it could produce a protein. Usually, there are few clues about what the protein actually does, although its structure can be deduced (and its shape inferred) from the order of the DNA letters which make it.

Many functioning genes are arranged in groups making related products. There are about a thousand of these "gene families" altogether. The best known is involved in the manufacture of the red pigment of the blood. Nearly all the DNA in the bone-marrow cells which produce the red cells of the blood is switched off. However, one small group of genes is tremendously busy. As a result they are better known than any other. Much of human molecular biology is based on this particular genetic industrial center, the beta-globin genes.

They are about halfway along the total length of the DNA; to follow the American analogy, about halfway to New York—roughly speaking at Cape Hatteras. They make some of the proteins involved in carrying oxygen. The globin industrial estate contains about half a dozen sections of DNA making related products. That for making part of the hemoglobin molecule, the red blood pigment, is quite small: about three feet long on this map's scale. A few feet away is another one which makes a globin found only in the embryo. Close to that is the rusting hulk of some equipment which stopped working years ago. The globin factory covers about a hundred feet altogether, most of which seems to be unused space between functional genes. It cooperates with another one about the same size a long way away (near Charleston on this mythical map) which produces a related protein. Joined together, the two products make the red blood pigment. Most working genes are arranged in families, either close together or scattered all over the genome.

Another surprising aspect of the map of ourselves is that genes are of very different size, from about five hundred letters long to more than two million. . . . [N]early all the working segments are interrupted by lengths of non-coding DNA. In very large genes (such as the one which goes wrong in muscular dystrophy) the great majority of the DNA codes for nothing. The non-coding DNA participates in the first part of the production process, but this segment of the genetic message is snipped out of the messenger RNA before the protein is assembled. This seems an extraordinary way to go about things, but it is the one which evolution has come up with.

Source: From pp. 64–66 in S. Jones, 1993, *The Language of Genes.* Copyright © 1994 Steve Jones. Used by permission of Doubleday, a division of Random House, Inc.

the introns are snipped out, and the mRNA molecule is spliced back together. Only then is the mRNA exported outside of the nucleus where protein synthesis takes place.

The function of introns is unclear. Some biologists think that introns facilitate the evolution of novel, useful proteins. According to this view, complex proteins are constructed from a number of standardized components—analogous to pillars, beams, and arches in buildings. Each component has its own exon because this makes it easier for exons to recombine to form novel proteins. Adherents of this hypothesis think that introns predate the origin of eukaryotes, which arose about 2 billion years ago, and that most of the modern proteins in prokaryotes and eukaryotes result from the recombination of exons. Prokaryotes have subsequently lost their introns, perhaps because DNA synthesis is more costly for such small organisms. Other biologists think that introns provide no benefit to the organism as a whole, but are bits of "selfish" DNA that replicate themselves by jumping from one place in the genome to another. According to this view, introns arose after the origin of the eurkaryotes.

Chromosomes also contain long strings of simple repeated sequences.

Introns are not the only kind of DNA that is not translated into proteins. Chromosomes also contain a lot of DNA that is composed of simple repeated patterns. For example, in fruit flies there are long segments of DNA composed of adenine and thymine in the following monotonous five-base pattern:

. . . ATAATATAATATAATATAATATAATATAATATAATATAAT . . .

In all eukaryotes, simple repeated sequences of DNA are found in particular sites on particular chromosomes.

Only DNA that is translated into protein is subject to natural selection.

DNA found in introns and in simple repeated sequences is not translated into proteins and doesn't have any direct effect on phenotype. Consequently, this DNA is not subject to natural selection. Fruit flies with the repeated sequence ATATT would be no more likely to survive and reproduce than those with ATAAT. It turns out that the bulk of the DNA in eurkaryotes is found in introns and simple repeated sequences. Molecular geneticists estimate that only about 5% of human DNA codes for proteins. Thus, the evolution of 95% of human DNA is not controlled by natural selection, but by random processes like mutation. As we will see, this fact is very important because it allows us to date distant evolutionary events using genetic information.

At this point, there is some danger of losing sight of the genes amid the discussion of introns, exons, and repeat sequences. A brief reprise may be helpful.

In summary, then, chromosomes contain an enormously long molecule of DNA. Genes are short segments of this DNA. A gene's DNA is, after suitable editing, transcribed into mRNA, which in turn is translated into a protein whose structure is determined by the gene's DNA sequence. Proteins determine the properties of living organisms by selectively catalyzing some chemical reactions and not others, and by forming some of the structural components of cells, organs, and tissues. Genes with different DNA sequences lead to the synthesis of proteins with different catalytic behavior and structural characteristics. Many changes in the morphology or behavior of organisms can be traced back to variations in the proteins and genes that build them. It is important to remember that evolution has a molecular basis. The changes in the genetic constitution of populations that we will explore in Chapter 3 are grounded in the physical and chemical properties of molecules and genes discussed here.

Further Reading

Berg, P., and M. Singer. 1992. *Dealing with Genes: The Language of Heredity*. University Science Books, Mill Valley, Calif.

Lodish, H., D. Baltimore, A. Berk, S. L. Zipursky, P. Matsudaira, and J. Darnell. 1995. *Molecular Cell Biology*. 3d ed. W. H. Freeman and Company, New York.

Maynard Smith, J. 1989. *Evolutionary Genetics*. Oxford University Press, New York.

Olby, R. C. 1966. *Origins of Mendelism*. Schocken Books, New York.

Ridley, M. 1996. *Evolution*. 2d ed. Blackwell Scientific, Oxford.

Study Questions

1. Explain why Mendel's principles follow from the mechanics of meiosis.
2. Mendel did experiments in which he kept track of the inheritance of seed texture (wrinkled or smooth). First, he created true breeding lines: parents with smooth seeds produced offspring with smooth seeds, and parents with wrinkled seeds produced offspring with wrinkled seeds. When he crossed wrinkled and smooth peas from these true-breeding lines, all of the offspring were smooth. Which trait is dominant? What happened when he crossed members of the F_1 generation? What would have happened had he backcrossed members of the F_1 generation with individuals from a true-breeding line (like their parents) with smooth seeds? What about crosses between the F_1 generation and the wrinkled true-breeding line?
3. Mendel also did experiments in which he kept track of two traits. For example, he created true-breeding lines of smooth-green and wrinkled-yellow, and then crossed these lines to produce an F_1 generation. What were the F_1 individuals like? He then crossed the members of the F_1 generation to form an F_2 generation. Assuming that the seed-color locus and the seed-texture locus are on different chromosomes, calculate the ratio of each of the four phenotypes among the F_2 generation. Calculate the ratios (approximately), assuming that the two loci are very close together on the same chromosome.
4. Many animals, like birds and mammals, have physiological mechanisms that enable them to maintain their body temperature well above the air temperature. They are called homeotherms. Poikilotherms, like snakes and other reptiles, regulate their body temperatures by moving closer to or further away from sources of heat. Use what you have learned about chemical reactions to develop a hypothesis that explains why animals have mechanisms for controlling body temperature.
5. Why is DNA so well suited for carrying information?
6. Recall the discussion about crossing over, and try to explain why introns might increase the rate of recombination between surrounding exons.

C H A P T E R   3

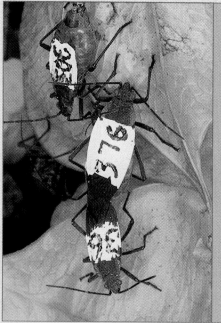

The Modern Synthesis

Population Genetics

Evolutionary change in a phenotype reflects change in the underlying genetic composition of the population.

When we discussed Mendel's experiments in Chapter 2, we briefly introduced the distinction between genotype and phenotype: phenotypes are the observable characteristics of organisms, and genotypes are the underlying genetic compositions. There need not be a one-to-one correspondence between genotype and phenotype. For example, there were only two seed-color phenotypes in Mendel's

Figure 3.1
Peas were a useful subject for Mendel's botanical experiments because they have a number of dichotomous traits. For example, pea seeds are either yellow or green but not an intermediate color.

pea populations (yellow and green), but there were three genotypes (*AA, Aa,* and *aa*) (Figure 3.1).

It is also clear from what we now know about the genetic nature of inheritance that evolutionary processes must entail changes in the genetic composition of populations. When evolution alters the morphology of a character like the finch's beak, there must be a corresponding change in the distribution of genes that control beak development within the population. To understand how Mendelian genetics solves each of Darwin's difficulties, we need to look more closely at what happens to genes in populations that are undergoing natural selection. This is the domain of **population genetics**.

Genes in Populations

Biologists describe the genetic composition of a population by specifying the frequency of alternate genotypes.

We can see what happens to genes in populations more readily if we consider a trait that is controlled by one gene operating at a single locus on a chromosome. Phenylketonuria (PKU), a potentially debilitating, genetically inherited disease in humans, is determined by the substitution of one allele for another at a single locus. Individuals who are homozygous for the PKU allele are missing a crucial enzyme in the biochemical pathway that allows people to metabolize the amino acid phenylalanine. If the disease is not treated, phenylalanine builds up in the bloodstream of children with PKU and leads to severe mental retardation. Fortunately, treatment is possible: people with PKU raised on a special low-phenylalanine diet can develop normally and lead normal lives as adults.

How do evolutionary processes control the distribution of this allele in a population? The first step in answering this question is to characterize the distribution of the harmful allele. Population geneticists do this by specifying the **genotypic frequency**, which is simply the fraction of the population that carries that genotype. Let's label the normal allele *A* and the deleterious PKU allele *a*. Suppose we perform a census of a population of 10,000 individuals and determine the number of individuals with each genotype. Geneticists can do this using established biochemical methods for determining an individual's genotype for a specific genetic locus. Table 3.1 shows the number of individuals with each genotype and the frequencies of each genotype in this hypothetical population. (In most real pop-

Table 3.1   The distribution of individuals with each of the three genotypes in a population of 10,000.

GENOTYPE	NUMBER OF INDIVIDUALS	FREQUENCY OF GENOTYPE
aa	2000	freq*(aa)* = 2000/10,000 = 0.2
Aa	4000	freq*(Aa)* = 4000/10,000 = 0.4
AA	4000	freq*(AA)* = 4000/10,000 = 0.4

ulations, the frequency of individuals homozygous for the PKU allele is only about 1 in 10,000. We have used larger numbers here to make the calculations simpler.)

Genotypic frequencies must add up to 1.0 because every individual in the population has to have a genotype. We keep track of the frequencies of genotypes, rather than the numbers of individuals with each genotype, because the frequencies provide a description of the genetic composition of populations that is independent of population size. This makes it easy to compare populations of different sizes.

One goal of evolutionary theory is to determine how genotypic frequencies change through time.

A variety of events in the lives of plants and animals act to change the frequency of alternative genotypes in populations from generation to generation. Population geneticists categorize these processes into a number of evolutionary mechanisms, or "forces." The most important mechanisms are sexual reproduction, natural selection, mutation, and genetic drift. In the remainder of this section we will see how sexual reproduction and natural selection alter the frequencies of genes and genotypes, and later in the chapter we will return to consider the effects of mutation and genetic drift.

How Random Mating and Sexual Reproduction Change Genotypic Frequencies

The events that occur during sexual reproduction can lead to changes in genotypic frequencies in a population.

First, let's consider the effects of the patterns of inheritance that Mendel observed. Imagine that men and women do not choose their mates according to whether or not they are afflicted with PKU, but mate randomly with respect to the individual's genotype for PKU. It is important to study the effects of random mating because for most genetic loci mating is random. Even though humans may choose their mates with care, and might even avoid mates with particular genetic characteristics like PKU, they cannot choose mates with a particular allele at each locus because there are about 100,000 genetic loci in humans. Random mating between individuals is equivalent to the random union of gametes. In this sense it's not really different from oysters shedding their eggs and sperm into the ocean, where chance dictates which gametes will form zygotes.

The first step in determining the effects of sexual reproduction on genotypic frequencies is to calculate the frequency of the PKU allele in the pool of gametes.

We can best understand how segregation affects genotypic frequencies by breaking the process into two steps. In the first step we determine the frequency of the

PKU allele among all the gametes in the mating population. Remember that the *a* allele is the PKU allele and the *A* allele is the "normal" allele. Table 3.1 gives the frequency of each of the three genotypes among the parental generation. We want to use this information to determine the genotypic frequencies among the F1 generation. First we calculate the frequencies of the two types of alleles in the pool of gametes. (The frequency of an allele is also referred to as its **gene frequency**.) Let's label the frequency of *A* as *p* and the frequency of *a* as *q*. (Since there are only two alleles, $p + q = 1$.) If all individuals produce the same number of gametes, then we can calculate *q* as follows:

$$q = \frac{\text{no. of } a \text{ gametes}}{\text{total no. of gametes}}$$

Note that this is simply the definition of a frequency. We can calculate the values of the numerator and denominator of this fraction from information we already know. Since *a* gametes can only be produced by *aa* and *Aa* individuals, the total number of *a* gametes is simply the sum of the number of *Aa* and *aa* parents multiplied by the number of *a* gametes that each parent produces. The denominator is the product of the sum of the total number of parents and the number of gametes per parent. Hence,

$$q = \frac{\left(\begin{array}{c}\text{no. of } a \\ \text{gametes per} \\ aa \text{ parent}\end{array}\right)\left(\begin{array}{c}\text{no. of } aa \\ \text{parents}\end{array}\right) + \left(\begin{array}{c}\text{no. of } a \\ \text{gametes per} \\ Aa \text{ parent}\end{array}\right)\left(\begin{array}{c}\text{no. of } Aa \\ \text{parents}\end{array}\right)}{(\text{no. of gametes per parent})(\text{total no. of parents})} \qquad (1)$$

To simplify this equation, let's first examine those terms involving numbers of individuals. Remember that the population size is 10,000 individuals. This means that the number of *aa* parents is equal to the frequency of *aa* parents times 10,000. Similarly, the number of *Aa* parents is equal to the frequency of *Aa* parents times 10,000. Now we examine the terms involving numbers of gametes. Suppose each parent produces two gametes. For *aa* individuals, both gametes contain the *a* allele, so the number of *a* gametes per *aa* parent is 2. For *Aa* individuals, half of the gametes will carry the *a* allele and half will carry the *A* allele, so here the number of *a* gametes per *Aa* parent is 0.5×2. Now we substitute all these values into Equation (1) to get

$$q = \frac{2[\text{freq}(aa) \times 10,000] + (0.5 \times 2)[\text{freq}(Aa) \times 10,000]}{2 \times 10,000}$$

We can reduce this fraction by dividing the top and bottom by $2 \times 10,000$, which yields the following formula for the frequency of the *a* allele in the pool of gametes:

$$q = \text{freq}(aa) + 0.5 \times \text{freq}(Aa) \qquad (2)$$

Notice that this form of the formula contains neither the population size nor the average number of gametes per individual in the population. Under normal circumstances the population size and the average number of gametes per individuals do not matter. This means that you can use this expression as a general formula for calculating gene frequencies among the gametes produced by any population of individuals as long as there are only two alleles at the genetic locus of interest.

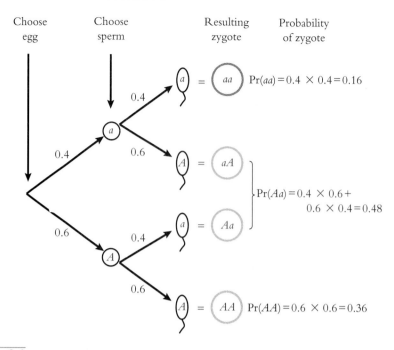

Figure 3.2

This is an event tree that shows how to calculate the frequency of each genotype among the zygotes created by the random union of gametes. In this pool of gametes, the frequency of a is 0.4 and the frequency of A is 0.6. The first node represents the choice of an egg. There is a 40% chance that the egg will carry the a allele, and a 60% chance it will carry the A allele. The second node represents the choice of the sperm. Once again there is a 40% chance of drawing an a-bearing sperm and a 60% chance of drawing an A-bearing sperm. The probability of each genotype can be calculated by computing the probability of taking a path through the tree. For example, the probability of getting an aa zygote is 0.4 × 0.4 = 0.16.

It is important to keep in mind that the formula results from applying Mendel's laws and from counting the number of a gametes that are produced.

By using Equation (2) and values from Table 3.1, for this particular population, we get q = 0.2 + (0.5 × 0.4) = 0.4. Since p + q must sum to 1.0, p (the frequency of the A gametes) must be 0.6. Notice that the frequency of each allele in the pool of gametes is the same as the frequency of the same allele among parents.

The next step is to calculate the frequencies of all the genotypes among the zygotes.

Now that we have calculated the frequency of the PKU *allele* among the pool of gametes, we can determine the frequencies of the *genotypes* among the zygotes. If each zygote is the product of the random union of two gametes, the process of zygote formation can be schematically represented in an event tree (Figure 3.2) similar to the one we used to represent Mendel's crosses (for example, Figure 2.9). First, we select a gamete, say an egg. The probability of selecting an egg carrying the *a* allele is 0.4 because this is the frequency of the *a* allele in the population. Now, we randomly draw a second gamete, the sperm. Again, the probability of getting a sperm carrying an *a* allele is 0.4. The chance that these two randomly chosen gametes are both *a* is 0.4 × 0.4, or 0.16. Figure 3.2 also shows how to

Table 3.2 The distribution of genotypes in the population of zygotes in the F_1 generation.

freq(aa)	=	0.4×0.4	=	0.16
freq(Aa)	=	$(0.4 \times 0.6) + (0.4 \times 0.6)$	=	0.48
freq(AA)	=	0.6×0.6	=	0.36

compute the probabilities of the other two genotypes. Note that the sum of the three genotypes always equals 1.0, since all individuals must have a genotype.

If we form a large number of zygotes by randomly drawing gametes from the gamete pool, we will obtain the genotypic frequencies shown in Table 3.2. Compare the frequencies of each genotype in the parental population (Table 3.1) with the frequencies of each genotype in the F_1 generation: the genotypic frequencies have changed. This is because the processes of independent segregation of alleles into gametes and random mating alter the distribution of alleles in zygotes. This in turn alters genotypic frequencies between the F_0 generation and the F_1 generation. (Note, however, that there has been no change in the frequencies of the two *alleles, a* and *A*; you can check this out using Equation 2.)

When no other forces (such as natural selection) are operating, genotypic frequencies reach stable proportions in just one generation. These proportions are called the Hardy-Weinberg equilibrium.

If no other processes act to change the distribution of genotypes, the set of genotypic frequencies in Table 3.2 will remain unchanged in subsequent generations. That is, if members of the F_1 generation mate at random, the distribution of genotypes among the F_2 generation will be exactly the same as the distribution of genotypes in the F_1 generation. The fact that genotypic frequencies remain constant was recognized independently by British mathematician G.H. Hardy and German physician W. Weinberg in 1908, and these constant frequencies are now called the **Hardy-Weinberg equilibrium**. As we will see later in this chapter, the realization that sexual reproduction alone does not alter phenotypic and genotypic frequencies was the key to understanding how variation is maintained.

In general, the Hardy-Weinberg proportions for a genetic locus with two alleles are

$$\text{freq}(aa) = q^2$$
$$\text{freq}(Aa) = 2pq \qquad (3)$$
$$\text{freq}(AA) = p^2$$

where q is the frequency of allele a, and p is the frequency of allele A. Figure 3.3 shows how to calculate these frequencies using a Punnett square.

If no other processes act to change genotypic frequencies, the Hardy-Weinberg equilibrium frequencies will be reached after only one generation and will remain unchanged thereafter. Moreover, if the Hardy-Weinberg proportions are altered by chance, the population will return to Hardy-Weinberg proportions in one generation. If this seems like an unlikely conclusion, work through the calculations

Figure 3.3

This Punnett square shows a second way to calculate the frequency of each type of zygote when there is random mating. The horizontal axis represents the proportion of A and a eggs, and is thus divided into fractions p and q; where $q = 1 - p$. The vertical axis is divided according to the proportion of sperm carrying each allele, and again it is divided into fractions p and q. The areas of the rectangles formed by the intersection of the vertical and horizontal dividing lines give the proportion of zygotes that result from each of the four possible fertilization events. The area of the square containing aa zygotes is q^2, the area of the square containing AA zygotes is p^2, and the total area of the two rectangles containing Aa zygotes is $2pq$.

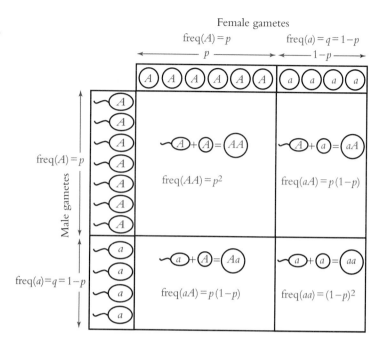

in Box 3.1, which shows how to calculate the genotypic frequencies after a second episode of segregation and random mating. As you can see, genotypic frequencies remain constant.

We have seen that sexual reproduction and random mating can change the distribution of genotypes, which will reach equilibrium after one generation. We have also seen that neither process changes the frequencies of alleles. Clearly sexual reproduction and random mating alone cannot lead to evolution over the long run. We now turn to a process that can produce changes in the frequencies of alleles—natural selection.

How Natural Selection Changes Gene Frequencies

If different genotypes are associated with different phenotypes and those phenotypes differ in their ability to reproduce, then the alleles that lead to the development of the favored phenotype will increase in frequency.

Genotypic frequencies in the population will remain at the Hardy-Weinberg proportions (Equation 3) as long as all genotypes are equally likely to survive and produce gametes. In wealthy countries in which PKU can be treated, this is approximately correct. However, this assumption will not be valid in environments in which PKU is not treated. Suppose some catastrophe occurs that interrupts the supply of modern medical care to our hypothetical population just after a new generation of zygotes is formed. In this case, virtually none of the PKU individuals (genotype aa) would survive to reproduce. Suppose the frequency of the PKU allele is 0.4 among zygotes and the population is in Hardy-Weinberg equilibrium. Let's compute q', the frequency of the PKU allele among adults, which is given by

$$q' = \frac{\text{no. of } a \text{ gametes produced by adults in the next generation}}{\text{total no. of gametes produced by adults in the next generation}} \quad (4)$$

BOX 3.1

Genotypic Frequencies after Two Generations of Random Mating

From Equation (2) given in the text, we know that the frequency of the gametes produced by adults in the second generation carrying the *a* allele will be

$$q = \text{freq}(aa) + 0.5 \times \text{freq}(Aa)$$
$$= 0.16 + 0.5 \times 0.48$$
$$= 0.16 + 0.24$$
$$= 0.4$$

Since there are only two alleles, the frequency of *A* is 0.6. As we asserted in the text, the frequency of the two alleles remains unchanged, and the frequencies of the gametes produced by random mating among members of this generation are the same as in the previous generation. The calculations for these frequencies are the same as those shown in Table 3.2.

We have already calculated the frequency of each of the genotypes among zygotes just after conception (Table 3.2), but if PKU is a lethal disease without treatment, not all of these individuals will survive. Now we need to calculate the frequency of *a* gametes after selection, which can be found by expanding Equation (4) as follows:

$$q' = \frac{\begin{pmatrix}\text{no. of } a \\ \text{gametes per} \\ aa \text{ parent after} \\ \text{selection}\end{pmatrix}\begin{pmatrix}\text{no. of } aa \\ \text{parents} \\ \text{after} \\ \text{selection}\end{pmatrix} + \begin{pmatrix}\text{no. of } a \\ \text{gametes per} \\ Aa \text{ parent after} \\ \text{selection}\end{pmatrix}\begin{pmatrix}\text{no. of } Aa \\ \text{parents} \\ \text{after} \\ \text{selection}\end{pmatrix}}{(\text{no. of gametes per parent})(\text{total no. of parents after selection})} \quad (5)$$

Since none of the *aa* individuals will survive, the first term in the numerator will be equal to zero. If we assume that all of the *AA* and *Aa* parents survive, then the number of parents after selection is $10{,}000 \times [(\text{freq}(AA) + \text{freq}(Aa)]$ and thus Equation (5) can be simplified to

$$q' = \frac{(0.5 \times 2)(0.48 \times 10{,}000)}{2 \times (0.36 + 0.48) \times 10{,}000} = 0.2857$$

The frequency of the PKU allele in adults is now 0.2857, a big decrease from 0.4. Several important lessons can be drawn from this example:

1. Selection cannot produce change unless there is variation in the population. If all the individuals were homozygous for the normal allele, there would be no change in gene frequencies from one generation to the next.
2. Selection does not operate directly on genes and does not change gene frequencies directly. Instead, natural selection changes the frequency of different phenotypes. In this case, individuals with PKU cannot survive without treatment. Selection decreases the frequency of the PKU allele because it is more likely to be associated with the lethal phenotype.

3. The strength and direction of selection depend on the environment. In an environment with medical care, the strength of selection against the PKU allele is negligible.

It is also important to see that, while this example shows how selection can change gene frequencies, it does not yet show how selection can lead to the evolution of new adaptations. Here, all phenotypes were present at the outset, and all selection did was to change their relative frequency.

The Modern Synthesis

The Genetics of Continuous Variation

When Mendelian genetics was rediscovered, biologists at first thought Mendelian genetics was incompatible with Darwin's theory of evolution by natural selection.

Darwin believed that evolution proceeded by the gradual accumulation of small changes. But Mendel and the biologists who elucidated the structure of the genetic system around the turn of the century were dealing with genes that had a noticeable effect on the phenotype. The substitution of one allele for another in Mendel's peas changed pea color. Genetic substitutions at other loci had visible effects on the shape of the pea seeds and the height of the pea plants. Genetics seemed to prove that inheritance was fundamentally discontinuous, and turn-of-the-century geneticists like de Vries and William Bateson argued that this fact could not be reconciled with Darwin's idea that adaptation occurs through the accumulation of small variations. If one genotype produces short plants and the other two genotypes produce tall plants, then there will be no intermediate types, and the size of pea plants cannot change in small steps. In a population of short plants, tall ones must be created all at once by mutation, not gradually lengthened over time by selection. These arguments convinced most biologists of the time, and consequently Darwinism was in decline during the early part of the 20th century.

Mendelian genetics and Darwinism were eventually reconciled, resulting in a body of theory that solved the problem of explaining how variation is maintained.

In the early 1930s, the British biologists R. A. Fisher and J. B. S. Haldane and the American biologist Sewall Wright showed how Mendelian genetics could be used to explain continuous variation (Figure 3.4). We will see how their insights led to the resolution of the two main objections to Darwin's theory—the absence of a theory of inheritance, and the problem of accounting for how variation is maintained in populations. When the theory of Wright, Fisher, and Haldane was combined with Darwin's theory of natural selection, and with modern field studies by biologists such as Theodozius Dobzhansky, Ernst Mayr, and George Gaylord Simpson, a powerful explanation of organic evolution emerged. This body of theory and the supporting empirical evidence is now called the **modern synthesis**.

Continuously varying characters are affected by genes at many loci, each locus having only a small effect on phenotype.

To see how the theory of Wright, Fisher, and Haldane works, let's start with an unrealistic, but instructive, case. Suppose that there is a measurable, continu-

(a) (b) (c)

Figure 3.4

Shown here are the three architects of the modern synthesis, which showed how Mendelian genetics could be used to account for continuous variation: (a) Ronald A. Fisher, a British biologist; (b) J. B. S. Haldane, another British biologist; (c) Sewall Wright, an American biologist.

ously varying character, such as beak depth, and suppose that two alleles, $+$ and $-$, operating at a single genetic locus control the character. We'll assume that the gene at this locus influences the production of a hormone that stimulates beak growth, and that each allele leads to production of a different amount of the growth hormone. Let's say that each "dose" of the $+$ allele increases the beak depth, while a $-$ dose decreases it. Thus $++$ individuals have the deepest beaks, $--$ individuals have the shallowest beaks, and $+-$ individuals have intermediate beaks. In addition, suppose the frequency of the allele $+$ in the population is 0.5. Now, we use the Hardy-Weinberg rule (Equation 3) to calculate the frequencies of different beak depths in the population. A quarter of the population will have deep beaks ($++$), half will have intermediate beaks ($+-$), and the remaining quarter will have shallow beaks ($--$) (Figure 3.5).

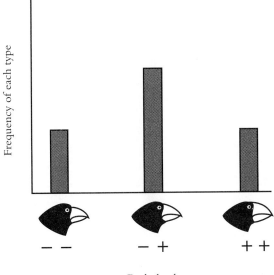

Beak depth

Figure 3.5

This figure shows the hypothetical distribution of beak depth assuming beak depth is controlled by a single genetic locus with two alleles that occur at equal frequency. The height of the bars represents the fraction of birds in the population with a given beak depth. The height of the bars is computed using the Hardy-Weinberg formula. The birds with the smallest beaks are homozygous for the $-$ allele, and have a frequency of $0.5 \times 0.5 = 0.25$. The birds with intermediate beaks are heterozygotes, and have a frequency of $2(0.5 \times 0.5) = 0.5$. The birds with the deepest beaks are homozygous for the $+$ allele, and thus have a frequency of $0.5 \times 0.5 = 0.25$.

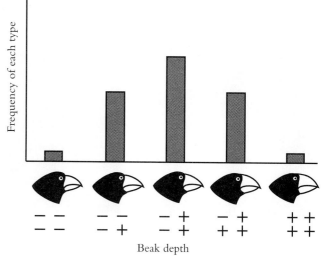

Figure 3.6

When beak depth is controlled by two loci, each locus having two alleles that occur with equal frequency, intermediate types are observed. The frequencies are calculated using a more advanced form of the Hardy-Weinberg formula, assuming that the genotypes at each locus are independent.

This does not look like the smooth bell-shaped distribution of beak depths that Grant observed on Daphne Major. If beak depth were controlled by a single locus, natural selection could not increase beak depth in small increments. But imagine what would happen if there were other genes at a second locus on a different chromosome that also affect beak depth, perhaps because they control the synthesis of the receptors for the growth hormone. As before, we assume there is a + allele that leads to larger beaks, and a − allele that leads to smaller beaks. Using the Hardy-Weinberg proportions and assuming the independent segregation of chromosomes, we can show that there are now more types of genotypes, and the distribution of phenotypes begins to look somewhat smoother (Figure 3.6). Now imagine that there is a third locus that controls the calcium supply to the growing beak, and that once again there is a + allele leading to larger beaks and a − allele leading to smaller beaks. As we see in Figure 3.7, the distribution of beak depths is even more like the bell-shaped distribution seen in nature. However, small gaps still exist in this distribution of bills. If genes were the only influence on bill depths, we might expect to see a more broken distribution than the Grants actually observed in the Galápagos.

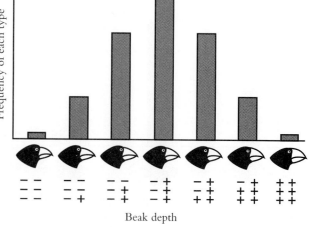

Figure 3.7

When beak depth is controlled by three loci, each locus having two alleles that occur with equal frequency, the distribution of phenotypes begins to resemble a bell-shaped curve.

The observed distribution of phenotypic values is a smooth bell-shaped curve because of the effect of **environmental variation**. The phenotypic expression of all characters, whether affected by one locus or many loci, will depend at least to some extent on the environment in which the organism develops. For example, the size of a bird's bill will depend on how well it is nourished during its development. When there is only one locus, and the effect of an allelic substitution at that locus is large, environmental variation is not important. It is easy to distinguish the phenotypes associated with different genotypes. This was the case for seed color in Mendel's peas. But when there are many loci and each has a small effect on the phenotype, environmental variation tends to blur together the phenotypes associated with different genotypes. You can't be sure whether you are measuring a +++−−− bird who matured in a good environment or a ++++−− bird who grew up in a poor environment. The result is that we observe a smooth curve of variation as shown in Figure 3.8.

Darwin's view of natural selection is easily incorporated into the genetic view that evolution typically results from changes in gene frequencies.

Darwin knew nothing about genetics, and his theory of adaptation by natural selection was framed in purely phenotypic terms: there is a struggle for existence, there is phenotypic variation that affects survival and reproduction, and this phenotypic variation is heritable. In Chapter 1 we saw how this theory could explain the adaptive changes in beak depth in a population of Darwin's finches on the Galápagos Islands. As we saw earlier in this chapter, population geneticists take the seemingly different view that evolution means changes in allelic frequencies by natural selection. However, these two views of evolution are easily reconciled. Suppose Figure 3.8 gave the distribution of beak depths before the drought on Daphne Major. Remember that individuals with deep beaks were more likely to survive and reproduce than individuals with smaller beaks. Figure 3.8 illustrates that individuals with deep beaks are more likely to have + alleles at the three loci assumed to affect beak depth. Thus at each locus, ++ individuals have deeper beaks on average than +− individuals, who in turn have deeper beaks on average than −− individuals. Since individuals with deeper beaks had higher fitness, natural selection would favor the + alleles at each of the three loci affecting beak depth, and thus the + alleles would increase in frequency.

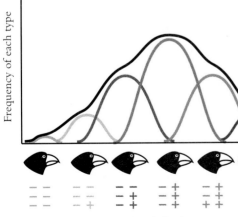

Figure 3.8

Environmental variation influences the distribution of beak depths. Again there are three loci affecting bill depths, as shown in Figure 3.7. However, now we assume that environmental conditions cause individuals with the same genotype to develop beaks of different depths. Each colored curve gives the distribution of phenotypes associated with the genotype of the same color, and the black curve gives the overall distribution of phenotypes.

DNA Is Digital

It is raining DNA outside. On the bank of the Oxford canal at the bottom of my garden is a large willow tree, and it is pumping downy seeds into the air. There is no consistent air movement, and the seeds are drifting outwards in all directions from the tree. Up and down the canal, as far as my binoculars can reach, the water is white with floating cottony flecks, and we can be sure that they have carpeted the ground to much the same radius in other directions too. The cotton wool is mostly made of cellulose, and it dwarfs the tiny capsule that contains the DNA, the genetic information. The DNA content must be a small proportion of the total, so why did I say that it was raining DNA rather than cellulose? The answer is that it is the DNA that matters. The cellulose fluff, although more bulky, is just a parachute, to be discarded. The whole performance, cotton wool, catkins, tree and all, is in aid of one thing and one thing only, the spreading of DNA around the countryside. Not just any DNA, but DNA whose coded characters spell out specific instructions for building willow trees that will shed a new generation of downy seeds. Those fluffy specks are, literally, spreading instructions for making themselves. They are there because their ancestors succeeded in doing the same. It is raining instructions out there; it's raining programs; it's raining tree-growing, fluff-spreading, algorithms. That is not a metaphor, it is the plain truth. It couldn't be any plainer if it were raining floppy discs.

It is plain and it is true, but it hasn't long been understood. A few years ago, if you had asked almost any biologist what was special about living things as opposed to nonliving things, he would have told you about a special substance called protoplasm. Protoplasm wasn't like any other substance; it was vital, vibrant, throbbing, pulsating, 'irritable' (a schoolmarmish way of saying responsive). If you took a living body and cut it up into ever smaller pieces, you would eventually come down to specks of pure protoplasm. At one time in the last century, a real-life counterpart of Arthur Conan Doyle's Professor Challenger thought that the 'globigerina ooze' at the bottom of the sea was pure protoplasm. When I was a schoolboy, elderly textbook authors still wrote about protoplasm although, by then, they really should have known better. Nowadays you never hear or see the word. It is as dead as phlogiston and the universal aether. There is nothing special about the substances from which living things are made. Living things are collections of molecules, like everything else.

What is special is that these molecules are put together in much more complicated patterns than the molecules of nonliving things, and this putting together is done by following programs, sets of instructions for how to develop, which the organisms carry around inside themselves. Maybe they do vibrate and throb and pulsate with 'irritability', and glow with 'living' warmth, but these properties all emerge incidentally. What lies at the heart of every living thing is not a fire, not warm breath, not a 'spark of life'. It is information, words, instructions. If you want a metaphor, don't think of fires and sparks and breath. Think, instead, of a billion discrete, digital characters carved in tablets of crystal. If you want to understand life, don't think about vibrant, throbbing gels and oozes, think about information technology. . . .

The basic requirement for an advanced information technology is some kind of storage medium with a large number of memory locations. Each location must be capable of being in one of a discrete number of states. This is true, anyway, of the *digital* information technology that now dominates our world of artifice. There is an alternative kind of information technology based upon *analogue* information. The information on an ordinary gramophone record is analogue. It is stored in a wavy groove. The information on a modern laser disc (often called 'compact disc', which is a pity, because the name is uninformative and also usually mispronounced with the stress on the first syllable) is digital, stored in a series of tiny pits, each of which is either definitely there or definitely not there: there are no half measures. That is the diagnostic feature of a digital system: its fundamental elements are either definitely in one state or definitely in another state, with no half measures and no intermediates or compromises.

The information technology of the genes is digital. This fact was discovered by Gregor Mendel in the last century, although he wouldn't have put it like that. Mendel showed that we don't blend our inheritance from our two parents. We receive our inheritance in discrete particles. As far as each particle is concerned, we either inherit it or we don't. Actually, as R. A. Fisher, one of the founding fathers of what is now called neo-Darwinism, has pointed out, this fact of particulate inheritance has always been staring us in the face, every time we think about sex. We inherit attributes from a male and a female parent, but each of us is either male or female, not hermaphrodite. Each new baby born has an approximately equal *probability* of inheriting maleness or femaleness, but any one baby inherits only one of these, and doesn't combine the two. We now know that the same goes for all our particles of inheritance. They don't blend, but remain discrete and separate as they shuffle and reshuffle their way down the generations. Of course there is often a powerful appearance of blending in the effects that the genetic units have on bodies. If a tall person mates with a short person, or a black person with a white person, their offspring are often intermediate. But the appearance of blending applies only to effects on bodies, and is due to the summed small effects of large numbers of particles. The particles themselves remain separate and discrete when it comes to being passed on to the next generation.

Source: From pp. 111–113 from *The Blind Watchmaker*, by R. Dawkins. Copyright © 1996, 1987, 1986 by Richard Dawkins. Reprinted by permission of W. W. Norton & Company and Sterling Lord Literistic, Inc..

How Variation Is Maintained

Genetics provides a ready explanation for why the phenotypes of offspring tend to be intermediate between those of their parents.

Recall from Chapter 1 that the blending model of inheritance appealed to 19th-century thinkers because it explained the fact that for most continuously varying characters, offspring are intermediate between their parents. However, the genetic model developed by Fisher, Wright, and Haldane is also consistent with this fact. To see why, consider a cross between the individuals with the biggest and smallest beaks: $(+++++)\times(-----)$. All of the offspring will be $(+-+-+-)$, intermediate between the parents because during development the effects of $+$ and $-$ alleles are averaged. Other matings will produce a distribution of different kinds of offspring, but intermediate types will tend to be most common.

There is no blending of genes during sexual reproduction.

We know from population genetics that, in the absence of selection (and other factors to be discussed later in this chapter), genotypic frequencies reach equilibrium after one generation and the distribution of phenotypes does not change. Furthermore, we know that sexual reproduction produces no blending in the genes themselves, despite the fact that offspring may appear to be intermediate between their parents. This is because genetic transmission involves faithful copying of the genes themselves and reassembling them in different combinations in zygotes. The only blending that occurs takes place at the level of the expression of genes in phenotypes. The genes themselves remain distinct physical entities (Figure 3.9).

These facts do not completely solve the problem of the maintenance of variation because selection tends to deplete variation. When selection favors birds with deeper bills, we might expect $-$ alleles to be replaced at all three loci affecting the trait, leaving a population in which every individual has the genotype $+++++$. There would still be phenotypic variation due to environmental effects, but without genetic variation there can be no further adaptation.

Mutation slowly adds new variation.

Genes are copied with amazing fidelity, and their messages are protected from random degradation by a number of molecular repair mechanisms. However, every once in a while a mistake in copying is made that goes unrepaired, and a new

Figure 3.9

The blending model of inheritance assumes that the hereditary material is changed by mating. When red and white parents are crossed to produce a pink offspring, the blending model posits that the hereditary material has mixed, so that when two pink individuals mate they produce only pink offspring. According to Mendelian genetics, however, the effects of genes are blended in their *expression* to produce a pink phenotype, but the genes themselves remain unchanged. Thus, when two pink parents mate, they can produce white, pink, or red offspring.

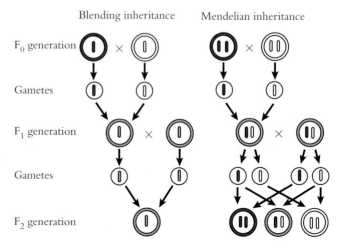

allele is introduced into the population. In Chapter 2 we learned that genes are pieces of DNA. Certain forms of ionizing radiation (such as X rays) and certain kinds of chemicals damage the DNA and alter the message that it carries. These changes are called **mutations**, and they add variation to a population by continuously introducing new alleles, some of which may produce novel phenotypic effects that selection can assemble into adaptations. Although rates of mutation are very low—ranging from 1 in 100,000 to 1 in 10 million per locus per gamete in each generation—this process plays an important role in generating variation.

Low mutation rates can maintain variation because a lot of variation is protected from selection.

For characters that are affected by genes at many different loci, low rates of mutation can maintain variation in populations. This is possible because many different genotypes generate intermediate phenotypes that are favored by stabilizing selection. If individuals with a variety of genotypes are equally likely to survive and reproduce, a considerable amount of variation is protected (or hidden) from selection. To make this concept more clear, let's consider the action of stabilizing selection on beak depth in the medium ground finch. Remember that stabilizing selection occurs when birds with intermediate-sized bills have higher fitness than birds with deeper or shallower bills. If beak depth is affected by genes at three loci, as shown in Figure 3.8, individuals with several different genotypes may develop very similar, intermediate-sized beaks. For example, individuals that are $++$ at one locus affecting the trait may be $--$ at another locus, and thus have the same phenotype as an individual that is $+-$ at both loci. Thus, when there are a large number of loci affecting a single trait, only a small fraction of the genetic variability present in the population is expressed phenotypically. As a result, selection removes variation from the population very slowly. The processes of segregation and recombination slowly shuffle and reshuffle the genome, and thereby expose the hidden variation to natural selection in subsequent generations (Figure 3.10). This process provides the solution to Darwin's other dilemma—because a considerable amount of variation is protected from selection, a very low mutation rate can maintain variation despite the depleting action of selection.

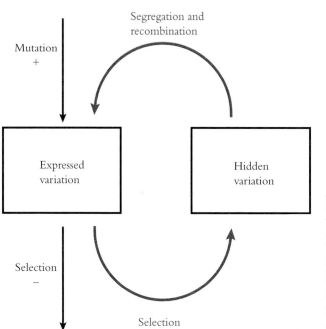

Figure 3.10

There are two pools of genetic variation, hidden and expressed. Mutation adds new genetic variation, and selection removes it from the pool of expressed variation. Segregation and recombination shuffle variation back and forth between the two pools with each generation.

Hidden variation explains why selection can move populations far beyond their initial range of variation.

Remember from Chapter 1 Fleeming Jenkin's argument that Darwin's theory could not explain cumulative evolutionary change because it provided no account of how a population could evolve beyond the initial range of variation present. Selection would cull away all of the small-beaked finches, he would have argued, but it could never make the average beak bigger than the biggest beak initially present. If this argument were correct, then selection could never lead to cumulative, long-term change.

Jenkin's argument was wrong because it failed to take hidden variation into account. Hidden variation is always present in continuously varying traits. Let's suppose that environmental conditions favor larger beaks. When beak depth is affected by genes at many loci, the birds in a population with the deepest beaks do not carry all + alleles. They carry a lot of + alleles and some − alleles. When the finches with the shallowest beaks die, alleles leading to small beaks are removed from the breeding population. This increases the frequency of + alleles at every locus, but because even the deepest-beaked individuals had some − alleles, a huge amount of variation remains. This variation is shuffled through the process of sexual reproduction. Because + alleles become more common, more of these alleles are likely to be combined in the genotype of a single individual. The greater the proportion of + alleles in an individual, the larger the beak will be. Thus the biggest beak will be larger than the biggest beak in the previous generation. In the next generation the same thing happens again: The deepest-beaked individuals still carry − alleles but fewer than before. This allows the biggest beaks to be bigger than any beaks in the previous generation.

This process can go on for many generations. An experiment on oil content in corn begun at the Illinois Experiment Station in 1896 provides a good example. Researchers selected for high and low oil content in corn and, as you can see from Figure 3.11, each generation showed significant change. In the initial population of 163 ears of corn, oil content ranged from 4% to 6%. After about 70 generations

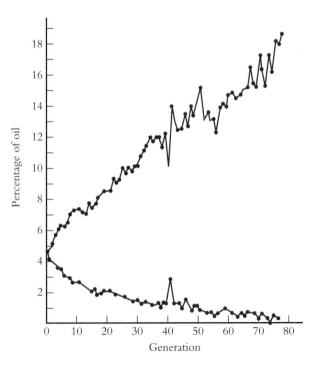

Figure 3.11

Selection can carry a population far beyond the original range of variation because at any given time a lot of genetic variation is not expressed as phenotypic variation. The range of variation in oil content at the beginning of the experiment was 4% to 6%. After about 80 generations, oil content in the line selected for high oil content had increased to 19%, and oil content in the low oil line had decreased to less than 1%.

Figure 3.12

This Pekinese is smaller than the smallest wolf, although wolves are the ancestors of all domestic breeds of dogs. Some of Darwin's critics argued that natural selection could not produce this kind of transformation because it could not lead to changes that fell outside the range of variation found in the original population. However, genes for very small size are present in wolves, but their effects are hidden by other genes.

of selection, both the high and low values for oil content far exceeded the initial range of variation. Researchers were then even able to reverse the direction of adaptation by taking plants from the high-oil line and selecting for low oil content.

An even more dramatic example is the Pekinese, whose very existence is a testament to the power of hidden variation (Figure 3.12). In a few thousand generations, a mere instant on the evolutionary time scale, wolves were transformed into Pekinese, as well as into bulldogs, great Danes, dachshunds, Chihuahuas, Irish wolfhounds, and toy poodles. This is far too little time for a large number of new mutations to accumulate. Instead, most of the genes necessary for the development of Irish wolfhounds and Pekinese must have been present in the ancestral population of wolves from which dogs were selected, their effects hidden by other genes. Dog breeders, acting as agents of selection, chose the genes leading to small size, long hair, and funny little faces, and made them more common in some populations; segregation and recombination brought them together over time in the Pekinese. The same process created dogs that are well adapted for herding sheep, ferreting out badgers, retrieving game, chasing mechanical rabbits around a track, and warming the laps of Chinese emperors.

Natural Selection and Behavior

The evolution of mate guarding in the soapberry bug illustrates how flexible behavior can evolve.

So far we have considered the evolution of morphological characters like beak depth and eye morphology that do not change once an individual reaches adulthood. In much of this book we will be interested in the evolution of the behavior of humans and other primates. Behavior is different from morphology in an important way—it is flexible, and individuals adjust their behavior in response to their circumstances. Some people think natural selection cannot account for flexible responses to environmental contingencies because natural selection only acts on phenotypic variation that results from genetic differences. Although this view is very common, particularly among social scientists, it is incorrect. To see why, let's consider an elegant empirical study that illustrates exactly how natural selection can shape flexible behavioral responses.

The soapberry bug *(Jadera haematoloma),* a seed-eating insect found in the south-

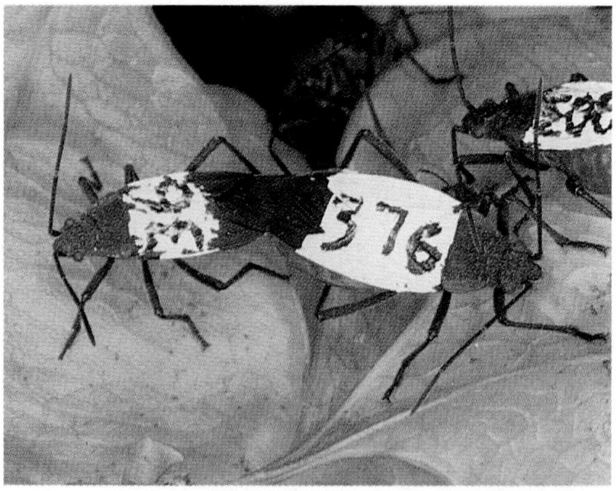

Figure 3.13
The male soapberry bug on the left is guarding the (larger) female below him. Carroll painted numbers on the bugs in order to identify individuals. (Photograph courtesy of Scott Carroll.)

eastern part of the United States, has been studied by biologist Scott Carroll of the University of New Mexico. (It's okay to call them bugs because they are members of the insect order *Hemiptera,* the true bugs.) Adult soapberry bugs are bright red and black and about 1 to 1.5 cm (0.5 in.) long (Figure 3.13). They gather together in huge groups near the plants that they eat. During mating, males mount females and copulate. The transfer of sperm typically takes about 10 minutes. However, males remain in the copulatory position, sometimes for hours, securely anchored to females by large genital hooks. This behavior is called **mate guarding**. Biologists believe the function of mate guarding is to prevent other males from copulating with the female before she lays her eggs. When a female mates with several males, they share in the paternity of the eggs that she lays. By guarding his mate and preventing her from mating with other males, a male can increase his reproductive success. However, mate guarding also has a cost: a male cannot find and copulate with other females while mate guarding. The relative magnitude of the costs and benefits of mate guarding depends on the **sex ratio**, the relative numbers of males and females. When the ratio of males to females is high, males have little chance of finding another female, and guarding is the best strategy. When females are more common than males, the chance of finding an unguarded female increases, so males may benefit more from looking for additional mates than from guarding.

Behavioral plasticity allows male soapberry bugs to vary their mate-guarding behavior adaptively in response to variation in the local abundance of females.

In populations of soapberry bugs in western Oklahoma, the sex ratio is quite variable. In some places there are equal numbers of males and females, while in others there are twice as many males as females. Males guard their mates more often where females are rare than where they are common (Figure 3.14). There are two possible mechanisms that might produce this pattern: (1) males in populations with high sex ratios might differ genetically from males in populations with low sex ratios, or (2) males might adjust their behavior in response to the local sex ratio. To distinguish between these two possibilities, Carroll brought soapberry bugs into the laboratory and created populations with different sex ratios. Then he watched males mate in each of these populations. If the mate-guarding trait were **canalized**, or showing the same phenotype in a wide range of environments, then a male would behave the same way in each population; if the trait were **plastic**, then a male would adjust his behavior in relation to the local sex ratio.

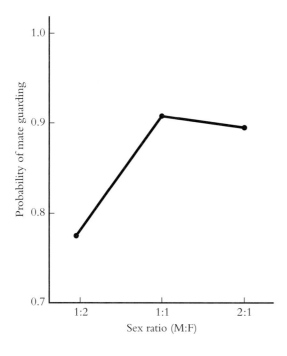

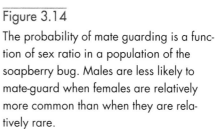

Figure 3.14
The probability of mate guarding is a function of sex ratio in a population of the soapberry bug. Males are less likely to mate-guard when females are relatively more common than when they are relatively rare.

Carrol's data confirmed that soapberry bug males in Oklahoma have a plastic behavioral strategy that causes them to modify their mating behavior in response to current social conditions.

There is evidence that the soapberry bug's plasticity has evolved in response to the variability in conditions in Oklahoma.

Most soapberry bugs live south of Oklahoma, in warmer, more stable habitats like the Florida Keys, and the sex ratio in these areas is always close to even. (Hence females are relatively rare.) Carroll subjected male soapberry bugs from the Florida Keys to the same experimental protocol as the bugs from Oklahoma. As is shown in Figure 3.15, the Florida males did not change their behavior in response to changes in sex ratio. They mate-guarded about 90% of the time regardless of the abundance of females. In a stable environment like the Florida Keys, sex ratios do not vary much from time to time, and the ability to adjust mate-guarding behavior provides no advantage. Behavioral flexibility is costly in a number of ways. For example, flexible males must spend time and energy assessing the sex ratio before they mate, males will sometimes make mistakes about the local sex ratio and behave inappropriately, and flexibility probably requires a more complex nervous system. Thus a simple, fixed behavioral rule is likely to be best in stable environments. In the variable climate of Oklahoma, however, the ability to adjust mate-guarding behavior provides enough in fitness benefits to compensate for the costs of maintaining behavioral flexibility.

A comparison of the mate-guarding behavior in two different populations of soapberry bugs indicates that behavioral plasticity evolves when the nature of the behavioral response to the environment is genetically variable.

How did behavioral flexibility evolve in the Oklahoma bugs? It evolved like any other adaptation—by the selective retention of beneficial genetic variants. For any character to evolve, (1) the character must vary, (2) the variation must affect

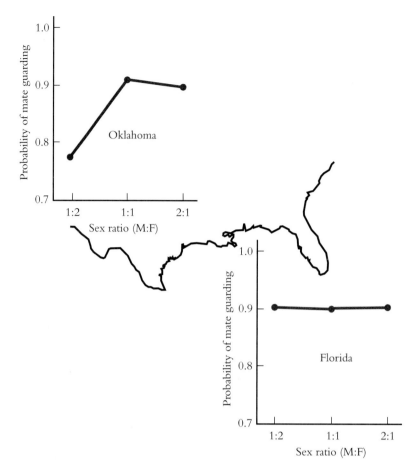

Figure 3.15

The average probability of mate guarding varies between populations of soapberry bugs in Oklahoma and Florida. In Oklahoma, males are more likely to guard when females are relatively rare, while in Florida males do not vary their mate-guarding behavior in relation to the local sex ratio.

reproductive success, and (3) the variation must be heritable. Mate guarding in soapberry bugs satisfies each of these conditions: First, there is variation, and this variation affects fitness. By bringing individual bugs into the laboratory and observing their behavior at different sex ratios, Carroll showed that individual males in Oklahoma have different behavioral strategies. Figure 3.16 plots the probability of mate guarding for four representative individuals numbered 1 to 4. Males 1 and 4 both have fixed behavioral strategies. Male 1 guards about 90% of the time, and male 4 guards about 80% of the time. Males 2 and 3 have variable behavioral strategies: male 2 is very sensitive to changes in sex ratio, while male 3 is less sensitive. Notice that both the amount of mate guarding and the amount of flexibility vary among the bugs in Oklahoma.

Second, it seems likely that this variation would affect male reproductive success. In Oklahoma, males would experience a range of sex ratios, so it seems likely that males with flexible strategies, like male 2, would tend to have the most offspring. In Florida, males with inflexible strategies, like male 1, would have the most offspring.

Third, the character is heritable. By doing controlled matings in the laboratory, Carroll was able to show that males tended to have the same strategies as their fathers. Bugs like male 1 had sons with fixed strategies, and bugs like male 2 had sons with flexible strategies. Thus, the Oklahoma bugs would come to have a variable strategy, while the Florida bugs would come to have a fixed strategy.

Behavior in the soapberry bug is relatively simple. Mate guarding depends on the sex ratio in the local population. The behavior of humans and other primates

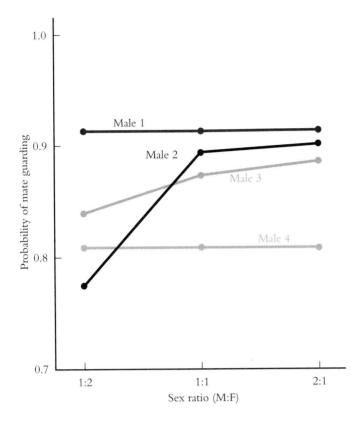

Figure 3.16

There is genetic variation in the behavioral rules of individual soapberry bugs, illustrated here by four representative individuals. There is variation in both the level of mate guarding (for example, male 1 guards more than male 4 for all sex ratios) and in the amount of plasticity (e.g., the behavior of males 1 and 4 does not change, male 3 changes some, and male 2 changes a lot). If this variation is heritable, the rule that works best, averaged over all of the environments of the population, will tend to increase.

is much more complex, but the principles that govern the evolution of complex forms of behavior are the same as the principles that govern the evolution of more simple forms of behavior. That is, individuals must differ in the ways they respond to the environment, these differences must affect their ability to survive and reproduce, and at least some of these differences must be heritable. Then individual responses to environmental circumstances will evolve in much the same way as finch beaks evolve.

Constraints on Adaptation

Natural selection plays a central role in our understanding of evolution because it is the only mechanism that can explain adaptation. However, evolution does not always lead to the best possible phenotype. In this section we consider five reasons why this is the case.

Correlated Characters

When individuals that have particular variants of one character also tend to have particular variants of a second character, the two characters are said to be correlated.

So far, we have only considered the action of natural selection on one character at a time. This approach is misleading if natural selection acts on more than one

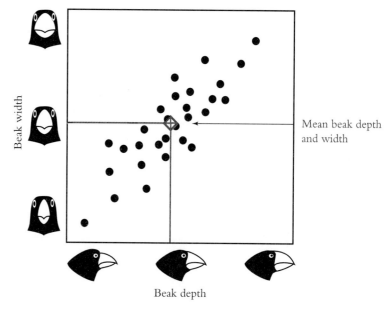

Figure 3.17

Beak depth and beak width are correlated in the medium ground finch on Daphne Major. The vertical axis gives the difference between the individual's beak width and the mean width in the population, and the horizontal axis gives the difference between the individual's beak depth and the population mean. Each point represents one individual. Birds with deep beaks are likely to have wide beaks, and birds with shallow beaks generally have narrow beaks.

character simultaneously and the characters are nonrandomly associated, or **correlated**. It's easiest to grasp the meaning and importance of correlated characters in the context of a now-familiar example, Darwin's finches. When the Grants and their colleagues captured medium ground finches on Daphne Major, they measured beak depth, beak width, and a number of other morphological characters. Beak depth is measured as the top-to-bottom dimension of the beak, while beak width is the side-to-side dimension. As is common for morphological characters like these, the Grants found that beak depth and beak width are positively correlated: birds with deep beaks also have wide beaks (Figure 3.17). Each point in Figure 3.17 represents one individual. Beak depth is plotted on the vertical axis, and beak width is plotted along the horizontal axis. If the cloud of points were round or the points were randomly scattered in the graph, it would mean these characters were uncorrelated. Then information about an individual's beak depth would tell us nothing about its beak width. However, the cloud of points forms an ellipse with the long axis oriented from the lower left to the upper right, so we know that the two characters are **positively correlated**: deep beaks also tend to be wide. If birds with deeper beaks tended to have narrow beaks, and birds with shallow beaks tended to have wide beaks, the two characters would be **negatively correlated**, and the long axis of the cloud of points would run from the upper left to the lower right.

Correlated characters occur because some genes affect more than one character.

Genes that affect more than one character are said to have **pleiotropic effects**, and most genes probably fall into this category. Often genes that are expressed early in development and affect overall size also influence a number of other discrete morphological traits. In Darwin's finches, individuals who have alleles for bigness at loci affecting overall size will have deeper beaks and wider beaks than individuals who have alleles for smallness. The PKU locus provides another good example of pleiotropy. Untreated PKU homozygotes are phenotypically different from heterozygotes or normal homozygotes in several ways. For example, they have lower IQs and different hair color from individuals with other genotypes.

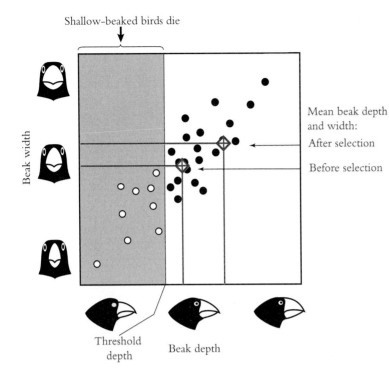

Figure 3.18

Selection on one character affects the mean value of correlated characters even if the other characters have no effect on fitness. The distribution of beak depth and beak width in the population is plotted here, assuming that only birds whose beaks are greater than a threshold depth survive. Beak width is assumed to have no effect on survival. Selection leads to an increase in both average beak depth and average beak width. Mean beak width is increased by the correlated response of selection on beak depth.

Thus, the substitution of another PKU allele for the normal allele in a heterozygote would affect a wide variety of phenotypic characters.

When two characters are correlated, selection that changes the mean value of one character in the population also changes the mean value of the other correlated character.

Returning to the finches, suppose that there is selection for individuals with deep beaks, and that beak width has no effect on survival (Figure 3.18). As we would expect, the mean value of beak depth increases. Notice, however, that the mean value of beak width also increases even though beak width has no effect on the probability that an individual will survive. Selection on beak depth affects the mean value of beak width because the two traits are correlated. This effect is called the **correlated response** to selection. It results from the fact that selection increases the frequency of genes that increase both beak depth and beak width.

A correlated response to selection can cause other characters to change in a maladaptive direction.

To understand how selection on one character can cause other characters to change in a **maladaptive** (less fit) direction, let's continue with our example. It turns out that beak width did affect survival during the Galápagos drought. By holding beak depth constant, the Grants showed that individuals with *thinner* beaks were more likely to survive during the drought, probably because birds with thinner beaks were able to generate more pressure on the tough seeds that predominated during the drought. If selection were acting only on bill width, then we would expect the mean beak width in the population to decrease. However, the correlated response to selection on beak depth also acts to increase mean beak width. If the correlated response to selection on beak depth were stronger than the effect of selection directly on beak width, mean beak width would increase

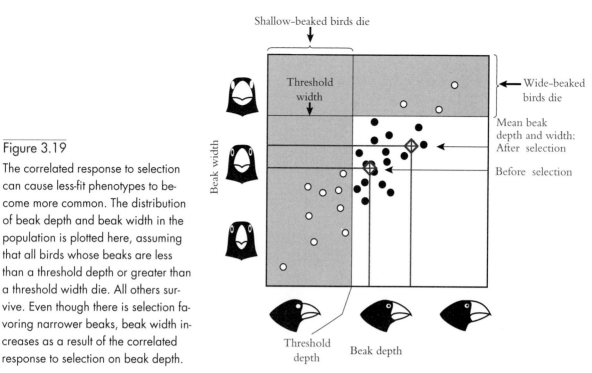

Figure 3.19

The correlated response to selection can cause less-fit phenotypes to become more common. The distribution of beak depth and beak width in the population is plotted here, assuming that all birds whose beaks are less than a threshold depth or greater than a threshold width die. All others survive. Even though there is selection favoring narrower beaks, beak width increases as a result of the correlated response to selection on beak depth.

even though selection favors thinner beaks. This is exactly what happened on Daphne Major (Figure 3.19).

Disequilibrium

Selection produces optimal adaptations only at equilibrium.

In Chapter 1 we saw how natural selection could gradually increase beak depth generation by generation until an equilibrium was reached, a point where stabilizing selection maintained the average beak depth in the population at the optimum size. This example illustrates the principle that selection keeps changing a population until an adaptive equilibrium is reached. It is easy to forget that the populations being observed have not necessarily reached equilibrium. If the environment has changed recently, there is every reason to suspect that the morphology or behavior of residents is not adaptive under current conditions. How long does it take for populations to adapt to environmental change? This depends on how fast selection acts. As we have seen, enormous changes can be created by artificial selection in a few dozen generations.

Disequilibrium is particularly important for some human characters because there have been big changes in the lives of humans during the last 10,000 years. It seems likely that many aspects of the human phenotype have not had time to catch up with recent changes in our subsistence strategies and living conditions. Our diet provides a good example. Five thousand years ago, most people hunted wild game and gathered wild plants for food, and typically had little access to sugar, fat, and salt (Figure 3.20). These are essential dietary requirements for the proper functioning of the human body, and so it was probably always good for people to eat as much of these things as they could find. In an adaptive response to human dietary needs, evolution equipped people with a nearly insatiable appetite for fat,

Figure 3.20

For most of our evolutionary history, humans have subsisted on wild game and plant foods. Sugar, salt, and fat were in short supply. Here a Hadza woman digs up a tuber. (Photograph courtesy of Nick Blurton Jones.)

salt, and sugar. With the advent of agriculture and trade, however, these substances became readily available and our evolved appetites for them became more problematic. Today, eating too much fat, salt, and sugar is associated with a variety of health problems including bad teeth, obesity, diabetes, and high blood pressure.

Genetic Drift

When populations are small, genetic drift may cause random changes in gene frequencies.

So far, we have assumed that evolving populations are always very large. When populations are small, however, random effects caused by **sampling variation** can be important. To see what this means, consider a statistical analogy. Suppose that we have a huge urn like the one in Figure 3.21. The urn contains 10,000 balls—half of them black and half red. Suppose we also have a collection of small urns and each holds 10 balls. We draw 10 balls at random from the big urn to put in each small urn. Not all of the little urns will have five red balls and five black balls. Some will have four red balls, some three red balls, and a few may even have no red balls. The fact that the distribution of black and red balls among the small urns varies is called sampling variation.

The same thing happens during genetic transmission in small populations (Figure 3.22). Suppose there is an organism that has only one pair of chromosomes with two possible alleles, *A* and *a*, at a particular locus and that selection does not act on this trait. In addition, a population of five individuals of this species is newly isolated from the rest. In this small population (generation 1), each allele has a frequency of $\frac{1}{2}$, so five chromosomes carry *A* and five carry *a*. These five individuals mate at random and produce five surviving offspring (generation 2). To keep things

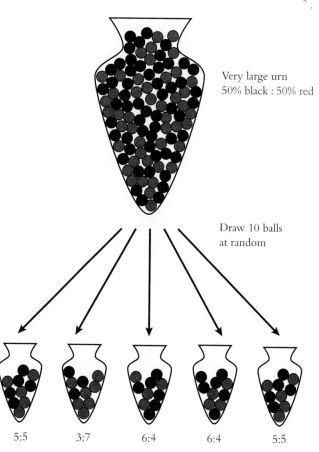

Very large urn
50% black : 50% red

Draw 10 balls
at random

Figure 3.21

A large urn has 10,000 balls, half of them black and half red. Ten balls are drawn from the large urn at random and placed in each of five small urns. Not all of the urns contain five red balls and five black balls. This phenomenon is called sampling variation.

5:5   3:7   6:4   6:4   5:5

simple, assume that the gamete pool is large, so there is no sampling variation. This means that half of the gametes produced by generation 1 will carry *A* and half will carry *a*. However, only 10 gametes will be drawn from this gamete pool to form the next generation of five individuals. These gametes will be sampled in the same way the balls were sampled from the urn (Figure 3.21). The most likely result is that there are five *A* alleles and five *a* alleles in generation 2 as there were in generation 1. However, just as there is some chance of drawing three, four, six, seven, or even zero dark balls from the urn, there is some chance that generation 2 will not carry equal numbers of *A* and *a* alleles. Suppose the five individuals in generation 2 have six *A* alleles and four *a* alleles. The frequency of the *A* allele is now 0.6, so when these individuals mate, the frequency of the *A* allele in the pool of gametes will be 0.6 as well. This means that the frequency of the allele in the population has changed by chance alone. In the third generation the gene frequencies change again, this time in the opposite direction. This phenomenon is called **genetic drift**. In small populations, genetic drift causes random fluctuations in genetic frequencies.

The way alleles are sampled in real populations depends on the biology of the evolving species and the nature of the selective forces that it faces. For instance, population size in some bird species might be limited by nesting sites, and success in obtaining nest sites might be related to body size. The larger the bird, the more chance it will have success in battling others for a prime nesting spot. Although body size is partly influenced by genes, and selection would favor genes for large body size, many genes don't affect body size, and they may be sampled randomly in this particular population.

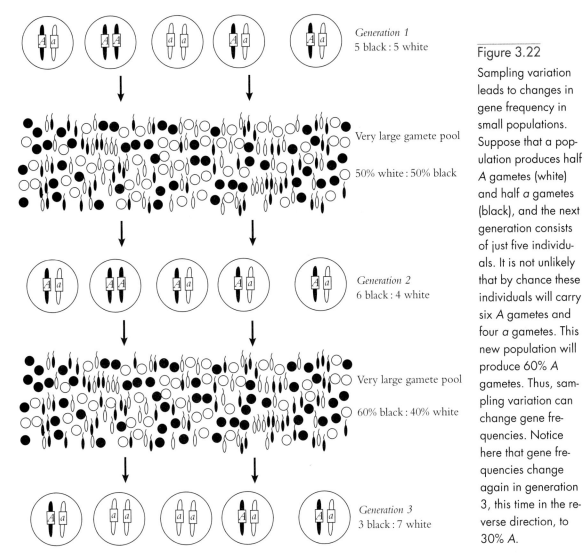

Generation 1
5 black : 5 white

Very large gamete pool

50% white : 50% black

Generation 2
6 black : 4 white

Very large gamete pool

60% black : 40% white

Generation 3
3 black : 7 white

Figure 3.22
Sampling variation
leads to changes in
gene frequency in
small populations.
Suppose that a pop-
ulation produces half
A gametes (white)
and half *a* gametes
(black), and the next
generation consists
of just five individu-
als. It is not unlikely
that by chance these
individuals will carry
six *A* gametes and
four *a* gametes. This
new population will
produce 60% *A*
gametes. Thus, sam-
pling variation can
change gene fre-
quencies. Notice
here that gene fre-
quencies change
again in generation
3, this time in the re-
verse direction, to
30% *A*.

The rate of genetic drift depends on population size.

Genetic drift causes more rapid change in small populations than in large ones because sampling variation is more pronounced in small samples. To see why this is true, consider a simple example. Suppose a company is trying to predict whether a new product, say low-fat peanut butter, will succeed in the marketplace. The company commissions two different survey firms to find out how many peanut butter fans would switch to the new product. One firm conducts interviews with five people, four of whom say that they would switch immediately to the low-fat variety. The other company polls 1000 people, and 50% of them say they would prefer low-fat peanut butter. Which survey should the company trust? Clearly, the second survey is more credible than the first one. It is easy to imagine that by chance we might find four fans of low-fat peanut butter in a sample of five people, even if the actual frequency of such preferences in the population were only 50%. In contrast, in a sample of 1000 people, we would be very unlikely to find such a large discrepancy between our sample and the population.

Exactly the same principle applies to genetic drift. In a population of five in-dividuals (with 10 chromosomes) in which two alleles are equally common (freq

= 0.5) there is a good chance that in the next generation the frequency of one of the alleles will be greater than or equal to 0.8. But in a population of 1000 individuals, there is virtually no chance that such a large deviation from the initial frequency will occur.

Genetic drift causes isolated populations to become genetically different from each other.

Genetic drift leads to unpredictable evolution because the changes in gene frequency caused by sampling variation are random. As a result, drift causes isolated populations to become genetically different from one another over time. Results from a computer simulation illustrate how this works for a single genetic locus with two alleles (Figure 3.23). Initially there is a large population in which each allele has a frequency of 0.5. Then four separate populations of 20 individuals (each individual carrying two sets of chromosomes) are created. These four populations are maintained at a constant size and remain isolated from each other during all subsequent generations. During the first generation the frequency of *A* in the gamete pool is 0.5 in each population. However, the unpredictable effects of sampling variation alter the frequencies of *A* among adults in each population. From the first generation we sample 40 gametes from each of the four populations. In

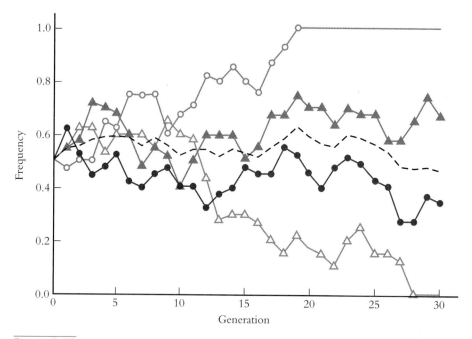

Figure 3.23

This computer simulation demonstrates that genetic drift causes isolated populations to become genetically different from one another. Four populations are initially formed by drawing 20 individuals from a single large population. In the original population, there is a gene with two alleles that each have a frequency of 0.5. The graph tracks the frequency of one of these alleles in each of these populations over time. Population 1 is in red, population 2 in blue, population 3 in green, and population 4 in purple. In each population, the frequency of the allele fluctuates randomly under the influence of genetic drift. Notice that populations 2 and 4 eventually reach fixation, meaning that one of the two alleles is lost.

population 1 (marked in red) the frequency of A increases dramatically, in populations 2 (blue) and 3 (green) it increases a little, while in population 4 (purple) it decreases. In each case, the change is created by chance alone.

During the next generation the same process is repeated, except that the gamete pools of each of the four populations now differ. Once again, 40 gametes are sampled from each population. This time the frequency of A in populations 2, 3, and 4 increases, while the frequency of A in population 1 decreases substantially. As these random and unpredictable changes continue, the four populations become more and more different from one another. Eventually, one of the alleles is lost entirely in two of our populations (A in population 2 and a in population 4), making all individuals in these populations identical at the locus in question. Such populations have reached **fixation**. In general, the smaller the population, the sooner it reaches fixation. Theoretically, if genetic drift goes on long enough, all populations eventually reach fixation. When populations are large, however, this may take such a long time that the species will become extinct before fixation occurs. Populations will remain at fixation until mutation introduces a new allele.

As you might expect, the rate at which populations become genetically different is strongly affected by their size—small populations differentiate rapidly, while larger populations differentiate more slowly. We will see in Chapter 14 that the relationship between population size and the rate of genetic drift allows us to make interesting and important inferences about the size of human populations tens of thousands of years ago.

Populations must be quite small for drift to lead to significant maladaptation.

Random changes produced by genetic drift will create adaptations only by chance, and we have seen that the probability of assembling complex adaptations in this way is small indeed. This means that genetic drift usually leads to maladaptation. The importance of genetic drift in evolution is a controversial topic. Nearly everyone agrees that populations must be fairly small (say less than 100 individuals) for drift to have an important effect when it is opposed by natural selection. There is a lot of debate about whether populations in nature are usually that small. Most scientists agree, however, that genetic drift is not likely to generate significant maladaptation in traits that vary continuously and are affected by many genetic loci. Thus, a trait like beak size in Darwin's finches is unlikely to be influenced much by genetic drift.

Local versus Optimal Adaptations

Natural selection may lead to an evolutionary equilibrium at which the most common phenotype is not the best possible phenotype.

Natural selection acts to increase the adaptedness of populations, but it does not necessarily lead to the best possible phenotypes. The reason for this is that natural selection is myopic—it favors small improvements to the existing phenotype, but it does not take into account the long-run consequences of these alterations. Selection is like a mountaineer who tries to climb a peak cloaked in dense cloud; she cannot see the surrounding country, but she figures she will reach her goal if she keeps climbing uphill. The mountain climber will eventually reach the top of something if she continues climbing, but if the topography is the least bit complicated, she won't necessarily reach the summit. Instead, when the fog clears, she is likely to find herself on the top of a small, subsidiary peak. In the same way, natural selection keeps changing the population until no more small improvements are possible, but there is no reason to believe that the end product is the optimal phenotype. The phenotype arrived at in such cases is called a local adaptation; it is analogous to the subsidiary peak reached by the mountaineer.

As an example of this effect, let's consider the evolution of eyes again (Figure 3.24). Humans, other vertebrates, and some invertebrates such as octopi have **camera-type eyes**. In this type of eye there is a single opening in front of a lens, which projects an image on photoreceptive tissue. Insects, in contrast, have **compound eyes**, in which a large number of very small separate photoreceptors build up an image composed of a grid of dots, something like a television image. Compound eyes are inferior to camera-type eyes in most ways. For example, they have lower resolution and less light-gathering power than similarly sized camera-type eyes. If this is so, why haven't insects evolved camera-type eyes?

The most likely answer is that once a species has evolved complex, compound eyes, selection cannot favor intermediate types, even though this might eventually allow superior camera-type eyes to evolve. Consider the early lineages in which eyes were first evolving. There may have come a time when selection favored greater light sensitivity. In the vertebrates and mollusks, this was achieved by increasing the area of sensitive tissue within each eye. In insects, it seems likely that increased sensitivity was achieved by multiplying the number of small photoreceptors, each in its own cup. At this early stage, these alternatives yielded equally useful eyes. However, once the insect lineage has evolved a visual system based on many small eyes, images could not be formed the same way as in camera-type

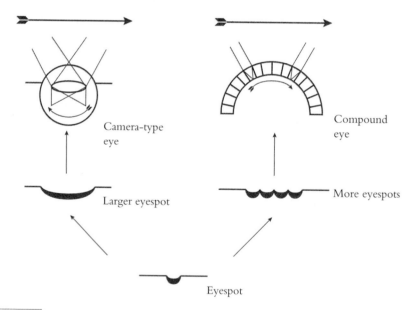

Camera-type eye

Compound eye

Larger eyespot

More eyespots

Eyespot

Figure 3.24

Researchers believe that different evolutionary pathways led to the development of camera-type and compound eyes. Selection for greater light sensitivity in a simple organism could favor the development of a larger eyespot or of multiple eyespots. The first pathway could lead to camera-type eyes, and the second to compound eyes. Images are formed very differently in compound and camera-type eyes, making it virtually impossible for compound eyes to evolve into camera-type eyes.

eyes because the camera-type eye inverts images while the compound eye does not. It does not seem structurally possible to evolve from a compound eye to a camera-type eye in a number of small steps, each favored by selection.

Some local adaptations are called developmental constraints.

Most kinds of organisms begin life as a single fertilized cell, a zygote. As an organism grows, this cell divides many times giving rise to specialized nerve cells, liver cells, and so on. This process of growth and differentiation is called **development**, and the development of complex structures like eyes involves many different interdependent processes. Developmental changes that would produce desirable modifications are often selected against because these alterations have many other negative effects. For example, as we will see later in the text, there is good reason to believe that it would be adaptive for males in some primate species to be able to **lactate** (produce milk for their young). However, there are no primate males who can lactate, and most biologists believe that the developmental changes that would allow males to lactate would also make them sterile. Thus, developmental processes constrain evolution, but such constraints are not absolute. They result from the particular phylogenetic history of the lineage. If primate reproductive biology had evolved along a different pathway so that the development of mammary glands and other elements of the primate reproductive system were independent, then there would be no constraints on the evolution of male lactation.

Other Constraints on Evolution

Evolutionary processes are also constrained by the laws of physics and chemistry.

The laws of physics and chemistry place additional constraints on the kinds of adaptations that are possible. For example, the laws of mechanics predict that the strength of bones is proportional to their cross-sectional area. This fact constrains the evolution of morphology. To see why, suppose that an animal's size (that is, its *linear* dimensions) doubles as the result of natural selection. This means that the animal's weight (which is proportional to its *volume*) will increase by a factor of 8 (see Box 3.2). The strength to bear an animal's weight comes from its muscles and bones, and is determined largely by their cross-sectional areas. However, if the linear dimensions of bones and muscles double, their cross-sectional areas will increase by only a factor of 4. This means that the bones and muscles will be only half as strong relative to the animal's weight as they were before. Thus, if the bones are to be as strong as they were before, they must be more than four times greater in cross section, which means that they must become proportionately thicker. Of course, thicker bones are heavier, and this imposes other constraints on the animal. If it is heavier, it may not be able to move as fast or suspend itself from branches. This constraint (and a closely related one that governs the strength of muscles) explains why big animals like elephants are typically slow and ponderous, while smaller animals like squirrels move quickly and are often agile leapers. It also explains why all flying animals are relatively light. Natural selection might well favor an elephant that could run like a cheetah, vault obstacles like an impala, and climb like a monkey. But because of the tradeoffs imposed by the laws of physics, there

BOX 3.2

The Geometry of Area–Volume Ratios

When an animal becomes larger without otherwise changing shape, the ratio of any fixed area measurement to its volume decreases. This is easiest to understand by computing how the ratio of an animal's entire surface area to its volume changes as the animal becomes larger. To see why, suppose that the animal is a cube x centimeters on a side. Then its volume is x^3, and its surface area is $6x^2$. Thus, the ratio of its surface area to its volume is

$$\frac{\text{surface area}}{\text{volume}} = \frac{6x^2}{x^3} = \frac{6}{x}$$

This means that an animal measuring 1 cm on a side has 6 cm^2 of surface area for each cubic centimeter of volume. An animal 2 cm on a side has only 3 cm^2 of surface area for each cubic centimeter of volume. Of course, there aren't any cubical animals, but it turns out that the shape doesn't alter this relationship. When an animal's linear dimension is doubled without the animal changing shape, the ratio of surface area to volume is halved. We will see in Chapters 14, 15, and 16 that this fact has consequences for how temperature affects the size of animals.

The same geometrical principle governs the relationship between any area measure and volume. Suppose the cubical animal has a vertical bone running through its center that supports the weight of the rest of the animal. For simplicity, we make the bone square in cross section, and assume it has the dimensions $\frac{1}{2}x$ by $\frac{1}{2}x$ as shown in Figure 3.25. The cross-sectional area of this bone is thus $\frac{1}{2}x \times \frac{1}{2}x = \frac{1}{4}x^2$. Thus the ratio of the volume of the animal to the cross-sectional area of the bone is then

$$\frac{\frac{1}{4}x^2}{x^3} = \frac{1}{4x}$$

Thus if the linear dimensions of an animal are doubled, its weight will increase eightfold, but the cross-sectional area of its bones—and therefore their strength—only increase fourfold.

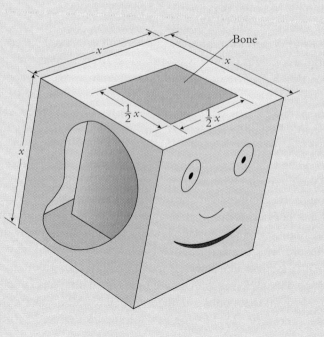

Figure 3.25

A cutaway diagram of the cubical animal discussed.

Figure 3.26

It might be advantageous for animals to be large enough to be relatively invulnerable to predators, but agile enough to leap considerable distances. But physical constraints limit evolution—not all things are possible.

are no mutant pachyderms whose bones are both light and strong enough to permit these behaviors (Figure 3.26).

Such constraints are closely related to genetic correlations. Selection eliminates mutants with heavier, weaker bones, and the tradeoffs inherent in the design of structures mean that there are also no possible mutant animals that have lighter, stronger bones. The genotypes that remain in the population either lead to the development of lighter bones or stronger bones, but not both. Thus, bone weight and bone strength typically show a positive genetic correlation.

Further Reading

Crow, J. F. 1986. *Basic Concepts in Population, Quantitative, and Evolutionary Genetics.* W. H. Freeman, New York.

Dawkins, M. S. 1986. *Unraveling Animal Behavior.* Longham, Essex.

Dawkins, R. 1996. *The Blind Watchmaker.* W.W. Norton, New York.

Falconer, D. S. 1960. *Quantitative Genetics.* Ronald, New York.

Hedrick, P. 1983. *Genetics of Populations.* Science Books International, Boston.

Maynard Smith, J. 1989. *Evolutionary Genetics.* Oxford University Press, Oxford.

Ridley, M. 1985. *The Problems of Evolution.* Oxford University Press, New York.

Ridley, M. 1996. *Evolution.* 2d ed. Blackwell Scientific, Oxford.

Study Questions

1. In human populations in Africa, two common alleles affect the structure of hemoglobin, the protein that carries oxygen on red blood cells. What is regarded as the normal allele (the allele common in European populations) is usually labeled A and the sickle-cell allele is designated S. There are three hemoglobin genotypes, and the phenotypes associated with these genotypes can be distinguished in a variety of ways. In an African population of 10,000 adults there are 3000 AS individuals, 7000 AA individuals, and 0 SS individuals.

(a) Suppose these individuals were to mate at random. What would be the frequency of the A and S alleles among the gametes that they produce?

(b) What would be the frequency of the three genotypes that result from random mating?

(c) Is the original population of adults in Hardy-Weinberg equilibrium?

2. The three common genotypes at the hemoglobin locus have very different phenotypes: SS individuals suffer from severe anemia, AS individuals have a relatively mild form of anemia but are resistant to malaria, and AA individuals have no anemia but are susceptible to malaria. The frequency of the S allele among the gametes produced by the first generation of a Central African population is 0.2.

(a) Assuming that mating occurs at random, what are the frequencies of the three genotypes among zygotes produced by this population?

(b) In this area, no SS individuals survive to adulthood, 70% of the AA individuals survive, and all of the AS individuals survive. What is the frequency of each of the three genotypes among the second generation of adults?

(c) What is the frequency of the S allele among gametes produced by these adults?

3. Tay-Sachs disease is a lethal genetic disease controlled by two alleles, the Tay-Sachs allele T, and the normal allele N. Children who have the TT genotype suffer from mental deterioration, blindness, paralysis, and convulsions, and they die sometime between three and five years of age. The proportion of infants afflicted with Tay-Sachs disease varies in different ethnic groups. The frequency of the Tay-Sachs *allele* is highest among descendants of Central European Jews. Suppose that in a population of 5000 people descended from Central European Jews the frequency of the Tay-Sachs allele is 0.02.

(a) Assuming that members of this population mate at random, and each adult has four children on average, how many gametes from each person are successful at being included in a zygote?

(b) Calculate the frequency of each of the three genotypes among the population of zygotes.

(c) Assume that no TT zygotes survive to become adults, and that 50% of the TN and NN individuals survive to become adults. What is the frequency of the TT and TN genotypes among these adults?

(d) When the adult individuals resulting from zygotes in part 3b produce gametes, what is the frequency of the T allele among their gametes? Compare your answer to the frequency of the T allele among gametes produced by their parents. Think about your result, given that these numbers are roughly realistic and that there have been no substantial medical breakthroughs in the treatment of Tay-Sachs disease. What is the paradox here?

4. Consider a hypothetical allele that is lethal, like the Tay-Sachs allele, but is dominant rather than recessive. Assume that this allele has a frequency of 0.02, like the Tay-Sachs allele. Recalculate the answers to parts a, b, and c in question 3. Is selection stronger against the recessive allele or the dominant one? Why? Assuming that mutation to the deleterious allele occurs at the same rate at both loci, which allele will occur at a higher frequency? Why?

5. Consider the Rh blood group. For this example we will assume there are only two alleles, R and r, where r is recessive. Homozygous rr individuals produce certain nonfunctional proteins and are said to be Rh-negative (Rh^-); while Rr and RR individuals are Rh-positive (Rh^+). When an Rh^- female mates with an Rh^+ male, their offspring may suffer serious anemia while still in the

uterus. The frequency of the *r* allele in a hypothetical population is 0.25. Assuming that the population is in Hardy-Weinberg equilibrium, what fraction of the offspring in each generation will be at risk for this form of anemia?

6. The blending model of inheritance was attractive to 19th-century biologists because it explained why offspring tend to be intermediate in appearance between their parents. However, this model fails because it predicts a loss in variation, but variation is maintained. How does Mendelian genetics explain the intermediate appearance of offspring without the loss of variation? Be sure to explain why the properties of the Hardy-Weinberg equilibrium are an important part of this explanation.

7. In a particular species of fish, egg size and egg number are negatively correlated. Draw a graph that illustrates this fact. What will happen to egg size if selection favors larger numbers of eggs?

8. A plant breeder trying to increase the yield of wheat selects plants with larger kernels. After a number of generations of selection, the size of kernels produced has increased substantially. However, the number of kernels per plant has decreased so that the total yield remains constant. What two sources of maladaptation best explain this result?

9. The compound eyes of insects provide poor image clarity compared with the images produced by the camera-type eyes of vertebrates. Explain why insects don't evolve camera-type eyes.

10. Explain why genetic drift has no effect on genetic loci that are not variable. (When all individuals in the population are homozygous for the same allele, that genetic locus is not variable.) What does this mean about the long-run outcome of genetic drift?

11. Why does natural selection produce adaptations only at equilibrium?

12. Two dog breeders are using artificial selection to create new breeds. Both begin with dachshunds. One breeder wants to create a breed with longer forelimbs and hindlimbs. The second breeder wants to create a breed with shorter forelimbs and longer hindlimbs. Which breeder will make more rapid progress? Why?

13. When selection makes animals larger overall, their bones also get thicker. The increase in bone thickness could be due to the correlated response to selection, or to independent selection for thicker bones, or some combination of the two factors. How could you determine the relative importance of these two processes?

CHAPTER 4

Speciation and Phylogeny

What Are Species?

Microevolution refers to how populations change under the influence of natural selection and other evolutionary forces, while macroevolution refers to how new species and higher taxa are created.

So far, we have focused on how natural selection, mutation, and genetic drift cause populations to change through time. These are mechanisms of **microevolution**, and they affect the morphology, physiology, and behavior of particular species in particular environments. For example, microevolutionary processes are responsible for variation in the size and shape of the beaks of medium ground finches on Daphne Major.

However, this is not all there is to the process of evolution. Darwin's major work was entitled *The Origin of Species* because he was interested in how new species are created, as well as how natural selection operates within populations. Evolutionary theory tells us how new species, genera, families, and higher groupings come into existence. These processes are mechanisms of **macroevolution**. Macroevolutionary processes play an important role in the story of human evolution. To properly interpret the fossil record and reconstruct the history of the human lineage, we need to understand how new species and higher groupings are created and transformed over time.

Species can usually be distinguished by their behavior and morphology.

Organisms cluster into distinct types called **species**. The individual organisms that belong to a species are similar to each other, and are usually quite distinct from the members of other species. For example, in Africa some tropical forests house two species of apes: chimpanzees and gorillas. These two species are similar in many ways: both are tailless, both bear weight on their knuckles when they walk, and both defend territories. Nonetheless, these two species can easily be distinguished on the basis of their morphology: chimpanzees are smaller than gorillas; male gorillas have tiny **testes** (the organ that produces sperm), while male chimpanzees have quite large ones; and gorillas have a fin of bone on their skull, while chimpanzees have more rounded skulls. Gorillas and chimpanzees also differ in their behavior: chimpanzees make and use tools in foraging, while gorillas do not; male gorillas beat their chests when they perform displays, while male chimpanzees flail branches and charge about; and gorillas live in smaller groups than chimpanzees do. These two species are easy to distinguish because there are no animals that are intermediate between them—no "gimps" or "chorillas" (Figure 4.1).

Species are not abstractions created by scientists; they are real biological categories. People all over the world name the plants and animals around them, and biologists use the same kinds of phenotypic characteristics to sort animals into species as do other people. For the most part there is little problem in identifying any particular specimen from its phenotype.

Although nearly everyone agrees that species exist and can recognize species in nature, biologists are much less certain about how species should be defined. This uncertainty arises from the fact that evolutionary biologists do not agree about why

(a)

(b)

Figure 4.1

(a) Chimpanzees and (b) gorillas sometimes occupy the same forests, and share a number of traits. However, these two species are readily distinguished because there are no animals that are intermediate between them.

species exist. There is now a considerable amount of controversy about the processes that give rise to new species and the processes that maintain established ones. Although there are many different views on these topics, we will concentrate on two of the most widely held points of view, the biological species concept and the ecological species concept.

The Biological Species Concept

The biological species concept defines a species as a group of interbreeding organisms that are reproductively isolated from other organisms.

Most zoologists believe in the **biological species concept**, which defines a biological species as a group of organisms that interbreed in nature and are reproductively isolated. **Reproductive isolation** means that members of a given group of organisms do not mate successfully with organisms outside the group. For example, there is just one species of gorilla, *Gorilla gorilla,* and this means that all gorillas are capable of mating with one another and they do not breed with any other kinds of animals in nature. According to adherents of the biological species concept, this is why there are no gimps or chorillas.

The biological species concept defines a species in terms of the ability to interbreed because successful mating leads to **gene flow**, the movement of genetic material within parts of a population or from one population to another. Gene flow tends to maintain similarities among members of the same species. To see how gene flow preserves homogeneity within species, consider the hypothetical situation diagrammed in Figure 4.2a. Imagine a population of finches living on a small island in which there are two habitats, moist and dry. Natural selection favors different-sized beaks in each habitat. Large beaks are favored in the dry habitat, and small beaks are favored in the moist habitat. However, because the island is small, birds fly back and forth between the two environments and mate at random, so there is a lot of gene flow between habitats. Unless selection is very strong, gene flow will swamp its effects. On average, birds in both habitats will have medium-sized beaks, a compromise between the optimal phenotypes for each habitat. In this way, gene flow tends to make the members of a species evolve as a unit.

Now, suppose there are finches living on two different islands, one wet and one dry, and that the islands are far enough apart that the birds are unable to fly from one island to the other (Figure 4.2b). This means that there will be no interbreeding and no gene flow will occur. With no genetic exchange between the two independent groups to counter the effects of selection, the two populations will diverge genetically and become less similar phenotypically. Birds on the dry island will develop large beaks, and birds on the wet island will develop small beaks.

Reproductive isolation prevents species from genetically blending together.

Reproductive isolation is the flip side of interbreeding. Suppose chimpanzees and gorillas could and did interbreed successfully in nature. This would lead to gene flow between the two kinds of apes, and this phenomenon would produce animals that were genetically intermediate between chimpanzees and gorillas. Eventually the two species would merge into one. There are no chorillas or gimps in nature because the two species are reproductively isolated.

Reproduction is a complicated process, and anything that alters the process can act as an isolating mechanism. Even subtle differences in activity patterns, courtship

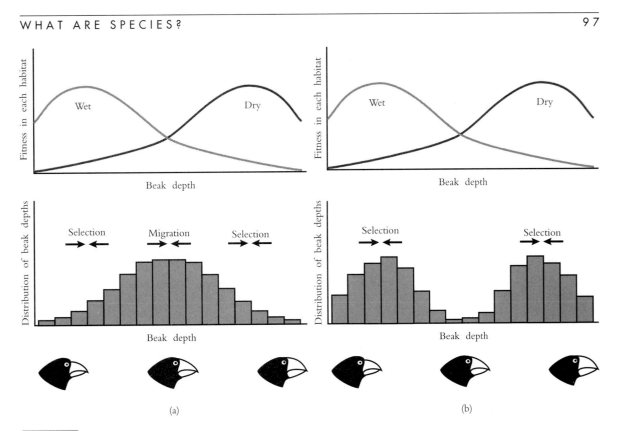

Figure 4.2
Gene flow among populations destroys differences between them. (a) Suppose a population of finches lives on an island with both wet and dry habitats. In the dry habitat, selection favors deep beaks, while selection in the wet habitat favors shallow beaks. Because
interbreeding leads to extensive gene flow between populations in the two habitats, the
population beak size responds to an average of the two environments. (b) Now suppose
two populations of finches live on different islands, one wet and one dry, with no gene
flow between the two islands. Now the deep beaks are common on the dry island, and
shallow beaks are common on the wet island.

behavior, or appearance may prevent individuals of different types from mating.
Moreover, even if a mating among individuals of different types does takes place,
the egg may not be fertilized or the zygote may not survive.

The Ecological Species Concept

*The ecological species concept emphasizes the role of selection in maintaining species
boundaries.*

Critics of the biological species concept point out that gene flow is neither
necessary nor sufficient to maintain species boundaries in every case, and argue
that selection plays an important role in preserving the boundaries between species.
The view that emphasizes the role of natural selection in creating and maintaining
species is called the **ecological species concept**.

In nature, species boundaries are often maintained even though there are substantial amounts of gene flow between species. For example, the medium ground
finch readily breeds with the large ground finch on islands where they coexist.
Peter Grant and his colleagues have estimated that approximately 10% of the time,

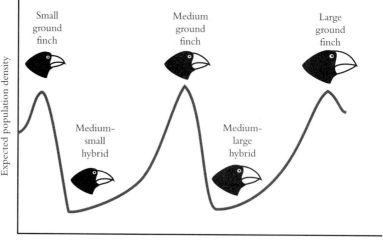

Figure 4.3

In the Galápagos, selection maintains three species of ground finches despite the fact that there is substantial gene flow among them. The red line represents the amount of food available in the environment for birds with different size beaks. Each peak in this curve represents a different species.

medium ground finches mate with large ground finches. Although there is considerable gene flow between these two species, they have not merged into a single species. Grant and his colleagues have concluded that the medium ground finch and the large ground finch have remained distinct because these two species represent two of the three optimal beak sizes for ground finches (Figure 4.3). These three optimal sizes are based on the availability of seeds of different size and hardness and the ability of birds with different-sized beaks to harvest these seeds. According to their calculations, the three optimal beak sizes for ground finches correspond to the average beak sizes of the small *(G. fuliginosa)*, medium *(G. fortis)*, and large *(G. magnirostris)* ground finch species (Figure 4.4). These researchers suggest that hybrids are selected against because their beaks fall in the "valleys" between these selective "peaks." The kind of interbreeding seen in the Galápagos finches is not particularly unusual; interspecific matings occur in a sizable minority of bird and mammal species.

There are also a number of species that have maintained their coherence with no gene flow between isolated subpopulations. For example, the checkerspot butterfly (Figure 4.5) is found in scattered populations throughout California. Members of different populations are very similar morphologically and are all classified

(a)   (b)   (c)

Figure 4.4

(a) The small ground finch, *Geospiza fuliginosa;* (b) the medium ground finch, *G. fortis;* (c) the large ground finch, *G. magnirostris.*

Figure 4.5
Checkerspot butterflies living in localized populations throughout California are all members of the same species, but there is probably very little gene flow among different populations.

as members of the species *Euphydryas editha*. However, careful studies by Stanford University biologist Paul Ehrlich have shown that these butterflies rarely disperse more than 100 m from their place of birth. Given that populations are often separated by several kilometers and sometimes by as much as 200 km, it seems unlikely that there is enough gene flow to unify the species.

The existence of strictly asexual organisms provides additional evidence that species can be maintained without gene flow. Asexual species, which reproduce by budding or fission, simply replicate their own genetic material and produce exact copies of themselves. Thus, gene flow cannot occur in asexual organisms and it does not make sense to think of them as being reproductively isolated. Nonetheless, many biologists who study purely asexual organisms contend that it is just as easy to classify them into species as to classify organisms that reproduce sexually.

What maintains the coherence of species in these cases? Most biologists think the answer is natural selection. To see why, let's return to the Galápagos once more. Imagine a situation in which selection favored finches with small bills in one habitat and finches with big bills in another habitat, but did not favor birds with medium-sized bills in either habitat. If birds with medium-sized beaks consistently failed to survive or reproduce successfully, natural selection could maintain the difference between the two bird species even if they interbred freely. In the case of the checkerspot butterflies and asexual species, we can imagine the opposite scenario: selection favors organisms with the same morphology and physiology and thus maintains their similarity even in the absence of gene flow.

Today most biologists concede that selection maintains species boundaries in a few odd cases like Darwin's finches, but insist that reproductive isolation plays the major role in most instances. However, there is a growing minority of researchers who contend that selection plays an important role in maintaining virtually all species boundaries.

The Origin of Species

Speciation is difficult to study empirically.

The species is one of the most important concepts in biology, so it would be helpful to know how new species come into existence. However, despite huge amounts of hard work and many heated arguments, there is still uncertainty about what Darwin called the "mystery of mysteries"—the origin of species. The mystery persists because the process of speciation is very difficult to study empirically. Unlike microevolutionary change within populations, which researchers are sometimes able to study in the field or laboratory, new species usually evolve too slowly for any single individual to study the entire process. On the other hand, speciation usually occurs much too rapidly to be detected in the fossil record. Nonetheless, biologists have compiled a substantial body of evidence that provides important clues about the processes that give rise to new species.

Allopatric Speciation

If geographic or environmental barriers isolate part of a population, and selection favors different phenotypes in these regions, then a new species may evolve.

Allopatric speciation occurs when a population is divided by some type of barrier, and different parts of the population adapt to different environments. The following hypothetical scenario captures the essential elements of this process. Mountainous islands in the Galápagos contain dry habitats at low elevations and moist habitats at higher elevations. Suppose that a finch species that lives on mountainous islands has a medium-sized beak, which represents a compromise between the best beak for survival in the wet areas and the best beak for survival in the dry areas. Further suppose that a number of birds are carried on the winds of a severe storm to another island that is uniformly dry, like the small, low-lying islands in the Galápagos archipelago. On this new, dry island, only large-beaked birds are favored because large beaks are best for processing the large, hard seeds that predominate there. As long as there is no competition from some other small bird adapted to the dry habitat and there is very little movement between the two islands, the population of finches on the dry island will rapidly adapt to their new habitat and the birds' average beak size will increase (Figure 4.6).

Now, let's suppose that after some time, finches from the small, dry island are blown back to the large island from which their ancestors came. If the large-beaked newcomers successfully mate with the medium-beaked residents, then gene flow between the two populations will rapidly eliminate the differences in beak size between them, and the recently created large-beaked variety will disappear. If, on the other hand, large-beaked immigrants and small-beaked residents cannot successfully interbreed, then the differences between the two populations will persist. As we noted earlier, there are a variety of mechanisms that can prevent successful interbreeding, but it seems that the most common obstacle to interbreeding is that hybrid progeny are less viable than other offspring. In this case, we might imagine that when the two populations of finches became isolated, they diverged genetically due to natural selection and genetic drift. The longer they remain isolated, the greater the genetic difference between the populations becomes. When the two distinct types then come into contact, hybrids may have reduced viability, either because genetic incompatibilities have arisen during their isolation or because hybrid birds are unable to compete successfully for food. If these processes cause complete reproductive isolation, then a new species has been formed.

Even if there is some gene flow after the members of the two populations initially come back into contact, two additional processes may increase the degree of reproductive isolation and facilitate the formation of a new species. The first process, **character displacement**, may occur if competition over food, mates, or other resources increases the morphological differences between the immigrants and the residents. In our example, the large-beaked immigrants will be better suited

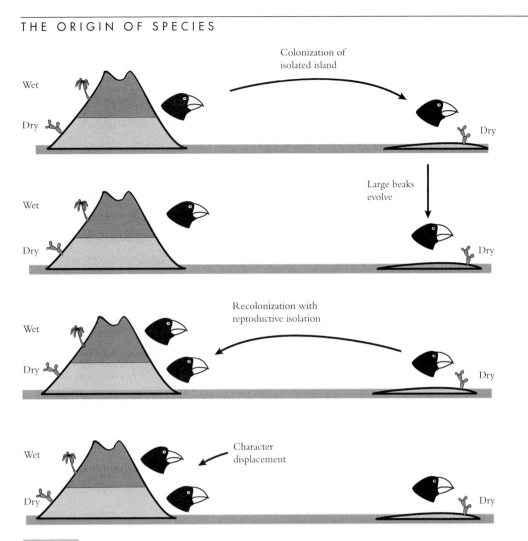

Figure 4.6

A likely sequence of events in a hypothetical allopatric speciation event in the Galápagos is depicted here. Initially, one population of finches occupies an island with both wet and dry habitats. Because there is extensive gene flow between habitats, these finches have intermediate-sized beaks. By chance, some finches disperse to a dry island where they evolve deeper beaks. Then by chance some of these birds are reintroduced to the original island. If the two populations are reproductively isolated, then a new species has been formed. Even if there is some gene flow, competition between the two populations will cause the beaks of the residents and immigrants to diverge further, a process called character displacement.

to the dry parts of the large island and will be able to out-compete the original residents of these areas. Resident birds will be better off in the wet habitats, where they face less competition from the immigrants. Since small beaks are advantageous in wet habitats, residents with smaller-than-average beaks will be favored by selection. At the same time, since large beaks are advantageous in dry habitats, natural selection will favor increased beak size among the immigrants. This process will cause the beaks of the competing populations to diverge. There is good evidence that character displacement has played an important role in shaping the morphology of Darwin's finches (Figure 4.7).

A second process, called **reinforcement**, may act to reduce the extent of gene flow between the populations. Since hybrids have reduced viability, selection will

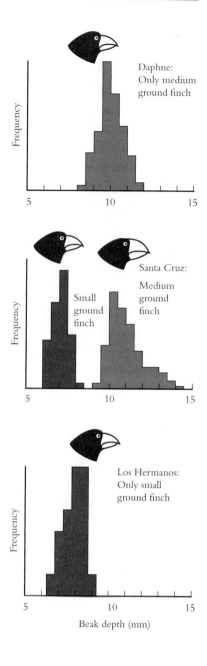

Figure 4.7

The distribution of beak sizes of the small ground finch *(G. fuliginosa)* and the medium ground finch *(G. fortis)* illustrates the effects of character displacement. The middle figure shows the distribution of beak depth for each species on Santa Cruz Island, where the two species compete. The top and bottom figures show the distributions of beak depth for the small and medium ground finches on islands where they do not compete with each other *(G. fortis* on Daphne Major and *G. fuliginosa* on Los Hermanos). The beaks of the two species are most different where there is direct competition. Careful measurements of seed density and other environmental conditions suggest that competition, not environmental differences, is responsible for the difference between the populations.

favor behavioral or morphological adaptations that prevent matings between members of the two populations (Figure 4.8). This process will further increase the reproductive isolation between the two populations. Thus, character displacement and reinforcement may amplify the initial differences between the populations and lead to two new species.

Allopatric speciation requires a physical barrier that initially isolates part of a population, interrupts gene flow, and allows the isolated subpopulation to diverge from the original population under the influence of natural selection. In our example, the physical barrier is the sea, but mountains, rivers, and deserts can also restrict movement and interrupt gene flow. Character displacement and reinforcement may work to increase differences when members of the two populations renew contact. However, these processes are not necessary elements in allopatric speciation. Many species become completely isolated while they are separated by a physical barrier.

Figure 4.8

Finch courtship displays can be quite elaborate. Here, a male finch displays to attract the female's attention. Subtle variations in courtship behavior among members of different populations may prevent mating and increase reproductive isolation.

Parapatric and Sympatric Speciation

New species may also be formed if there is strong selection that favors two different phenotypes.

Biologists who endorse the biological species concept usually argue that allopatric speciation is the most important mechanism for creating new species in nature. For these scientists, gene flow "welds" a species together, and species can be split only if gene flow is interrupted. Biologists who think that natural selection plays an important role in maintaining species often contend that selection can lead to speciation even when there is interbreeding. There are strong and weak versions of this hypothesis. The weak version, called **parapatric speciation**, holds that selection alone is not sufficient to produce a new species, but new species can be formed if selection is combined with partial genetic isolation. For example, baboons range from Saudi Arabia to the Cape of Good Hope and occupy an extremely broad range of environments. Some baboons live in moist tropical forests, some live in arid deserts, and some live in high-altitude grasslands (Figure 4.9). Different behaviors and morphological traits may be favored in each of these environments, and this may cause baboons in different regions to vary. At habitat boundaries, animals that come from different habitats and have different characteristics may mate and create a **hybrid zone**. Study of hybrid zones in a wide variety of species suggests that hybrids are usually less fit than nonhybrids. When this is the case, selection should favor behavior or morphology that prevents mating between members of individuals from different habitats. If such reinforcement does occur, gene flow will be reduced, and eventually two new, reproductively isolated species will evolve.

The strong version of the hypothesis, called **sympatric speciation**, contends that strong selection favoring different phenotypes can lead to speciation even when there is no geographic separation, and therefore initially there is extensive gene flow among individuals in the population. Sympatric speciation is theoretically possible, and this form of speciation has been induced in laboratory popula-

(a)

(b)

(c)

(d)

Figure 4.9

Baboons are distributed all over Africa, and occupy a very diverse range of habitats: (a) In the Drakensburg Mountains in South Africa, winter temperatures drop below freezing and snow sometimes falls. (b) In the Motopos of Zimbabwe, woodlands are interspersed with open savanna areas. (c) Amboseli National Park lies at the foot of Mt. Kilimanjaro just inside the Kenyan border. (d) The Gombe Stream National Park lies on the hilly shores of Lake Tanganyika in Tanzania.

tions, but it is uncertain whether it occurs in nature. The three speciation mechanisms are diagrammed in Figure 4.10.

Adaptive radiation occurs when there are many empty niches.

Ecologists use the term **niche** to refer to a particular way of "making a living," which includes the kinds of food eaten and when, how, and where the food is

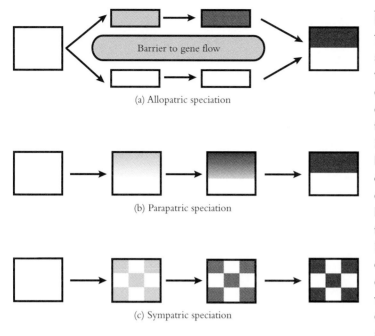

(a) Allopatric speciation

(b) Parapatric speciation

(c) Sympatric speciation

Figure 4.10

There are three different mechanisms of speciation. (a) Allopatric speciation occurs when a species is divided into two reproductively isolated populations that subsequently diverge. If the divergent populations regain contact, they cannot interbreed and two new species exist. (b) Parapatric speciation occurs when a species experiences different environments in different parts of its geographical range. Natural selection causes different populations of the species to diverge while adapting to these different surroundings, and eventually reproductive isolation is achieved. (c) Sympatric speciation occurs when selection strongly favors different adaptations to a similar environment within a single species.

acquired. One consequence of all three models of speciation is that the rate of speciation depends on the number of available ecological niches. Once again, Darwin's finches provide a good example. When the first finches migrated to the Galápagos from the mainland of South America (or perhaps from the Cocos Islands) about half a million years ago, all of the niches for small birds in the Galápagos were empty. There were opportunities to make a living as a seed eater, a cactus eater, and so on. The finches' ancestors diversified to fill all of these ecological niches, and eventually became 14 distinct species (Figure 4.11). An even more spectacular example of this process occurred at the end of the Cretaceous era. As you probably know, dinosaurs dominated the earth during the Cretaceous era but disappeared suddenly 65 mya (million years ago). The mammals that coexisted with the dinosaurs were mostly small, nocturnal, and insectivorous. But when the

(a)   (b)   (c)

Figure 4.11

When the first finches arrived in the Galápagos 500,000 years ago, there were many empty niches. (a) Some finches became cactus eaters, (b) others became seed eaters, and (c) some became predators on insects and other arthropods.

Evolution Is Not Always So Slow

When Darwin looked at the fossil record, he saw it static and frozen for long stretches. In retrospect that is not surprising. If selection events do not show up among a tribe of sparrows when summed over the course of a single generation, consider how much less visible these events will be in the strata of rock beneath our feet, in which the generations have been summed for many millions of generations.

One of the most famous examples of "rapid" evolution in the fossil record is the ancestral lineage of the modern horse. That was the subject that Huxley chose for his lecture "The Demonstrative Evidence of Evolution." The transmutation of an upper molar in *Mesohippus* from *Hyracotherium* is one of the most rapid changes whose sequences are preserved like a snip of flicker-frame silent movie in the rock record. This event took about one and a half million years, or about half a million generations.

Even that rate is molasses and stasis compared with the stutter step of live finches and sparrows. Compared with life above the ground, the record of life in the stones flows almost as slowly as stone itself. The apparent lack of action in the fossil record confirmed for Darwin his belief that evolution by natural selection must be a rare event, and that any and all action must be unendurably, geologically slow. It is because the changes in the record are invisible, and the changes around us were at the time invisible, that he says in the *Origin* we "see nothing of these slow changes in progress, until the hand of time has marked the lapse of ages."

In 1949, the evolutionist J. B. S. Haldane suggested that the rate of evolution be described in terms of a universal unit, whether the change took place in the animal or vegetable kingdoms, among the living or the long extinct. This unit he named in the same spirit of tribute that we call continents and oceans after their discoverers. He called his unit the darwin.

To keep things simple, Haldane said, let us define the darwin as a change in the length of some character. Since we want a universal unit, and not one that depends on the size of the creature, let us use the percentage change. The beak of a duck-billed dinosaur might lengthen by a few meters in a million years, and a bird's beak by a few millimeters in a million years, Haldane said, and yet both could represent a change of 10 percent. A percentage change will mean the same thing in myriad living and once-living forms, from the beaks of birds to the teeth of hyenas, from the skulls and cannon bones of horses to the inner whorls of ammonites. Haldane defined one darwin as a change of 1 percent per million years.

When Haldane looked at a few representative fossil records of evolution he found that the rate of change was very slow, on the order of 1 percent per million years, or a rate of change of a single darwin. Other investigators since Haldane have borne out his guess that this snail's pace is typical of the fossil record.

Haldane concluded as Darwin had that the rate of evolution by natural selection in the world around us must be infinitesimally slow, far too slow to watch, that it could only be watched in the long, slow additions of the fossil record. Rates in the living world would have to be measured in millidarwins. In artificial selection, he said, you could get rates of thousands of darwins, but that is not something you would see in the wild: "Rates of one darwin would be exceptional in nature."

"Such calculations are extremely rough," Haldane said, "but they suggest the remarkably small order of magnitude of the selective 'forces' which are at work if natural selection is largely responsible for evolution, and the extreme difficulty of demonstrating them in action."

Now it is possible to translate the evolution of Darwin's finches in the drought and the flood into Haldane's whimsically named unit, the darwin. In the drought the change was 25,000 darwins. After the flood the change was about 6,000 darwins.

So there is an enormous gulf between what we see when we take the time to watch the living world in action, and what we see when we look at the world recorded in stone. To set this discrepancy in a broad perspective, an evolutionist not long ago compiled more than five hundred cases of evolutionary change, from short and fast experiments in artificial selection (events that took months or at most a year and a half) to evolutionary experiments in the fossil record (events that took millions of years). The evolutionist, Philip Gingerich, translated all of these events into darwins. He discovered a simple pattern, a pattern that is just the opposite of what earlier evolutionists—the whole lineage of evolutionists from Darwin to Haldane—would have expected. The closer you look at life, the more rapid and intense the rate of evolutionary change. The farther back in time you stand, the less you see.

In a single year, you can find rates of change as high as 60,000 darwins. But in the fossil record the average is only a tenth of a darwin.

The reasons for the discrepancy are not far to seek. If at any time in those millions of years a species changed swiftly, but the rest of the time it changed slowly, that start-and-stop motion would average out into a very sluggish movement. What is more, if the species changed first one way and then the other way, over and over again, as Darwin's finches did in the first decade of the Grants' watch, then the fossil record would register virtually no change, a near equilibrium. Yet the beak of the finch is in fact in so much evolutionary motion that as soon as people started watching closely enough, they saw it change right before their eyes.

"There's quite a bit of wobble in the fossil record too," says Peter Grant. "But you don't normally see it. When you look at a series of fossils, the usual practice is to take 3,000 or 5,000 or 10,000 years as the unit. You take the average position at the beginning and the end, and average the rate of change in between.

"Even then, you see wobbles in the record. It's one in a thousand wobbles, if you like. You rarely get successive generations preserved in fossils. It's like looking at specimens of Darwin's finches from 1874, 1932, and 1987. You may get some wobbles there—but you miss all sorts of action in between."

The fossil record is just too primitive a motion-picture camera to capture the fast-moving life. Rapid motion disappears like the whir of a humming-bird's wings. In such a record, the two wonder years of Darwin's finches would disappear as surely as a wing-beat up and a wing-beat down, can-celing out in the blur.

If we look at the billowing smoke of a volcano from close up, we see in-tense and rapid motion, enormous and dangerous turbulence. If we look at the eruption from far, far away (a safe distance that puts it almost to the ho-rizon), the smoke seems to hang in the air almost motionless: we have to watch a long while to see any change at all. The evolution of life turns out to be rather like the eruption of a volcano. The closer you look, the more turbu-lent and dangerous the action; the farther your remove, the more the living world seems fixed and stable, hardly moving at all.

It is amazing to think of all these species around us not fixed but in jittery motion. It is like the difference between the old view of solid physical matter around us—the view in the time of Newton—and the view now, of infinite motion down to the level of each atom and molecule, and below, in the ceaseless assault and battery of elementary particles. The beak of the finch is an icon of evolution the way the Bohr atom is an icon of modern physics, and the study of either one shows us more primal energy and eternal change than our minds are built to take in. Yet like the vista of the atoms, the vista of evolution in action, of evolution in the flesh, has enormous implica-tions for our sense of reality, of what life is, and also for our sense of power, of what we can do with life.

"This jittering is an aspect of all populations everywhere," says Dolph Schluter. "It is demonstrating that populations are currently dynamic—still wobbling—so that a larger change in the environment at any moment might push them one way or the other." If they weren't jittering, that would sug-gest that the processes that brought them here had finished, that the creation was over, just as the universe would be moribund or dead if there were no motions to be found in its atoms. But these motions are to be found every-where, Dolph says. "They are ever-present."

Having seen even this much of the view from the rim of Daphne Major, we can no longer picture the story of life as slow and almost static, a world view for which the chief emblem of evolutionary change is a fossil in a stone. What we must picture instead is an emblem of life in motion. For all species, including our own, the true figure of life is a perching bird, a pas-serine, alert and nervous in every part, ready to dart off in an instant. Life is always poised for flight. From a distance it looks still, silhouetted against the bright sky or the dark ground; but up close it is flitting this way and that, as if displaying to the world at every moment its perpetual readiness to take off in any of a thousand directions.

Source: From pp. 109–112 in *The Beak of the Finch,* by Jonathan Weiner. Copyright © 1994 by Jonathan Weiner. Reprinted by permission of Alfred A. Knopf, Inc.

Figure 4.12

Darwin's finches are not the only example of adaptive radiations that occur when immigrants encounter a range of empty ecological niches. In the Hawaiian archipelago, the adaptive radiation of honeycreeper finches produced an even greater diversity of species.

dinosaurs became extinct, these humble creatures diversified to fill a broad range of ecological niches, evolving into elephants, killer whales, buffalo, wolves, bats, gorillas, humans, and the other kinds of mammals. When a single kind of animal or plant diversifies to fill many available niches, the process is called **adaptive radiation** (Figure 4.12).

The Tree of Life

Organisms can be classified hierarchically based on similarities. Many such similarities are unrelated to adaptation.

In Chapter 1 we saw how Darwin's theory of evolution explains the existence of adaptation. Now we can show how the same theory explains why organisms can be classified into a hierarchy based on their similarities, a fact that puzzled 19th-century biologists much more than the existence of adaptations. Richard Owen, a famous anatomist and one of Darwin's principal opponents in the debates following the publication of *The Origin of Species,* used dugongs (aquatic mammals much like manatees), bats, and moles as an example of this phenomena. All three of these creatures have the same kind and number of bones in their forelimbs, even though the shapes of these bones are quite different (Figure 4.13). The forelimb of the bat is adapted to flying, and that of the dugong to paddling (Figure 4.14). Nonetheless, the basic structure of a bat's forelimb is much more similar to that of a dugong than to the forelimb of a swift, even though the swift's forelimb is also designed for flight.

Such patterns of similarity make it possible to cluster species hierarchically—like a series of nested boxes. This remarkable property of life is the basis of the system for classifying plants and animals devised by the 18th-century Swedish bi-

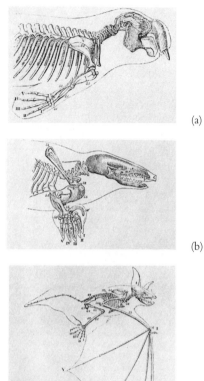

(a)

(b)

(c)

Figure 4.13

Organisms show patterns of similarity that have nothing to do with adaptation. Richard Owen's drawings of the forelimbs of three mammalian species illustrate this fact. The dugong (a), the mole (b) and the bat (c) all have the same basic bone structure in their forelimbs; dugongs use their forelimbs to swim in the sea, moles use theirs to burrow into the earth, and bats use their forelimbs to fly.

ologist Linnaeus (Figure 4.15). All species of bats share many similarities and are grouped together in one box. Bats can be clustered with dugongs and moles in a larger box that contains all of the mammals. Mammals in turn are classified together with birds, reptiles, and amphibians in an even larger box. Sometimes the similarities that lead us to classify animals together are functional similarities. Many of the features shared by different species of bats are related to the fact they fly at night. However, many shared features bear little relation to adaptation: bats, tiny aerial

Figure 4.14

The dugong is an aquatic animal. Its forelimb is adapted for paddling through the water.

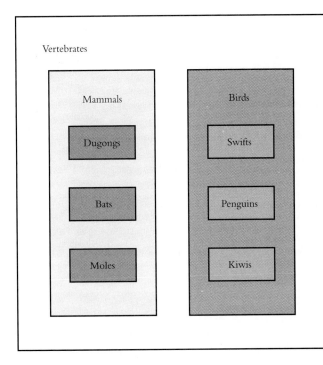

Vertebrates

Mammals

Dugongs

Bats

Moles

Birds

Swifts

Penguins

Kiwis

Figure 4.15

Patterns of similarity allow organisms to be classified hierarchically into a series of nested boxes. All mammals share more similarities with each other than they do with birds, even though some mammals, like bats, must solve the same adaptive problems that face birds.

acrobats, are grouped with the enormous, placid dugong, not the small, acrobatic swift.

The process of speciation explains why organisms can be classified hierarchically.

The fact that new species derive from existing species accounts for the existence of the patterns of nonadaptive similarity that allow organisms to be classified hierarchically. Clearly, new species originate by splitting off from older ones, and we can arrange a group of species that share a common ancestor into a family tree, or **phylogeny**. Figure 4.16 shows the family tree of the **hominoids**—the superfamily that includes apes and humans—and Figure 4.17 shows what some of these present-day apes look like. At the root of the tree is an unknown ancestral species

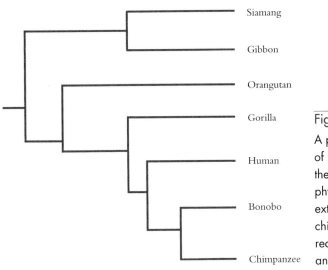

Siamang

Gibbon

Orangutan

Gorilla

Human

Bonobo

Chimpanzee

Figure 4.16

A phylogeny for the hominoids. The names of living species are given at the ends of the branches. The branching pattern of the phylogenetic tree reflects the ancestry of extant (still-living) lineages. For example, chimpanzees and bonobos have a more recent common ancestor than chimpanzees and orangutans do.

Figure 4.17

(a) Gibbons and siamangs are called lesser apes, and are classified in the family Hylobatidae; a gibbon is shown here. (b) Orangutans, (c) gorillas, and (d) chimpanzees are called great apes, and are classified in the family Pongidae.

from which all hominoids evolved. Each branch represents a speciation event in which one species split into two daughter species, and one or both of the new daughter species diverged in morphology or behavior from the parent species.

It is important to realize that when two daughter species diverge from each other, they don't differ in all phenotypic details. A few traits may differ, while most others retain their original form. For example, the hominoids share many common features, including an unspecialized digestive system, five toes on each foot, and no tail. At the same time, these animals have diverged in a number of ways. Gorillas live in groups that contain one adult male and several adult females, while orangutans are mainly solitary. Male and female gibbons are about the same size, while among the other apes, males are larger than females. In general, we expect to see the greatest divergence in those traits that are related to making a living in different habitats and to choosing mates.

Now, let us return to the family tree. Each time one species splits to become two new species, the new daughter species will differ in some way. They will continue to diverge through time because once speciation occurs, the two lineages evolve independently. Some differences will arise because the two species adapt to different habitats, while other differences may result from random processes like genetic drift. In general, species that have recently diverged will have more characters in common with one another than with species that diverged in the more distant past. For example, chimpanzees and gorillas share more traits with one another than they do with orangutans or gibbons because they share a more recent common ancestor. This is the source of the hierarchical nature of life.

Why Reconstruct Phylogenies?

Phylogenetic reconstruction plays three important roles in the study of organic evolution.

We have seen that descent with modification explains the hierarchical structure of the living world. Since new species always evolve from existing species, and species are reproductively isolated, all living organisms can be placed on a single phylogenetic tree, which we can then use to trace the ancestry of all living species. In the remainder of this chapter, we will see how the pattern of similarities and differences observed in living things can be used to construct phylogenies and help to establish the evolutionary history of life.

There are several reasons that reconstructing phylogenies plays an important role in the study of evolution:

1. *Phylogeny is the basis for the identification and classification of organisms.* In the latter part of this chapter we will see how scientists use phylogenetic relationships to name organisms and arrange them into hierarchies. This endeavor is called **taxonomy**.

2. *Knowing phylogenetic relationships often helps to explain why a species evolved certain adaptations and not others.* Natural selection creates new species by modifying existing body structures to perform new functions. To understand why a new organism evolved a particular trait, it helps to know what kind of organism it evolved from, and this is what phylogenetic trees tell us. The phylogenetic relationships among the apes provide a good example of this point. Most scientists used to believe that chimpanzees and gorillas shared a more recent common ancestor than either of them shared with humans. This view influenced their interpretation of the evolution of locomotion, or forms of movement, among the apes. All of the great apes are **quadrupedal**, which means that they walk on their hands and feet. However, gorillas and chimpanzees curl their fingers over their palms and bear weight on their knuckles, a form of locomotion called **knuckle walking**, while orangutans bear weight on their palms (Figure 4.18). Humans, of course, stand upright on two legs. Knuckle walking involves distinctive modifications of the anatomy of the hand, and since human hands show none of these anatomical features, most scientists believed that humans did not evolve from a knuckle-walking species. Because both chimpanzees and gorillas are knuckle walkers, it was generally assumed that this trait evolved in their common ancestor (Figure 4.19). However, recent measurements of genetic similarity now lead many scientists to believe that humans and chimpanzees are more closely related to one another than either species is to the gorilla. If this is correct, then the old account of the evolution of locomotion in apes must be wrong. Two accounts are consistent with the new phylogeny. It is possible that the common ancestor of humans, chimpanzees, and gorillas was a knuckle walker and that knuckle walking was retained in the common ancestor of humans and chimpanzees. This would mean that knuckle walking evolved only once and that humans are descended from a knuckle-walking

Figure 4.18

Knuckle walking among the great apes: (a) Chimpanzees and gorillas are knuckle walkers; this means that when they walk, they bend their fingers under their palms and bear weight on their knuckles. (b) In contrast, orangutans are not knuckle walkers; when they walk, they bear their weight on their palms.

(a)   (b)

species (Figure 4.20a). Alternatively, the common ancestor of humans, chimpanzees, and other apes may not have been a knuckle walker. If this was the case, then knuckle walking evolved independently in chimpanzees and gorillas (Figure 4.20b). Each of these scenarios raises interesting questions about the evolution of locomotion in humans and apes. If humans evolved from a knuckle walker, then perhaps more careful study will reveal the traces of our former mode of adaptation. If chimpanzees and gorillas evolved knuckle walking independently, then a close examination should reveal subtle differences in their morphology.

3. *We can deduce the function of morphological features or behaviors by comparing the traits of different species.* This technique is called the **comparative method**. As we will see in Part II, most primates live in groups. Some scientists have argued that **terrestrial** (ground-dwelling) primates live in larger groups than **arboreal** (tree-dwelling) primates because terrestrial species are more vulnerable to pred-

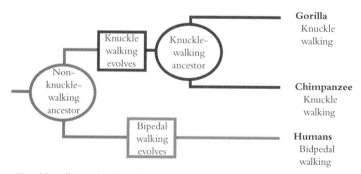

Knuckle walking evolved in the common ancestor of chimpanzees and gorillas

Figure 4.19

If chimpanzees are more closely related to gorillas than to humans, as morphological evidence suggests, then this is the most plausible scenario for the evolution of locomotion in gorillas, chimpanzees, and humans. The fossil record suggests that the ancestor of chimpanzees, gorillas, and humans was neither knuckle walking nor bipedal. The simplest account of the current distribution of locomotion is that knuckle walking evolved once in the common ancestor of chimpanzees and gorillas.

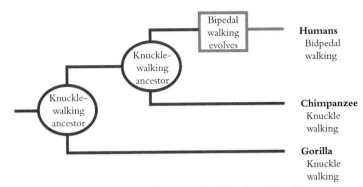

(a) Bipedal locomotion in humans evolved from knuckle walking

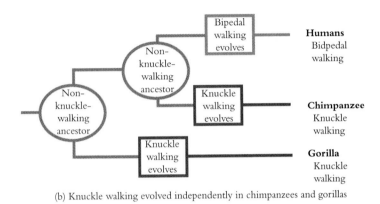

(b) Knuckle walking evolved independently in chimpanzees and gorillas

Figure 4.20

If humans are more closely related to chimpanzees than to gorillas, as the genetic data suggest, then there are two possible scenarios for the evolution of locomotion in these three species. (a) If the common ancestor of chimpanzees and gorillas was a knuckle walker, then it follows that human bipedal locomotion evolved from knuckle walking. (b) If the common ancestor of all three species was not a knuckle walker, then knuckle walking must have evolved independently in chimpanzees and gorillas.

ators and animals are safer in larger groups. To test the relationship between group size and terrestiality using the comparative method, we would collect data on group size and lifestyle (arboreal/terrestrial) for many different primate species. However, most biologists believe that only independently evolved cases should be counted in comparative analyses, so we must take the phylogenetic relationships among species into account. Box 4.1 provides a hypothetical example of how phylogenetic information can alter our interpretations of comparative data.

For many years, scientists constructed phylogenies only for the purpose of classification. The terms *taxonomy* and *systematics* were used interchangeably to refer to the construction of phylogenies and the use of the such phylogenies for naming and classifying organisms. With the recent realization that phylogenies have other important uses like the ones just described, there is a need for terms that distinguish phylogenetic construction from classification. Here we adopt the suggestion of University of Zurich anthropologist Robert Martin to employ the term **systematics** to refer to the construction of phylogenies, and the term *taxonomy* to mean the use of phylogenies in naming and classification. Although this distinction may not seem important now, it will become more relevant as we proceed.

BOX 4.1

The Role of Phylogeny in the Comparative Method

To understand why it is important to take phylogeny into account when using the comparative method, consider the phylogeny shown in Figure 4.21, which shows the pattern of relationships for eight hypothetical primate species.

As you can see, there are three terrestrial species that live in large groups and three arboreal species that live in large groups. There is only one terrestrial species that lives in small groups and only one arboreal species that lives in large

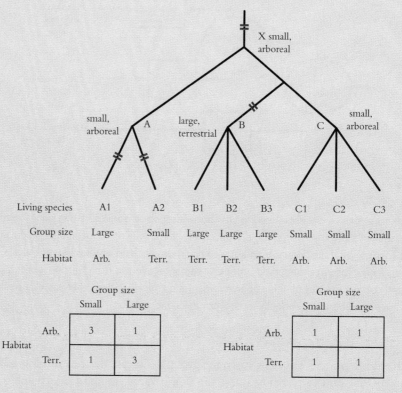

Living species	A1	A2	B1	B2	B3	C1	C2	C3
Group size	Large	Small	Large	Large	Large	Small	Small	Small
Habitat	Arb.	Terr.	Terr.	Terr.	Terr.	Arb.	Arb.	Arb.

Group size

		Small	Large
Habitat	Arb.	3	1
	Terr.	1	3

Counting living species implies that habitat predicts group size.

Group size

		Small	Large
Habitat	Arb.	1	1
	Terr.	1	1

Counting evolutionary events implies that habitat does *not* predict group size.

Figure 4.21

The phylogenetic relationships among eight hypothetical primate species are shown in this figure. Living species lie at the ends of branches, and their ancestors are identified at the branching points of the tree. These species vary in group size (small or large) and lifestyle (arboreal or terrestrial). The lineages in which novel associations between group size and lifestyle first evolved are marked with a double red bar. The matrix on the left tallies the association between group size and lifestyle among living species, and suggests that arboreal species live in small groups while terrestrial species live in large ones. The matrix on the right tallies the number of times the association between group size and lifestyle changed in the course of the evolution of these species. In this case, no relationship between group size and lifestyle is evident. This example demonstrates why it is important to keep track of independent evolutionary events in comparative analyses.

groups. Thus, if we based our conclusions on living species, we would conclude that there is a statistical relationship between group size and lifestyle. However, if we count independent evolutionary events, we get a very different answer. Large group size and terrestriality are found together only in species B and its descendants, B1, B2, and B3. This combination evolved only once, though we now observe this combination in three living species. There is also only one case of selection creating an arboreal species that lives in small groups (species C and its descendants, C1, C2, and C3). Note that each of the other possible combinations has also evolved once. When we tabulate independent evolutionary events, we find no consistent relationship between lifestyle and group size. Clearly, phylogenetic information is crucial to making sense of the patterns we see in nature.

How to Reconstruct Phylogenies

We reconstruct phylogenies by assuming species that are phenotypically similar are more closely related than are species that are less phenotypically similar.

To see how systematists reconstruct a phylogeny, consider the following example. We begin with three species, named for the moment A, B, and C, whose myoglobin (a cellular protein involved in oxygen metabolism) differs in the ways shown in Table 4.1. Remember from Chapter 2 that proteins are long chains of amino acids; the letters in Table 4.1 stand for the amino acids present at various positions along the protein chain.

A systematist wants to find the pattern of descent that is most likely to have produced these data. The first thing she would notice is that all three types of myoglobin have the same amino acid at many positions, exemplified here by the positions 1 (all G), 2 (all L), and 3 (all S). At positions 5, 9, 30, 34, and 13, species A and B have the same amino acids, while species C has a different set. At position 48, species B and C have the same amino acid, and at position 59 A and C have the same amino acid. At position 66, all three species have different amino acids. The systematist would see that there are fewer differences between species A and B than between A and C or B and C. She would infer that fewer genetic changes

Table 4.1 The amino acid sequence for the protein myoglobin is shown for three different species. The numbers refer to positions on the myoglobin chain, and the letter in each cell stands for the particular amino acid found at that position in each kind of myoglobin. All three species have the same amino acids at all of the positions not shown here, as well as at positions 1 to 3. However, at eight positions there is at least one discrepancy among the three species (shaded).

	AMINO ACID NUMBER										
SPECIES	1	2	3	5	9	30	34	13	48	59	66
A	G	L	S	G	L	I	K	V	H	E	A
B	G	L	S	G	L	I	K	V	G	A	I
C	G	L	S	Q	Q	M	H	I	G	E	Q

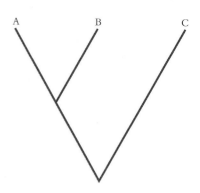

Figure 4.22

The phylogenetic tree for species A, B, and C derived from their myoglobin amino acid sequences.

have accumulated since A and B shared a common ancestor sometime in the past than since A and C or B and C shared a common ancestor. Since fewer evolutionary changes are necessary to convert A to B than A to C or B to C, A and B are assumed to have a more recent common ancestor than either species shares with C (Figure 4.22). In other words, species that are more similar are assumed to be more closely related.

Let's end the suspense and reveal the identities of the three species. Species A is our own species, *Homo sapiens;* species B is the duck-billed platypus (Figure 4.23); and species C is the domestic chicken. The phylogeny shown in Figure 4.22 suggests that humans and duck-billed platypuses are more closely related to each other than either of them is to chickens. However, this phylogeny is not based on nearly enough data to be conclusive. To be convinced of the relationships among these three organisms, we would need to gather and analyze data on many more characters, and generate the same tree each time. In fact, many other characters show the same pattern as the myoglobin. Humans and duck-billed platypuses, for example, both have hair and mammary glands, while chickens lack these structures. The phylogeny in Figure 4.22 is the currently accepted tree for these three species.

Problems Due to Convergence

In constructing phylogenies, we must avoid basing decisions on characters that are similar because of convergent evolution.

However, everything is not completely hunky-dory. Two amino acid positions, 48 and 59, are not consistent with the phylogeny shown in Figure 4.22. Moreover,

Figure 4.23

Duck-billed platypuses illustrate the importance of distinguishing between derived and ancestral traits. These creatures lay eggs and have horny bills like birds, but they lactate like mammals.

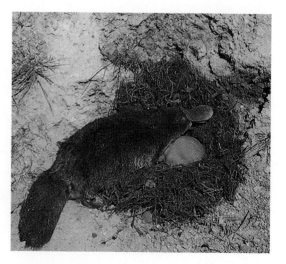

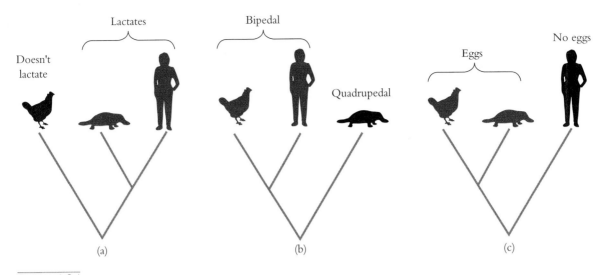

Figure 4.24

(a) Many traits, like lactation, generate the same pattern of relationships among humans, chickens, and platypuses as myoglobin does (see Figure 4.22). (b) There are other characters, like bipedal locomotion, that suggest a closer relationship between humans and chickens than between humans and platypuses or platypuses and chickens. (c) Some traits, like egg laying, suggest a closer relationship between chickens and platypuses than between chickens and humans or platypuses and humans.

there are other characters that don't fit this tree. For example, both humans and chickens are bipedal while platypuses are not (Figure 4.24b). Platypuses and chickens both lay eggs (Figure 4.24c), have a feature of the gut called a cloaca, and sport horny bills. Why don't these characters fit neatly into our phylogeny?

One reason for these anomalies is convergent evolution. Sometimes traits shared by two species are not the result of common ancestry. Instead, they are separate adaptations independently produced by natural selection. Chickens and humans are bipedal—*not* because they are descended from the same bipedal ancestor, but rather, because they *each* evolved this mode of locomotion independently. Similarly, the horny bills of chickens and platypuses are not similarities due to descent; they are independently derived characters. Systematists say that characters similar because of convergence are **analogous**, while characters whose similarity is due to descent from a common ancestor are **homologous**. It is important to avoid using convergent traits in reconstructing phylogenetic relationships. While it is fairly obvious that bipedal locomotion in humans and chickens is not homologous, convergence is sometimes very difficult to detect.

Problems Due to Ancestral Characters

It is also important to ignore similarity based on ancestral characters, traits that also characterized the common ancestor of the species being classified.

There are also homologous traits that do not fit neatly into correct phylogenies. For example, chickens and platypuses both reproduce by laying eggs, while humans do not. It seems likely that both chickens and platypuses lay eggs because they are descended from a common egg-laying ancestor. If these characters are homologous, why don't they allow us to generate the correct phylogeny (see Figure 4.24c)?

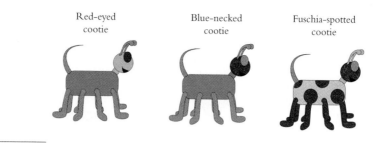

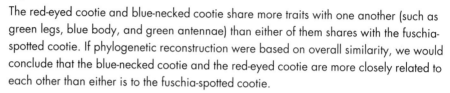

Red-eyed
cootie

Blue-necked
cootie

Fuschia-spotted
cootie

Figure 4.25

The red-eyed cootie and blue-necked cootie share more traits with one another (such as green legs, blue body, and green antennae) than either of them shares with the fuschia-spotted cootie. If phylogenetic reconstruction were based on overall similarity, we would conclude that the blue-necked cootie and the red-eyed cootie are more closely related to each other than either is to the fuschia-spotted cootie.

Egg laying is an example of what systematists call an **ancestral trait**, one that characterized the common ancestor of chickens, platypuses, and humans. Egg laying has been retained in the chicken and the platypus, but lost in humans. It is important to avoid using ancestral characters when constructing phylogenies. Only **derived traits**—features that have evolved since the time of the last common ancestor of the species under consideration—can be used in constructing phylogenies.

To see why we need to distinguish between ancestral and derived traits, consider the three hypothetical species of "cooties" pictured in Figure 4.25. It seems at first glance that the red-eyed cootie and the blue-necked cootie are more similar to one another than either is to the fuschia-spotted cootie, and a careful count of characters will show that they do share more traits with one another than with the fuschia-spotted variety. But consider Figure 4.26, which shows the phylogeny for

Figure 4.26

The phylogenetic history of the three cootie species indicates that the blue-necked cootie is actually more closely related to the fuschia-spotted cootie than to the red-necked cootie. This is because the blue-necked cootie and the fuschia-spotted cootie have a more recent common ancestor with each other than with the red-eyed cootie.

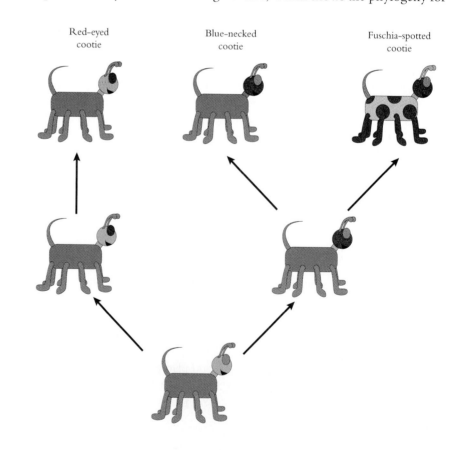

Red-eyed
cootie

Blue-necked
cootie

Fuschia-spotted
cootie

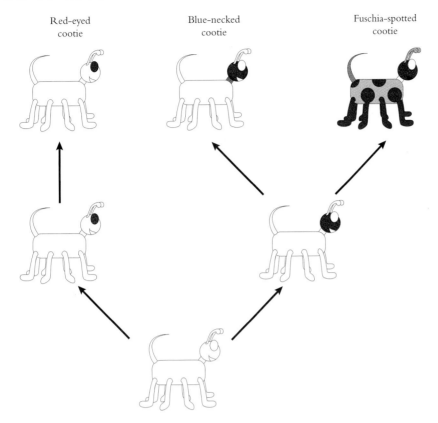

Red-eyed   Blue-necked   Fuschia-spotted
cootie   cootie   cootie

Figure 4.27
In this phylogeny, only derived characters are shown: red eyes in the red-eyed cootie; fuschia face in the blue-necked and fuschia-spotted cooties; blue neck in the blue-necked cootie; and red legs, orange tail, yellow antennae, and fuschia spots in the fuschia-spotted species. The correct phylogeny is based on similarity in shared, derived characters.

the three species. You can see that the blue-necked cootie is actually more closely related to the fuchsia-spotted cootie because they share a more recent common ancestor. The blue-necked cootie seems more similar to the red-eyed cootie because they share many ancestral characters, while the spotted cootie has undergone a period of rapid evolution that has eliminated most ancestral characters. Looking at ancestral traits will not generate the correct phylogeny if rates of evolution differ among species.

If we base our assessment of similarity only on the number of derived characters each species displays (as shown in Figure 4.27), then the most-similar species are the ones most closely related. The blue-necked cootie and fuschia-spotted cootie share one derived character (a fuschia face) with each other, but share no derived characters with the red-eyed cootie. Thus, if we avoid ancestral characters, we can construct the correct phylogeny.

Systematists distinguish ancestral from derived characters by the following criteria: ancestral characters (1) appear earlier in organismal development, (2) appear earlier in the fossil record, and (3) are seen in outgroups.

It is easy to see that distinguishing ancestral characters from derived characters is important, but hard to see how to do it in practice. If you can only observe living organisms, how can you tell whether a particular character in two species is ancestral or derived? There is no surefire solution to this problem, but biologists use three rules of thumb:

1. The development of a multicellular organism is a complex process, and it seems logical that modifications early in the process are likely to be more disruptive than modifications occurring later. As a result, evolution often (but not always) proceeds by modifying the ends of existing developmental pathways. To the

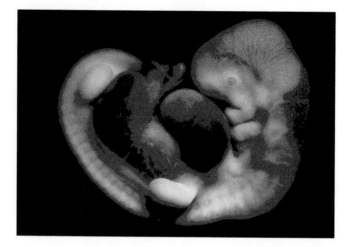

Figure 4.28

At an early stage of development, the human embryo has a tail. This feature disappears as development continues.

extent that this generalization is true, characters that occur early in development are ancestral. For example, humans and other apes do not have tails, but the fact that a tail appears in the development of the human embryo and then disappears is evidence that the tail is an ancestral character (Figure 4.28). We conclude from this that the absence of a tail in humans is derived. This reasoning will not be helpful if ancestral traits have been completely lost during development or if traits (like egg laying) are expressed only in adults.

2. In many cases, fossils provide information about the ancestors of modern species. If we see in the fossil record that all of the earliest likely ancestors of apes had tails, and that primates without tails only appear later in the fossil record, then it is reasonable to infer that having a tail is an ancestral character. This criterion may fail if the fossil record is incomplete, and therefore a derived character appears in the fossil record before an ancestral one. As we will see in Chapter 11, new fossil finds can sometimes lead to radical revisions of existing phylogenies.

3. Finally, we can determine which characters are ancestral in a particular group by looking at neighboring groups, or **outgroups**. Suppose we are trying to determine whether having a tail is ancestral in the primates. We know that monkeys have tails and that apes do not, but we don't know which state is ancestral. To find out, we look at neighboring mammalian groups, such as insectivores or carnivores. Since the members of these outgroups typically have tails, it is reasonable to infer that the common ancestor of all primates also had a tail.

Reconstructing Phylogenies Using Genetic Distance Data

Phylogenetic reconstruction can also be based on genetic distance.

Molecular data can be used to construct phylogenies, as we did when referring to the structure of myoglobin to construct the phylogeny of humans, chickens, and duck-billed platypuses. Molecular data are useful because they provide a large number of easily comparable characters. However, molecular data are also used in a completely different way to reconstruct phylogenies. Instead of looking for shared derived characters, biologists compute a measure of overall genetic similarity, called the **genetic distance**, between each pair of species. Scientists then reconstruct the phylogenies based on the assumption that pairs of genetically similar organisms have a more recent common ancestor than pairs of species less similar genetically.

The molecular technique called **DNA hybridization** is an important example of how genetic distance measures can be used to reconstruct phylogenetic relationships. In this process, an investigator takes tissue from a pair of organisms, extracts the DNA, and cuts it into small fragments. Then she mixes the DNA of the two organisms together, which allows the fragments to bond to each other. When the DNA sequences from the two species are similar, any bonds created will be strong; when the DNA sequences are very different, the bonds will be relatively weak. Finally, the researcher gradually raises the temperature until the bonded pairs of DNA dissociate. The more strongly bonded (and hence similar) the molecules are, the higher the temperature needed to separate them. Specifically, the temperature at which half of the bound DNA has dissociated is a measure of how many DNA bases are identical in the strands from the two species, and provides a measure of the genetic distance between the organisms. The investigator repeats this process for all of the pairs of species in the taxa she wants to classify. Table 4.2 gives the genetic distances between humans and several species of apes. Box 4.2 shows how the genetic distance data can be used to construct a phylogeny. Other genetic distance techniques use molecular methods to assess the genotypes of DNA sequences of different species, and then the genetic distance is computed using these data.

These data are consistent with the hypothesis that genetic distance, as measured by DNA hybridization, changes at approximately a constant rate.

Rearranging the rows and columns of the genetic distance data so that the most closely related species are next to each other produces the matrix shown in Table 4.3, p. 126. Notice that gorillas, humans, chimpanzees, and orangutans are all essentially the same genetic distance from gibbons. This is evidence that genetic distance changes at an approximately constant rate. To see why, think of genetic distance accumulating along the path leading from the last common ancestor of all of the hominoids to each living species. The total distance between any pair of living species is the sum of their paths from the last common ancestor. For example, the genetic distance between humans and gibbons is the accumulated distance between the last common ancestor of all hominoids and gibbons plus the accumulated distance between the last common ancestor of all hominoids and humans. Now notice that the distance between gibbons and humans is the same as the distance between gibbons and orangutans. This means that the genetic distance

Table 4.2 Genetic distance data for all pairs of species in the superfamily Hominoidea using DNA hybridization. The smaller the genetic distance, the more closely related two species are assumed to be. Genetic distance units are based on the temperature at which fixed fractions of bound DNA dissociates. (From C. G. Sibley and J. E. Alquist, 1987, DNA hybridization evidence of hominoid phylogeny: results from an expanded data set, *Journal of Molecular Evolution*, 26:99–121.)

	CHIMPANZEE	**GIBBON**	**GORILLA**	**HUMAN**	**BONOBO**
GIBBON	4.76	—			
GORILLA	2.37	4.75	—		
HUMAN	1.63	4.78	2.27	—	
BONOBO	0.69	5.00	2.37	1.64	—
ORANGUTAN	3.58	4.74	3.55	3.60	3.56

BOX 4.2

Phylogeny Reconstruction Using Genetic Distance

To construct a phylogeny using the genetic distance data in Table 4.2, we proceed as follows. First, notice that the smallest measured genetic distance is between chimpanzees and bonobos (0.69). If the most genetically similar species are the most closely related, then these two species must share a more recent common ancestor than do any other pair of hominoids. In Figure 4.29, the genetic distance between any two species is plotted as the horizontal distance between them on the graph. If ge-

netic distance changes at a constant rate, the genetic distance between chimpanzees and the common ancestor must be equal to the genetic distance between bonobos and the common ancestor (Figure 4.29a). Since the total distance between chimpanzees and bonobos is 0.69, the distance from the common ancestor to each species must be half that, or 0.345. Thus, we place the common ancestor of chimpanzees and bonobos 0.345 genetic distance units from both chimpanzees and bonobos. Notice that

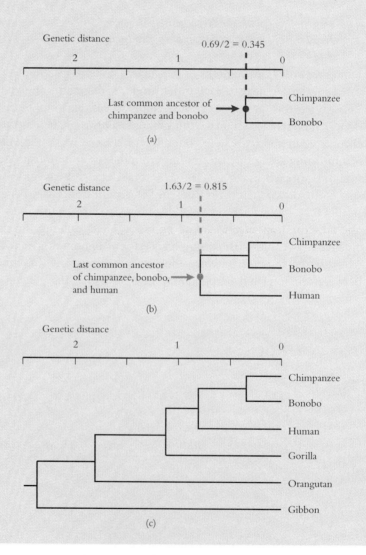

the next-shortest genetic distances are between chimpanzees and humans (1.63) and between bonobos and humans (1.64). Thus, we place the last common ancestor of humans, chimpanzees, and bonobos half this distance from these three species (Figure 4.29b). If we continue using this approach to construct the phylogeny for all six hominoid species, the final result is that shown in Figure 4.29c.

Note that according to genetic distance measures, chimpanzees are more closely related to humans than they are to gorillas, and the majority of phylogenies based on molecular data produce these results. However, phylogenies based on morphological and chromosomal data indicate that chimpanzees and gorillas are more closely related to each other than either is to humans. Thus, the phylogenetic relationships between chimpanzees, gorillas, and humans remain controversial.

←

Figure 4.29

(a) The DNA hybridization genetic distance between bonobos and chimpanzees, 0.69, is smaller than the genetic distance between any other pair of hominoids. This indicates that bonobos and chimpanzees are more closely related to one another than either species is to other apes. If genetic distance changes at a constant rate, then the distance between the last common ancestor and the two living species is $0.69/2 = 0.345$. Thus we locate the branch point between chimpanzees and bonobos 0.345 genetic distance units to the left in the horizontal direction. (b) The next branch in the phylogenetic tree links bonobos and chimpanzees with humans. The distance from humans to either chimpanzees or bonobos is 1.63, so the last common ancestor of all three living species is located $1.63/2 = 0.815$ genetic distance units to the left. (c) Proceeding in this way yields the complete phylogeny for the hominoids. To find the genetic distance between any two species, add the genetic distances along the evolutionary pathway that joins them. For example, the genetic distance between chimpanzees and bonobos is the distance from chimpanzees to the last common ancestor of chimpanzees and bonobos, 0.345, plus the distance from the last common ancestor to bonobos, also 0.345. Thus the total distance between chimpanzees and bonobos is $0.345 + 0.345 = 0.69$.

Table 4.3   The data from Table 4.2 have been rearranged so that the most closely related species are placed next to each other.

	BONOBO	CHIMPANZEE	HUMAN	GORILLA	ORANGUTAN
CHIMPANZEE	0.69	—			
HUMAN	1.64	1.63	—		
GORILLA	2.37	2.21	2.27	—	
ORANGUTAN	3.56	3.58	3.60	3.55	—
GIBBON	5.00	4.76	4.78	4.75	4.74

between the last common ancestor of all hominoids and humans is the same as the distance between the last common ancestor and orangutans. Since the time elapsed since the branching off from the last common ancestor is the same for all hominoids, it follows that the rate of change of genetic distance along both of these paths through the phylogeny must be approximately the same, and that the genetic distance between two species is a measure of the time elapsed since both had a common ancestor. Evolutionists refer to genetic distances with a constant rate of change as **molecular clocks** because genetic change acts like a clock that measures the time since two species shared a common ancestor. Data from other groups of organisms suggest that genetic distances based on DNA hybridization and other methods often have this clocklike property, but there are also important exceptions.

A majority of biologists agree that as long as genetic distances are not too big or too small, the molecular-clock assumption is a useful approximation. However, there is intense controversy about *why* genetic distance changes in a clocklike way. Advocates of the **neutral theory** believe that most changes in DNA sequences have little or no effect on fitness, so the evolution of this neutral DNA must be controlled by drift and mutation. The neutral theory suggests that under the right circumstances mutation and drift will produce clocklike change. Other biologists argue that the molecular clock is the product of natural selection in variable environments.

Phylogenies based on genetic distance combine ancestral and derived traits.

DNA hybridization (and other genetic distance methods) measure the overall genetic similarity between two organisms and, as a result, genetic distance includes both derived characters and ancestral ones. If genetic distance changes in a clocklike way, however, only the differences between lineages will remain and the inclusion of ancestral traits does not cause any problems.

If the molecular-clock hypothesis is correct, then knowing the genetic distance between two living species allows us to estimate how long ago the two lineages diverged.

The data from contemporary species indicate that genetic distance changes at a constant rate, but it doesn't provide a clue as to what that rate might be. However, dated fossils allow us to estimate when the splits between lineages occurred. By plotting these dates against the genetic distances between the same pairs of species, we can estimate the rate at which genetic distance changes through time. For example, one point in Figure 4.30 plots the date of the last common ancestor of humans and orangutans, which lived about 12 mya, against the genetic distance

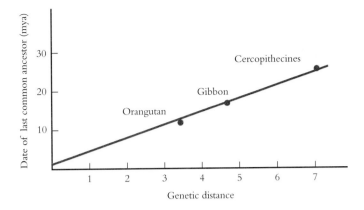

Figure 4.30

Genetic distance measures are calibrated using well-dated fossil material. Among the primates, certain divergence dates can be established from the fossil record. To estimate the speed of the molecular clock, or the rate of genetic change, the genetic distance among pairs of species whose divergence dates are known is plotted against the time since they diverged. Each data point here represents the genetic distance between humans and the named group. The slope of the line gives the rate of change in genetic distance per unit time.

between humans and orangutans, about 3.60 units. The other two points represent the corresponding data for humans and gibbons, and for humans and cercopithecines (the Old World monkeys such as baboons and macaques). The slope of the line through these points gives the rate of change in genetic distance with time. Charles Sibley and his colleagues at Yale University applied the same techniques to measure genetic distances in birds, for which the value of the slope is about the same as for primates, about 4.2. To convert the genetic distances given in Figure 4.29 to the number of years since the divergence of the lineages, we can just multiply the genetic distances by 4.2 million.

Taxonomy—Naming Names

The hierarchical pattern of similarity created by evolution provides the basis for the way science classifies and names organisms.

Putting names on things is to some extent arbitrary. We could give species names like Sam or Ruby, or perhaps use numbers like the social security system. This is, in fact, the way common names work. The word *lion* is an arbitrary label, as is "Charles" or "550-72-9928." The problem for scientists is that there are a very large number of animals and plants, far too many species for any individual to keep track of. One way to cope with this massive complexity is to devise a system in which organisms are grouped together in a hierarchical system of classification. Once again there are many possible systems. For example, we could categorize organisms alphabetically, grouping alligators and apricots with the *a*'s, barnacles and baboons with the *b*'s, and so on. This approach has little to recommend it because knowing how an animal is classified tells us nothing about the organism besides its location in the alphabet. An alternative approach would be to adopt a system of classification that groups together organisms with similar char-

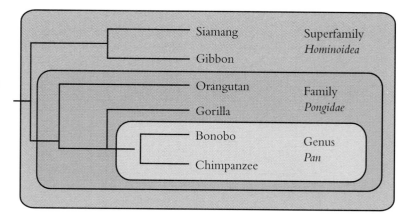

Figure 4.31

Bonobos and chimpanzees are classed together in the genus *Pan;* these species are classified with gorillas and orangutans in the family Pongidae; and all of the apes are grouped together in the superfamily Hominoidea. Note that humans are missing from this phylogenetic tree.

acteristics, analogous to the Library of Congress system used by libraries to classify books. The Qs might be predators, the QHs aquatic predators, the QPs aerial predators, and so on. Thus, knowing that the scientific name of the red-tailed hawk is QP604.4 might tell you that the hawk is a small, aerial predator that lives in North America. The problem with such a system is that not all organisms would fall into a single category. Where would you classify animals that eat both animals and plants, such as bears, or amphibious predators, such as frogs?

The scientific system for naming animals is based on the hierarchy of descent: species that are closely related are classified together. Closely related species are grouped together in the same **genus** (Figure 4.31). For example, the genus *Pan* contains two closely related species of chimpanzee: the common chimpanzee, *Pan troglodytes,* and the bonobo, *Pan paniscus.* Closely related genera (the plural of genus) are usually grouped together in a higher unit, often the **family**. Chimpanzees are in the family *Pongidae* along with the other great apes, the orangutan *(Pongo pygmaeus),* and the gorilla *(Gorilla gorilla).* Closely related families are then grouped together in a more inclusive unit, often a **superfamily**. The *Pongidae* belong to the superfamily *Hominoidea,* along with the gibbons, siamangs, and humans.

Taxonomists disagree about whether overall similarity should also be used in classifying organisms.

The majority of taxonomists agree that descent should play a major role in classifying organisms. However, they vehemently disagree about whether descent should be the *only* factor used to classify organisms. The members of a relatively new school of thought, called **cladistic systematics**, argue that only descent should matter. Adherents to an older school of thought, called **evolutionary systematics**, believe classification should be based both on descent *and* overall similarity. To understand the difference between these two philosophies, consider Figure 4.32, in which humans have been added to the phylogeny of the apes. Evolutionary taxonomists would say that humans are qualitatively different from other apes and so deserve to be distinguished at a higher taxonomic level (Figure 4.32a). Accordingly, these taxonomists classify humans in a family of their own, the *Hominidae.* For a cladist this is unacceptable because humans are descended from the same common ancestor as other members of the family *Pongidae.* This means that humans *must* be classified in the same family as chimpanzees, bonobos, and gorillas (Figure 4.32b). It is not just chauvinism about our own place in the primate phylogeny that causes discrepancies between these classification schemes. The same problem arises in many other taxa. For example, it turns out that croc-

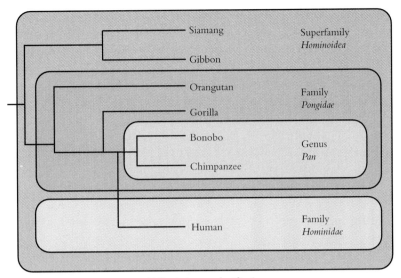

(a) Evolutionary classification

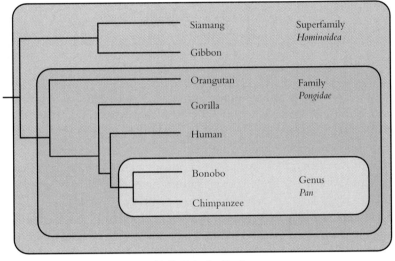

(b) Cladistic classification

Figure 4.32

Cladistic and evolutionary systematic schemes generate two different phylogenies for the hominoids. (a) The evolutionary classification classifies humans in a different family from other apes because apes are more similar to each other than they are to humans. (b) According to the cladistic classification, humans must be classified in the same family as the other great apes because they share a common ancestor with these creatures.

odiles and birds share a more recent common ancestor than either does with lizards. For a cladist, this means that birds and crocodiles must be classified together, and lizards must be classified separately. Evolutionary taxonomists argue that birds are obviously distinctive and deserve a separate taxonomic grouping.

In theory, cladistic systematics is both informative and unambiguous. It is informative because knowing an organism's name and its position in the hierarchy of life tells us how it is related to other organisms. It is unambiguous because the position of each organism is given by the actual pattern of descent. Once you are confident that you understand the phylogenetic relationships within a group, there is no doubt about how any organism in that group should be classified. Cladists believe that evolutionary systematics is ambiguous because judgments of overall similarity are necessarily subjective. On the other hand, evolutionary taxonomists complain that the advantages of cladistics are mainly theoretical. In real life, they argue, uncertainty about phylogenetic relationships introduces far more ambiguity and instability in classification than do judgments about overall similarity. For example, we are not completely certain whether chimpanzees are more closely related to humans or to gorillas. Morphological data suggest that chimpanzees are more closely related to gorillas. While most genetic distance measures indicate that chimpanzees are more closely related to humans, some genetic distance data suggest the opposite. Given such uncertainty, how would cladists name and classify these species?

It is important to keep in mind that this controversy is not about what the world is like, or even about how evolution works. Instead it is a debate about how we should name and classify organisms. Thus, there is no experiment or observation that can prove either school right or wrong. Instead, scientists must determine which system is more useful in practice.

Further Reading

Dawkins, R. 1996. *The Blind Watchmaker.* W. W. Norton, New York, chap. 11.
Ridley, M. 1986. *Evolution and Classification.* Longman, London.
Ridley, M. 1996. *Evolution.* 2d ed. Blackwell, Oxford.

Study Questions

1. There are two large plants in the authors' yard: a yucca tree and an enormous prickly pear cactus. Both have rough scaly bark, but it is known that these plants are not closely related.
   (a) What are two different explanations for the similarity of the bark?
   (b) It is also known that yuccas are descended from a grass species instead of a tree species. Which of the two explanations given in part a is consistent with this fact?
2. Chimpanzees and gorillas more closely resemble each other anatomically than either resembles humans. For example, the hands of chimpanzees and gorillas are structurally similar, and quite different from human hands. Genetic distance data suggest, however, that humans and chimpanzees are more closely related to each other than either is to gorillas. Assuming the genetic distance data are correct, give two different explanations for the observed anatomical similarity between chimpanzees and gorillas.
3. Use the genetic distance matrix given below to establish the taxonomic relationships between the species listed. (*Hint:* Draw a phylogenetic tree to illustrate these taxonomic relationships.)

	A	B	C	D
B	4.8	—		
C	0.7	5.0	—	
D	3.6	4.7	3.6	—

4. According to the biological species concept, what is a species? Why do some biologists define species in this way?
5. What is the ecological species concept? Why have some biologists questioned the biological species concept?
6. Molecular methods allow biologists to measure the amount of gene flow among populations that make up a species. When such methods first became available, systematists were surprised to find that many morphologically indistinguishable populations seem to be reproductively isolated from each other and thus, ac-

cording to the biological species concept, are considered entirely different species. Is it possible to account for the existence of such cryptic species by allopatric speciation? by parapatric speciation? by sympatric speciation?

7. New plant species are sometimes formed by the hybridization of existing species. The new species retains all of the genes of each parent. For example, the variety of wheat used to make bread is a hybrid of three different grass species. Explain how such hybridization affects the family tree of these plants.

Primate Ecology and Behavior

CHAPTER 5

Introduction to the Primates

Two Reasons to Study the Primates

The chapters in Part Two focus on the behavior of living nonhuman primates. Studies of nonhuman primates help us understand human evolution for two complementary, but distinct, reasons. First, closely related species tend to be similar morphologically. As we saw in Chapter 4, this similarity is due to the fact that closely related species retain and share traits acquired through descent from a common ancestor. Thus, **viviparity** (bearing live young) and lactation are traits that all placental and marsupial mammals share, and these traits distinguish mammals from other taxa, such as reptiles. The existence of such similarities means that studies of living primates often give us more insight about the behavior of our ancestors than do studies of other organisms. This approach is called "reasoning by homology." The second reason we study primates is based on the idea that natural selection leads to similar organisms in similar environments. By assessing the patterns of diversity in behavior and morphology of organisms in relation to their environments, we can see how evolution shapes adaptation in response to different selective pressures. This approach is called "reasoning by analogy."

Primates Are Our Closest Relatives

The fact that humans and other primates share many characteristics means other primates provide valuable insights about early humans.

We humans are more closely related to nonhuman primates than we are to any other animal species. The anatomical similarities among monkeys, apes, and humans led the Swedish naturalist Carl Linnaeus to place humans in the order Primates in the first scientific taxonomy, *Systema Naturae,* published in 1758. Later, naturalists such as Georges Cuvier and Johann Blumenbach placed humans in their own order because of our distinctive mental capacities and upright posture. However, in *The Descent of Man,* Charles Darwin firmly advocated reinstating humans in the order Primates; he cited biologist T. H. Huxley's essay enumerating the many anatomical similarities between us and apes, and suggested that "if man had not been his own classifier, he would never have thought of founding a separate order for his own reception." Modern systematics unambiguously confirms that humans are more closely related to other primates than to any other living creatures.

Because we are closely related to other primates, we share with them many aspects of morphology, physiology, and development. For example, like other primates, we have well-developed visual abilities and grasping hands and feet. We share certain features of our life history with other primates, including an extended period of juvenile development, and primates as a whole have larger brains in relation to body size than the members of most other taxonomic groups do. Homologies between humans and other primates also extend to behavior, since the physiological and cognitive structures that underlie human behavior are more similar to those of other primates than to members of other taxonomic groups. The existence of this extensive array of homologous traits, the product of the common evolutionary history of the primates, means that nonhuman primates provide useful models for understanding the evolutionary roots of human morphology and for unraveling the origins of human nature.

Primates Are a Diverse Order

Diversity within the primate order helps us to understand how natural selection shapes behavior.

During the last 30 years, hundreds of researchers from several academic disciplines have spent thousands of hours observing many different species of nonhuman primates in the wild, in captive colonies, and in laboratories. All primate species have evolved adaptations that enable them to meet the basic challenges of life, such as finding food, avoiding predators, obtaining mates, rearing young, and coping with competitors. At the same time, there is great morphological, ecological, and behavioral diversity among species within the primate order. For example, they range in size from the tiny mouse lemur, which weighs less than 30 g (less than one ounce) to male gorillas weighing 160 kg (352 lb). Some species live in dense tropical forests, while others are at home in open woodlands and savannas. Some subsist almost entirely on leaves, while others rely on an omnivorous diet that includes fruit, leaves, flowers, seeds, gum, nectar, insects, and small prey. Some species are solitary, and others are highly gregarious. Some are active at night (nocturnal); others are active during daylight hours (diurnal). One primate, the fat-tailed dwarf lemur, enters a torpid state and sleeps for six months each year. Some species actively defend territories from incursions by other members of their own species (conspecifics); others do not. In some species, females provide all care of their young, while in others males participate actively in this process.

This variety is inherently interesting in and of itself. Researchers who study primates are motivated by absorption in the lives of their subjects to endure the hardships of fieldwork, the frustrations of attempting to obtain a share of ever-shrinking research funds, and the puzzlement of family and friends who wonder why they have chosen such an odd occupation. However, evidence of diversity among closely related organisms living under somewhat different ecological and social conditions also helps researchers to understand how evolution shapes behavior. Animals that are closely related to one another phylogenetically tend to be very similar in morphology, physiology, life history, and behavior. Thus differences observed among closely related species are likely to represent adaptive responses to specific ecological conditions. At the same time, similarities among more distantly related creatures living under similar ecological conditions are likely to be the product of convergence.

This approach, sometimes called the comparative method, has become an important form of analysis as researchers attempt to explain the patterns of variation in morphology and behavior observed in nature. The same principles have been borrowed to reconstruct the behavior of extinct hominids, early members of the human lineage. Since behavior leaves virtually no trace in the fossil record, the comparative method provides one of our only objective means of testing hypotheses about the lives of our hominid ancestors. For example, the observation that there are substantial differences in male and female body size, a phenomenon called **sexual dimorphism**, in species that form nonmonogamous groups suggests that highly dimorphic hominids in the past were not monogamous. We will see in Part III how the data and theories about behavior produced by primatologists have played an important role in reshaping our ideas about human origins.

Features That Define the Primates

The primate order is generally defined by a number of shared, derived characters, but not all primates share all of these traits.

The animals pictured in Figure 5.1 are all members of the primate order. These animals are similar in many ways: they are covered with a thick coat of hair, they have four limbs, and they have five fingers on each hand. However, these ancestral features are shared with all mammals. Beyond these ancestral features, it is hard to see what this group of animals has in common that makes them distinct from other mammals. What distinguishes a ring-tailed lemur from a mongoose or raccoon? What features link the langur and the aye-aye?

In fact, primates are a rather nondescript mammalian order that cannot be unambiguously characterized by a single derived feature shared by all members. However, in his extensive treatise on primate evolution, University of Zurich biologist Robert Martin defines the primate order in terms of the derived features listed in Table 5.1.

The first three traits in Table 5.1 are related to the flexible movement of hands

(a)   (b)   (c)

Figure 5.1

All of these animals are primates (a = aye-aye, b = ring-tailed lemur, c = langur, d = howler, e = gelada baboon). Primates are a diverse order, and primates do not possess a suite of traits that unambiguously distinguish them from other animals.

(d)   (e)

and feet. Primates can grasp with their hands and feet (Figure 5.2a), and most monkeys and apes can oppose their thumb and forefinger in a precision grip (Figure 5.2b). The flat nails, distinct from the claws of many animals, and the tactile pads on the tips of primate fingers and toes further enhance their dexterity. These traits enable primates to use their hands and feet differently than most other animals. Primates are able to grasp fruit, squirming insects, and other small items in their

Table 5.1   Definition of the primate order. See the text for more complete descriptions of these features.

1. The big toe on the foot is **opposable**, and hands are **prehensile**. This means that primates can use their feet and hands for grasping. The opposable big toe has been lost in humans.
2. There are flat nails on the hands and feet in most species, instead of claws, and there are sensitive tactile pads with "finger prints" on fingers and toes.
3. Locomotion is **hindlimb dominated**, meaning the hindlimbs do most of the work, and the center of gravity is nearer the hindlimbs than the forelimbs.
4. There is an unspecialized **olfactory** (smelling) apparatus that is reduced in diurnal primates.
5. The visual sense is highly developed. The eyes are large and moved forward in the head, providing stereoscopic vision.
6. Females have small litters, and gestation and juvenile periods are longer than in other mammals of similar size.
7. The brain is large compared with the brains of similarly sized mammals, and it has a number of unique anatomical features.
8. The **molars** are relatively unspecialized, and there is a maximum of two **incisors**, one **canine**, three **premolars**, and three molars on each half of the upper and lower jaw.
9. There are a number of other subtle anatomical characteristics that are useful to systematists but are hard to interpret functionally.

worth reading

(a)

(b)

(c)

Figure 5.2

(a) Primates have grasping feet, which they use to climb, cling to branches, hold food, and scratch themselves. (b) Primates can oppose the thumb and forefinger in a precision grip, a feature that enables them to hold food in one hand while they are feeding, to pick small ticks and bits of debris from their hair while grooming, and (in some species) to use tools. (c) Most primates have flat nails on their hands and sensitive tactile pads on the tips of their fingers.

Figure 5.3

In most primates, the eyes are moved forward in the head. The field of vision of the two eyes overlaps, creating binocular stereoscopic vision.

hands and feet, and they can grip branches with their fingers and toes. During grooming sessions, they delicately part their partner's hair and use their thumb and forefinger to remove small bits of debris from the skin (Figure 5.2c).

The next two traits in Table 5.1 are related to a shift in emphasis among the sense organs. Primates are generally characterized by a greater reliance on visual stimuli and a reduced reliance on olfactory stimuli than other mammals are. Many primate species can perceive color, and their eyes are set forward in the head, providing them with binocular stereoscopic vision (Figure 5.3). **Binocular vision** means that the fields of vision of the two eyes overlap so that both eyes perceive the same image. **Stereoscopic vision** means that each eye sends a signal of the visual image to both hemispheres in the brain to create an image with depth. These trends are not uniformly expressed within the primate order, as olfactory cues play a more important role in the lives of prosimian primates than of anthropoid primates. As we will explain shortly, the **prosimian** primates include the lorises and lemurs, while the **anthropoid** primates include the monkeys and apes.

The next two features in Table 5.1 result from the distinctive life history of primates. As a group, primates have longer pregnancies, mature at later ages, live longer, and have larger brains than other animals of similar body size do. These features reflect a progressive trend toward increased dependence on complex behavior, learning, and behavioral flexibility within the primate order. As the noted primatologist Alison Jolly points out, "If there is an essence of being a primate, it is the progressive evolution of intelligence as a way of life." As we will see in the chapters that follow, these traits have a profound impact on the mating and parenting strategies of males and females and the patterns of social interaction among members of the order Primates.

The eighth feature in Table 5.1 concerns primate dentition. Teeth play a very important role in the lives of primates and in our understanding of their evolution. The utility of teeth to primates themselves is straightforward: teeth are necessary for processing food and are also used as weapons in conflicts with other animals. Teeth are also useful features for those who study living and fossil primates. Primatologists sometimes rely on tooth wear to gauge the age of individuals, and they use features of the teeth to assess the phylogenetic relationships among species. As we will see, paleontologists often rely on teeth, which are hard and preserve well, to identify the phylogenetic relationships of extinct creatures and to make inferences about their developmental patterns, their dietary preferences, and their social structure. Box 5.1 describes primate dentition in greater detail.

BOX 5.1

What's in a Tooth?

To appreciate the basic features of primate dentition, you can consult Figure 5.12, or you can simply look in a mirror since your teeth are much like those of other primates. Teeth are rooted in the jaw. The jaw holds four different kinds of teeth; in order they are first the incisors at the front, then the canines, premolars, and the molars in the rear. To understand what each kind of teeth does, imagine yourself eating a sandwich. You bite into the sandwich with your incisors and canines and use your front teeth to detach a piece. The incisors are relatively small, peg-shaped teeth. The canines are dagger-shaped. The upper canines are usually considerably longer than the other teeth, and the upper canine is sharpened on the lower premolar. The incisors and canines are mainly involved in getting food into the mouth and preparing food for further processing by the molars and premolars. The molars and premolars have broad surfaces, covered by a series of enamel bumps, or **cusps,** connected by crests or ridges. The molars and premolars are mainly used to crush, shred, and chew food before it is swallowed. In Chapter 6 we will show how the size and shape of the teeth are related to the types of foods primates eat.

Although all primates have the same kinds of teeth, species vary in how many of each kind of tooth they have. For convenience, these combinations are expressed in a standard format called the **dental formula,** which is commonly written in the following form:

$$\frac{2.1.3.3}{2.1.3.3}$$

The numerals separated by periods tell us how many of each of the four types of teeth a particular species has (or had) on one side of its jaw. Left to right, the four types are given for the front of the mouth (incisors) to the rear (molars). The top line represents the teeth on one side of the upper jaw **(maxilla),** and the bottom line represents the teeth on the corresponding side of the lower jaw **(mandible).** Hence this species— which happens to be the common ancestor of all primates—had two incisors, one canine, three premolars, and three molars each on one side of the upper and lower jaws. Usually, but not always, the formula is the same for both upper and lower jaws. Like most other parts of the body, our dentition is **bilaterally symmetric,** which means that the left side is identical to the right side. The ancestral pattern shown above has been modified in various primate taxa, as the total number of teeth has been reduced (Table 5.2).

The dental formulas among living primates vary. Prosimians have the most variable dentition. While the lorises, pottos, galagos, and a number of lemurids have retained the primitive dental formula, other groups have lost incisors, canines, or premolars. Tarsiers have lost one incisor on the mandible but retained two on the maxilla. Dentition is generally less variable among the anthropoid primates than among the prosimians. All of the New World monkeys, except the marmosets and tamarins, have retained the primitive dental formula; the marmosets and tamarins have lost one molar. The Old World monkeys, apes, and humans have reduced the number of premolars from 3 to 2.

Table 5.2 Primates vary in the numbers of each type of tooth that they have. The dental formulas here give the number of incisors, canines, premolars, and molars on each side of the upper and lower jaw. For example, lorises have two incisors, one canine, three premolars, and three molars on each side of their upper and lower jaw.

Primate Group	Dental Formula
Prosimians	
Lorises, pottos, and galagos	$\dfrac{2.1.3.3}{2.1.3.3}$
Dwarf lemurs, mouse lemurs, and true lemurs	$\dfrac{2.1.3.3}{2.1.3.3}$
Indriis	$\dfrac{2.1.2.3}{2.0.2.3}$
Aye-aye	$\dfrac{1.0.1.3}{1.0.0.3}$
Tarsiers	$\dfrac{2.1.3.3}{1.1.3.3}$
New World monkeys	
Most species	$\dfrac{2.1.3.3}{2.1.3.3}$
Marmosets and tamarins	$\dfrac{2.1.3.2}{2.1.3.2}$
All Old World monkeys, apes, and humans	$\dfrac{2.1.2.3}{2.1.2.3}$

Although these traits are generally characteristic of primates, you should keep two points in mind. First, none of them makes primates unique. Dolphins, for example, have large brains and extended periods of juvenile dependence, and their social behavior may be just as complicated and flexible as that of any nonhuman primate (Figure 5.4). Second, not every primate possesses all of these traits. Humans

Figure 5.4

A high degree of intelligence characterizes some animals besides primates. Dolphins, for example, have very large brains in relation to their body size, and behavior that is quite complex.

have lost the grasping big toe that characterizes other primates, and some prosimians have claws on some of their fingers and toes.

Primate Biogeography

Primates are generally restricted to tropical regions of the world.

The continents of Asia, Africa, and South America and the islands that lie near their coasts are home to most of the world's primates (Figure 5.5). Just a few species are currently found in southern Europe, the southernmost part of North America, and Central America, areas where primates were formerly more abundant. There are no natural populations of primates in Australia or Antarctica, and none occupied these continents in the past.

Primates are mainly found in tropical regions, where the fluctuations in temperature from day to night greatly exceed fluctuations in temperature over the course of the year. In the tropics, the distribution of resources that primates rely on for subsistence is more strongly affected by seasonal changes in rainfall than by seasonal changes in temperature. Some primate species extend their ranges into temperate areas of Africa and Asia, where they manage to cope with substantial seasonal fluctuations in environmental conditions.

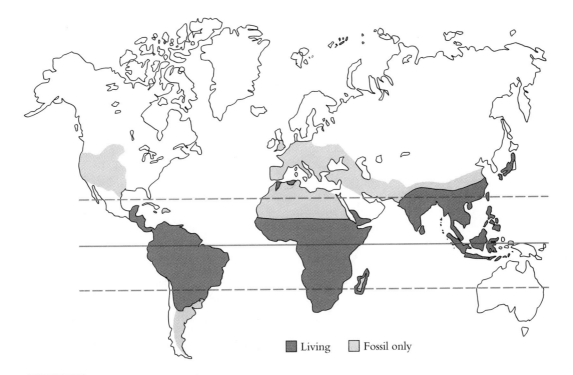

■ Living   ■ Fossil only

Figure 5.5

The distribution of living and fossil primates is shown here. Primates are now found in Central America, South America, Africa, and Asia. Primates are mainly found in tropical regions of the world. Primates were formerly found in southern Europe and northern Africa. There have never been indigenous populations of primates in Australia or Antarctica.

Within their ranges, primates occupy an extremely diverse set of habitats that includes all types of tropical forests, savanna woodlands, mangrove swamps, grasslands, high-altitude plateaus, and deserts. The vast majority of primates, however, are found in forested areas, where they travel, feed, socialize, and sleep in a largely arboreal world.

A Taxonomy of Living Primates

The primates are divided into two groups, the prosimians and the anthropoids.

Scientists classify primates into two suborders, the Prosimii and Anthropoidea (Table 5.3). Many of the primates included in the suborder Prosimii are nocturnal, and like some of the earliest primates that lived 50 mya, they have many adaptations to living in darkness, including a well-developed sense of smell, large eyes, and independently movable ears. By contrast, monkeys and apes, which make up the suborder Anthropoidea, evolved adaptations more suited to a diurnal lifestyle early in their evolutionary history. In the Anthropoidea, the traits related to increased complexity of behavior are most fully developed.

The classification of the primates into prosimians and anthropoids does not strictly reflect the patterns of genetic relationships among the animals in the suborders. Tarsiers are included in the prosimians because, like the lorises and lemurs, they are nocturnal creatures that have retained many ancestral characters. However, both genetic and morphological data suggest that tarsiers are more closely related to monkeys and apes than to prosimians. Thus, a purely cladistic classification would place tarsiers in the same **infraorder** (the taxonomic level immediately below suborder) as the monkeys. In fact, many primate taxonomists advocate a taxonomy in which the lemurs and lorises are classified together as **strepsirhines** and the rest of the primates are classified together as the **haplorhines**. The more traditional division into prosimians and anthropoids is an example of evolutionary systematics in which overall similarity and relatedness are used to classify species.

The Prosimians

The prosimians are divided into three infraorders: Lemuriformes, Lorisiformes, and Tarsiiformes.

The Lemuriformes, or lemurs, are found only on Madagascar and the Comoro Islands, off the southeastern coast of Africa. These islands have been separated from Africa for 120 million years. The primitive prosimians that reached Madagascar evolved in total isolation from primates elsewhere in the world, as well as from many of the predators and competitors that primates confront in other places. Faced with a diverse set of available ecological niches, the lemurs underwent a spectacular adaptive radiation. When humans first colonized Madagascar about 2000 years ago, there were approximately 44 species of lemurs, ranging in size from mouse lemurs that weigh less than 30 g (< 1 oz) to lemurs that were as large as gorillas (200 kgs, 440 lbs). Within the next few centuries, all of the larger lemur species became extinct, probably the victims of human hunters. The extant lemurs are mainly small or medium-sized arboreal residents of forested areas; they travel quadrupedally or by jumping in an upright posture from one tree to another, a form of

Table 5.3   A taxonomy of the living primates. (From R. Martin, 1992; Classification of Primates, p. 20–21 in S. Jones, R. Martin, and D. Pilbeam, 1994, *The Cambridge Encyclopedia of Human Evolution*, Cambridge University Press, Cambridge.)

Suborder	Infraorder	Superfamily	Family	Subfamily
Prosimii (prosimians)	Lemuriformes	Lemuroidea (lemurs)	Cheirogaleidae (dwarf and mouse lemurs)	
			Lemuridae	Lemurinae (true lemurs)
				Lepilemurinae (sportive lemurs)
			Indiidae (indriis)	
			Daubentoniidae (aye-ayes)	
	Lorisiformes	Lorisoidea (loris group)	Lorisidae	Lorisinae (lorises)
				Galaginae (galagos)
	Tarsiiformes	Tarsioidea	Tarsiidae (tarsiers)	
Anthropoidea (anthropoids)	Platyrrhini	Ceboidea (New World monkeys)	Cebidae	Cebinae (e.g., capuchins)
				Aotinae (e.g., owl monkeys)
				Atelinae (e.g., spider monkeys)
				Alouattinae (howling monkeys)
				Pithecinae (e.g., sakis)
				Callimiconinae (Goeldi's monkeys)
			Callitrichidae (marmosets and tamarins)	
	Catarrhini	Cercopithecoidea (Old World monkeys)	Cercopithecidae	Cercopithecinae (e.g., macaques and vervets)
				Colobinae (e.g., langurs)
		Hominoidea (apes and humans)	Hylobatidae	Hylobatinae (gibbons and siamangs)
			Pongidae	Ponginae (great apes)
			Hominidae	Homininae (humans)

locomotion known as **vertical clinging and leaping**. Activity patterns of lemurs are quite variable: about half are primarily diurnal, others are nocturnal, and some are active during both day and night. One of the most interesting aspects of lemur behavior is that females routinely dominate males. In most lemur species, females are able to supplant males from desirable feeding sites, and in some lemur species females regularly defeat males in aggressive encounters. Alison Jolly, one of the first observers of free-ranging lemurs, noted ". . . at any time a female may casually supplant any male or irritably cuff him over the nose and take a tamarind pod from his hand." While such behavior may seem unremarkable in our own liberated times, female dominance is very rare in other primate species.

Preserving Primates

At the end of the twentieth century, the future of primate ecology is interlocked with that of the world's tropical forests. Most primate species occur in rainforests, which are disappearing at an astounding rate. It is clear that conservation of tropical forests and their rich faunas must take place within the context of the economic development of the countries in which they are situated. Further, the conservation of these resources must prove valuable to the peoples who own them. The international community, including World Bank and USAid, have made it a priority to fund projects combining rural dvelopment programs linked with park management. In 1987, the Malagasy government asked me to help them establish a national park at Ranomafana, in southeast Madagascar. This article provides a brief description of the route we followed establishing this park. Undoubtedly, many of the problems and opportunities we faced, and the solutions we arrived at, were unique to Madagascar at this particular time; others, I hope, will be broadly applicable to conservation efforts throughout the world.

The rainforest habitat in the southeastern part of Madagascar has one of the richest and most endangered biotas in the world. So far, twelve species of primates are known to be at Ranomafana. This number includes three species that were found within the last five years. One, the golden bamboo lemur, *Hapalemur aureus*, was a previously unknown species; another, the greater bamboo lemur, *Hapalemur simus*, was assumed to be extinct; the third, the aye-aye, *Daubentonia madagascariensis*, was not known to occur in this region. Much of the park remains unexplored, making it possible that one or two additional species exist within its boundaries.

A UNIQUE PRIMATE FAUNA

Madagascar has twenty-nine species and more than fifty distinct taxa of endemic primates ranging in size from the 40-gram mouse lemur to the 7 to 8 kg indris. In the recent past there were an additional fourteen or more species that are now extinct. Less than 1,000 years ago, Madagascar contained more primate species than any region of comparable size in the world. One explanation for this diversity is that many other groups of vertebrates either never occurred on the island or are poorly represented. For example, ungulates [hoofed mammals] are absent from Madagascar. Some of the largest lemurs (the size of chimpanzees) and the extinct sloth-like lemurs filled the role of browsers and even grazers. Today, some of the largest lemurs extensively browse in the forest. There are few bird species, but some primates seem to fill niches that birds fill in other ecosystems. For example, the aye-aye eats insects from dead wood—woodpeckers, which do this in other habitats, are absent from Madagascar. Other roles in which extant

Agriculture Sustainable alternatives to slash-and-burn farming have been recommended as a long-term strategy for the tropics. Agricultural experts from North Carolina State University have designed a development program for villages around the national park. The village projects planned include creating local gardens, setting up bee-keeping, making fish ponds, and building dams to encourage growing rice in paddies.

Forestry The forest currently serves as a source of fuel, construction woods, and medicinal plants. Tree ferns, which are a cash crop, are sold for use as planters. Current policies for maintaining tropical forests no longer suggest planting eucalyptus and pine stands; rather, they concentrate on planting endemic trees and fruit-bearing trees. Tropical forestry experts from North Carolina State University helped with the plan to make a vegetation and habitat map of the forest, establish tree nurseries, encourage planting of endemic species for firewood and construction. In addition, we plan to establish ceramic workshops to produce pots and thereby replace the tree fern industry.

Socioeconomic Studies and Health Studies in other areas of Madagascar have shown that parasites and intestinal worms are prevalent and that nutrition is inadequate in villages. Although protected-area authorities have traditionally focused on biological issues, there is a growing realization that a social impact assessment—the systematic gathering and analysis of social data—can be useful in planning successful rural development programs.

In 1989, the RNPP health team, with the collaboration of the Ministry of Health, began to visit all the remote villages to do a socioeconomic and health survey. Questions addressed to the villagers dealt with their perceptions of the national park, use of forest resources, and needs. During 1990 and 1991 these visits were continued with the addition of a Malagasy medical doctor to provide medical assistance and advice, a Malagasy botanist to find out about the use of native plants as medicines, and an anthropological economist. The plan is to continue this monitoring of the villages throughout the life of the project, with the help of the Ministry of Research and Development. UNICEF has also established a collaborative program with the RNPP that assists with health care, encourages women's cooperatives, and provides special assistance to women and children.

Education This has been a major priority for local villages and for us as well. Education is critical not only for developing the economy and well-being of the people, but must form the backbone of any long-term conservation strategy. The World Wildlife Fund planned a nationwide conservation-education program for all primary and secondary schools in Madagascar. However, provisions for remote villages were not yet installed (nearly all villages surrounding the park are off-road). With the assistance of the World Wildlife Fund's in-country program, we planned a primary school conservation-education program for the remote villages. USAid has begun constructing several local schools and the Ministry of Education has agreed to provide teachers. A public awareness and museum center was funded by the MacArthur Foundation. The displays about the park and biodiversity are enjoyed by local villagers, tourists, and school children.

Park Management Legal establishment of the park is only the first step; the protected area must be managed. During recent decades, the International Union for the Conservation of Nature and Natural Resources (IUCN, World Conservation Union) has been developing and refining management categories for the world's protected areas. The establishment of multiple-use buffer zones around national park boundaries is encouraged.

In Madagascar, park management is the jurisdiction of the Department of Water and Forests. In addition to the National Director of that department, the RNPP, has a conservation technical advisor. One of the first steps in management is to understand the resource and its potential. The park ranges in altitude from 400 to 1500 meters and contains diverse habitats, including high plateau forest, lowland rainforest, swamps, bamboo stands, and patches of savannah. We know from preliminary surveys that more than half of this park has been selectively logged, browsed by cattle, or cut for farmland within the past sixty years. During the next two years, forest transects and inventories, soil mapping, annual rainfall, temperature information, and biodiversity surveys will be combined with satellite image information and aerial photographs to produce a database and detailed maps. These maps will generate exact information about the location and condition of different habitats. Pristine areas will give us baseline data about plant and animal distribution that can be compared with the disturbed areas. Restoration of the disturbed areas will be monitored over the years, providing important recovery data. All of this baseline data will be shared with other projects and become part of a national biodiversity database.

A buffer zone larger than the park itself has been delineated around it. Most of this area is secondary forest or agricultural land. Ten villages in this zone have been selected for pilot projects involving the planting of endemic tree species or the implementation of new agricultural techniques. These projects will be extended to all villages in the buffer zone in the second phase of this project. A one-kilometer forested border around the perimeter of the buffer zone has been discussed. This area would be used by the villagers for the extraction of construction woods and medicinal plants.

Ranomafana National Park. (Photo by Steve Zack)

Ecotourism The assistance in agriculture, forestry, health and education will help improve the well-being and lifestyle of the region at a local level in conjunction with development of the park. However, it is clear that in the long run the park has to generate income and help pay for itself. Local economies have benefited from the encouragement of wildlife tourists in many other areas, including Rwanda and the Galapagos. However, it is important to make sure that tourism provides funds for maintenance of reserves, provides economic opportunities for local people, and helps justify reserves to local governments.

Ranomafana is one of the few accessible rainforests where habituated lemurs can be seen. A rustic and charming hotel was built fifty years ago to accommodate tourists to the hot mineral springs, but it has only fifteen rooms. The project began to train twenty tourist guides, encourage the upgrading of the existing hotel and the establishment of tourist bungalows. The museum and public awareness center, a snack bar, marked trails, camping sites, and information about the park infrastructure facilitate visits by tourists. However, it was critical to acknowledge at the planning stage that the park has a maximum carrying capacity for tourists and to build these limitations into the plans.

INAUGURATION AND FUTURE

On May 31, 1991, the inauguration of the Ranomafana National Park was held at the park entrance. More than 1,000 international visitors and local villagers attended. At the morning ceremony, the village elders sacrificed four cows and shared a traditional rum drink with the officials. The villagers then invoked their ancestors' protection of the park. The ceremony ended with the performance of traditional dances by people from different villages. In the afternoon ceremony, representatives of the governments and institutions that had collaborated in the park project made speeches. Afterward, villagers and officials shared a celebratory dinner.

Ranomafana is the fourth National Park created in Madagascar. The programs relating to health, socioeconomic studies, and biodiversity research and training have now been in place for three years and, we hope, will continue for the next three to ten years. A three-year USAid grant has been awarded to establish the park infrastructure and to study needs. The programs in agriculture, forestry, park management, and ecotourism are beginning. So far, the Ranomafana Project has received extraordinary support, both internationally and locally. However, the future of this integrated conservation and development project rests on the abilities of the local people to accept and continue this program.

Source: From pp. 25–33 in Patricia C. Wright, 1992, Primate ecology, rainforest conservation, and economic development: building a national park in Madagascar, *Evolutionary Anthropology* 1:25–33. Reprinted by permission of Wiley-Liss, Inc., a division of John Wiley & Sons, Inc.

Figure 5.6

Galagos are small, arboreal, nocturnal animals who can leap great distances. They are mainly solitary, though residents of neighboring territories sometimes rest together during the day.

The Lorisiformes are small, nocturnal, arboreal residents of the forests of Africa and Asia (Figure 5.6). These animals include two subfamilies with different types of locomotion and activity patterns. Galagos are active and agile, leaping through the trees and running quickly along branches. The lorises move with ponderous deliberation, and their wrists and ankles have a specialized network of blood vessels that allows them to remain immobile for long periods of time. These traits may be adaptations that help them avoid detection by predators. Traveling alone, the Lorisiformes generally feed on fruit, gum, and insect prey. The Lorisiformes leave their dependent offspring in nests built in the hollows of trees or hidden in masses of tangled vegetation. During the day, females sleep, nurse their young, and groom, sometimes in the company of mature offspring or familiar neighbors.

The Tarsiiformes, or tarsiers, are enigmatic primates who live in the rain forests of Borneo, Sulawesi, and the Philippines (Figure 5.7). Like many prosimians, they are small, nocturnal, and arboreal and move by vertical clinging and leaping. Some tarsiers live in monogamous family groups, but many have more than one breeding female. Female tarsiers give birth to infants who weigh 25% of their own weight; mothers leave their bulky infants behind in safe hiding places when they forage for insects. Tarsiers are unique among primates because they are the only primate who relies exclusively on animal matter, feeding on insects and small vertebrate prey.

Figure 5.7

Tarsiers are small, insectivorous primates who live in Asia. Some tarsiers form monogamous pairs.

The Anthropoids

The suborder Anthropoidea contains the infraorders Platyrrhini and Catarrhini.

The two infraorders Platyrrhini and Catarrhini are commonly referred to as the New World monkeys and the Old World monkeys and apes, respectively, because platyrrhine monkeys are found in South and Central America, while catarrhine monkeys and apes are found in Africa and Asia. This geographic dichotomy breaks down with humans, however—we are catarrhine primates, but we are spread over the globe.

Platyrrhini, or New World monkeys, is divided into two families: Callitrichidae and Cebidae.

Although the New World monkeys encompass considerable diversity in size, diet, and social organization, they do share some basic features. All but those in one genus *(Aotus)* are diurnal, all live in forested areas, and all are mainly arboreal. Most New World monkeys are quadrupedal, moving along the tops of branches and jumping between adjacent trees. However, some species within the family Cebidae can suspend themselves by their hands, feet, or tail and can move by swinging by their arms beneath branches.

The family Callitrichidae is composed of the marmosets and tamarins. These species share several morphological features that distinguish them from other anthropoid primate species: they are extremely small, the largest weighing less than 1 kg (2.2 lb), they have claws instead of nails, they have only two molars while all other monkeys have three, and they frequently give birth to twins and sometimes triplets (Figure 5.8). Marmosets and tamarins are also notable for their domestic arrangements: most species seem to be monogamous, and some may be polyandrous. **Polyandry**, which occurs when two or more males simultaneously form

Figure 5.8

Marmosets are small-bodied South American monkeys who form monogamous or polyandrous social groups. Males and older offspring participate actively in the care of infants.

pair-bonds with a single female, is extremely uncommon among mammalian species. Marmoset and tamarin mothers receive a considerable amount of assistance in caring for their young from their mates and older offspring.

The Cebidae are generally larger than the marmosets and tamarins, ranging in size from the 600-g (21-oz) squirrel monkey to the 9.5-kg (21-lb) muriqui (Figure 5.9). Although many people think that all monkeys can swing by their tails, prehensile tails are actually restricted to only the largest species of the Cebidae.

The family Cebidae is divided into six subfamilies which encompass considerable diversity in social organization, feeding behavior, and ecology. The subfamily Alouattinae is composed of several species of howlers, named for the long-distance roars they give in intergroup interactions. Howlers live in small one-male or multimale groups, defend their home ranges, and feed mainly on leaves. The subfamily Atelinae includes spider monkeys and woolly monkeys (Figure 5.9a). These species subsist mainly on fruit and leaves, and live in multimale, multifemale groups of 15 to 25. Spider monkeys, which rely heavily on ripe fruit, typically break up into

Figure 5.9

Portraits of some cebid monkeys: (a) Muriquis, or woolly spider monkeys, are large-bodied, arboreal monkeys. They are extremely peaceful creatures, rarely fighting or competing over access to resources. (b) Spider monkeys rely heavily on ripe fruit and travel in small parties. They have prehensile tails that they can use much like an extra hand or foot. (c) Capuchin monkeys have larger brains in relation to their body sizes than any of the other nonhuman primates. (d) Squirrel monkeys form large multimale, multifemale groups. In the mating season, males gain weight and become "fatted," and then compete actively for access to receptive females. (Photographs courtesy of: a, Sue Boinski; c, Susan Perry; d, Carlão Limeira.)

(a)

(c)

(b)

(d)

small parties for feeding (Figure 5.9b). The subfamily Cebinae includes capuchins and squirrel monkeys. Capuchins are best known to the public as the clever creatures dressed in red caps and jackets that retrieve coins for organ grinders (Figure 5.9c). To primatologists, capuchins are notable in part because they have very large brains in relation to their body size (see Chapter 9). Squirrel monkeys and capuchins live in multimale, multifemale groups of 10 to 50 individuals and forage for fruit, leaves, and insects (Figure 5.9d). The subfamily Aotinae includes the diurnal titi monkey and the only nocturnal anthropoid primate, the owl monkey. The subfamily Callimiconinae is composed of just one species, Goeldi's monkey. Aotinae and Callimiconae are small-bodied fruit eaters and live in monogamous groups. Finally, Pithecinae is a subfamily composed of the uakaris and sakis. Uakaris have little hair around their faces, making them look like wizened old men.

The infraorder Catarrhini contains the monkeys and apes of the Old World and humans.

As a group, the catarrhine primates share a number of anatomical and behavioral features that distinguish them from the New World primates. For example, most Old World monkeys and apes have narrow nostrils that face downward, while New World monkeys have round nostrils. Old World monkeys have two premolars on each side of the upper and lower jaws, while New World monkeys have three. Most Old World primates are larger than most New World species, and Old World monkeys and apes occupy a wider range of habitats than New World species do.

The catarrhine primates are divided into two superfamilies, the Cercopithecoidea (Old World monkeys) and Hominoidea (apes and humans). Cercopithecoidea contains one extant (still-living) family, which is further divided into two subfamilies of monkeys, Cercopithecinae and Colobinae.

The superfamily Cercopithecoidea encompasses great diversity in social organization, ecological specializations, and biogeography.

Colobine monkeys are found in the forests of Africa and Asia, and are collectively perhaps the most elegant of the primates (Figure 5.10). They have slender bodies, long legs, long tails, and their coats are often beautifully colored. The black-and-white colobus, for example, has a white ring around its black face, a striking white cape on its black back, and a bushy white tail that flies out behind as it leaps from tree to tree. Colobines are mainly leaf and seed eaters, and most species spend the majority of their time in trees. They have complex stomachs, almost like the chambered ones of cows, that allow them to maintain bacterial colonies that facilitate the digestion of cellulose. Colobines are most often found in groups composed of one adult male and a number of adult females. As in many other vertebrate taxa, the replacement of resident males in one-male groups is often accompanied by lethal attacks on infants by new males. Infanticide under such circumstances is believed to be favored by selection because it improves the relative reproductive success of infanticidal males. This issue is discussed more fully in Chapter 7.

Most cercopithecine monkeys are found in Africa, though one successful genus (*Macaca*) is widely distributed through Asia and part of Europe (Figure 5.11). The cercopithecines are more variable in size and diet than the colobines are. The social behavior, reproductive behavior, life history, and ecology of a number of cercopithecine species (particularly, baboons, macaques, and vervets) have been exten-

(a)   (b)

Figure 5.10

(a) African colobines, like this black-and-white colobus monkey, are arboreal and feed
mainly on leaves. These animals are sometimes hunted for their spectacular coats.
(b) Hanuman langurs are native to India and have been the subject of extensive study
during the last three decades. In some areas, Hanuman langurs form one-male, multifemale groups, and males engage in fierce fights over membership in bisexual groups. In
these groups, infanticide often follows when a new male takes over the group.

Baboons, Macaques, Vervets

sively studied. Cercopithecines typically live in medium or large bisexual (multimale, multifemale) groups. Females typically remain in their **natal groups** (the
groups into which they are born) throughout their lives and establish close and
enduring relationships with their maternal kin, while males leave their natal groups
and join new groups when they reach sexual maturity.

*The superfamily Hominoidea includes three families of apes: Hylobatidae (gibbons),
Pongidae (orangutans, gorillas, and chimpanzees), and Hominidae (humans).*

The Hominoidea are different from the Cercopithecoidea in a number of ways.
The most readily observed difference between monkeys and apes is that apes lack
tails. But there are many other more subtle differences between monkeys and apes.
For example, the apes share some derived traits, including broader noses, broader
palates, and larger brains, and they retain some primitive traits, such as relatively
unspecialized molars. In Old World monkeys the prominent anterior and posterior
cusps are arranged to form two parallel ridges. In apes, the five cusps on the lower

(a)

(b)

(c)

Figure 5.11

Some representative cercopithecines: (a) Bonnet macaques are one of several species of macaques that are found throughout Asia and North Africa. Like other macaques, bonnet macaques form multimale, multifemale groups, and females spend their entire lives in their natal (birth) groups. (b) Vervet monkeys are found throughout Africa. Like macaques and baboons, females live among their mothers, daughters, and other maternal kin. Males transfer to nonnatal groups when they reach maturity. Vervets defend their ranges against incursions by members of other groups. (c) Blue monkeys live in one-male, multifemale groups. However, during the mating season, one or more unfamiliar males may join bisexual groups and mate with females. (Photographs courtesy of: a, Kathy West; c, Marina Cords.)

molars are arranged to form a side-turned Y-shaped pattern of ridges (Figure 5.12).

The Hylobatidae are called lesser apes, while the larger-bodied Pongidae (orangutans, gorillas, and chimpanzees) are called great apes. Humans are traditionally placed in their own family, the Hominidae, but many taxonomists believe that humans belong with the other large-bodied apes, in the Pongidae. Gibbons, siamangs, and orangutans are found in Asia, while chimpanzees, bonobos, and gorillas are restricted to Africa.

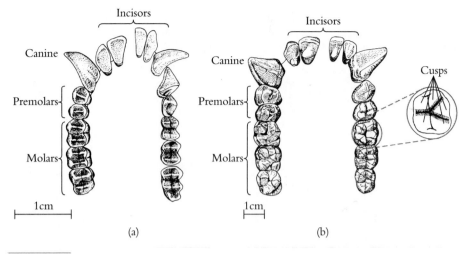

Incisors

Canine

Premolars

Molars

1cm

(a)

Incisors

Canine

Premolars

Molars

1cm

(b)

Cusps

Figure 5.12

The upper jaw (left) and lower jaw (right) are shown here for (a) a male colobine and (b) a male gorilla. In Old World monkeys the prominent anterior and posterior cusps of the lower molars form two parallel ridges. In apes the five cusps of the lower molar form a Y-shaped pattern.

The lesser apes (Figure 5.13) are slightly built creatures with extremely long arms in relation to their body size. Gibbons and siamangs are strictly arboreal, and they use their long arms to perform spectacular acrobatic feats, moving through the canopy with grace, speed, and agility. Gibbons and siamangs are the only true **brachiators** among the primates, which means they propel themselves by their arms alone and are in free flight between handholds. (To picture this, think about swinging on monkey bars in your elementary school playground.) All of the lesser apes live in monogamous family groups, vigorously defend their **home ranges**

(a)

(b)

Figure 5.13

(a) Gibbons live in monogamous groups and actively defend their territories against intruders. They have extremely long arms, which they use to propel themselves from one branch to another as they swing hand over hand through the canopy, a form of locomotion called brachiation. Gibbons and siamangs (b) are confined to the tropical forests of Asia. Like other residents of tropical forests, their survival is threatened by the rapid destruction of tropical forests. (Photographs courtesy of John Mitani.)

(a)

(b)

Figure 5.14

(a) Orangutans are large, ponderous, and mostly solitary creatures. Male orangutans often descend to the ground to travel, while lighter females often move through the tree canopy. (b) Today, orangutans are found only on the islands of Borneo and Sumatra, in tropical forests like this one.

(the areas they occupy), and feed on fruit, leaves, flowers, and insects. Siamang males play an active role in caring for young, frequently carrying them during the day, while male gibbons are less attentive fathers. In territorial displays, mated pairs of siamangs perform coordinated vocal duets that can be heard over long distances.

Orangutans, now found only on the Southeast Asian islands of Sumatra and Borneo, are among the largest and most solitary species of primates (Figure 5.14). Orangutans have been most extensively studied by Biruté Galdikas, who has monitored their behavior and ecology in Sumatra for more than 20 years. Orangutans feed primarily on fruit, but also eat some leaves and bark. Adult females associate mainly with their own infants and immature offspring, and do not often meet or interact with other orangutans. Adult males spend the majority of their time alone. A single adult male may defend a home range that encompasses the home ranges of several adult females, while other males wander over larger areas and mate opportunistically with receptive females. When resident males encounter these nomads, fierce and noisy encounters may take place.

Gorillas, the largest of the apes, existed in splendid isolation from Western science until the middle of the 19th century (Figure 5.15). Today, our knowledge of the behavior and ecology of gorillas is based mainly on detailed long-term studies of one subspecies, the mountain gorilla, at the Karisoke Research Center in Rwanda founded by the late Dian Fossey. Mountain gorillas live in small groups that contain one or two adult males and a number of adult females and their young. Each day, mountain gorillas ingest great quantities of various herbs, vines, shrubs, and bamboo. They eat little fruit because fruiting plants are scarce in their mountainous habitat. Adult male mountain gorillas, called **silverbacks** because the hair on their backs and shoulders turns a striking silver-gray when they mature, play a central role in the structure and cohesion of their social groups. Males sometimes remain in their natal groups to breed, but most males leave their natal groups and acquire females by drawing them away from other males during intergroup encounters. The silverback largely determines the timing of group activity and the

(a)   (b)

Figure 5.15

(a) Gorillas are the largest of the primates. Mountain gorillas usually live in one-male, multifemale groups, but some groups contain more than one adult male. (b) Most behavioral information about gorillas comes from observations of mountain gorillas who live in the Virunga Mountains of central Africa pictured here. The harsh montane habitat may influence the nature of social organization and social behavior in these animals, and the behavior of gorillas living at lower elevations may differ. (Photographs courtesy of John Mitani.)

direction of travel. As data from newly established field studies of lowland gorilla populations become available, we may have reason to revise some elements of this view of gorilla social organization. For example, lowland gorillas seem to eat substantial amounts of fruit, spend more of their time in trees, and form larger and less cohesive social groups than mountain gorillas do.

As humankind's closest living relatives, chimpanzees (Figure 5.16a) have played a uniquely important role in the study of human evolution. Whether reasoning by homology or by analogy, researchers have found observations about chimpanzees to be important bases for hypotheses about the behavior of early hominids.

Most of our detailed knowledge of chimpanzee behavior and ecology comes from three long-term field studies of chimpanzees: one conducted by Jane Goodall and her associates at the Gombe Stream National Park in Tanzania (Figure 5.16b); another by Toshisada Nishida, Yukimaru Sugiyama, and their colleagues in the nearby Mahale Mountains; and a third conducted by Christophe and Hedwige Boesch in the Taï Forest of the Ivory Coast. Bonobos (Figure 5.16c), another member of the genus *Pan,* have also been studied for shorter periods at two sites in the Democratic Republic of the Congo, Wamba and Lamako. Chimpanzees and bonobos form large multimale, multifemale communities. These communities differ from the social groups formed by most other species of primates in two important ways. First, female chimpanzees usually disperse from their natal groups when they reach sexual maturity, while males remain in their natal groups throughout their lives. Secondly, the members of chimpanzee communities are rarely found together in a unified group. Instead, they split up into smaller parties that vary in size and composition from day to day. In chimpanzees, the strongest social bonds among adults are formed among males, while bonobo females form stronger bonds with one another and with their adult sons than males do. Chimpanzees are

(a) (b) (c)

Figure 5.16

(a) Chimpanzees live in multimale, multifemale social groups. In this species, males form the core of the social group and remain in their natal groups for life. Many researchers believe that chimpanzees are our closest living relatives. (b) Like other apes, chimpanzees are found mainly in forests like this area on the shores of Lake Tanganyika in Tanzania. However, chimpanzees sometimes range into more open areas as well. (c) Bonobos are members of the same genus as chimpanzees, and are similar in many ways. Bonobos are sometimes called pygmy chimpanzees, but this is a misnomer, because bonobos and chimpanzees are about the same size. This infant bonobo is sitting in a patch of terrestial herbaceous vegetation, one of the staples of the bonobo's diet. (b, c, photographs courtesy of John Mitani.)

the only primates that routinely modify natural objects for use as tools in the wild. At several sites, chimpanzees strip twigs and poke them into termite mounds and ant nests to extract insects, a much-prized delicacy. In the Taï Forest, chimpanzees crack hard-shelled nuts using one stone as a hammer and a heavy, flat stone or a protruding root as an anvil. At Gombe, chimpanzees wad leaves in their mouths, and then dip these "sponges" into crevices to soak up water.

Primate Conservation

Many species of primates are endangered by habitat destruction, hunting, or live capture for trade and export.

Sadly, no introduction to the primate order would be complete without acknowledging that the prospects for the continued survival of many primate species are grim. Several species, and some entire genera, are considered by conservationists to be **endangered**, meaning they are in immediate danger of extinction and will not survive in the wild unless the factors that have caused their predicament are altered. The populations of some of the most gravely endangered species and populations, such as the mountain gorilla, number in the hundreds. In Africa and Asia, one-third to one-half of all primate species are endangered, while in some parts of South America an even larger fraction of species are at risk (Figure 5.17). At least two-thirds of the lemur species in Madagascar are in immediate danger of extinction.

(a)   (b)

Figure 5.17

Many populations of primates are in danger of becoming extinct in the wild. (a) This is a free-ranging member of a subspecies of squirrel monkeys that may number only 200 in the wild. (b) This infant of the same subspecies is being kept as a pet. Its mother was killed to capture the infant. (Photographs courtesy of Sue Boinski.)

Primate populations are endangered by three major factors: habitat destruction, hunting, and capture for trade and export. As arboreal residents of the tropics, many primate populations are directly affected by the rapid and widespread destruction of the world's forests. Their habitats are vulnerable for several different reasons. Rapid increases in the population of underdeveloped countries in the tropics have created intense demand for additional agricultural land (Figure 5.18). For example, in West Africa, Asia, and South America, vast areas of forest and other habitats have been cleared to accommodate the demands of subsistence farm–

Figure 5.18

Mountain gorillas live in one of the most densely populated regions on earth. Their existence is threatened because their needs conflict with the needs of local people, who need more land to grow crops to feed their families.

ers as well as the needs of large-scale agricultural projects (Figure 5.19). In Central and South America, massive tracts of forest have been cleared for large cattle ranches. A considerable amount of the wood that is removed from tropical forests is taken for fuel, while industrial logging operations also reduce the size and diversity of tropical forests.

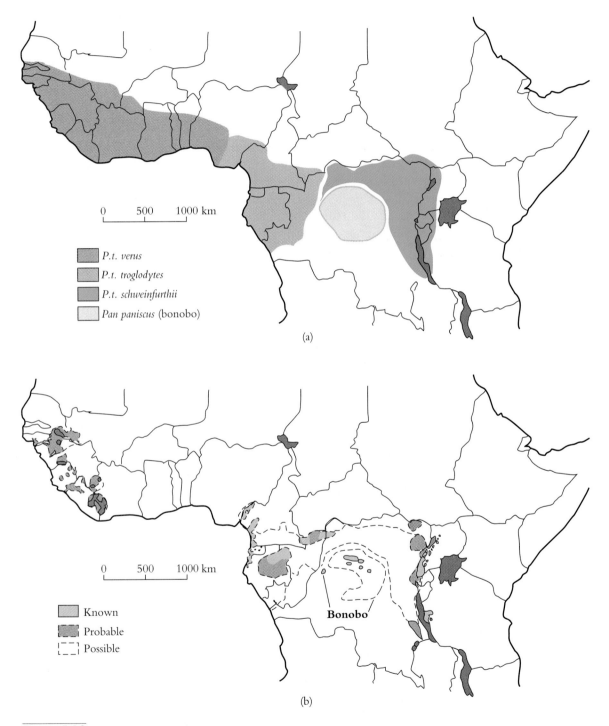

Figure 5.19

(a) Chimpanzees and bonobos once ranged throughout much of central Africa.

(b) Today, they are restricted to much smaller pockets within their former ranges.

Primates are also directly endangered due to hunting by humans. Primates are killed for meat, to obtain trophies, and to keep them from damaging crops in agricultural areas. The impact of hunting for meat varies among species and among regions of the world; primates are most endangered by hunting in the Amazon, West Africa, and central Africa. In contrast, primates in India were traditionally protected by their religious association with the Hindu monkey god, Hanuman. Where primates are hunted for food, larger-bodied species are killed more often than small ones. Some species, such as the beautiful black-and-white colobus monkey, are hunted to obtain their skins and other body parts for decoration, medicine, or trophies. In some areas, where primates range close to agricultural settlements, primates are killed because they are considered agricultural pests.

The third major source of danger to primate populations is capture for sale as pets or as research subjects (Figure 5.17b). The species affected in the largest numbers are those most often used in biomedical research, such as rhesus and squirrel monkeys. International trade in primates has a potentially devastating effect on species that are vulnerable to extinction in the wild. Happily, this is one area in which real progress has been made over the last few years. International trade in primates has been significantly reduced since the Convention of Trade in Endangered Species of Wild Flora and Fauna (CITES) was drafted in 1973. The parties to CITES, which now number nearly 100 countries, ban commercial trade in all endangered species and monitor trade of those that are likely to become endangered. Although not all countries are party to this agreement, and not all individuals or institutions around the world honor its policies, CITES has been an effective weapon in protecting primate populations. The United States imported more than 100,000 monkeys and apes per year before ratifying CITES, but reduced this number to approximately 13,000 a decade after signing the international agreement. More recently, the Rio Biodiversity Treaty established international guidelines for preserving biodiversity.

Although much remains to be done, conservation efforts have significantly improved the survival prospects for a number of primate species. These efforts have helped to preserve muriquis and golden lion tamarins in Brazil, golden bamboo lemurs in Madagascar, orangutans in Borneo, and mountain gorillas in Rwanda.

Those who study human evolution have a special interest in promoting the survival of primates, since living primates provide important clues about early human behavior and ecology.

While there are many different reasons to conserve primates and the habitats in which they live, those of us who are interested in human evolution have a very special reason to favor such efforts. We can gain valuable insights about our own evolutionary history by studying the behavior and ecology of living primates. If these species disappear, an important source of insight about ourselves will vanish. As we describe the ecology, mating systems, and social behavior of primates in the next few chapters, keep in mind that some of the species you read about here will not survive long in the wild unless rapid and effective action is taken to conserve their habitats, promote their survival, and safeguard their future. Conservation projects to save endangered primates, such as the mountain gorilla, will be successful only if endorsed by local residents, facilitated by the government of the countries in which the animals live, and supported and funded by the international community.

Further Reading

Fleagle, J. G. 1998. *Primate Adaptation and Evolution*. 2d ed. Academic Press, San Diego, Calif., chap. 1.

Martin, R. D. 1990. *Primate Origins and Evolution: A Phylogenetic Analysis*. Princeton University Press, Princeton, N.J.

Mittermeier, R. A. 1986. A global overview of primate conservation. Pp. 325–340 in *Primate Ecology and Conservation,* ed. by J. G. Else and P. C. Lee. Cambridge University Press, Cambridge.

Study Questions

1. What is the difference between homology and analogy? What are the evolutionary processes that correspond to these terms?
2. Suppose a group of extraterrestrial scientists landed on earth and enlisted your help in identifying animals. How would you help them recognize members of the primate order?
3. What primitive characters do modern prosimians retain?
4. How can we balance the needs and rights of people living in developing nations with the needs of the primates who live around them?

C H A P T E R 6

Primate Ecology

The Distribution of Food
Activity Patterns
Ranging Behavior
Predation

In the silver light of the moon, a solitary aye-aye quietly searches a promising tree trunk for its primary prey, insect larvae. It stops frequently to smell and to listen for the muted sounds of insect larvae writhing beneath the tree's bark. Locating a promising site along the trunk, the aye-aye strips off a layer of bark with its teeth and repeatedly probes the hole with its long, slender middle finger. As it does so, insect larvae are mashed into a thick paste that the aye-aye extracts and licks from its long finger.

A group of savanna baboons is spread out over more than a kilometer of open ground. The earth is parched by the sun and sparsely dotted with the browned stalks of grasses long gone to seed. The baboons sit on their haunches and painstakingly excavate corms, the bulbous stems of grasses buried beneath the hard surface of the ground (Figure 6.1) Periodically, each baboon raises its head, and scans the surrounding area. At a sudden rustle from a thick clump of bushes, the baboons call in alarm and race away. From the safety of a nearby acacia tree, they stare intently at the bushes for several minutes.

Figure 6.1
A female baboon feeds on corms in Amboseli National Park, Kenya.

Seeing nothing more threatening than a warthog trotting away, they descend from the tree and continue feeding.

Much of the day-to-day life of primates is driven by two concerns: getting enough to eat and avoiding being eaten. Food is essential for growth, survival, and reproduction, and it should not be surprising that primates spend much of every day finding, processing, consuming, and digesting a wide variety of foods. At the same time, primates must always guard against predators like lions, pythons, and eagles, which hunt them by day, and leopards, which stalk them by night. As we will see in the chapters that follow, both the distribution of food and the threat of predation influence the extent of sociality among primates, and shape the patterning of social interactions within and between primate groups.

In this chapter we describe the basic features of primate ecology. Later, we will draw on this information to explore the relationships among ecological factors, social organization, and primate behavior. It is important to understand the nature of these relationships because the same ecological factors are likely to have influenced the social organization and behavior of our earliest ancestors.

The Distribution of Food

Food provides energy that is essential for growth, survival, and reproduction.

Like all animals, primates need energy to maintain normal metabolic processes, to regulate essential bodily functions, and to sustain growth, development, and reproduction. The total amount of energy that an animal requires depends on four components: basal metabolism, active metabolism, growth rate, and reproductive effort.

1. **Basal metabolic rate** is the rate at which an animal expends energy to maintain life when at rest. As you can see from Figure 6.2, larger animals have higher basal metabolic rates than smaller ones do. However, large animals require relatively fewer calories *per unit* of body weight.

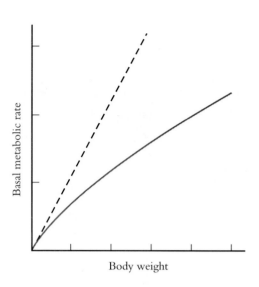

Figure 6.2

Average basal metabolism is affected by body size. The dashed line represents a direct linear relationship between body weight and basal metabolic rate. The solid line represents the actual relationship between body weight and basal metabolic rate. The fact that the curve "bends over" means that larger animals use relatively less energy per unit body weight.

2. When animals become active, their energy needs rise above baseline levels. The number of additional calories required depends on how much energy the animal expends. This, in turn, depends on the size of the animal and how fast it moves. In general, to sustain a normal range of activities, an average-sized primate like a baboon or macaque requires enough energy per day to maintain a rate about twice its basal metabolic rate.

3. Growth imposes further energetic demands on organisms. Infants and juveniles, who are gaining weight and growing in stature, require more energy than would be expected on the basis of their body weight and activity levels alone.

4. In addition for female primates, the energetic costs of reproduction are also substantial. During the latter stages of their pregnancies, for example, primate females require about 25% more calories than usual, and during lactation they require about 50% more calories than usual.

A primate's diet must satisfy the animal's energy requirements, provide specific types of nutrients, and minimize exposure to dangerous toxins.

The food that primates eat provides them with energy and essential nutrients, such as amino acids and minerals, that they cannot synthesize themselves. Proteins are essential for virtually every aspect of growth and reproduction, and for the regulation of many bodily functions. As we saw in Chapter 2, proteins are composed of long chains of amino acids. Primates cannot synthesize amino acids from simpler molecules, and in order to build many essential proteins, they must ingest foods that contain sufficient amounts of a number of amino acids. Fats and oils are important sources of energy for animals, and provide about twice as much energy as equivalent volumes of **carbohydrates**. Vitamins, minerals, and trace amounts of certain elements play an essential role in regulating many of the body's metabolic functions. Although they are needed in only small amounts, deficiencies of specific vitamins, minerals, or trace elements can cause significant impairment of normal body function. For example, trace amounts of the elements iron and copper are important in the synthesis of hemoglobin, while vitamin D is essential for the metabolism of calcium and phosphorous, and sodium regulates the quantity and distribution of body fluids. Primates cannot synthesize any of these compounds and must acquire them from the foods they eat. Water is the major constituent of the bodies of all animals and most plants. For survival, most animals must balance their water intake with their water loss; moderate dehydration can be debilitating, and significant dehydration can be fatal.

At the same time that primates attempt to obtain nourishment from food, they must also take care to avoid **toxins**, substances in the environment that are harmful to them. Many plants produce toxins called **secondary compounds** in order to protect themselves from being eaten. Thousands of these secondary compounds have been identified: caffeine and morphine are among the secondary compounds most familiar to us. Some secondary compounds, such as **alkaloids**, are toxic to consumers because they pass through the stomach into various types of cells, where they disrupt normal metabolic functions. Common alkaloids include capsicum (the compound that brings tears to your eyes when you eat red peppers) and chocolate. Other secondary compounds, such as tannins (the bitter-tasting compound in tea), act in the consumer's gut to reduce the digestibility of plant material. Secondary compounds are particularly common among tropical plant species, and are often concentrated in mature leaves and seeds. Young leaves, fruit, and flowers tend to have lower concentrations of secondary compounds, making them relatively more palatable to primates.

Primates obtain nutrients from many different sources.

Primates obtain energy and essential nutrients from a variety of sources (Table 6.1). Carbohydrates are obtained mainly from the simple sugars in fruit, but animal prey, such as insects, also provides a good source of fats and oils. **Gum**, a substance that plants produce in response to physical injury, is an important source of carbohydrates for some primates, particularly galagos, marmosets, and tamarins. Primates get most of their protein from insect prey or from young leaves. Some species have special adaptations that facilitate the breakdown of cellulose, enabling them to digest more of the protein contained in the cells of mature leaves. While seeds provide a good source of vitamins, fats, and oils, many plants package their seeds in husks or pods that shield their contents from seed predators. Many primates

Table 6.1 Sources of nutrients for primates. (x) indicates that the nutrient content is generally accessible only to animals that have specific digestive adaptations.

Source	Protein	Carbohydrates	Fats and Oils	Vitamins	Minerals	Water
Animals	x	(x)	x	x	x	x
Fruit		x				x
Seeds	x		x	x		
Flowers		x				x
Young leaves	x			x	x	x
Mature leaves	(x)					
Woody stems	x					
Sap		x			x	x
Gum	x	(x)			x	
Underground parts	x	x				x

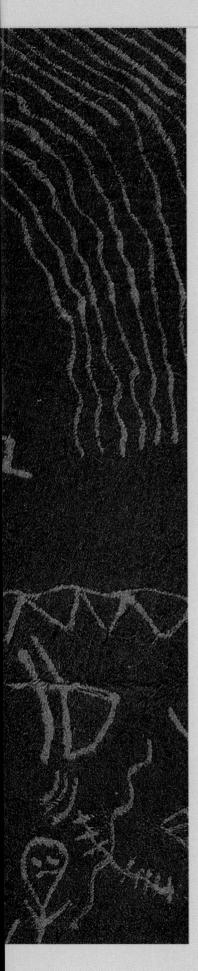

Monkeys Can Be Picky Eaters

[*Author's note:* Muriquis, or woolly spider monkeys, are large, arboreal South American monkeys. They live in large multimale, multifemale groups and are remarkably peaceful.]

Energy from fruits and flowers, and protein from leaves, are the principal constituents of the muriqui's diet. But the forest is full of plants that produce chemical and physical deterrents to avoid being eaten, and muriquis must distinguish between edible and inedible ones. Mature leaves of most woody plants contain high levels of tannins, which bind with the leaf's proteins and make it difficult to digest. The presence of tannins, as well as alkaloids and other secondary compounds, probably explains why muriquis rarely feed on any particular leaf species for any length of time. Several individuals will feed in sequence from the same tree, but each one feeds only briefly before moving to another food source. Immature leaves, however, generally contain fewer tannins, and muriquis spend more time eating larger quantities of these when they are available.

Unripe fruits tend to have more toxic compounds than the ripe fruits that muriquis prefer. When ripe fruit is plentiful in a single large tree, or food patch, the number of muriquis that feed together and the time they spend eating increases with the size of the patch. Unlike mature leaves, which must be eaten in only small quantities at a time, mature fruits can be eaten until the monkeys get bored or the food runs out. . . .

Even seeds contain toxins, or are wrapped in thick coats for protection from predators. Muriquis chew small seeds along with the fruit's flesh, but larger seeds are often dropped after the flesh has been removed. Some

A muriqui feeds on fruit. (Photograph courtesy of C. P. Nogueira.)

seeds, however, are covered with a smooth, hard coat, a design that makes them easy to swallow along with the fruit. These seeds glide through the muriquis' digestive tracts, and appear intact in their feces. . . .

Muriquis serve an important ecological function in dispersing these large, smooth seeds. Because they usually leave a fruit tree before their meal has been digested, the muriquis effectively transport the seeds to another part of the forest, away from the shade and competitive environment of the mother tree. When my students collected seeds from muriqui feces and planted them elsewhere, they almost always germinated. In fact, some muriqui-dispersed seeds actually germinate faster after they have been eaten than they do without the benefit of having passed through the muriquis' digestive system.

Muriquis eat only a few species of seeds that are not surrounded by fleshy fruit. The seeds of one of these, *Mucuna* (Leguminosae), are contained in large pods which are covered with a hard shell bristling with minuscule spines that break off and implant themselves in any hand or mouth that touches them. Adult and subadult muriquis are able to open these pods and eat the large, fatty seeds, but smaller muriquis avoid *Mucuna*, perhaps because they are not strong enough to open the pods, or because the fleshy pads of their fingers are not tough enough to deflect the spines. The *Mucuna*'s spiny defenses protect it from juvenile muriquis and other potential predators, but its fatty seeds are too important for larger muriquis to resist.

Leaves from the fig-like *Cecropia* (Moraceae) are another example of a food that muriquis go to great lengths to get. *Cecropia* have co-evolved with ants, which live in their trunks and defend them by attacking and stinging any animal that tries to enter the trees to feed. To get around this defense, muriquis eat *Cecropia* leaves by suspending themselves from neighboring trees. I once witnessed an adult male attempting to reach a *Cecropia* from a small palm tree growing next to it. He rocked back and forth in the crown of the palm causing it to swing. Finally, the momentum of his weight brought the palm close enough to the *Cecropia* for him to grab a leaf before the palm swung back. He then sat in the center of the still-swaying palm tree to eat the leaf, his reward for a 10 minute effort. Biochemical analyses indicate that these *Cecropia* are especially high in protein and low in other toxic compounds, so the nutritional pay-offs appear to compensate for the time and ingenuity required to outsmart the tree's defense system.

The trade-off between nutritional gains and plant defenses may not be the only basis for muriqui food choices. Some of the secondary compounds that plants produce may also have important medicinal properties. Many indigenous human populations throughout the tropics exploit different plants and plant parts for pharmaceutical purposes, and chimpanzees and baboons are known to eat certain plants for their antiparasitic and antibacterial compounds. I suspected that the muriquis at Fazenda Montes Claros might find similarly important compounds in some of their foods, so in 1989 I began a study of the parasites in their gastrointestinal tracts. I relied on the information that could be gleaned from noninvasive examination of their feces, which [was] conducted by Dr. Michael Stuart and his students at the University of North Carolina–Asheville. Looking at over 80 fecal samples collected from 32 different monkeys, the results were astonishing. The muriquis' feces were completely devoid of parasites, and our comparative analyses of howler monkey feces at this site show that they, too, are parasite-free. Most other primates, including howler monkeys and muriquis at other forests, carry in their gastrointestinal tracts a number of parasites that are passed along in their feces, so the fact that neither of these primates at Fazenda Montes Claros [is] infected suggests that something unusual is going on.

Parasites have evolved complex ecological relationships with their hosts, and the absence of parasites in both muriquis and howlers could mean that these ecological—and evolutionary—relationships have been disrupted at this forest. Perhaps a key secondary host has become locally extinct. On the other hand, it is possible that both primate species at Fazenda Montes Claros have discovered antiparasitic agents in their diets. Muriquis and howler monkeys eat many of the same plants, and by analyzing the ones that are found only at this forest we may be able to identify bioactive plants with value, not only for monkeys, but for humans as well.

Source: From pp. 53–56 in K. B. Strier, 1992, *Faces in the Forest: The Endangered Muriqui Monkeys of Brazil,* Oxford University Press, New York. Reprinted by permission of Oxford University Press.

Figure 6.3

Savanna baboons drink from a pool of rainwater. Most primates must drink every day.

drink daily from streams, water holes, springs, or puddles of rainwater (Figure 6.3). Primates can also obtain water from fruit, flowers, young leaves, animal prey, and the underground storage parts (roots and tubers) of various plants. This is particularly important for arboreal animals that do not descend from the canopy of tree branches and for terrestrial animals during times of the year when surface water is scarce. Vitamins, minerals, and trace elements are obtained in small quantities from many different sources.

Although there is considerable diversity in diet among primates, some generalizations are possible:

1. All primates rely on at least one type of food high in protein and another high in carbohydrates. Prosimians generally obtain protein from insects and carbohydrates from gum and fruit. Monkeys and apes usually obtain carbohydrates from fruit and protein from insects or young leaves.
2. Most primates rely more heavily on some types of foods than on others. Moreover, most species are fairly conservative in their dietary preferences, even though the availability of specific types of food varies over the course of the year in many habitats, and the species of food plants eaten often varies between habitats. Chimpanzees, for example, feed mainly on ripe fruit throughout their range from Tanzania to the Ivory Coast. Scientists use the terms **frugivore, folivore, insectivore**, and **gummivore** to refer to primates that rely most heavily on fruit, leaves, insects, and plant gum, respectively (Figure 6.4). Box 6.1 examines some of the adaptations in morphology among primates with different diets.
3. In general, insectivores are smaller than frugivores and frugivores are smaller than folivores (Figure 6.6). This is related to the fact that small animals have relatively higher energy requirements than larger animals do, and require relatively small amounts of high-quality foods that can be processed quickly. Larger animals are less constrained by the quality of their food than by the quantity, as they can afford to process lower-quality foods more slowly.

The nature of dietary specializations and the challenge of foraging in tropical forests influence ranging patterns.

Nonhuman primates do not have the luxury of shopping in supermarkets where abundant supplies of food are concentrated in a single location and are constantly

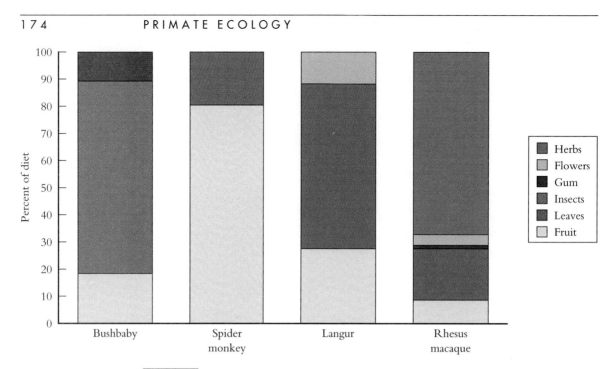

Figure 6.4

The diets of primates are quite variable. The black-handed spider monkey is one of the most dedicated frugivores in the primate order—80% of its diet consists of fruit. Langurs are among the most folivorous primate species; more than half of the purple-faced langur's diet is composed of leaves, while fruit and flowers play a much less important role. The bushbaby is primarily insectivorous, relying on insects for 70% of its food supply. Macaques have more eclectic diets, feeding on herbs, fruit, gum and sap, flowers, and leaves.

replenished. Instead, the availability of their preferred foods varies widely in space and time, making their food sources patchy and often unpredictable. Most primate species live in tropical forests. Although such forests, with their dense greenery, seem to provide abundant supplies of food for primates, appearances can be deceiving. Tropical forests contain a very large number of tree species, and individual trees of any particular species are few in number. Katherine Milton, a primatologist at the University of California at Berkeley, has conducted detailed studies of the feeding ecology of howler monkeys in a lowland tropical forest in central Panama (Figure 6.7). In the course of her work, she carefully surveyed the composition of the forest in which the howlers live. She discovered that 65% of all tree species occur less than once per hectare. (A hectare is a square, one hundred meters on a side, or about 2.5 acres.) This means that potential food sources are patchily distributed in space. Moreover, Milton's studies showed there is considerable variation in the production of new leaves, flowers, and fruit over the course of the year. In the forests of central Panama, young leaves are available on a single species for an average of seven months, green and ripe fruits are available for four months, and flower parts are available for three months. On individual trees, these items are available for an even shorter period. Ripe fruit, for example, is available on individual trees for less than one month on average. In some cases food items sustain an optimal nutritional content and edibility for only a few days. Thus, foraging in a tropical forest is not an easy task.

Primates with different dietary specializations confront different foraging challenges (Figure 6.8). Plants generally produce more leaves than flowers or fruit, and

BOX 6.1

Dietary Adaptations of Primates

Primates have evolved a number of adaptations to enhance their ability to process and digest certain types of foods. Their morphological dietary specializations include specific adaptations of the teeth and gut (Figure 6.5).

TEETH Primates that rely heavily on gum tend to have large and prominent incisors, which they use to gouge holes in the bark of trees. In some prosimian species the incisors and canines are projected forward in the jaw, and are used to scrape hardened gum off the surface of branches and tree trunks. Dietary specializations are also reflected in the size and shape of the molars. Insectivores and folivores have molars with well-developed shearing crests that permit them to cut their food into small pieces when they chew. Insectivores tend to have higher and more pointed cusps on their molars, which are useful for puncturing and crushing the bodies of their prey. The molars of frugivores tend to have flatter, more rounded cusps, with broad and flat areas used to crush their food. Primates that rely on hard seeds and nuts have molars with very thick enamel that can withstand the heavy chewing forces needed to process these types of food.

GUT Primates that feed principally on insects or animal prey have relatively simple digestive systems that are specialized for absorption. They generally have a simple, small stomach, a small caecum (a pouch located at the upper end of the large intestine), and a small colon relative to the rest of the small intestine. Frugivores also tend to have simple digestive systems, but frugivorous species with large bodies have capacious stomachs to hold large quantities of leaves they consume along with the fruit in their diet. Folivores have the most specialized digestive systems because they must deal with large quantities of cellulose and secondary plant compounds. Since primates cannot digest cellulose or other structural carbohydrates directly, folivores maintain colonies of microorganisms in their digestive systems that break down these substances. In some species these colonies of microorganisms are housed in an enlarged caecum, while in other species the colon is enlarged for this purpose. Colobines, for example, have an enlarged and complex stomach divided into a number of different sections where microorganisms help process cellulose.

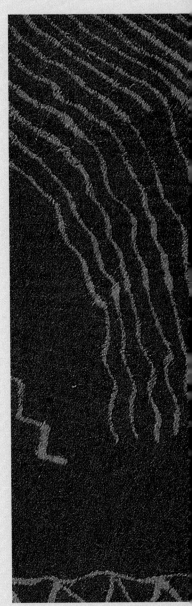

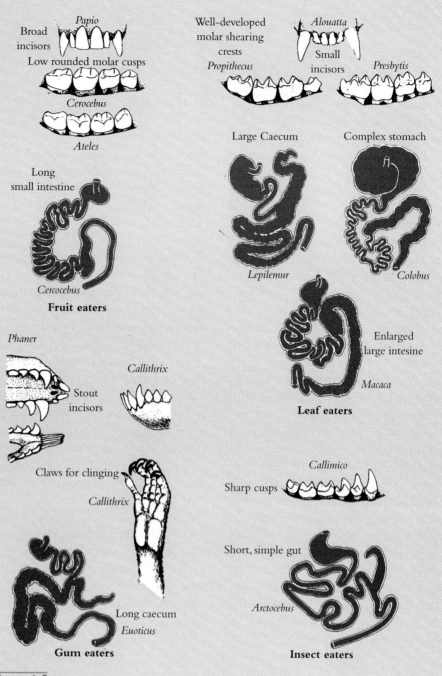

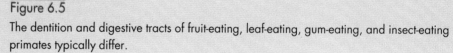

Figure 6.5

The dentition and digestive tracts of fruit-eating, leaf-eating, gum-eating, and insect-eating primates typically differ.

bear leaves for a longer period during the year than they bear flowers and fruit. As a result, foliage is normally more abundant than fruit or flowers are at a given time during the year, and mature leaves are more abundant than young leaves. Insects and other suitable prey animals occur at even lower densities than plants. This means that folivores can generally find more food in a given area than frugivores or insectivores can. However, the high concentration of toxic secondary compounds in mature leaves complicates the foraging strategies of folivores. Some leaves must be avoided altogether, and others can be eaten only in small quantities.

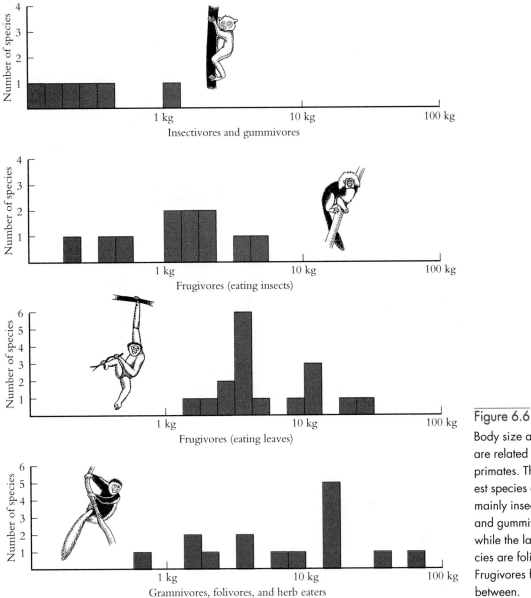

Figure 6.6

Body size and diet are related among primates. The smallest species are mainly insectivores and gummivores, while the largest species are folivores. Frugivores fall in between.

Figure 6.7

This is the forest of Barro Colorado Island, where howler monkeys have been studied for many years.

(a)

(b)

(c)

(d)

(e)

Figure 6.8

(a) Some primates feed mainly on leaves, though many leaves contain toxic secondary plant compounds. These primates are red colobus monkeys in the Kibale Forest of Uganda. (b) Some primates include a variety of insects and other animal prey in their diet. This capuchin monkey in Costa Rica is feeding on a wasp's nest. (c) Mountain gorillas are mainly vegetarians. They consume vast quantities of plant material, like this fibrous stem. (d) A vervet monkey feeds on grass stems. (e) Although many primates feed mainly on one type of food, such as leaves or fruit, none rely exclusively on one type of food. Thus, the main bulk of the muriqui diet comes from fruit, but muriquis also eat leaves, as you see here. (Photographs courtesy of: a, Lynne Isbell; b, Susan Perry; c, John Mitani; e, Carlão Limeira.)

Nonetheless, the food supplies of folivorous species are generally more uniform and predictable in space and time than are the food supplies of frugivores or insectivores. Thus, it is not surprising to find that folivores generally have smaller home ranges than frugivores or insectivores.

Activity Patterns

The fact that many prosimian species are nocturnal suggests that primates evolved from a nocturnal ancestor.

Most primate species are active either during the day (diurnal) or at night (nocturnal), although a few species are active at intervals throughout the day and night **(cathemeral)**. Since all monkeys and apes (except the owl monkey) are diurnal, and a substantial number of prosimians are nocturnal, it seems likely that primates originally evolved from a nocturnal ancestor. Nocturnal primate species tend to be smaller, more solitary, and more exclusively arboreal than diurnal species. They also rely more heavily on olfactory signals than do diurnal species.

Primate activity patterns show regularity in seasonal and daily cycles.

Primates spend the majority of their time feeding, moving around their home ranges, and resting (Figures 6.9 and 6.10). Relatively small portions of each day are spent grooming, playing, fighting, or mating (Figure 6.11). The proportion of time devoted to various activities is influenced to some extent by ecological conditions. For primates living in seasonal habitats, for example, the dry season is often a time of scarce resources, and it is harder to find adequate amounts of appropriate types of food. In some cases, this means that the proportion of time spent feeding and traveling increases during the dry season, while the proportion of time spent resting and socializing decreases.

Primate activity patterns also show regular patterns over the course of the day. When they wake up, their stomachs are empty, so the first order of the day is to

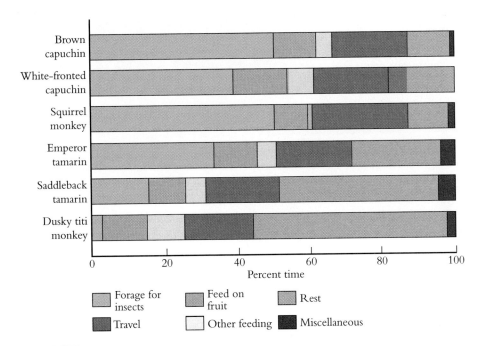

Figure 6.9

The amount of time that animals devote to various types of activities is called a time budget. Time budgets of different species vary considerably. These six monkey species all live in a tropical rain forest in Manu Park in Peru.

Figure 6.10

All diurnal primates, like this capuchin monkey, spend some part of each day resting. (Photograph courtesy of Susan Perry.)

Figure 6.11

Immature monkeys spend much of their ''free'' time playing. These patas monkeys are play wrestling. (Photograph courtesy of Lynne Isbell.)

visit a feeding site. Much of the morning is spent eating and moving between feeding sites. As the sun moves directly overhead and the temperature rises, most species settle down in a shady spot to rest, socialize, and digest their morning meals (Figure 6.12). Later in the afternoon, they resume feeding. Before dusk, they move to the night's sleeping site; some species habitually sleep in the same trees every night, while others have multiple sleeping sites within their ranges (Figure 6.13).

Ranging Behavior

In the hours after dawn in the Malaysian rain forest, the calls of gibbons reverberate through the canopy. The vocal performance starts as one male begins to sing in short bursts interspersed with intervals of silence. The male's mate joins and they perform an elaborate vocal duet. The gibbon pair's song rouses their neighbors a kilometer

Figure 6.12
Gorillas often rest in close proximity to other group members and socialize during a midday rest period.

Figure 6.13
Like most other primates, baboons sleep perched on the branches of trees at night. These baboons are descending from their sleeping trees shortly after dawn.

Figure 6.14

(a) Some primates defend the boundaries of territories (a schematic map of three territories is shown here), and do not tolerate intrusions from neighbors or strangers of the same species.
(b) Other primates occupy fixed home ranges (three shown here), but do not defend their borders against members of the same species.
When two groups meet in areas of home range overlap, one may defer to the other, they may fight, or they may mingle peacefully together.

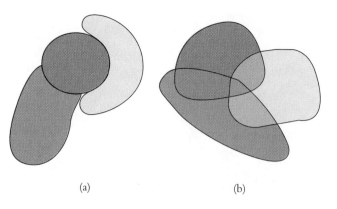

(a)   (b)

away who begin their own chorus of calls. The vocal skirmish continues as the sun climbs over the horizon and light begins to filter through the canopy. Eventually, both pairs become quiet, and move off to feed in their respective ranges.

In Manu Park in Peru, a noisy group of squirrel monkeys swarms through a fig tree, plucking ripe figs and popping them into their mouths. Partially eaten fruits rain down onto the ground, as the monkeys taste and discard several figs for each one that they consume. The animals return to the same site for several successive days, sometimes meeting the members of neighboring groups in the same tree. They mingle peacefully while they feed.

All primates have home ranges, but only some species are territorial—defending their home range against incursions by other members of their species.

In virtually all primate species, groups range over a relatively fixed area, and members of a given group can be consistently found in a particular area over time. These areas are called home ranges, and they contain all of the resources that group members exploit in feeding, resting, and sleeping. However, the extent of overlap among adjacent home ranges and the nature of interactions with members of neighboring groups or strangers varies considerably among species. Some primate species, like gibbons, maintain exclusive access to fixed areas, called **territories** (Figure 6.14a). Territory residents regularly advertise their presence by vocalizing, and they aggressively protect the boundaries of their territories from encroachment by outsiders (Figure 6.15). While some territorial birds defend only their nest sites,

Figure 6.15

Gibbons perform complex vocal duets as part of territorial defense.

primate territories contain all of the sites at which the residents feed, rest, and sleep and the areas in which they travel. Thus, among territorial primates, the boundaries for the territory are essentially the same as for their home range, and territories do not overlap.

Nonterritorial species, like squirrel monkeys and long-tailed macaques, establish home ranges that overlap considerably with those of neighboring groups (Figure 6.14b). When members of neighboring nonterritorial groups meet, they may fight (Figure 6.16), exclude members of lower-ranking groups from resources, avoid one another, or mingle peacefully together. This last option is unusual, but in some species, adult females sexually solicit males from other groups, males attempt to mate with females from other groups, and juveniles from neighboring groups play together when their groups are in proximity.

The two main functions suggested for territoriality are resource defense and mate defense.

In order to understand why some primate species defend their home ranges from intruders and others do not, we need to think about the possible costs and benefits associated with defending resources from conspecifics. Costs and benefits are measured in terms of the impact on the individual's ability to survive and reproduce successfully. Territoriality is beneficial because it prevents outsiders from exploiting the limited resources within a territory. At the same time, however, territoriality is costly because the residents must be constantly vigilant against intruders, regularly advertise their presence, and be prepared to defend their ranges against encroachment. Territoriality is expected to occur only when the benefits of maintaining exclusive access to a particular piece of land outweigh the costs of protecting these benefits.

When will the benefits of territoriality exceed the costs? The answer to this question depends in part on the kind of resources that individuals need in order to survive and reproduce successfully, and in part on the way these resources are distributed spatially and seasonally. For reasons we will discuss more fully in Chapter 7, the reproductive strategies of mammalian males and females generally differ. In most cases, female reproductive success depends mainly on getting enough to eat for themselves and their dependent offspring, while males' reproductive success depends mainly on their ability to mate with females. As a consequence, females

Figure 6.16

Rhesus monkeys don't defend fixed territories, but they sometimes react aggressively when they meet members of neighboring groups. Here, members of two groups meet on Cayo Santiago Island. (Photograph courtesy of Susan Perry and Joseph Manson.)

are more concerned about access to food, while males are more interested in access to females. Thus, territoriality has two different functions. Sometimes females defend food resources, or males defend food resources on their behalf. Other times, males defend groups of females against incursions by other males. In primates, both resource defense and mate defense seem to have influenced the evolution of territoriality.

Resource-defense territoriality occurs when resources are limited, but also clumped and defendable.

Territoriality occurs when resources are economically defensible, meaning that resources are limited in abundance but occur within an area that can be defended with a reasonable amount of effort. When food resources are distributed over a wide area, it is costly to detect and evict intruders, so territoriality does not pay. Similarly, when resources are readily available, it makes little sense to defend them (Figure 6.17a). If, on the other hand, valuable resources are clumped within a small area, territoriality can be favored by natural selection (Figure 6.17b).

Mate defense also plays a role in the evolution of territoriality in some primate species.

Although territoriality seems to be linked to the defendability of resources in many species, there are also species in which resource defense does not seem to be the primary factor favoring territoriality. This conclusion is based on the observation that in a number of species, males are active in intergroup encounters while females seem largely indifferent to the presence of intruders. This pattern is characteristic of gorillas, red colobus monkeys, many Southeast Asian langurs, and some species of lesser apes. Since males' reproductive success is influenced by their access to females, when males are the principal actors in territorial encounters, it seems likely that the primary function of territoriality is to defend access to mates, not food resources. By the same token, since females' ability to reproduce is related to their access to food resources, it seems likely that if females don't participate in territorial disputes, then resource defense is not the central factor determining the nature of intergroup encounters.

(a)   (b)

Figure 6.17

(a) When resources are evenly distributed, like these *Ramphicarpa montana* flowers, there is little point in defending access to them. (b) When resources are clumped together, like these acacia flowers, then animals may profit from driving away competitors.

Predation

The members of Camp Troop, a group of chacma baboons in the Okavango Delta of Botswana, settle one evening among the branches of trees in one of their regular sleeping sites. Late that night, the baboons erupt in a frantic chorus of wahoos, loud calls of alarm. The next day an adult female named Sugar is missing from the group. Near the sleeping site there is a long, shallow depression in the soft sand where Sugar's body had been dragged along the ground. The footprints of a leopard straddle the depression at regular intervals. The trail leads to a tree about 20 m away from the sleeping site. At the base of the tree there are tufts of Sugar's coarse, gray-brown hair, the bloody remains of her upper jaw, and small fragments of her skull. The bark in the crook of the tree is stained with Sugar's blood (Figure 6.18).

Predation is believed to be a significant source of mortality among primates, but direct evidence of predation is normally difficult to obtain.

Primates are hunted by a wide range of predators, including pythons, raptors, crocodiles, leopards, lions, tigers, and humans (Figure 6.19). In Madagascar, large lemurs are preyed upon by fossas, puma-like carnivores. Primates are also preyed on by other primates. Chimpanzees, for example, hunt red colobus monkeys, and baboons prey on infant vervet monkeys.

In primate populations, the estimated rates of predation vary from less than 1% of the population per year to more than 15% of the population per year. These figures must be treated with some caution because few kills are actually seen by humans. This is partly because most predators avoid close contact with humans, so observers of primate groups see few predatory attacks. It is also due to the fact

(a)

(b)

(c)

Figure 6.18
(a) This depression in the sand was made when a leopard dragged the body of an adult female baboon out of the sleeping tree and across a small, sandy clearing.
(b) The leopard's footprints were clearly visible next to the drag marks.
(c) The next morning, all that remained of Sugar's body was her jaw and skull.

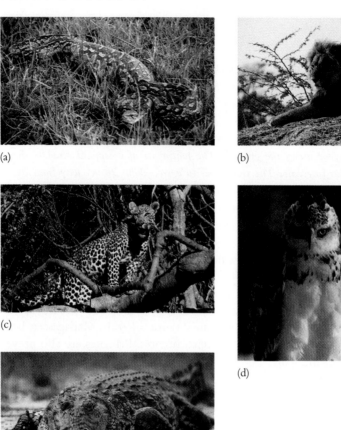

(a)

(b)

(c)

(d)

Figure 6.19

Primates are preyed upon by a variety of predators, including (a) pythons, (b) lions, (c) leopards, (d) Martial eagles, and (e) crocodiles.

(e)

that some predators, such as leopards, often hunt at night when most researchers are asleep themselves. In most cases, predation is inferred when a healthy animal, who is unlikely to have left the group, abruptly disappears.

University of Pennsylvania biologist Dorothy Cheney and Harvard anthropologist Richard Wrangham have reviewed the data on predation on primates. These researchers found that small-bodied primate species are more vulnerable to predation than larger ones (Figure 6.20). Moreover, young primates are generally more susceptible to predation than adults. They did not find, however, any consistent associations between the rate of predation, habitat use (arboreal versus terrestrial), or the composition of social groups.

Primates have evolved an array of defenses against predators.

Many primates give alarm calls when they sight potential predators, and some species have specific vocalizations for particular predators. Vervet monkeys, for example, give different calls when they are alerted to the presence of leopards, small carnivores, eagles, snakes, baboons, and strange humans. In many species, the most common response to predators is to flee or take cover. Small primates sometimes try to conceal themselves from predators, while larger ones may confront potential predators. For example, slow-moving pottos encountering snakes, fall to the ground, move a short distance, and freeze. At some sites, adult red colobus monkeys aggressively attack chimpanzees who hunt their infants.

(a)   (b)

Figure 6.20

Small primates are more vulnerable to predators but even large primates are vigilant.
(a) This squirrel monkey is scanning the canopy for potential predators. (b) A male rhesus
monkey scans the ground below for possible dangers. (a, photograph courtesy of Sue
Boinski.)

Another antipredator strategy that primates adopt is to associate with members
of other primate species. In the Taï Forest of Ivory Coast, a number of monkey
species share the canopy and form regular associations with one another. For example,
groups of red colobus monkeys spend approximately half their time with
groups of Diana monkeys. Ronald Noë and Rdeouan Bshary of the Max-
Planck-Institut für Verhaltenphysiologie in Seewiesen, Germany, have shown that
both species benefit when they are together. Red colobus are less vigilant about
dangers from terrestrial predators when they are with Diana monkeys—they can
spend more time at lower heights in trees, where they would normally be exposed
to threats from the ground. At the same time, Diana monkeys are less vigilant
about aerial predators when they are with red colobus monkeys—they spend more
time at higher positions in trees where they are more often exposed to predators
that might strike them from in front, above, or behind.

Red colobus and Diana monkeys can afford to be less vigilant when they are
together because they are safer. Diana monkeys benefit from their associations with
red colobus monkeys, who sometimes detect eagles before they do and may even
launch attacks on eagles. Red colobus monkeys benefit because Diana monkeys
are better at detecting terrestrial predators than they are. This is probably why red
colobus monkeys associate with Diana monkeys most often during times of the
year when they are at greatest danger from their most important terrestrial predators,
chimpanzees.

Although empirical information about predation on primates is incomplete,
many researchers believe that predation has played an important role in the evolution
of primate social organization and behavior. This is one of the topics that
we will consider in the next chapter.

Further Reading

Isbell, L. A. 1994. Predation on primates: ecological patterns and evolutionary
      consequences. *Evolutionary Anthropology* 3:61–71.

Janson, C. H. 1992. Evolutionary ecology of primate social structure. Pp. 95–130 in *Evolutionary Ecology and Human Behavior,* ed. by E. A. Smith and B. Winterhalder. Aldine de Gruyter, New York.

Oates, J. F. 1986. Food distribution and foraging behavior. Pp. 197–209 in *Primate Societies,* ed. by B. B. Smuts, D. L. Cheney, R. M. Seyfarth, R. W. Wrangham, and T. T. Struhsaker. University of Chicago Press, Chicago.

Richard, A. 1985. *Primates in Nature.* W. H. Freeman, New York, chaps. 4 and 5.

Terborgh, J. 1983. *Five New World Primates.* Princeton University Press, Princeton, N.J.

Study Questions

1. Large primates often subsist on low-quality food such as leaves, while small primates specialize in high-quality foods such as fruit and insects. Why is body size associated with dietary quality in this way?

2. For folivores, tropical forests seem to provide an abundant and constant supply of food. Why is this not an accurate assessment?

3. Territorial primates do not have to share access to food, sleeping sites, mates, and other resources with members of other groups. Given that territoriality reduces the extent of competition over resources, why aren't all primates territorial?

4. Territoriality is often linked to group size, day range, and diet. What is the nature of the association, and why does the association occur?

5. Most primates specialize in one type of food, such as fruit, leaves, or insects. What benefits might such specializations have? What costs might be associated with specialization?

6. Nocturnal primates are smaller, more solitary, and more arboreal than diurnal primates. What might the reason(s) be for this pattern?

C H A P T E R   7

Primate Mating Systems

Reproduction is the central act in the life of every living thing. Primates perform a dizzying variety of behaviors—gibbons fill the forest with their haunting duets, baboons threaten and posture in their struggle for dominance over other members of their group, and chimpanzees use carefully selected stone hammers to crack open tough nuts. But all of these behaviors evolved for a single ultimate purpose—reproduction. According to Darwin's theory, complex adaptations exist because they evolved step by step by natural selection. At each step, only those modifications that increased reproductive success were favored and retained in subsequent generations of offspring. Thus, every morphological feature and every

189

behavior exists only because it was part of an adaptation that contributed to reproduction in ancestral populations. As a consequence, **mating systems** (the way animals find mates and care for offspring) play a crucial role in our understanding of primate societies.

Understanding the diverse reproductive strategies of nonhuman primates illuminates human evolution because we share many elements of our reproductive physiology with other species of primates.

To understand the evolution of primate mating systems, we must take into account that the reproductive strategies of living primates are influenced by their phylogenetic heritage as mammals. Mammals reproduce sexually. After conception, mammalian females carry their young internally. After they give birth, females suckle the young for an extended period of time. The mammalian male's role in the reproductive process is more variable than that of the female. In some species, males contribute little to their offspring's development besides a single sperm at the moment of conception, while in other species males defend territories; provide for their mates, and feed, carry, and protect their offspring.

Although mammalian physiology imposes some bounds on the nature of primate reproductive strategies, there is considerable room for diversity in primate mating systems and in reproductive behavior. Patterns of courtship, mate choice, and parental care vary greatly within the primate order. In some species, male reproductive success is determined mainly by success in competition with other males, while in others it is strongly influenced by female preferences. In many monogamous species, both males and females care for their offspring, while in most nonmonogamous species, females care for offspring and males compete with other males to inseminate females. To fully understand the evolution of primate mating systems and reproductive strategies, we must first consider the factors that sustain this diversity, which we will explore later in this chapter.

What aspects of mating do humans share with other primates? Until very recent times, all pregnant women nursed their offspring for an extended period, as with other primates. In nearly all traditional human societies, fathers contribute extensively to their children's welfare, providing resources, security, and social support. An understanding of the phylogenetic and ecological factors that shape the reproductive strategies of other primates may help us to understand how evolutionary forces shaped the reproductive strategies of our hominid ancestors, and the reproductive behavior of men and women in contemporary human societies.

The Language of Adaptive Explanations

In evolutionary biology, strategy is used to refer to behavioral mechanisms that lead to particular courses of behavior in particular functional contexts, such as foraging or reproduction.

Biologists often use the term **strategy** when describing certain aspects of the behavior of animals. For example, folivory is characterized as a foraging strategy, and monogamy is described as a mating strategy. When evolutionary biologists use these terms, they mean something very different than we normally mean when we use the term *strategy* to describe, say, a general's military maneuvers or a baseball manager's tactics. In common usage, *strategy* implies a conscious plan and a goal-directed course of action. Evolutionary biologists do not think other animals consciously decide to defend their territories, wean their offspring at a particular age, monitor their ingestion of secondary plant compounds, and so on. Instead, the term is used to refer to a set of behaviors occurring in a specific functional context, such as mating, parenting, or foraging. Strategies are the product of natural selection acting on individuals to shape the motivations, reactions, preferences, capacities, and choices that influence behavior. Predispositions that produce behaviors that led to greater reproductive success in ancestral populations have been favored by natural selection and represent adaptations.

For example, howler monkeys, who are folivorous, behave as though they know that some leaves are good for them in small quantities but harmful in large quantities and that young leaves contain more protein than older ones. Moreover, in different habitats they adjust the mix of plants in their diet in what appears to be a deliberate attempt to balance nutrients and to minimize toxins. But biologists doubt very much that howlers have any conscious knowledge of the nutritional content of the foods they eat or the components of an optimal diet. Instead, the underlying mechanisms that influence their decisions about what to eat, how much to eat, and what to avoid have been shaped by natural selection over many generations, producing the foraging behavior that we observe in nature.

Cost *and* benefit *refer to the effect of particular behavioral strategies on reproductive success.*

Different behaviors have different impacts on an animal's genetic fitness. Behaviors are said to be beneficial if they increase the genetic fitness of individuals, and costly if they reduce the genetic fitness of individuals. Thus, we argued in Chapter 6 that ranging behavior reflects a tradeoff between the benefits of exclusive access to a particular area and the costs of territorial defense. Reproductive success is the ultimate currency in which these tradeoffs are measured. Although this is a simple concept in principle, it is often very difficult to measure the costs and benefits associated with individual behavioral acts, particularly in long-lived animals like primates. Instead, researchers rely on indirect measures, such as foraging efficiency (measured as the quantity of nutrients obtained per unit time), and assume that, all other things being equal, behavioral strategies that increase foraging efficiency also enhance genetic fitness and will be favored by natural selection. We will encounter many other examples of this type of reasoning in the chapters that follow.

The Evolution of Reproductive Strategies

Primate females always provide extensive care for their young, while males do so in only a few species.

The amount of parental care varies greatly within the animal kingdom. In most species, parents do little for their offspring. For example, most frogs lay their eggs and never see their offspring again. In such species, the nutrients that females leave in the egg are the only form of parental care. In contrast, primates like almost all birds and mammals and like some invertebrates and fish, provide much more than

Figure 7.1

In all primate species, females nurse their young. In baboons, and many other species, females provide most of the direct care that infants receive.

just the resources included in gametes. At least one parent, and sometimes both, shelter their young from the elements, protect them from predators, and provide them with food.

The *relative* amount of parental care provided by mothers and fathers also varies within the animal kingdom. In species without parental care, females produce large, nutrient-rich gametes, while males produce small gametes and supply only genes. Among species with parental care, however, all possible arrangements occur. Among primates, mothers always nurse their offspring and often provide extensive care (Figure 7.1). The behavior of fathers is much more variable. In most species, fathers give nothing to their offspring other than the genes contained in their sperm. However, in a minority of species, males are devoted fathers. In other taxa, patterns differ. For example, in most bird species, males and females form monogamous pairs and raise their young together (Figure 7.2), and there are even some bird species in which only males care for their chicks.

The amount of time, energy, and resources that the males and females of a species invest in their offspring has profound consequences for the evolution of virtually every aspect of that species' social behavior. As we shall see, species in which both parents care for their offspring are subject to radically different selection pressures than those species in which only one parent invests. We will also discover that species with equal parental investment differ from species with unequal parental care in morphology, behavior, and social organization. Thus, it is important to understand why the amounts and patterns of parental investment differ among species.

Figure 7.2

In most species of birds, the male and female form a pair-bond and jointly raise their young. Here a bald eagle carries food to its hungry brood.

Males do not care for their offspring when (1) they can easily use their resources to acquire many additional matings or (2) when caring for their offspring would not appreciably increase the offspring's fitness.

At first glance, it seems odd that most primate males fail to provide any care for their offspring. Surely, if the males helped their mates, they would increase the chances of their offspring surviving to adulthood. Therefore, we might expect paternal care to be favored by natural selection (Figure 7.3).

If time, energy, and other resources were unlimited, this reasoning would be correct. But in real life, time, energy, and material resources are always in short supply. The effort that an individual devotes to caring for offspring is effort that cannot be spent competing for prospective mates. Natural selection will favor individuals that allocate effort among these competing demands so as to maximize the number of surviving offspring that they produce.

To understand the evolution of asymmetries in parental investment, we must identify the conditions under which one sex can profitably reduce its parental effort at the expense of its partner. Consider a species in which most males help their mates feed and care for their offspring. There are a few males, however, with a heritable tendency to invest less in their offspring. We will refer to these two types as "investing" and "noninvesting" fathers. Since time, energy, and resources are always limited, males that devote more effort to caring for offspring must allocate less effort to competing for access to females. On average, the offspring of noninvesting males will receive less care than will the offspring of investing males, making them less likely to survive and to reproduce successfully when they mature. On the other hand, since noninvesting males are not kept busy caring for their offspring, they will on average acquire more mates than will investing males. Mutations favoring the tendency to provide less parental care will increase in frequency when the benefits to males, measured in terms of the increase in fitness gained from additional matings, outweigh the costs to males, measured as the decrease in offspring fitness due to a reduction in paternal care.

Figure 7.3

In most nonmonogamous species, males have relatively little contact with infants. Although males, like this bonnet macaque, are sometimes quite tolerant of infants, they rarely carry, groom, feed, or play with infants. (Photograph by Kathy West.)

This reasoning suggests that unequal parental investment will be favored when one or both of the following are true:

1. Acquiring additional mates is relatively easy, so considerable gains are achieved by allocating additional effort to attracting mates.
2. The fitness of offspring raised by only one parent is high, so the payoff for additional parental investment is relatively low.

Thus, the key factors are the costs of finding additional mates and the benefits associated with incremental increases in the amount of care that offspring receive. When females are widely separated, for example, it may be difficult for males to locate them. In these cases, males may profit more from helping their current mates and investing in their offspring than from searching for additional mates. If females are capable of rearing their offspring alone and need little help from males, then investing males may be at a reproductive disadvantage compared with males that abandon females after mating and devote their efforts to finding eligible females.

The mammalian reproductive system commits primate females to invest in their offspring.

So far, there is nothing in our reasoning that says that if only one sex invests, it should always be females. Why aren't there primate species in which males do all the work, and females compete with each other for access to males? This is not simply a theoretical possibility—female sea horses deposit their eggs in their mate's brood pouch, and then swim away and look for a new mate (Figure 7.4). There are whole families of fish in which male parental care is more common than female parental care; and in several species of birds, including rheas, spotted sandpipers, and jacanas, females abandon their clutches after the eggs are laid, leaving their mates to feed and protect the young.

In primates and other mammals, the bias toward female investment arises because females lactate and males do not. Among primates, pregnancy and lactation commit mammalian females to invest in their young, and limit the benefits of male investment in offspring. Since offspring depend on their mothers for nourishment during pregnancy and after birth, females cannot abandon their young without greatly reducing the offspring's chances of surviving. On the other hand, males are never capable of rearing their offspring without help from females. Therefore,

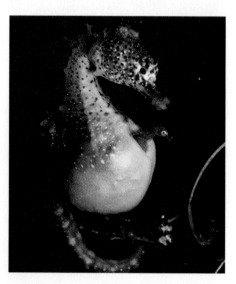

Figure 7.4

In sea horses, males carry fertilized eggs in a special pouch and provide care for their young as they grow.

when only one sex invests in offspring, it is invariably females that do so. Sometimes males can help females by defending territories or by carrying infants so that the mother can feed more efficiently, as siamangs and owl monkeys do. In most cases, however, these benefits are relatively insignificant, and selection favors males that allocate more time and energy to mating than to caring for their offspring.

You may be wondering why selection has not produced males that are able to lactate. As we noted in Chapter 3, most biologists believe the developmental changes that would enable males to lactate would also make them sterile. This is an example of a developmental constraint.

Reproductive Strategies of Primate Females

So far, we have established that female primates often invest more heavily in their offspring than do males. The next step is to consider the reproductive strategies of primate females in more detail.

Female primates invest heavily in each of their offspring.

Pregnancy and lactation are time-consuming and energetically expensive activities for female primates. The duration of pregnancy plus lactation ranges from 102 days in the tiny mouse lemur to 1839 days in the hefty gorilla. In primates, as in most other animals, larger animals tend to have longer pregnancies than do smaller animals, but primates have considerably longer pregnancies than we would expect based on their body sizes alone. The extended duration of pregnancy in primates is related to the fact that brain tissue develops very slowly. Primates have very large brains in relation to their body sizes, so additional time is needed for fetal brain growth and development during pregnancy. Primates also have an extended period of dependence after birth, further increasing the amount of care mothers must provide. Throughout this period, mothers must meet not only their own nutritional requirements but also those of their growing infants. In some species, offspring may weigh as much as 30% of their mother's body weight at the time of weaning.

The energy costs of pregnancy and lactation impose important constraints on female reproductive behavior. Since it takes so much time and energy to produce an infant, each female can rear a relatively small number of surviving infants during her lifetime (Figure 7.5). For example, a female toque macaque who gives birth for the first time when she is five years old and survives to the age of 20 would produce 15 infants if she gave birth annually and all of her infants survived. This number undoubtedly represents an upper limit; in the wild a substantial fraction of all toque macaque infants die before they reach reproductive age, intervals between successive live births often last two years, and some females die before they reach old age. Thus, most toque macaque females will produce a relatively small number of surviving infants over the course of a lifetime, and each infant represents a substantial proportion of a female's lifetime fitness. Therefore, we would expect mothers to be strongly committed to the welfare of their young.

A female's reproductive success depends largely on her ability to obtain enough resources to support herself and her offspring.

In most species of primates, including humans, females must achieve a minimum nutritional level in order to ovulate and to conceive. This means that females may

Figure 7.5

A female bonnet macaque produced twins, but only one survived beyond infancy. Twins are common among marmosets and tamarins, but are otherwise uncommon among monkeys and apes. (Photograph by Kathy West.)

sometimes be unable to conceive because their nutritional status is poor, they may have long periods of infertility while they recover from the rigors of their last pregnancy, and they may not always be able to nourish themselves or their newborns adequately.

For animals living in the wild, without take-out pizza or 24-hour grocery stores, getting enough to eat each day is often a serious dilemma. There is considerable evidence that female reproductive success is limited by the availability of resources within their local habitat. For example, there are a number of sites in Japan where free-ranging monkeys' natural diets have been supplemented with wheat, sweet potatoes, rice, and other foods by humans for many years (Figure 7.6). When their

Figure 7.6

At a number of locations in Japan, indigenous monkeys are fed regularly. The size of these artificially fed groups has risen rapidly, indicating that population growth is limited by the availability of resources.

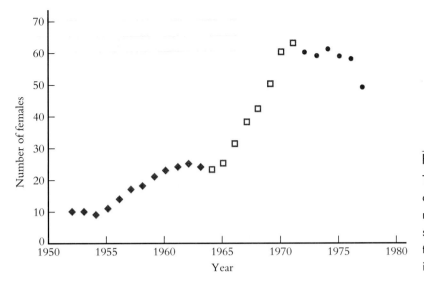

Figure 7.7

The size of the Koshima troop of Japanese monkeys grew rapidly when they were provisioned intensively (□), and then dropped when provisioning was restricted (•).

diet is supplemented in this way, females grow faster, mature earlier, survive longer, and produce infants at shorter intervals. Figure 7.7 shows changes in the size of one such group in Japan before, during, and after a period of intense supplementation. Similar effects have been documented under natural conditions. For example, a long period of environmental deterioration led to a drastic decline in the size and number of baboon groups in Amboseli National Park, Kenya. In 1963 and 1964, biologists Jeanne and Stuart Altmann of the University of Chicago counted 2500 baboons in 51 groups, but in 1979 there were only 123 baboons in five groups. During this period, female birth rates declined and mortality among immature animals increased.

Female Dominance Hierarchies

Females compete with each other for access to resources. Such competition often results in stable and predictable dominance relationships among them.

In many primate species, females compete with one another over access to resources. The frequency and intensity of such competition varies considerably both within and between species. For example, female mountain gorillas rarely confront one another directly over access to food, but female rhesus macaques frequently do (Figure 7.8). In some species, particularly the terrestrial Old World monkeys, the same females win all types of contests: over food, grooming partners, resting spots, and fights. In fact, many confrontations among females do not involve access to resources and have no obvious precipitating cause. Instead, females seem to be concerned with establishing and maintaining their ability to dominate, or intimidate, other females. In some species, **dominance** relationships are very stable over time. This means that if a female defeats another female on one day, she is likely to be able to defeat her again the next day, the next week, and sometimes even years later. Females can then be assigned ranks in a **dominance hierarchy** (Box 7.1).

There are some groups in which high-ranking females invariably defeat lower-ranking ones. For example, in one group of savanna baboons in Kenya, Barbara Smuts of the University of Michigan observed nearly 3000 dominance

Figure 7.8

Female primates sometimes fight over access to food and other resources.

interactions among females over a 16-month period. In less than 1% of these interactions, a lower-ranking female defeated a higher-ranking one. Highly predictable dominance relationships generally characterize macaques, baboons, vervets, some mangabeys, and cebus monkeys. In some other species, such as langurs and some species of squirrel monkeys, dominance hierarchies are less clearly defined, and the outcome of aggressive interactions is much less certain from day to day.

Lynne Isbell of the University of California, Davis, noted that clear-cut dominance relationships among females characterize primate species where competition over access to food within groups plays an important role. In some of these species it has been shown that high-ranking females are able to obtain greater access to preferred, limited, or clumped foods, or are able to feed more efficiently than can low-ranking females. Among brown capuchins, for example, dominance rank was directly related to the intake of foods that the monkeys most frequently fought over; for foods that created little competition, dominance rank was uncorrelated with food intake. Similarly, Patricia Whitten of Emory University observed that when female vervet monkeys fed on desirable, clumped foods like acacia flowers, high-ranking females had higher feeding rates than did low-ranking ones.

Females with high dominance rank tend to have higher reproductive success than do lower-ranked females.

Differences in dominance rank among females are sometimes reflected in differences in various components of their reproductive performance. In marmoset and tamarin groups, for example, there are sometimes several mature females, but only the dominant female reproduces. Among both brown capuchins and vervets, high-ranking females have greater success in producing surviving offspring than do low-ranking ones. Anne Pusey and her colleagues at the University of Minnesota have shown that at Gombe, infants of high-ranking female chimpanzees are more likely to survive to the age of weaning than the infants of low-ranking females are (Figure 7.10a). Moreover, the daughters of high-ranking females grow faster and mature at earlier ages than the daughters of low-ranking females do (Figure 7.10b). Other studies have documented a positive correlation between female dominance rank and various parameters of reproductive performance, such as age at first birth, length of interbirth intervals, number of births per year, and

BOX 7.1

Dominance Hierarchies

When dominance interactions have predictable outcomes, we can assign dominance rankings to individuals. Consider the four hypothetical females which we shall call Blue, Turquoise, Green, and Lavender in Figure 7.9a. Blue always beats Turquoise, Green, and Lavender. Turquoise never beats Blue but always beats Green and Lavender. Green never beats Blue or Turquoise but always beats Lavender. Poor Lavender never beats anybody. We can summarize the outcome of these confrontations between pairs of females in a **dominance matrix** like the one in Figure 7.9b, and we can use the data to assign numerical ranks to the females. In this case, Blue ranks first, Turquoise second, Green third, and Lavender fourth. When females can defeat all the females ranked below them and none of the females ranked above them, dominance relationships are said to be **transitive**.

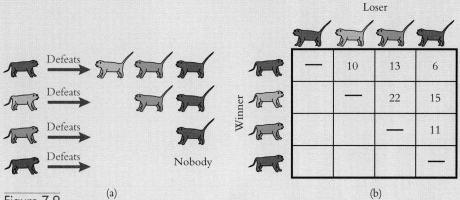

Figure 7.9

(a) Suppose that four hypothetical females, named Blue, Turquoise, Green, and Lavender, have the following transitive dominance relationships: Blue defeats the other three in dominance contests. Turquoise cannot defeat Blue, but can defeat Green and Lavender. Green loses to Blue and Turquoise, but is able to defeat Lavender. Lavender can't defeat anyone. (b) The results of data like those in part a are often tabulated in a dominance matrix, with the winners listed down the left side and the losers across the top. The value in each cell of the matrix represents the number of times one female defeated the other. Here, Blue defeated Turquoise 10 times, and Green defeated Lavender 11 times. There are no entries below the diagonal because females were never defeated by lower-ranking females.

infant survivorship. The importance of the correlation between dominance rank and reproductive success is sometimes questioned because there are some populations in which this pattern is not seen. For example, high-ranking female baboons at Gombe suffer higher rates of miscarriages and reproductive failure than lower-ranking females, resulting in lower lifetime reproductive success for the highest-ranking females. Moreover, there are a number of species in which dominance relationships among females are unstable or weakly expressed. However, correlations between female dominance rank and reproductive success are positive much more often than they are negative.

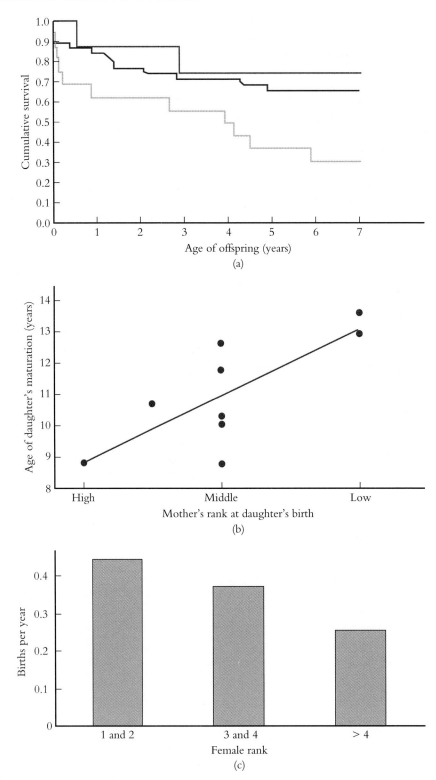

Figure 7.10

In many cases, there is a positive correlation between female dominance rank and reproductive performance. (a) Among chimpanzees at Gombe, offspring of high- (red line) and middle-ranking (purple line) females are more likely to survive to weaning age than are the offspring of low-ranking females (yellow line). (b) At Gombe, daughters of high-ranking females also mature at earlier ages than the daughters of lower-ranking females. (c) Among gelada baboons in Ethiopia, high-ranking females (low rank numbers) have more infants per year than lower-ranking females do.

Reproductive Tradeoffs

Females must make a tradeoff between the number of offspring they produce and the quality of care that they provide.

Just as both males and females must allocate limited effort to parental investment and mating, females must apportion resources among their offspring. All other

Figure 7.11

A female chimpanzee sits beside her youngest infant in the Gombe Stream National Park in Tanzania.

things being equal, natural selection will favor individuals that are able to convert effort into offspring most efficiently. Since mothers have a finite amount of effort to devote to offspring, they cannot maximize both the quality and the quantity of the offspring they produce. If a mother invests great effort in one infant, she must reduce her investment in others. If a mother produces many offspring, she will be unable to invest very much in any of them.

In nature, maternal behavior reflects this tradeoff when a mother modifies her investment in relation to an offspring's needs. Initially, infants spend virtually all of their time in contact with their mothers. The very young infant is entirely dependent on its mother for food and transportation and is unable to anticipate or to cope with environmental hazards. At this stage, mothers actively maintain close contact with their infants (Figure 7.11): retrieving them when they stray too far, signaling to them when they are about to depart, and scooping them up when danger arises.

As infants grow older, however, they become progressively more independent and more competent. They venture away from their mothers to play with other infants and to explore their surroundings. They begin to sample food plants, sometimes mooching scraps of their mother's food. They become aware of the dangers around them, attending to alarm calls given by other group members and reacting to disturbances within the group. Mothers use a variety of tactics to actively encourage their infants to become more independent. Initially, they may subtly resist their infants' attempts to suckle. At the same time, mothers may begin to encourage their infants to travel independently by moving off before the infant has climbed onto their backs, shrugging their infants off their backs while moving, or by simply failing to retrieve their infants when ready to depart. Although infants may protest these developments with tantrums or whimpers, and persistently attempt to ride on their mothers and to nurse, they eventually become independent of their mothers. Nursing is gradually limited to brief and widely spaced bouts that may provide the infant with more psychological comfort than physical nourishment. At this stage, infants are carried only when they are ill, injured, or in great danger.

The changes in maternal behavior reflect the shifting balance between the requirements of the growing infant and the energy costs to the mother of catering to her infant's needs. As infants grow older, they become heavier to carry and require more food, which imposes substantial burdens on mothers (Figure 7.12). As infants become more capable of feeding themselves and of traveling independently, mothers can gradually limit investment in their older infants without jeopardizing the infants' welfare. This enables mothers to conserve resources that can be allocated to subsequent infants. Moreover, since lactation inhibits ovulation in many primate species, mothers must wean their present infant before they are able

Figure 7.12

As infants become older, they become a bigger burden on their mothers. Here, a female baboon leaps across a ditch carrying her infant on her back. Later, the mother will encourage her infant to leap across the ditch on her own.

to conceive another. Phyllis Lee of Cambridge University and her colleagues have discovered that, although the duration of lactation varies within and among species, generally primate infants are weaned when they reach about four times their birth weight (Figures 7.13 and 7.14).

Male reproductive behavior in primates is strongly affected by the distribution of females.

Earlier we saw that male reproductive strategies are influenced by two factors: the costs of obtaining additional matings, and the benefits to infants of receiving care from both parents rather than from just one. We have established that primate females are generally prepared to raise their offspring with relatively little assistance from males. Thus, male mating strategies are likely to be determined largely by the costs of obtaining additional matings, and these costs will in turn depend mainly on the distribution of females. If females are sparsely dispersed in the environment, then it may be difficult and time-consuming for males to locate prospective mates. In such cases, males may benefit more from establishing a pair-bond with a single female and investing in their offspring than from seeking additional matings with

Figure 7.13

Primate infants are generally weaned when they reach about four times their birth weight, although the duration of lactation and infant growth rates vary considerably. The threshold weaning weight probably reflects constraints on the mother's ability to meet the energetic demands of growing infants. Each point represents the average value of birth weight and weaning weight for a given primate genus. ○ represents a human foraging group, the !Kung.

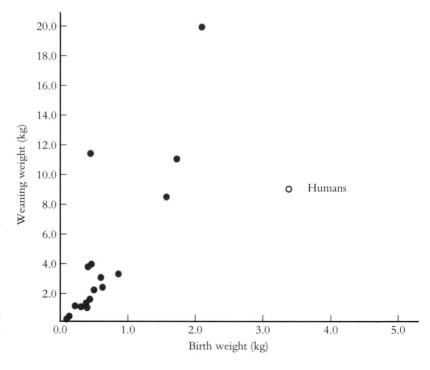

Figure 7.14

A large infant squirrel monkey is being carried by its mother. By the time they are weaned, infants in some species may weigh up to a third of their mother's weight. (Photograph by Sue Boinski.)

a number of females. If females are clumped together in groups, however, then the cost of finding mates is likely to be low. Under those circumstances, males are expected to focus on attempting to obtain additional matings, rather than investing in their offspring.

This line of reasoning raises two related questions. What factors determine whether females aggregate in social groups or remain separated from one another? How are male mating and parenting strategies influenced by the nature of female social arrangements? In the remainder of this chapter, we will explore theoretical and empirical evidence that bears on both of these questions.

Before considering these two questions, however, it is helpful to ask the larger question: why is understanding the origins and nature of sociality among nonhuman primates relevant to our understanding of human evolution? The evolution of sociality is important for understanding human evolution because sociality is a trait that contemporary humans share with other primates. Although the scale of contemporary human societies ranges from nuclear family units to industrialized nations, regular association among familiar individuals is a common feature of all human communities. Moreover, the fossil evidence suggests that humans and their hominid ancestors have always lived in social groups, although we do not know how these groups were structured. If we understand how ecological and evolutionary forces have shaped the patterns of group formation in other primate species, we may have better insights about the forces that shaped the evolution of sociality among ancestral hominids. With this information, we will be able to build reasonable models about the kinds of lives that early humans lived.

Sociality is an intriguing phenomenon because it creates opportunities for complex social interactions. For solitary individuals, social interactions are mainly limited to brief encounters with neighbors, potential mates, and offspring. For those that live in social groups, however, much more extensive opportunities for interaction arise. Group members encounter each other regularly and may interact when they feed, rest, mate, care for offspring, confront strangers, defend their territories, or attempt to avoid predators during the course of the day. The nature of their interactions in these contexts may be friendly, cooperative, competitive, or hostile. These contacts have a significant impact on the daily lives of individuals and on their reproductive strategies, and may ultimately affect their ability to reproduce successfully.

Primate Sociality

In the hours before sunrise, a pregnant baboon goes into labor. The delivery lasts only a few minutes, and the infant emerges bloody and wet. The mother licks her infant

Figure 7.15

A female baboon bends low to greet a newborn infant.

clean and cradles it to her chest. Soon after dawn breaks, the group descends from its sleeping trees, and other group members approach the weary mother and her newborn. Some glance at the mother and infant but make no further contact, while others give a low grunt, sit near the mother, and stare intently at the newborn. One female bends low in front of the infant, raising her eyebrows in a distinctive greeting reserved for infants (Figure 7.15). Others approach, grunt to the mother, reach out and pull the infant's legs away from the mother's stomach, and stare at its exposed genitals. The mother's older daughter, weaned only months before her mother became pregnant with this infant, grooms her mother assiduously, gradually working her way closer and closer to the infant, and cautiously touches her new sibling. By the end of the day, the entire group will be aware of the infant's presence, and the infant will be firmly embedded in the group's complex social network.

Most monkeys and apes live in social groups, although sociality is an unusual feature among mammalian species.

The infant baboon's experience reflects the fact that primates are, with few exceptions, gregarious creatures. Most monkeys and apes live in stable, social groups in which all members recognize one another as individuals (Figure 7.16). Group life is by no means unique to primates—killer whales, wolves, wild dogs, mongooses, and elephants are some of the many animals that live in cohesive social groups (Figure 7.17). However, group life is not common among mammals. In many mammalian species, adult males and females occupy separate home ranges or territories. They meet for brief periods to court and mate when females are sexually receptive, and then they part to resume their independent lives. Females raise their young without assistance from males.

Figure 7.16

Most primates live in stable and cohesive social groups. Baboons often live in groups of 50 to 75 individuals. All members of the group recognize each other as individuals.

Figure 7.17

Sociality is not restricted to primates. For example, the social organization of elephants is highly structured. The basic social unit is the family group, which consists of one or more related adult females and their offspring. Certain family groups are frequently found together and form bond groups. Members of certain bond groups tend to associate with one another and are clustered into clans.

What factors determine whether primate females aggregate in social groups or remain solitary? In order to answer this question, we need to consider the selective pressures that influence the evolution of sociality. Then, we will consider how these factors may influence the tendency to aggregate among primates and shape the form of primate social groups.

Why Do Primates Live in Groups?

This question has provoked a lively debate in primatology during the last 15 years, and there is no consensus about an answer. The fact that sociality is the exception, rather than the rule, among mammalian species is generally thought to be due to the fact that sociality is costly. Animals who live in groups face greater competition for resources and are more vulnerable to infectious diseases. At the same time, the fact that sociality has evolved in a number of primate species suggests that there must also be compensating advantages or benefits associated with it. In this section, we will briefly consider some of the advantages and disadvantages that confront primates living in social groups.

Most of the discussion among researchers has focused on the relative importance of two benefits of sociality: enhanced access to resources and reduced vulnerability to predation. Richard Wrangham, a primatologist at Harvard University, believes that sociality is beneficial because it enables individuals, particularly females, to defend access to food resources collectively. In contrast, Duke University primatologist Carel van Schaik argues that sociality creates substantial resource competition among group members. He suggests instead that primates form groups in order to reduce their vulnerability to predators. We will examine the logic and evidence related to each of these models.

According to Wrangham's resource-defense model, primates live in social groups because it allows them to better defend access to limited resources such as food.

The logic of Wrangham's resource-defense model is based on the simple idea that there is power in numbers: animals that band together to defend resources can exclude solitary individuals from those resources. Joint defense of food resources is most profitable when (1) food items are relatively valuable, (2) food sources are clumped in space and time, and (3) there is enough food within defended patches to meet the needs of several individuals. Wrangham notes that fruit

Figure 7.18

Female baboons feed together on *Salvadora persica* fruit in Amboseli National Park.

often meets these three criteria. Fruit is valuable because it is high in carbohydrates that can be quickly converted to energy to fuel the body's metabolic needs. Moreover, fruit grows in discrete patches on trees, bushes, and vines, and many fruiting species produce large enough crops to feed several individuals at the same time (Figure 7.18). Wrangham points out that many primates feed preferentially on fruit, and suggests that this observation may explain why so many primates live in social groups.

The fact that members of large primate groups usually defeat members of smaller groups in confrontations over access to food supports the resource-defense model.

The resource-defense model is consistent with the fact that primates sometimes compete with members of other groups. In some nonterritorial species, hostile intergroup interactions occur when a desirable food is located in an area where two groups' ranges overlap. The resource-defense model is further supported by the fact that larger groups are generally more successful in intergroup encounters than smaller groups are. For example, in capuchin monkeys larger groups win a greater fraction of encounters than smaller ones do (Figure 7.19). Even when groups of monkeys do not meet face to face, smaller groups may leave feeding trees when they hear larger groups approaching. In a long-term study of several groups of vervet monkeys occupying adjacent territories in Amboseli National Park in Kenya, Dorothy Cheney and Robert Seyfarth of the University of Pennsylvania found that small groups suffered more incursions into their ranges and

Figure 7.19

One possible benefit of living in groups is that group members can collectively defend access to resources. For example, large groups of capuchin monkeys are more successful in confrontations with neighboring groups of conspecifics than small groups are.

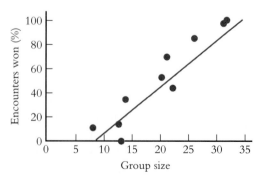

Figure 7.20

A brown capuchin monkey feeds in Peru's Manu National Park. (Photograph courtesy of Charles Janson.)

defended their ranges more aggressively than did larger groups, which probably means that members of smaller groups incurred greater costs in territorial defense than did members of larger groups. The larger groups also had bigger and more desirable territories than smaller groups did. These differences in territorial defense and territory quality were reflected in the females' life histories, as females in the small groups had lower survivorship than females in the large groups.

There are several problems with the resource-defense model: (1) it does not take into account tradeoffs between intragroup and intergroup competition over access to resources; (2) it does not explain the social organization of certain species; and (3) there is no direct evidence that living in larger groups enhances females' fitness.

The resource-defense model emphasizes the benefits that females in large groups gain in intergroup competition over access to resources. But sociality may also impose costs on group members. Carel van Schaik has pointed out that as groups grow, the extent of intragroup competition over access to food is likely to increase. While competition between groups may favor females that live in large groups, competition within groups may favor females who live in smaller ones.

The balance between intergroup and intragroup competition has been assessed carefully in one detailed study of feeding ecology among brown capuchins (Figure 7.20) conducted by Charles Janson of the State University of New York at Stony Brook. He measured the food intake of the members of two groups of brown capuchins in Peru's Manu National Park. In this population, monkeys competed actively and often with other group members over access to some types of foods. The largest groups had the highest rates of aggression, and an individual's access to food was strongly influenced by the rate of aggression it received from other group members (Figure 7.21). Energy intake among individuals within the group varied by 37%. By contrast, intergroup competition was limited to relatively brief and infrequent encounters over access to fruiting fig trees. While energy intake varied considerably within the group, it varied by only 3% between the groups that were most and least successful in intergroup competition. Studies of other

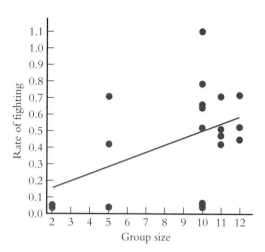

Figure 7.21

One possible cost of living in social groups is increased competition among group members over access to resources. For example, the rate of aggression increases as group size increases among capuchin monkeys in Peru's Manu National Park.

species and groups may reveal different balances between intragroup and intergroup competition.

Not surprisingly, Wrangham's simple model does not perfectly predict all primate social groupings. Not all frugivorous primates live in groups, and some species that form the most cohesive groups subsist mainly on leaves, not fruit. For example, although gibbons, chimpanzees, and orangutans all rely heavily on fruit, gibbons are monogamous, individual female chimpanzees maintain separate home ranges, and orangutans are largely solitary (Figure 7.22). Black-and-white colobus, on the other hand, live in cohesive social groups but are mainly folivorous. These exceptions suggest that the relationship between resource defense, diet, and social organization is quite complicated. A more specific shortcoming of the resource-defense model is that there is little evidence that membership in social groups increases female reproductive success by enhancing their access to resources.

 The predation model holds that the primary benefit of grouping is to reduce vulnerability to predators.

The predation model developed by Carel van Schaik offers the main alternative to the resource-defense model. As we noted in Chapter 6, a wide array of predators

Figure 7.22

Orangutans do not live in social groups. They are the only solitary anthropoid species.

hunt primates. Grouping may provide an effective means of reducing vulnerability to predation, and several lines of evidence suggest that grouping among primates is associated with overall predation risk. First, terrestrial species generally form larger groups than do arboreal species, perhaps because life on the ground is inherently riskier than life in the trees. Although we noted in Chapter 6 that there is no evidence that rates of predation are higher in terrestrial species, this may be due to the fact that large group size in terrestrial species provides an effective defense against predators. Second, several species that typically do not forage in groups (such as orangutans, chimpanzees, and spider monkeys) are large in size and apparently face little danger from predators. Third, juveniles, members of the age class most vulnerable to predation, suffer higher mortality in small groups than in large groups in habitats where predators are present. Finally, primates seem to adjust their behavior in response to the risk of predation. On islands without predators, macaques form smaller groups than more vulnerable conspecifics on the mainland do. In central Venezuela, capuchin monkeys in small groups scan for predators more often than monkeys in larger groups do (Figure 7.23). And some of the monkeys who regularly form mixed species groups in the Taï Forest (Chapter 6) do not associate at sites where predators are absent.

Like the resource-defense model, the predation model also has weaknesses.

There are several problems with the predation model. First, since acts of predation are rarely observed, it is difficult to establish whether predation is clearly linked to group size. Second, there is no consistent relationship among social organization, habitat use, and predation pressure. The rates of predation on species that form large multimale, multifemale groups are not consistently lower than rates of predation on species that live in smaller one-male groups, and terrestrial species do not seem more vulnerable to predators than arboreal species.

Both resource defense and predation may influence group sociality.

In summary, we have two plausible theories about the evolution of sociality among primates. With the evidence that is now available, it is impossible to draw

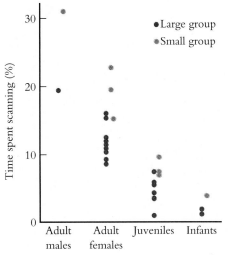

Figure 7.23

Another possible benefit of living in social groups is that grouping reduces vulnerability to predators. Among capuchin monkeys in Venezuela, the frequency of scanning for predators is related to group size. The members of the larger group, which consisted of 40 monkeys, scanned for predators less often than did members of the smaller group of eight individuals.

firm conclusions about the relative importance of resource defense and predation in the lives of contemporary primates. It is possible that more detailed empirical evidence will help resolve this question more clearly in specific cases. However, even if we find that grouping enhances defense of resources or reduces vulnerability to predation in contemporary populations, we cannot be sure that the same conditions existed in the past. This further complicates the problem of identifying the principle factor initially responsible for the evolution of sociality among primates. Finally, we should bear in mind the possibility that both resource defense and protection against predation may have jointly favored the evolution of sociality among primates. By grouping, primates may reduce the risk of predation and at the same time increase their ability to defend resources. Moreover, different species may be subject to different types of selective pressures.

How Big Should Groups Be?

It is difficult to predict exactly how big groups should be from theory, but comparative studies have established some patterns in the variation in group size among species.

In principle, we expect primates to form groups that minimize intragroup competition, lessen danger from predators, and maximize success in intergroup contests over resources. In practice, it is difficult to define the optimal group size since primate species vary in their nutritional requirements, and habitats vary in the abundance and distribution of resources and in the number of local predators. Nonetheless, comparative studies have established some broad connections between primate group size and ecological parameters:

- Terrestrial primates generally live in larger groups than arboreal species do (Figure 7.24). This may be related to the fact that terrestrial animals face greater danger from predators, and large group size provides a measure of protection against predators.
- Frugivores live in larger groups than folivores do. Perhaps this is because intergroup competition plays a more important role in the lives of frugivores than folivores, and large group size enhances intergroup competition.

Figure 7.24

A group of langur monkeys rests together. Arboreal species tend to live in smaller groups than terrestial species do.

- Diurnal species live in larger groups than nocturnal species do. This may be due to differences in predator avoidance strategies between diurnal and nocturnal species.

What Kinds of Groups Should Primates Form?

Both resource-defense and predation models predict that long-term, cooperative bonds among females should play an important role in primate groups.

Even if we agree that most primates are better off in groups than they are alone, and we have some idea of how large their groups should be, we still need to consider the kinds of groups that primates are expected to establish. In general, the form of competition over food resources is expected to influence the structure of primate groups and the nature of relationships among females.

Resource competition plays a central role in the evolution of primate groups because the reproductive performance of primate females is primarily influenced by their nutritional status. Females who are well nourished grow faster, mature earlier, and give birth at shorter intervals than do females who are poorly nourished. In contrast, the reproductive success of males is more directly influenced by their access to sexually receptive females than by their own nutritional status. If access to food resources limits female reproductive success, then the abundance and distribution of food is likely to shape the nature of female relationships and the composition of social groups. When food is limited, as it is under most circumstances, females are expected to compete over access to resources. The form of competition is expected to be influenced by how their food is distributed. Some species rely on foods that are typically clumped in space and time. Clumped food resources can be monopolized by one individual or a few, so animals that rely on clumped foods are likely to encounter competition over access to food resources. Regular competition over access to food may lead to the formation of strict, linear dominance hierarchies (see Box 7.1). Females may be able to enhance their access to clumped resources by forming alliances with other individuals in competitive encounters. These alliances are expected to be most effective when they are based on permanent relationships among familiar individuals. Kin are likely to be reliable allies, and so stable associations among related females will be favored. In the jargon of primatology, these are called **female-bonded kin groups**. But some species feed on foods that are widely dispersed and are not worth competing over directly. In these cases, there is little advantage gained from competing over access to food or from forming alliances with other females. In these species, dominance hierarchies are expected to be poorly defined. Social bonds between females are expected to be weak and unstable and may not be based upon maternal kinship relationships.

A key element of this argument is the premise that cooperation among kin is more effective and beneficial to individuals than cooperation among nonkin is. The theoretical rationale underlying this premise is that cooperation among genetic relatives is more likely to occur than cooperation among unrelated individuals because kin share some fraction of their genetic material. By helping a relative, a female thereby increases the probability that copies of genes that both animals share will be passed along to members of the next generation. So, under some circumstances females can indirectly increase their own fitness by helping their relatives,

Figure 7.25

Ring-tailed lemurs form female-bonded groups. Females live their entire lives in their natal group, the group in which they were born. Males leave their natal group at sexual maturity.

and this compensates them for the costs of participating in alliances. We will discuss the evolution of cooperation more fully in Chapter 8.

 Most primate groups are female philopatric and female-bonded.

Female **philopatry** (permanent residence in natal groups) characterizes most of the nonmonogamous, gregarious primate species. In most of the social prosimians in Madagascar, for example, females generally remain in their natal groups throughout their lives (Figure 7.25). The same pattern characterizes all of the Old World monkeys except hamadryas baboons, red colobus monkeys, and possibly black-and-white colobus monkeys. In the New World, several species of capuchins and squirrel monkeys show female philopatry, while howler monkeys, spider monkeys, and woolly spider monkeys do not. Among gorillas, chimpanzees, and bonobos, females do not remain in their natal groups. The remaining ape species are either solitary (orangutans) or monogamous (gibbons). Box 7.2 briefly outlines the range of social systems found among living primates.

In many cases, female philopatry is associated with female-bonded kin groups. For example, among baboons, macaques, and vervets, females form close bonds with one another (Figure 7.27). In these species, social ties among females are often stronger and more enduring than bonds among males or bonds among males and females. Again, however, there are exceptions to this pattern. Thus, among bonobos, females form close bonds with one another, although males are the philopatric sex.

The distribution of males is determined by the distribution of females.

Thus far we have focused on the behavioral strategies of female primates, largely ignoring males. This is because male options are significantly constrained by the distribution of females. Males go where their mating partners are.

The number of males in social groups is generally related to the number of females in them. When females do not associate with other females, they are usually associated with a single male. When two to four females are together, they are generally accompanied by one adult male, while groups with larger numbers of females contain multiple males. In cercopithecines (Old World monkeys such as macaques, baboons, vervets, and mangabeys), the number of adult males in social groups positively correlates with the number of females (Figure 7.28).

BOX 7.2

Forms of Social Groups among Primates

It should be obvious by now that while most primates are social, their groups vary in size and composition. One way to order this diversity is to classify primate social systems according to the mating and residence patterns of females and males. The major forms of primate social systems are described here and diagrammed in Figure 7.26:

Solitary Females maintain separate home ranges or territories and associate mainly with their dependent offspring. Males may establish their own territories or home ranges, or they may defend the ranges of several adult females from incursions by other males. With the exception of orangutans, all of the solitary primates are prosimians.

Monogamy One male and one female form a pair-bond and share a territory with their immature offspring. Monogamy is characteristic of gibbons, some of the small New World monkeys, and a few prosimian species.

Polyandry One female is paired with two or more males. They share a territory or home range with their offspring. Polyandry may occur among some of the marmosets and tamarins.

Polygyny, one-male Groups are composed of several adult females, a single resident male, and immature individuals. (We refer to them in the text as one-male, multifemale groups.) In species with this form of polygyny, males that do not reside in bisexual groups often form all-male groups. These all-male groups may mount vigorous assaults on resident males of bisexual groups in an attempt to oust residents. Dispersal patterns are variable in species that form one-male, multifemale groups. One-male, multifemale groups are common among howlers, langurs, and gelada baboons.

Polygyny, multimale, multifemale Groups are composed of several adult males, adult females, and immature animals. (We refer to them in the text as multimale, multifemale groups.) Most such groups are female-bonded. Dispersing males may move directly between bisexual groups, remain solitary for short periods of time, or join all-male groups. Multimale, multifemale groups are characteristic of macaques, baboons, vervets, squirrel monkeys, capuchin monkeys, and some colobines.

This taxonomy of social groups provides one way to classify primate social organization, but readers should recognize that these categories are idealized descriptions of residence and mating patterns. The reality is inevitably more complicated. For example, in some monogamous groups, partners sometimes copulate with nonresidents or neighbors, and in species that normally form multimale groups, there may be some groups that have only one adult male. Moreover, some species do not fit neatly within this classification system. Female chimpanzees and spider monkeys occupy individual home ranges, and spend much of their time alone or with their dependent offspring. In these species, a number of males jointly defend the home ranges of a number of females. Hamadryas and gelada baboons form one-male, multifemale units, but several one-male, multifemale units collectively belong to larger aggregations.

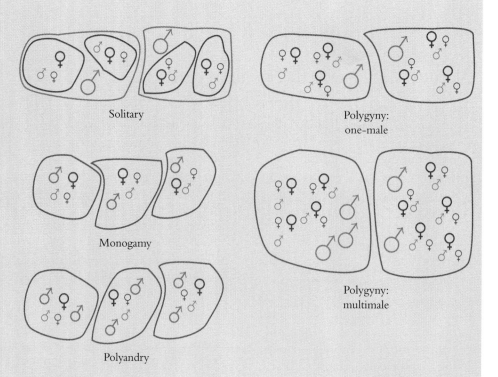

Solitary

Polygyny:
one-male

Monogamy

Polygyny:
multimale

Polyandry

Figure 7.26

The major types of social groups that primates form are diagrammed here. When males and females share their home ranges, their home ranges are drawn in lavender. When the ranges of the two sexes differ, male home ranges are drawn in blue and female home ranges are drawn in magenta. The size of the male and female symbols reflects the degree of sexual dimorphism among males and females.

Males avoid mating with genetically related females.

Male options are also influenced by the availability of unrelated females. The presence of unrelated females is important because in most species mating among close kin, or **inbreeding**, reduces the viability of offspring by increasing the probability that deleterious recessive alleles will be expressed. Surveys of captive primate populations have shown that the offspring of closely related parents are less likely

Figure 7.27

Two female baboons rest together. In many Old World monkey groups, females live in their natal groups throughout their lives and develop close bonds with other females.

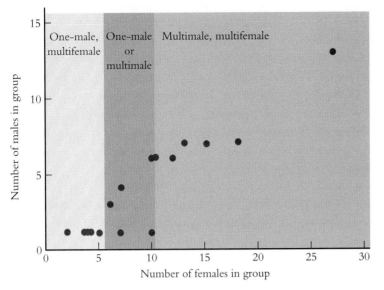

Figure 7.28

Among cercopithecine primate species, the number of females in social groups is positively correlated with the number of males in the same groups. Groups with only a few adult females generally contain only one adult male, while groups with many females contain several adult males.

to survive than the offspring of unrelated parents are, and close kin generally avoid mating with one another. In all primate species studied to date, members of one or both sexes typically disperse from their natal groups near the time of sexual maturity. In female-bonded groups, males are necessarily the dispersing sex.

Sexual Selection and Male Mating Strategies

Sexual selection leads to adaptations that allow males to compete more effectively with other males for access to females.

So far, we have seen that primate females invest heavily in each of their young and produce relatively few offspring over the course of their lives. Moreover, most primate females can raise their offspring without help from males. Female reproductive success is limited by access to food, not access to mates (Figure 7.29). Males

Figure 7.29

For female primates, access to resources has a bigger impact on reproduction than access to potential mates does.

(a)

(b)

Figure 7.30

Sexual selection can favor traits not favored by natural selection. (a) The peacock's tail hinders his ability to escape from predators, but it enhances his attractiveness to females. Female peahens are attracted to males with the most eyespots in their trains. (b) Male red deer use their antlers when they fight with other males. Red deer antlers are a good example of a trait that has been favored by sexual selection.

can potentially produce progeny from many females, and as a result males compete for access to females. Characteristics that increase male success in competition for mates will spread as a result of what Darwin called sexual selection.

It is important to understand the distinction between natural selection and sexual selection. Most kinds of natural selection favor phenotypes in both males and females that enhance their ability to survive and reproduce. Many of these traits are related to resource acquisition, predator avoidance, and offspring care. **Sexual selection** is a special category of natural selection that favors traits that increase success in competition for mates, and will be expressed most strongly in the sex whose access to members of the opposite sex is most limited. Sexual selection may favor traits like the peacock's tail, red deer's antlers, and the hamadryas baboon's mane, even if those traits reduce the ability of the animal to survive or acquire resources, outcomes not usually favored by natural selection (Figure 7.30).

Sexual selection is often much stronger than ordinary natural selection.

In mammalian males, sexual selection can have a greater effect on behavior and morphology than other forms of natural selection do. This is because male reproductive success usually varies much more than female reproductive success. Data from long-term studies of lions conducted by Craig Packer and Anne Pusey of the University of Minnesota demonstrate that the lifetime reproductive success of the most successful males is often much greater than that of even the most successful females (Figure 7.31). The same pattern is likely to hold for nonmonogamous primates. A primate male who succeeds in competition may sire scores of offspring, while a successful female might give birth to five or 10 offspring. Unsuccessful males and females will fail to reproduce at all. Since the strength of selection depends on how much variation in fitness there is among individuals, sexual selection acting on male primates can be much stronger than selective forces acting on female primates. (As an aside, in species like sea horses, in which males invest

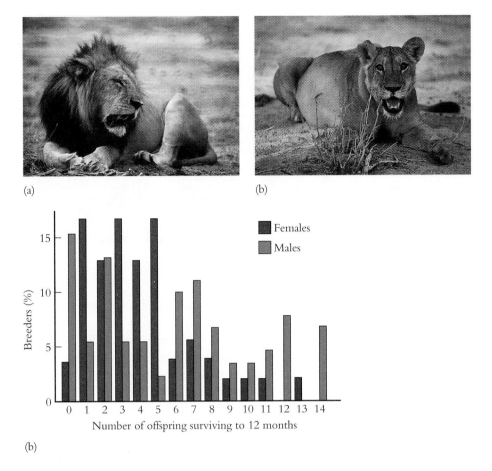

(a)

(b)

(b)

Figure 7.31

(a) The reproductive success of male lions is considerably more variable than (b) the reproductive success of female lions. (c) In Serengeti National Park and the Ngorongoro Crater of Tanzania, few female lions fail to produce any surviving cubs, but most females produce less than six surviving cubs over the course of their lives. Many males fail to produce any cubs, and a few males produce many cubs.

in offspring and females do not, the entire pattern is reversed. Sexual selection acts much more strongly on females than on males.)

> *There are two types of sexual selection: (1) intrasexual selection results from competition among males, and (2) intersexual selection results from female choice.*

Many students of animal behavior subdivide sexual selection into two categories: in species in which females can choose whom they mate with, selection favors traits that make males more attractive to females. This is **intersexual selection**. In species where females cannot choose their mates, access to females will be determined by competition among males. In such species, **intrasexual selection** favors traits that enhance success in male–male competition.

Intrasexual Selection in Primates

Sexual Dimorphism

> *Competition among males for access to females favors large body size, large canine teeth, and other weapons that enhance male competitive ability.*

For primates and most other mammals, intrasexual competition is most intense among males. Male–male competition can take different forms, but the simplest occurs when males drive other males away from females. Males that regularly win such fights have higher reproductive success than those that lose. Thus, intrasexual selection favors features such as large body size, horns, tusks, antlers, and large canine teeth that enable males to be effective fighters. For example, male gorillas

[handwritten margin notes: intersexual selection — making yourself as attractive to the opp. sex as possible]

Figure 7.32

Adult male baboons are nearly twice the size of adult females. The degree of sexual dimorphism in body size is most pronounced in species with the greatest competition among males over access to females.

compete fiercely over access to groups of females, and males weigh twice as much as females and have longer canine teeth.

As explained in Chapter 5, when the two sexes consistently differ in size or appearance, they are said to be sexually dimorphic (Figure 7.32). The body sizes of males and females represent compromises among many competing selective pressures. Larger animals are better fighters and are less vulnerable to predation, but they also need more food and take longer to mature. Intrasexual competition favors larger body size, larger teeth, and other traits that enhance fighting ability. Males compete over females, while females compete over resources but generally do not compete over mates. However, the effect of intrasexual competition among males is quantitatively greater than the effect of competition among females because the fitness payoff to a successful male is greater than it is to a successful female. Therefore sexual selection is much more intense than ordinary natural selection. As a result, intrasexual selection leads to the evolution of sexual dimorphism.

The fact that sexual dimorphism is greater in primate species forming one-male and multimale groups than in monogamous species indicates that intrasexual selection is the likely cause of sexual dimorphism in primates.

If sexual dimorphism among primates is the product of intrasexual competition among males over access to females, then we should expect to see the most pronounced sexual dimorphism in the species in which males compete most actively over access to females. One indirect way to assess the potential extent of competition among males is to consider the ratio of males to females in social groups. In general, male competition is expected to be most intense in social groups where males are most outnumbered by females. This prediction might seem paradoxical at first, since we might expect to have more competition when there are more males present. The key to resolving this paradox is to remember that there are approximately equal numbers of males and females at birth in most natural populations. In species that form one-male groups, there are many **bachelor males** (males who don't belong to social groups) who exert constant pressure on resident males. In species that form monogamous pair-bonds, each male is paired with a single female, reducing the intensity of competition among males over access to females.

Comparative analyses conducted by Paul Harvey of Oxford University and Tim Clutton-Brock of Cambridge University have demonstrated that the extent of sexual dimorphism in primates corresponds roughly to the form of social groups in which the males live (Figure 7.33). There is little difference in body weight or canine size between males and females in species that typically form monogamous groups, such as gibbons, titi monkeys, and marmosets. At the other extreme, the most pronounced dimorphism is found in species, such as gorillas and black-and-white colobus monkeys, that live in one-male, multifemale groups. And in species that form multimale, multifemale groups, the extent of sexual dimorphism is generally intermediate between these extremes. Thus, sexual dimorphism is most pronounced in the species in which the ratio of males to females living in bisexual groups is lowest (that is, with the highest relative number of females).

In multimale, multifemale groups, where females mate with several males during a given estrous period, sexual selection favors increased sperm production.

In most primate species, as with mammals in general, the female is receptive to mating mainly during the portion of her reproductive cycle when fertilization is possible. That period of time is called **estrus**. In primate species that live in multimale, multifemale groups, females can potentially mate with several different males during a single estrous period. In such species, sexual selection favors increased sperm production because males that deposit the largest volume of sperm in the female reproductive tract have the greatest chance of impregnating them. Competition in the quantity of sperm is likely to be relatively unimportant in monogamous species, because females mainly mate with their own partners. Since sperm production involves some cost to males, monogamous males may do better by guarding their partners when they are sexually receptive than by producing large quantities of sperm. Similarly, competition in sperm quantity probably does not play an important role in species that form one-male, multifemale groups. In these species, competition among males is over access to groups of females, which favors traits related to fighting ability. If resident males are able to exclude other males from associating with females in their groups, there may be little need to produce additional quantities of sperm.

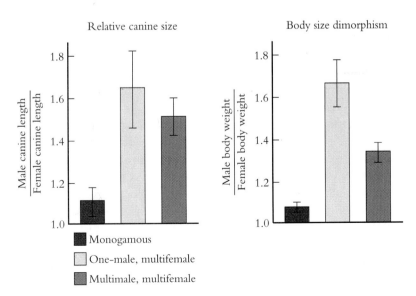

Figure 7.33

The degree of sexual dimorphism is a function of the ratio of males to females in social groups. Relative canine size (male canine length divided by female canine length) and body size dimorphism (male body weight divided by female body weight) are greater in species that form one-male, multifemale groups than in species that form multimale, multifemale groups or monogamous groups.

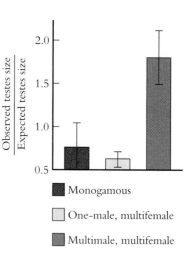

Figure 7.34

The average size of testes in species that typically form monogamous and one-male, multifemale groups is relatively smaller than the average size of testes in multimale, multifemale groups. Here observed testes' weight is divided by the expected testes' weight to produce relative testes' size. The expected testes' weight is derived from analyses that correct for the effects of body size.

Social organization is associated with testes' size, much as we would expect. Males with larger testes typically produce more sperm than males with smaller testes, and males that live in multimale groups have much larger testes in relation to their body size than males that live in either monogamous or in one-male, multifemale groups (Figure 7.34). Interestingly, in polyandrous groups where a single breeding female mates regularly with multiple males, males have relatively large testes.

Intersexual Selection in Primates

Intersexual selection favors three kinds of traits among males: (1) those that increase the fitness of their mates, (2) those that indicate good genes and thus increase the fitness of the offspring; and (3) nonadaptive traits that make males more conspicuous to females.

Darwin also realized that females might be attracted to males that exhibit particular traits. In this case, intersexual selection favors males that are pleasing to females rather than males that can defeat other males. Modern sexual selection theory now identifies three basic modes of female choice. First, if males with certain traits confer direct benefits on females, then selection will favor females that mate selectively with males possessing such traits. This will increase the frequency of such traits among males. Thus, if females mate preferentially with males that defend superior territories, protect offspring more vigorously, provision the young better, or confer more material benefits on females than other males do, these traits will be favored.

Second, sexual selection may also favor female preferences if females are able in some way to distinguish male genetic quality, and choose mates with desirable genes. These females are at an advantage because their offspring will carry certain genes that give them a greater chance of surviving and reproducing successfully. The gaudy peacock and the noisy sage grouse may be examples of such preferences. Males gather together in a small area called a lek, and each male displays his tail feathers to attract female attention (Figure 7.35). Marion Petrie of Oxford University has found that female peahens never mate with the first male they encounter on the lek, and show distinct preferences for the males with the most eyespots on their tail feathers. The offspring of the most richly ornamented males grow faster

Figure 7.35

Male sage grouse congregate on leks and display to attract females. (Photograph courtesy of Mark Chappell.)

and survive better than the offspring of other males. Thus, female mate preferences enhance offspring fitness. Again, it's important to remember that we need not imagine that peahens, or any other animals, assess male genetic quality *consciously*. Instead, selection may have favored preferences for specific traits that are reliably associated with genetic quality. By choosing males with the most eyespots, hens may be inadvertently and unconsciously choosing males with desirable genotypes.

Third, females may prefer males that exhibit distinctive traits (such as conspicuous coloration, exaggerated morphological characters, or elaborate courtship behaviors) for nonadaptive reasons. Even if such characters do not increase male fitness directly, they may be favored if females can discriminate and consistently select mates that most often display preferred traits. For example, male frogs call to attract mates, and the sensory system of female frogs seems to respond more strongly to some tones in male calls than to others. Researchers have discovered that male frogs whose calls are most readily heard attract the most mates.

Surprisingly, it turns out that female choice can also lead to the exaggeration of female preferences themselves. To see why, suppose there are two types of females: very picky females who prefer males with only the most exaggerated male trait—say, the deepest tones of a male frog's call or the longest tails of a male bird—and less picky females who will mate with a wider range of males. It is easy to see that the existence of more picky females will increase the reproductive success of males with the most exaggerated traits, and therefore increase the degree of exaggeration of the male trait in the population. However, the same process can also lead to the spread of more picky females. This is because the offspring of more picky females will tend to carry both the genes for pickiness and the genes for the exaggerated male trait, creating a positive correlation between the female preference and the exaggerated male trait (see Chapter 3). As a result of the correlated response to selection, the female choice that increases the reproductive success of exaggerated males will also increase the pickiness of females. If conditions are right, this can lead to a runaway process in which females become pickier and pickier about a male trait, and male traits become more and more exaggerated, even if they impair the male's ability to survive.

There are only a few morphological traits among primates that seem to have evolved because they help males to attract females.

The best candidates for traits that help primate males to attract females are found among mandrills and proboscis monkeys. Both the face and penis of the male mandrill are brilliantly colored. Their faces are striped red, white, and blue like an

Figure 7.36

Male mandrills have brightly colored faces. This trait may be favored by intersexual selection, but it is not known whether females actually prefer to mate with brightly colored males.

exotic mask (Figure 7.36), while the female mandrill's face is duller and less conspicuous. The proboscis monkey is named for its oddly shaped nose: males have quite long, pendulous noses, while females have much smaller noses that turn up at the end (Figure 7.37). There is pronounced sexual dimorphism in body size among both these species as well. Very little is known about mandrills, who live in dense primary rain forests, or about proboscis monkeys, who live in mangrove swamps, so we don't really know whether females actually prefer mandrill males with brightly colored faces or proboscis males with elongated noses. We also do not know which of the three modes of female choice discussed earlier may be

Figure 7.37

Male proboscis monkeys have long, pendulous noses, while females have much smaller upturned noses. This trait may be favored by intersexual selection, although we do not know whether females are attracted to males with long noses.

operating. However, since these traits do not seem to be directly related to male-male competition, and are more fully developed in males than in females, it seems possible these traits evolved in response to female preferences.

Sexual Selection and Primate Behavior

Like morphology, behavior in males is expected to be shaped by the extent of male-male competition over access to females and by female choice, which are in turn linked to social organization. We will now consider some elements of behavior in species that form three different kinds of social groups: monogamous pair-bonds; one-male, multifemale groups; and multimale, multifemale groups.

Monogamous Males

There is apparently little sexual selection in monogamous primate species because competition for females is minimal.

In monogamous groups, competition among males over access to females is greatly reduced. After they have established territories and found mates, males do not invest much energy in courtship or mating. Instead, they often invest heavily in their offspring and in guarding their mates. Male care of infants is most pronounced among the small New World monkey species, including owl monkeys, titi monkeys, marmosets, and tamarins (Figure 7.38), but it also occurs among siamangs. In these species, males carry their infants, sometimes for most of the day, play with them, share food with them, and protect them from predators. In titi monkeys, who have been studied in the forests of Peru by Patricia Wright of the State University of New York at Stony Brook, adult males carry their infants much more than other group members do. Extensive male investment in offspring may

Figure 7.38

In many monogamous species, like these golden lion tamarins, males participate actively in caring for infants and juveniles.

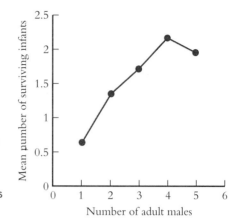

Figure 7.39

In tamarin groups, males clearly contribute to females' reproductive success. Tamarin groups that contain several adult males produce more surviving infants than groups that contain only one adult male.

enable females to recover quickly from the energy depletion of pregnancy and gestation, increasing the pair's joint reproductive capacity. Data compiled by Paul Garber of the University of Illinois at Urbana-Champaign show just how important males are in tamarin groups (Figure 7.39). Groups with multiple adult males reared more surviving infants than groups with only one male, while groups with multiple females produced slightly fewer surviving infants than groups with only one female resident. Monogamy is not associated with extensive male investment in every species. For example, male gibbons form monogamous pair-bonds but do not often carry their infants or share food with them.

Recently, primatologists began considering the potential importance of mate guarding in monogamous primate species. This was prompted in part by anecdotal reports that pair-bonded titi monkeys and gibbons sometimes copulate with individuals from neighboring groups. Moreover, numerous genetic studies of supposedly monogamous birds demonstrate that a surprisingly large fraction of the young are not sired by the female's mate. If females occasionally participate in extra-pair copulations, their partners may benefit from keeping close watch on them. Ryne Palombit of Rutgers University, who has studied the dynamics of pair bonding in gibbons, suspects that males do just that. Males are principally responsible for maintaining proximity to their female partners, and most males groom their mates more than they are groomed in return (Figure 7.40). It seems likely that males' attentiveness to females is a form of mate guarding and is favored by selection because it increases males' reproductive success.

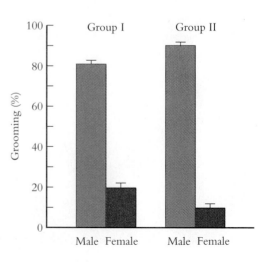

Figure 7.40

In white-handed gibbon groups, males groom their mates far more than they are groomed in return. Males' solicitous attention to females may be a form of mate guarding.

One-Male, Multifemale Groups

In species that form one-male groups, males compete actively to establish residence in groups of females.

Among primate species that form groups consisting of one male and several females, resident males face persistent pressure from nonresidents (bachelor males). In the highlands of Ethiopia, gelada baboons sometimes try to drive resident males out of social groups, leading to fierce confrontations that may last for several days. Robin Dunbar of Liverpool University estimates that half of the males involved in aggressive takeover attempts are seriously wounded. Among Hanuman langurs, males form all-male groups that collectively attempt to oust resident males from bisexual (coed) groups. Once they succeed in driving out the resident male, the members of the all-male group compete among themselves for sole access to the group of females. One consequence of this competition is that male tenure in such groups is often short.

The extent of male care of offspring varies among species that form one-male, multi-female groups.

In some species that form one-male, multifemale groups, like gorillas, males are immensely tolerant and protective of infants and juveniles within their groups. Silverbacks sometimes intervene in intragroup conflicts, and generally support the younger of two antagonists when they do so. If a mother dies, her juvenile offspring treat the resident silverback as their caretaker, even if they have other relatives within the group. In some other species that form one-male, multifemale groups, such as langurs, howler monkeys, and guenons, the resident male is more peripheral to the group and does not form close bonds with juveniles. It is not yet clear why the extent of male care varies among these species.

Infanticide is an important part of intrasexual competition in primates that live in one-male, multifemale groups.

The other extreme in the range of behaviors among males toward offspring is infanticide. Sarah Blaffer Hrdy, at the University of California, Davis, was the first to suggest that infanticide is an evolved male reproductive strategy. She based her hypothesis on the following reasoning. When a female gives birth to an infant, she nurses it for a number of months, and does not become pregnant again for an extended period of time. However, after the death of an infant, lactation ends abruptly, and females may become sexually receptive almost immediately. Thus, the death of nursing infants hastens the resumption of maternal receptivity. A male who takes over a group may benefit from killing nursing infants because their deaths cause their mothers to become sexually receptive much sooner than they would otherwise. Since male tenure in one-male, multifemale groups may be short, and interbirth intervals are relatively long, infanticide may substantially increase a new resident's mating opportunities.

This hypothesis was initially quite controversial. Some researchers who studied langur populations in which infanticide was not observed contended that infanticide was a pathological response to overcrowded conditions and did not occur in undisturbed populations of monkeys. Moreover, there are very few eyewitness accounts of infanticidal attacks; researchers rely mainly on circumstantial evidence that links male takeovers to infant disappearances. Nonetheless, reports of the dis-

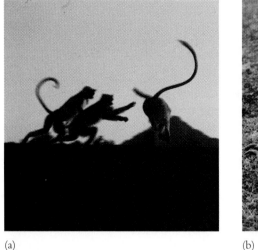
(a)

(b)

Figure 7.41

(a) In Hanuman langurs, and in many other species that form one-male groups, infanticide is an element of the male reproductive strategy. Infants are killed by new resident males to hasten the mothers' return to estrous cycling. (b) Hanuman langur females may lose many of their infants to infanticidal attacks.

appearance of infants immediately after male takeovers in a number of primate species, including gorillas, chimpanzees, howler monkeys, and sifakas, limited direct observations of immigrant males persistently stalking and attacking infants; and documentation of infanticide under similar circumstances in a wide range of other animal taxa have convinced most researchers that infanticide is not simply an aberrant response to crowding (Figure 7.41).

Hrdy's hypothesis was further strengthened as more detailed information about the patterning of infanticidal events became available. It was observed that males are quite selective in infanticidal attacks. The new resident male kills only the infants born before he arrived in the group; therefore he virtually never kills infants that he himself fathered. Moreover, males tend to kill only very young infants, the ones whose deaths will have the biggest impact on their mothers' sexual receptivity. Thus, males selectively kill infants whose deaths will increase their own reproductive success.

These points are clearly illustrated among red howler monkeys in Venezuela (Table 7.1). Carolyn Crockett and Ranka Sekulic monitored the composition of

Table 7.1 Among red howlers in Venezuela, infant mortality is much higher in groups that experience takeovers than in groups with stable male membership. (Adapted from Table IV in C. M. Crockett and R. Sekulic, 1984, Infanticide in red howling monkeys, pp. 173–191 in *Infanticide: Comparative and Evolutionary Perspectives*, ed. by G. Hausfater and S. B. Hrdy, Aldine, New York.

Group Status	Number of Births	Number of Deaths	Mortality
Stable	78	7	9%
Unstable	60	19	32%

43 groups of red howler monkeys, and documented the events that followed changes in adult male membership in these groups. They found that mortality among immature animals was greatly increased in unstable groups—those that experienced changes in male membership. Moreover, not all immature monkeys were at equal risk in unstable groups; the youngest infants were the ones most likely to be injured or to disappear (Figure 7.42). The infants at greatest risk were those who were young enough that their deaths would shorten their mother's interbirth interval. Extensive long-term data on Hanuman langurs in India shows much the same pattern.

Males in one-male, multifemale groups may not have exclusive access to females in their groups.

Even though we generally assume that males in one-male, multifemale groups face little competition for access to females within their groups during their tenure as residents, this may not always be the case. In patas monkeys and blue monkeys, for example, researchers have discovered that the resident male is sometimes unable to prevent other males from associating with the group and mating with sexually receptive females. Such incursions are concentrated during the mating season and may last for hours, days, or weeks and involve one male or several males.

Female choice may also play a role in male reproductive success in one-male, multi-female groups.

Although it might seem that there would be little opportunity for female mate choice in species that form one-male, multifemale groups, female preferences for particular males may influence male reproductive success. In experimental studies of captive hamadryas baboons, Hans Kummer and his colleagues at the University of Zürich have shown that unfamiliar males are less likely to attempt to disrupt a male-female pair if the female seems to be closely bonded to her mate than if she seems to be distant or uninterested in him. Female gorillas sometimes leave their groups to join strange males, and they may prefer males that are not closely related to themselves or males that can provide effective protection for infants.

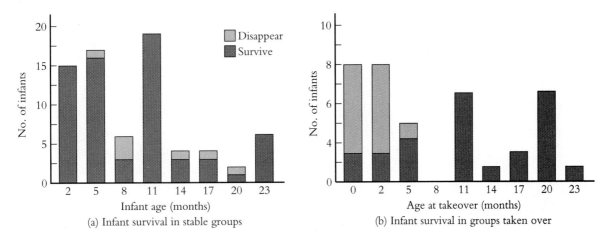

(a) Infant survival in stable groups (b) Infant survival in groups taken over

Figure 7.42

In red howlers in Venezuela, infant mortality is much greater in groups that are taken over by new males than in groups with stable male residents. In groups that are taken over by new males, it is the youngest infants who suffer the highest levels of mortality.

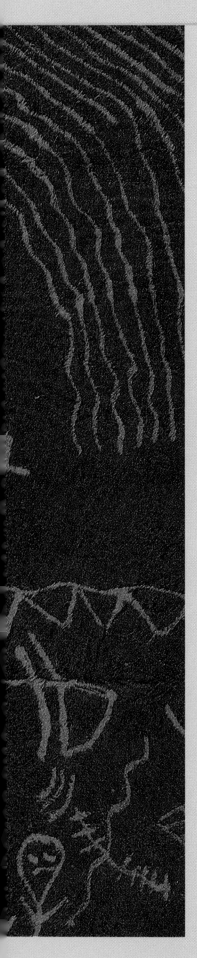

Dangers of Dispersal

Some primates, such as orangutans, lead solitary lives, meeting only for the occasional mating. But the average group of primates is supremely social, whether it is a family of a dozen gorillas living in a mountain rain forest, a band of twenty langur monkeys on the outskirts of an Indian village, or a troop of one hundred baboons in the African grasslands. In such groups an infant is born into a world filled with relatives, friends, and adversaries, surrounded by intrigue, double-dealing, trysts, and heroics—the staples of great small-town gossip. Pretty heady stuff for the kid that's learning whom it can trust and what the rules are, along with its species' equivalent of table manners.

Most young primates are socialized in this way quite effectively, and home must seem homier all the time. Then, inevitably, a lot of the young must leave the group and set out on their own. It is a simple fact driven by genetics and evolution: if everyone stayed on, matured, and reproduced there, and if their kids stayed on, and their kids' kids too, then ultimately everyone would be pretty closely related. You would have the classic problems of inbreeding—lots of funny-looking kids with six fingers and two tails (as well as more serious genetic problems). Thus essentially all social primates have evolved mechanisms for adolescent emigration from one group to another. Not all adolescents have to leave. The problem of inbreeding is typically solved so long as all the adolescents of one sex go and make their fortune elsewhere; the members of the other sex can remain at home and mate with the newcomers immigrating to the group. In chimpanzees and gorillas, it is typically the females who leave for new groups, while the males stay home with their mothers. But among most Old World monkeys—baboons, macaques, langurs—it is the males who make the transfer. Why one sex transfers in one species but not in another is a complete mystery, the sort that keeps primatologists happily arguing with each other ad nauseam.

This pattern of adolescent transfer solves the specter of inbreeding. To look at it just as a solution to an evolutionary problem is mechanistic, however. You can't lose sight of the fact that those are real animals going through the harrowing process. It is a remarkable thing to observe: every day, in the world of primates, someone young and frightened picks up, leaves Mommy and everyone it knows, and heads off into the unknown.

The baboons that I study each summer in East Africa provide one of the best examples of this pattern of adolescent transfer. Two troops encounter each other at midday at some sort of natural boundary—a river, for example. As is the baboon propensity in such settings, the males of the two troops carry on with a variety of aggressive displays, hooting and hollering with what they no doubt hope is a great air of menace. Eventually everyone gets bored and goes back to eating and lounging, ignoring the interlopers on the other side of the river. Suddenly you spot the kid—some adolescent in your troop. He stands there at the river's edge, absolutely riveted. New baboons, a whole bunch of 'em! He runs five steps toward them, runs four back, searches among the other members of his troop to see why no one else seems mesmerized by the strangers. After endless contemplation, he gingerly crosses the river and sits on the very edge of the other bank, scampering back down in a panic should any new baboon so much as glance at him.

A week later. When the troops run into each other again, the kid repeats the pattern. Except this time he spends the afternoon sitting at the edge of the new troop. At the next encounter he follows them for a short distance before the anxiety becomes too much and he turns back. Finally, one brave night, he stays with them. He may vacillate awhile longer, perhaps even ultimately settling on a third troop, but he has begun his transfer into adulthood.

And what an awful experience it is—a painfully lonely, peripheralized stage of life. There are no freshman orientation weeks, no cohorts of newcomers banding together and covering their nervousness with bravado. There is just a baboon kid, all alone on the edge of a new group, and no one there could care less about him. Actually, that is not true—there are often members of the new troop who pay quite a lot of attention, displaying some of the least-charming behavior seen among social primates, and most reminiscent of that of their human cousins. Suppose you are a low-ranking member of that troop, a puny kid a year or so after your own transfer. You spend most of your time losing fights, being pushed around, having food ripped off from you by someone of higher rank. You have a list of grievances a mile long and there's little you can do about it. Sure, there are youngsters in the troop that you could harass pretty successfully, but if they are pretransfer age, their mothers—and maybe their fathers and their whole extended family—will descend on you like a ton of bricks. Then suddenly, like a gift from heaven, a new, even punier kid shows up: someone to take it out on. (Among chimps, where females do the transferring, the same thing occurs; resident females are brutally aggressive toward the new female living on the group's edge.)

Yet that's only the beginning of a transfer animal's problems. When, as part of my studies of disease patterns among baboons, I anesthetize and examine transfer males, I find that these young animals are teeming with parasites. There is no longer anyone to groom them, to sit with them and methodically clean their fur, half for hygiene, half for friendship. And if no one is interested in grooming a recent transfer animal, certainly no one is interested in anything more intimate than that—it's a time of life where sex is mostly just practicing. The young males suffer all the indignities of being certified primate nerds.

They are also highly vulnerable. If a predator attacks, the transfer animal—who is typically peripheral and exposed to begin with—is not likely to recognize the group's signals and has no one to count on for his defense. I witnessed an incident like this once. The unfortunate animal was so new to my group that he rated only a number, 273, instead of a name. The troop was meandering in the midday heat and descended down the bank of a dry streambed into some bad luck: a half-asleep lioness. Panic ensued, with animals scattering every which way as the lioness stirred—while Male 273 stood bewildered and terribly visible. He was badly mauled and, in a poignant act, crawled for miles to return to his former home troop to die near his mother.

In short, the transfer period is one of the most dangerous and miserable times in a primate's life. Yet, almost inconceivably, life gets better. One day an adolescent female will sit beside our nervous transfer male and briefly groom him. Some afternoon everyone hungrily descends on a tree in fruit and the older adolescent males forget to chase the newcomer away. One morning the adolescent and an adult male exchange greetings (which, among male baboons, consists of yanking on each other's penis, a social gesture predicated on trust if ever I saw one). And someday, inevitably, a terrified new transfer male appears on the scene, and our hero, to his perpetual shame perhaps, indulges in the aggressive pleasures of finding someone lower on the ladder to bully.

In a gradual process of assimilation, the transfer animal makes a friend, finds an ally, mates, and rises in the hierarchy of the troop. It can take years. Which is why Hobbes was such an extraordinary beast.

Three years ago my wife and I spent the summer working with a baboon troop at Amboseli National Park in Kenya, a research site run by the behavioral biologists Jeanne and Stuart Altmann from the University of Chicago. When I'm in the field, I study the relationships between a baboon's rank and personality, how its body responds to stress, and what sorts of stress-related diseases it acquires. In order to obtain the physiological data—blood samples to gauge an animal's stress hormone levels, immune system function, and so on—you have to anesthetize the baboon for a few hours with the aid of a small aluminum blowgun and drug-filled darts. Fill the dart with the right amount of anesthetic, walk up to a baboon, aim, and blow, and he is snoozing five minutes later. Naturally it's not quite that simple—you can only dart someone in the mornings (to control for the effect of circadian rhythms on the animal's hormones). You must ensure there are no predators around to shred the guy, and you must make certain he doesn't climb a tree before passing out. And most of all, you have to dart and remove him from the troop when none of the other baboons are looking so that you don't alarm them and disrupt their habituation to the scientists. So essentially, what I do with my college education is creep around in the bushes after a bunch of baboons, waiting for the instant when they are all looking the other way so that I can zip a dart into someone's tush.

It was about halfway through the season. We were just beginning to know the baboons in our troop (named Hook's troop, for a long-deceased matriarch) and were becoming familiar with their daily routine. Each night they slept in their favorite grove of trees; each morning they rolled out of bed to forage for food in an open savannah strewn with volcanic rocks tossed up aeons ago by nearby Mount Kilimanjaro. It was the dry season, which meant the baboons had to do a bit more walking than usual to find food and water, but there was still plenty of both, and the troop had time to lounge around in the shade during the afternoon heat. Dominating the social hierarchy among the males was an imposing character named Ruto, who had joined the troop a few years before and had risen relatively quickly in the ranks. Number two was a male named Fatso, who had had the misfortune of being a rotund adolescent when he'd transferred into the troop years before and was named by a callous researcher. Fat he was no longer. Now a muscular prime-age male, he was Ruto's most obvious competitor, though still clearly subordinate. It was a fairly peaceful period for the troop; mail was delivered regularly and the trains ran on time.

Then one morning we arrived to find the baboons in complete turmoil. It is not an anthropomorphism to say that everyone was mightily frazzled. There was a new transfer male, and not someone meekly scurrying about the periphery. He was in the middle of the troop, raising hell—threatening and chasing everyone in sight. This nasty, brutish animal was soon named Hobbes (in deference to the seventeenth-century English philosopher who described the life of man as solitary, poor, nasty, brutish, and short).

This is an extremely rare, though not unheard of, event among baboons. The transfer male in such cases is usually a big, muscular, intimidating kid. Maybe he is older than the average seven-year-old transfer male, or perhaps this is his second transfer and he picked up confidence from his first emigration. Maybe he is the cocky son of a high-ranking female in his old troop. In any case, the rare animal with these traits comes on like a locomotive, and he often gets away with it for a time. As long as he keeps pushing, it will be a while before any of the resident males works up the nerve to confront him. Nobody knows him yet; thus no one knows if he is asking for a fatal injury by being the first fool to challenge this aggressive maniac.

This was Hobbes's style. The intimidated males stood around helplessly. Fatso discovered all sorts of errands he had to run elsewhere. Ruto hid behind females. No one else was going to take a stand. Hobbes rose to the number one rank in the troop within a week.

Despite his sudden ascendancy, Hobbes's success wasn't going to last forever. He was still a relatively inexperienced kid, and eventually one of the bigger males was bound to cut him down to size. Hobbes had about a month's free ride. At this point he did something brutally violent but which had a certain grim evolutionary logic. He began to selectively attack pregnant females. He beat and mauled them, causing three out of four to abort within a few days.

One of the great clichés of animal behavior in the context of evolution is that animals act for the good of the species. This idea was discredited in the 1960s but continues to permeate *Wild Kingdom*–like versions of animal behavior. The more accurate view is that animals usually behave in ways that maximize their own reproduction and the reproduction of their close relatives. This helps explain extraordinary acts of altruism and self-sacrifice in some circumstances, and sickening aggression in others. It's in this context that Hobbes's attacks make sense. Were those females to carry through their pregnancies and raise their offspring, they would not be likely to mate again for two years—and who knows where Hobbes would be at that point. Instead he harassed the females into aborting, and they were ovulating again a few weeks later. Although female baboons have a say in whom they mate with, in the case of someone as forceful as Hobbes they have little choice: within weeks Hobbes, still the dominant male in the troop, was mating with two of those three females. (This is not to imply that Hobbes had read his textbooks on evolution, animal behavior, and primate obstetrics and had thought through this strategy, any more than animals think through the logic of when they should reach puberty. The wording here is a convenient shorthand for the more correct way of stating that his pattern of behavior was almost certainly an unconscious, evolved one.)

As it happened, Hobbes's arrival on the scene—just as we had darted and tested about half the animals for our studies—afforded us a rare opportunity. We could compare the physiology of the troop before and after his tumultuous transfer. And in a study published with Jeanne Altmann and Susan Alberts, I documented the not very surprising fact that Hobbes was stressing the bejesus out of these animals. Their blood levels of cortisol (also known as hydrocortisone), one of the hormones most reliably secreted during stress, rose significantly. At the same time, their numbers of white blood cells, or lymphocytes, the sentinel cells of the immune system that defend the body against infections, declined markedly, another highly reliable index of stress. These stress-response markers were most pronounced in the animals receiving the most grief from Hobbes. Unmolested females had three times as many circulating lymphocytes as one poor female who was attacked five times during those first two weeks.

An obvious question: Why doesn't every new transfer male try something as audaciously successful as Hobbes's method? For one thing, most transfer males are too small at the typical transfer age of seven years to intimidate a gazelle, let alone an eighty-pound adult male baboon. (Hobbes, unusually, weighed a good seventy pounds.) Most don't have the personality needed for this sort of unpleasantry. Moreover, it's a risky strategy, as someone like Hobbes stands a good chance of sustaining a crippling injury early in life.

But there was another reason as well, which didn't become apparent until later. One morning, when Hobbes was concentrating on whom to hassle next and paying no attention to us, I managed to put a dart into his haunches. Months later, when examining his blood sample in the laboratory, we found that Hobbes had among the highest levels of cortisol in the troop and extremely low lymphocyte counts, less than one-quarter the troop average. (Ruto and Fatso, conveniently sitting on the sidelines, had three and six times as many lymphocytes as Hobbes had.) The young baboon was experiencing a massive stress response himself, larger even than those of the females he was traumatizing, and certainly larger than is typical of the other, meek transfer males I've studied. In other words, it doesn't come cheap to be a bastard twelve hours a day—a couple of months of this sort of thing is likely to exert a physiological toll.

As a postscript, Hobbes did not hold on to his position. Within five months he was toppled, dropping down to number three in the hierarchy. After three years in the troop, he disappeared into the sunset, transferring out to parts unknown to try his luck in some other troop.

Source: From pp. 78–86 in Robert M. Sapolsky, 1997, *The Trouble with Testosterone*, Scribner, New York. Reprinted by permission of Scribner, a division of Simon & Schuster.

Figure 7.43

Male baboons compete over access to an estrous female.

Multimale, Multifemale Groups

For males in multimale, multifemale groups, conflict arises both over membership in bisexual groups and over access to receptive females.

In some macaque species, males hover near the periphery of social groups, avoid aggressive challenges by resident males, and attempt to ingratiate themselves with females. In chacma baboons, immigrant males sometimes move directly into the body of the group and engage resident males in prolonged vocal duels and chases. Although there may be conflict when males attempt to join nonnatal groups, males spend most of their adult lives as members of multimale, multifemale groups. Thus, the ability to establish residence in social groups is unlikely to be a major source of variation in male reproductive success.

In multimale, multifemale groups, males sometimes compete directly over access to receptive females. Males may attempt to drive other males away from females, to interrupt copulations, and to prevent other males from approaching or interacting with females (Figure 7.43). More often, however, male-male competition is mediated through dominance relationships that reflect male competitive abilities. These relationships are generally established in spontaneous confrontations that are not linked to immediate access to resources or to females. Such contests may involve mild threats and stereotyped submissive gestures, but may also lead to escalated contests in which males chase, wrestle, and bite one another (Figure 7.44).

(a)   (b)

Figure 7.44

(a) A male baboon displays his canines as he threatens a rival. (b) Males' canines are formidable weapons, and males sometimes sustain serious injuries in fights with other males. This male baboon was injured in a fight with another male.

In many cases, the outcomes of encounters between particular pairs of males are relatively predictable from day to day, and males can be ordered in a linear dominance hierarchy. Male fighting ability and dominance rank are generally closely linked to physical condition, as prime-aged males in good physical condition are usually able to dominate others. However, the correlation between physical abilities and dominance rank is imperfect, as old males are sometimes able to dominate younger and stronger males.

It seems logical that male dominance rank would correlate with male reproductive success in multimale, multifemale groups, but this conclusion is energetically debated. The issue is difficult to resolve because unambiguous data on paternity are difficult to come by. It is difficult to determine which male fathered a particular infant in multimale, multifemale groups for several reasons: more than one male may mate with a given female near the time of conception, clandestine matings may be overlooked, and observations typically end at sunset although sexual activity may extend into the night. New genetic techniques make it possible to assess paternity with a much higher degree of precision. In some groups, genetic data reveal a strong relationship between male rank and reproductive success (Figure 7.45). As is the case for females, some studies don't find positive correlations

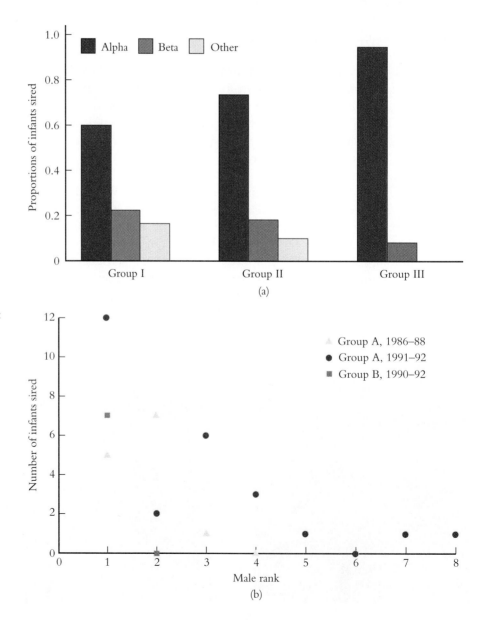

Figure 7.45

Genetic analyses of paternity have provided evidence that male rank is correlated with reproductive success in both free-ranging and captive groups. (a) In three groups of free-ranging long-tailed macaques, the highest-ranking male (alpha) fathered a much larger fraction of infants than the second-ranking male (beta) and other lower-ranking males did. (b) In two groups of captive sooty mangabeys, high-ranking males also produced more offspring than lower-ranking males did.

Figure 7.46

Here two middle-ranking males (on the right) challenge the dominant male (on the left), who has been consorting with a receptive female.

between male rank and reproductive success, but none shows significant negative correlations.

In some species, males form alliances in order to gain access to females.

In East Africa, male savanna baboons cooperate in contests over access to estrous females. Typically, one high-ranking male forms an exclusive bond, called a consortship, with a receptive female. Consortships may last hours or days. During the consortship, the male follows the female, grooms her, and copulates with her. When other males attempt to approach the pair, the consorting male drives them off with threat displays. Consortships are sometimes terminated when the consort male is simultaneously challenged by two less-dominant rivals. Usually, one male recruits another, and they jointly confront the consorting male (Figure 7.46). Sometimes these contests are prolonged and highly conspicuous events involving charges, long-distance chases, and loud vocalizations. In a substantial fraction of cases, the original consortship is terminated, and one of the challengers begins to consort with the female.

In multimale, multifemale groups, female preferences sometimes influence male mating success.

A growing body of work suggests that female primates exercise mating preferences (Figure 7.47), but there is no consensus about the types of traits females prefer. For example, free-ranging ring-tailed lemur females prefer to mate with males from neighboring groups and rebuff mating attempts by natal males. Among provisioned groups of rhesus macaques on Cayo Santiago and Japanese macaques,

Figure 7.47

Although male access to females is influenced by the outcome of male-male competition, female choice also plays an important role in some situations. Here a female rhesus monkey evades a male's attempt to copulate with her. (Photograph courtesy of Susan Perry and Joseph Manson.)

Figure 7.48

In a group of brown capuchins, the dominant male allows an infant to feed beside him undisturbed. (Photograph by Charles Janson.)

females seem to prefer unfamiliar, low-ranking males over high-ranking ones. Joseph Manson at the University of California, Los Angeles, notes that in these macaque populations, male tenure is typically long and male dominance hierarchies are very stable over time. Thus, female preferences for new males may be a means of avoiding inbreeding or of increasing genetic diversity among offspring. In other cases, females seem to prefer high-ranking males over low-ranking ones. For example, in one group of brown capuchins, estrous females associated selectively with the highest-ranking adult male, and he dominated matings with receptive females. These female preferences may be related to the fact that males generally allowed their own offspring to feed beside them, and monkeys that fed near the highest-ranking male had higher feeding rates than other monkeys (Figure 7.48).

Our difficulty in identifying the kinds of traits that females prefer may be due in part to the fact that females are not always able to choose their mating partners freely. When males compete for receptive females, subordinate males may be reluctant to approach receptive females, or they may be interrupted by higher-ranking males when they attempt to mate. Moreover, high-ranking males may coerce females by behaving aggressively when they approach other males, herding them away from other males and persistently pursuing them when they are receptive. Recently, Joseph Soltis from the University of California, Los Angeles, and his colleagues from the Primate Research Institute in Kyoto, Japan, have shown that both female choice and male dominance rank influenced mating behavior in one group of Japanese macaques. Females showed clear tendencies to associate with certain males (and to avoid others) while they were in estrous. Females' preferences were uncorrelated with male dominance rank, as females sometimes preferred low-ranking males over high-ranking males. At the same time, high-ranking males were more likely to mate with females than low-ranking ones were. DNA fingerprinting analyses revealed, however, that female preferences were more important than male rank in determining mating success—the males who were most attractive to females were most successful in siring their offspring.

In this chapter, we have seen that competition for food, scarce resources, mating partners, and reproductive opportunities play a pivotal role in the lives of primates. Dominance interactions and aggressive contests are common events in most primate groups. But this picture of primate social life is not quite complete—primates also interact peacefully, develop close bonds with other members of their groups, and cooperate in certain contexts. In the next chapter, we will consider how evolution shapes the pattern of these kinds of interactions.

Further Reading

Hausfater, G., and S. B. Hrdy, eds. 1984. *Infanticide: Comparative and Evolutionary Perspectives.* Aldine, Hawthorne, N.Y.

Kappeler, P. M., and J. G. Ganzhorn. 1993. *Lemur Social Systems and Their Ecological Basis.* Plenum, New York.

Krebs, J. R., and N. B. Davies. 1993. *An Introduction to Behavioural Ecology.* Sinauer Associates, Sunderland, Mass., chaps. 8 and 9.

Nicolson, N. A. 1986. Infants, mothers, and other females. Pp. 330–342 in *Primate Societies,* ed. by B. B. Smuts, D. L. Cheney, R. M. Seyfarth, R. W. Wrangham, and T. T. Struhsaker, University of Chicago Press, Chicago.

Palombit, R. In press. Infanticide and the evolution of pair bonds in nonhuman primates. *Evolutionary Anthropology.*

Strier, K. 1990. *Faces in the Forest: The Endangered Muriqui Monkeys in Brazil.* Oxford University Press, London.

Wrangham, R. W. 1986. Evolution of social structure. Pp. 282–296 in *Primate Societies,* ed. by B. B. Smuts, D. L. Cheney, R. M. Seyfarth, R. W. Wrangham, and T. T. Struhsaker. University of Chicago Press, Chicago.

Study Questions

1. Explain why reproductive success is a critical element of evolution by natural selection.

2. Why does frugivory play such an important role in the resource-defense model? What evidence supports the resource-defense model? What evidence supports the predation model?

3. What is the difference between polyandry and polygyny? Why is polygyny so much more common than polyandry among primates? Under what kinds of conditions might selection favor polyandry?

4. Having the members of one sex disperse from their natal groups after reaching adulthood helps to reduce inbreeding by reducing the possibility of matings between close kin. Explain why inbreeding has deleterious consequences in most species.

5. In many primate species, reproduction is seasonal. The females are sexually receptive for a few months during the year, and births are clumped during a few months of the year. How would this seasonality influence the extent of male-male competition and the intensity of sexual selection?

6. Infanticide by males is generally limited to species that form one-male, multi-female groups. However, it seems likely that males in multimale, multifemale groups would also profit from killing infants if it reduced the mother's interbirth interval. Why is infanticide more prevalent in species that form one-male, multifemale groups than in species that form multimale, multifemale groups?

7. Why is the variance in male fitness usually greater than the variance in female fitness among mammals? What effect does this have on sexual selection?

C H A P T E R   8

The Evolution of Social Behavior

Kinds of Social Interactions
Altruism: A Conundrum
Kin Selection
　　Hamilton's Rule
　　Evidence of Kin Selection in Primates
Reciprocal Altruism

A social primate lives in a group of known individuals. At one time or another, the other members of the group may become playmates, grooming partners, competitors for food, rivals for mates, allies in aggressive confrontations, caretakers for offspiring, collaborators in intergroup encounters, and so on. Even solitary primates, like galagos and orangutans, interact regularly with their neighbors, maturing offspring, and prospective mates.

There is great diversity in the form and frequency of social interactions that occur within and among groups of primates. Just as evolutionary theory provides the framework for understanding the patterning of mating and parenting behaviors in nature, it also provides an essential foundation for understanding the form and distribution of social interactions among individuals within social groups.

Kinds of Social Interactions

Social interactions are behaviors that affect the fitness of more than one individual.

When an animal feeds, travels, or sleeps, it has little impact on the fitness of others. But when two individuals interact, say in competition over a prized resource or in cooperative defense of valuable food items, their activities necessarily involve each other—as collaborators, rivals, or opponents. In these situations, the behavior of one individual directly affects the fitness of the other. Interactions that involve two individuals are called **dyadic** or **pairwise** interactions. Table 8.1 classifies four kinds of pairwise interactions according to their effects on the **actor** (the individual performing the behavior) and the **recipient** (the individual affected by the behavior). An act is said to be beneficial (+) if it increases fitness, and costly or detrimental (−) if it reduces fitness. It is evident that social interactions do not necessarily have the same kinds of effects on actors and on recipients. For example, **selfish** acts are beneficial to the actor and costly to the recipient, while the costs and benefits are reversed for **altruistic** interactions. On the other hand, **mutualistic** interactions are beneficial to both actor and recipient, and **spiteful** interactions are costly to both parties.

Before we go any further, two caveats are in order. First, this classification uses ordinary English words like *altruism* and *spite* because they provide a convenient, easily remembered shorthand for describing the fitness effects of different kinds of social behaviors. However, the technical definitions sometimes differ from the meanings these words have in ordinary usage. For example, an act that is beneficial to the recipient but has no adverse impact on the fitness of the actor might be considered altruistic in common usage but is not altruistic in the biological sense of the word. So, you should be careful to keep the technical definitions in mind when these terms are used.

Second, it is extremely difficult to measure the effects of particular behavioral acts on the fitness of individuals, particularly for long-lived animals like primates. A female may participate in hundreds of grooming sessions over the course of a year and in thousands during her lifetime. At the same time, she has a multitude of different experiences that may influence her reproductive career. As a result, it is virtually impossible to assess the effects of a single behavioral act, or even the effects of a class of behavioral acts, on her reproductive success. Nonetheless, we can make reasonable inferences about the immediate costs and benefits of particular acts based on more general considerations, such as the energy demands or the risks associated with specific behaviors or social interactions. For example, we could assess the caloric value of a food item that a female chimpanzee shares with her

Table 8.1 A classification of pairwise, or dyadic, social interactions. + indicates a positive effect on fitness, and − a negative impact on fitness.

Case	Actor	Recipient
Selfish	+	−
Mutualistic	+	+
Altruistic	−	+
Spiteful	−	−

daughter or measure the time required to replace the food item with another one of similar value. These kinds of assessments provide a basic estimate of the effects of particular kinds of social interactions on fitness.

Altruism: A Conundrum

Altruistic behavior cannot evolve by ordinary natural selection.

Notice from Table 8.1 that two of the four forms of interactions enhance the fitness of the actor: selfish and mutualistic acts increase the fitness of the actor, and will be favored by natural selection, all other things being equal. Thus, there is no problem in explaining why one female capuchin monkey supplants another from a patch of ripe fruit, or why two chimpanzees jointly corner a red colobus monkey and then share the carcass.

Altruism is a puzzle because it *decreases* the fitness of the individual performing the behavior. According to Darwin's theory, complex adaptations, including behavioral adaptations, must be assembled step by small step, each change favored by natural selection. Thus, while altruistic behaviors might initially arise by accident or as side effects of other behaviors, it seems impossible for complex altruistic behaviors to be assembled by natural selection. Each small genetic change that made it more likely for an individual to perform the behavior would be selected against because of the behavior's negative effects on genetic fitness. The same argument applies to spite. Thus, the existence of either spiteful or altruistic behavior would seem to contradict the fundamental logic of natural selection.

Primates perform altruistic behaviors in nature.

This theoretical conundrum would not be a problem if altruistic interactions were rare or unimportant, and this does seem to be the case for spite. However, there is a lot of evidence that primates, and many other social animals, regularly perform complex altruistic behaviors that play an important role in primate social life. Virtually all social primates groom other group members, picking parasites, scabs, and bits of debris from their hair. When one individual in a group sights a predator, she often gives a distinctive alarm call that alerts the rest of the group. Sometimes two individuals form an alliance against a higher-ranking individual. Occasionally, one individual allows another to share his food. All of these behaviors seem to meet the biological definition of altruism. When it is grooming, the donor expends valuable time and energy that could otherwise be used to look for food, to court prospective mates, or for other vital tasks. The recipient gets a thorough cleaning of parts of her body that she may find difficult to reach and may generally enjoy a period of pleasant relaxation (Figure 8.1). By giving an alarm call, the caller makes herself more conspicuous to predators, while the individuals hearing the warning are able to flee to safety. When animals form alliances, particularly when they come to the defense of other group members, they make themselves vulnerable to retaliation by the aggressor while the defended individual may have less chance of being injured or defeated during the encounter (Figure 8.2). Sharing food may reduce the amount of food that the owner eats while increasing the amount of food that the recipient obtains. Thus, in each of these cases, it seems

Figure 8.1
Gray langurs groom one another. Grooming is usually considered to be altruistic because the groomer expends time and energy when it grooms another animal, and the recipient benefits from having ticks removed from its skin, wounds cleaned, and debris removed from its hair.

likely that the actor suffers a decrease in fitness while the recipient incurs an increase in fitness, which satisfies the criteria for altruism.

Altruistic behaviors cannot be favored by selection just because they are beneficial to the group as a whole.

You might think that if the average effect of an act on all members of the group is positive, then it would be beneficial for all individuals to perform it. For example, suppose that when one monkey gives an alarm call, the other members of the group benefit, and the total benefits to all group members exceed the cost of giving the call. Then, if every individual gives the call when a predator is sighted, all members of the group will be better off than if no warning calls were ever given.

The problem with this line of reasoning is that it confuses the effect on the group with the effect on the actor. In most circumstances, the fact that alarm calls are beneficial to those hearing them doesn't affect whether the trait alarm calling evolves; all that matters is the effect of giving the alarm call on the caller. To see why, imagine a hypothetical monkey species in which some individuals give alarm calls when they are the first to spot a predator. Monkeys who are warned by the

Figure 8.2
Two capuchin monkeys form a coalition against an opponent who is not visible in the photograph. Coalitions are altruistic because the ally puts itself at risk and expends energy when it becomes involved in an ongoing dispute. The animal that receives support may benefit if the fight ends sooner or with less costly consequences. (Photograph by Susan Perry.)

(a) Altruist gives alarm call to group.   (b) Nonaltruist doesn't give alarm call to group.

Figure 8.3

Two groups of monkeys are approached by a predator. (a) In one group, there is an individual (pink) who has a gene that makes her call in this context. Giving the call lowers the caller's fitness, but increases the fitness of every other individual in the group. Like the rest of the population, one out of four of these beneficiaries also carry the genes for calling. (b) In the second group, the female who detects the predator does not carry the gene for calling, and remains silent. This lowers the fitness of all members of the group a certain amount because they are more likely to be caught unawares by the predator. Once again, one out of four are callers. Although members of the caller's group are better off on average than members of the noncaller's group, the gene for calling is not favored. This is because callers and noncallers in the caller's group both benefit from the caller's behavior, but callers incur some costs. Callers are at a disadvantage relative to noncallers. Thus, calling is not favored, even though the group as a whole benefits.

call have a chance to flee. Suppose that one-quarter of the population gives the call when they spot predators ("callers") and three-quarters of the individuals do not give an alarm call in the same circumstance ("noncallers"). (These proportions are arbitrary; we chose them because it's easier to follow the reasoning in examples with concrete numbers.) Let's suppose that in this species the tendency to give alarm calls is genetically inherited.

Now we compare the fitness of callers and noncallers. Since everyone in the group can hear the alarm calls and take appropriate action, alarm calls benefit everyone in the group to the same extent (Figure 8.3). Calling has no effect on the *relative* fitness of callers and noncallers because, on average, one-quarter of the beneficiaries will be callers and three-quarters of the beneficiaries will be noncallers, the same proportions we find in the population as whole. Calling reduces the risk of mortality for everyone who hears the call, but it does not change the frequency of callers and noncallers in the population because everyone gains the same benefits. However, callers are conspicuous when they call, which makes them more vulnerable to predators. While all individuals benefit from hearing alarm calls, callers are the only ones who suffer costs from calling. This means that, on average, noncallers will have a higher fitness than callers. Thus, genes that cause alarm calling will not be favored by selection even if the cost of giving alarm calls is small and the benefit to the rest of the group is large. Instead, selection will favor genes that suppress alarm calling because noncallers have higher fitness than callers. This simple example can be reformulated for grooming, food sharing, coalition formation, and so on, and in each case the conclusion is the same—individual selection is not expected to favor the evolution of these kinds of behaviors (Box 8.1).

BOX 8.1

Group Selection

Group selection was once thought to be the mechanism for the evolution of altruistic interactions. In the early 1960s a British ornithologist, V. C. Wynne-Edwards, contended that altruistic behaviors like those we have been considering here evolved because they enhanced the survival of whole groups of organisms. Thus, individuals gave alarm calls, despite the costs of becoming more conspicuous to predators, because calling protected the group as a whole from attacks. Wynne-Edwards reasoned that groups that contained more altruistic individuals would be more likely to survive and prosper than groups that contained fewer altruists, and the frequency of the genes leading to altruism would increase.

Wynne-Edwards' argument is logical, because Darwin's postulates logically apply to groups as well as individuals. However, group selection is not an important force in nature because there is generally not enough genetic variation among groups for selection to act on. Group selection can occur if groups vary in their ability to survive and to reproduce, and that variation is heritable. Then group selection may increase the frequency of genes that increase group survival and reproductive success. The strength of selection among groups depends on the amount of genetic variation among groups, just as the strength of selection among individuals depends on the amount of genetic variation among individuals. However, when individual selection and group selection are opposed, and group selection favors altruistic behavior while individual selection favors selfish alternatives, individual selection has a tremendous advantage. This is because the amount of variation among groups is much smaller than the amount of variation among individuals, unless groups are very small and there is very little migration among them. Thus, individual selection favoring selfish behavior will generally prevail over group selection, making group selection an unlikely source of altruism in nature.

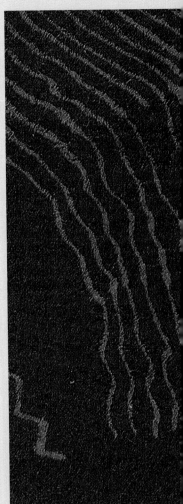

Kin Selection

Natural selection can favor altruistic behavior if altruistic individuals are more likely to interact with each other than chance alone would dictate.

If altruistic behaviors can't evolve by either ordinary natural selection or by group selection, then how do they evolve? A clear answer to this question did not come until 1964, when the young biologist W. D. Hamilton published a landmark paper. This paper was the first of a series of fundamental contributions Hamilton made to our understanding of the evolution of behavior. There are several different ways to conceptualize Hamilton's basic idea, and we will adopt the approach presented in his original paper.

The argument made in the previous section contains a hidden assumption—that altruists and nonaltruists are equally likely to interact with one another. Thus, we supposed that callers give alarm calls when they hear a predator, no matter who is nearby. This is why callers and noncallers were equally likely to benefit from hearing an alarm call. Hamilton's insight was to see that any process that causes altruists to be more likely to interact with other altruists than they would by chance could facilitate the evolution of altruism.

To see why this is such an important insight, let's modify the previous example by assuming that our hypothetical species lives in groups composed of full siblings—offspring of the same mother and father. The frequency of the calling and noncalling genes doesn't change, but their distribution will be affected by the fact that siblings live together (Figure 8.4). If an individual is a caller, then by the rules of Mendelian genetics there is a 50% chance that the individual's siblings will share the genes that cause calling behavior. This means that the frequency of the genes for calling will be higher in groups that contain callers than in the population as a whole, and therefore, more than a quarter of the beneficiaries will be callers themselves. When a caller gives an alarm call, the audience will contain a higher fraction of callers than the population at large. Thus, the caller raises the average fitness of callers relative to noncallers. Similarly, because the siblings of noncallers are more likely to be noncallers than chance alone would dictate, callers are less likely to be

(a) Altruist gives alarm call to siblings.   (b) Nonaltruist doesn't give alarm call to siblings.

Figure 8.4

Two groups of monkeys are approached by a predator. Each group is composed of nine sisters. (a) In one group there is a caller (pink), an individual who has a gene that makes her call in this context. Her call lowers her own fitness, but increases the fitness of her sisters. (b) In the second group, the female who detects the predator is not a caller, and does not call when she spots the predator. As in Figure 8.3, calling benefits the other group members but imposes costs on the caller. However, there is an important difference between the situations portrayed here and in Figure 8.3. Here the groups are made up of sisters, so five of the eight recipients of the call also carry the calling gene. In any pair of siblings, half of the genes are identical because the siblings inherited the same gene from one of their parents. Thus, on average, half of the caller's siblings also carry the calling allele. The remaining four siblings carry genes inherited from the other parent, and like the population as a whole, one out of four of them is a caller. The same reasoning shows that in the group with the noncaller, there is only one caller among the beneficiaries of the call. Half are identical to their sister because they inherited the same noncalling gene from one of their parents; one of the remaining four is a caller. In this situation, callers are more likely to benefit from calling than noncallers, and so calling alters the relative fitness of callers and noncallers. Whether the calling behavior actually evolves depends on whether these benefits are big enough to compensate for the reduction in the caller's fitness.

present in such groups than in the population at large. Therefore, the absence of a warning call lowers the relative fitness of noncallers relative to callers.

When individuals interact selectively with relatives, callers are more likely to benefit than noncallers, and the benefits of calling will, all other things being equal, favor the genes for calling. However, we must remember that calling is costly, and this will tend to reduce the fitness of callers. Calling will only be favored by natural selection if the benefits of calling are sufficiently greater than the costs. The exact nature of this tradeoff is specified by what we call Hamilton's rule.

Hamilton's Rule

Hamilton's theory of kin selection predicts that altruistic behaviors will be favored by selection if the costs of performing the behavior are less than the benefits discounted by the coefficient of relatedness between actor and recipient.

Hamilton's theory of **kin selection** is based on the idea that selection could favor altruistic alleles if animals interact selectively with their genetic relatives. Hamilton's theory also specifies the quantity and distribution of help among individuals. According to **Hamilton's rule**, an act will be favored by selection if

$$rb > c \qquad (1)$$

where

r = the average coefficient of relatedness between the actor and the recipients
b = the sum of the fitness benefits to all individuals affected by the behavior
c = the fitness cost to the individual performing the behavior

The **coefficient of relatedness**, r, measures the genetic relationship between interacting individuals. More precisely, r is the average probability that two individuals acquire the same allele through descent from a common ancestor. Figure 8.5 shows how these probabilities are derived in a simple genealogy. Female A obtains one allele at a given locus from her mother and one from her father. Her half-sister, female B, also obtains one allele at the same locus from each of her parents. The probability that both females obtain the same allele from their mother is obtained by multiplying the probability that female A obtains the allele (0.5) by the probability that female B obtains the same allele (0.5), and equals 0.25. Thus,

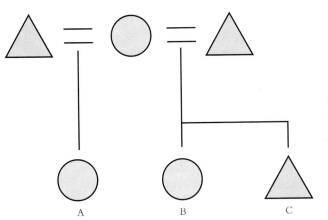

A   B   C

Figure 8.5

This genealogy shows how the value of r is computed. Triangles represent males, circles represent females, and the equals sign represents mating. The relationships between individuals labeled in the genealogy are described in the text.

Table 8.2 *The value of r for selected categories of relatives.*

RELATIONSHIP	r
Parent and offspring	$\frac{1}{2}$
Full siblings	$\frac{1}{2}$
Half-siblings	$\frac{1}{4}$
Grandparent and grandchild	$\frac{1}{4}$
Aunt-uncle and niece-nephew	$\frac{1}{4}$
First cousins	$\frac{1}{8}$
Unrelated individuals	0

half-sisters have, on average, a 25% chance of obtaining the same allele from their mothers. Now consider the relatedness between female B and her brother male C. In this case, note that female B and male C are full siblings: they have the same mother and the same father. The probability of each acquiring the same allele from their mother is still 0.25, but female B and male C might also share an allele from their father. The probability of this event is also 0.25. Thus, the probability that female B and male C share an allele is equal to the sum of 0.25 and 0.25, or 0.5. This basic reasoning can be extended to calculate the degrees of relatedness among various categories of kin (Table 8.2).

Hamilton's rule leads to two important insights: (1) altruism is limited to kin, and (2) closer kinship facilitates more costly altruism.

If you reflect on Hamilton's rule for a while, you will see that it produces two fundamental predictions about the conditions that favor the evolution of altruistic behaviors. First, altruism is not expected to be directed toward nonkin because the coefficient of relatedness between nonkin is $r = 0$. The condition for the evolution of altruistic traits will only be satisfied for interactions between kin, when $r > 0$. Thus, altruists are expected to be nepotistic, showing favoritism toward kin.

Second, close kinship is expected to facilitate altruism. If an act is particularly costly, it is most likely to be restricted to close kin. Figure 8.6 shows how the benefit/cost ratio scales with the degree of relatedness among individuals. Compare

Figure 8.6

As the degree of relatedness *(r)* between two individuals declines, the value of the ratio of benefits to costs *(b/c)* that is required to satisfy Hamilton's rule for the evolution of altruism rises rapidly.

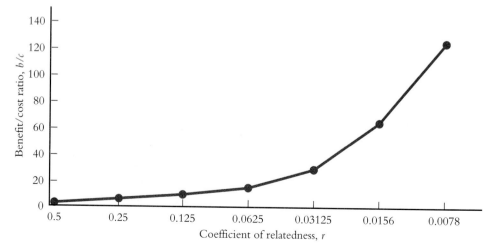

what happens when $r = \frac{1}{16}$ (or 0.0625) and when $r = \frac{1}{2}$ (or 0.5). When $r = \frac{1}{16}$, the benefits must be more than 16 times as great as the costs for Hamilton's inequality (Equation 1) to be satisfied. When $r = \frac{1}{2}$, the benefit needs to be just over twice as large as the costs. All other things being equal, altruism will be more common among close relatives than among distant ones.

It is important to understand that the fractions in Table 8.2 are not equivalent to the quantities in a recipe. If a recipe calls for 2 cups of flour and $\frac{3}{4}$ cup of sugar, the dutiful cook measures out these amounts. But monkeys don't necessarily apportion altruism in precise amounts according to the coefficients of relatedness. Reality is, as always, more complicated. There are sometimes dramatic asymmetries in the benefit/cost ratios within pairs of relatives which influence the distribution of altruism. For example, in many primate species, it may be much less costly for a mother to defend her infant than vice versa. The inability to recognize certain categories of kin, particularly distant kin or paternal kin, may also limit the distribution of altruism. Finally, it is important to remember that close kinship does not always prevent violence and aggression. In some species of birds, nestlings kill their own siblings to prevent them from competing for resources that parents bring back to the nest. Juvenile primates sometimes jostle with their newborn siblings for their mother's time and attention.

Evidence of Kin Selection in Primates

A considerable body of evidence suggests that the patterns of many forms of altruistic interactions among primates are largely consistent with predictions derived from Hamilton's rule. Later in this section we consider three examples: food sharing, grooming, and coalition formation. In each case, the behaviors are usually directed toward kin. First, however, we discuss how primates identify relatives.

Primates seem to know which animals are their relatives.

In order for kin selection to provide an effective mechanism for the evolution of cooperative behavior, animals must be able to distinguish relatives from nonrelatives and close relatives from distant ones. Is this a reasonable requirement for nonhuman primates? Current evidence suggests that unlike some other organisms, primates have no innate ability to distinguish kin from nonkin. Primates must learn to recognize relatives, and they probably base their knowledge of kinship on cues such as familiarity and proximity (Figure 8.7).

After they give birth, females repeatedly sniff and inspect their newborns. Nevertheless, if a strange infant is substituted for the female's own infant shortly after birth, most mothers accept the strange infant without any clear sign of having noticed the substitution, even if the foster infant is not the same sex as their own infant. But sometime within the first few weeks of life, mothers learn to recognize their infants. After this, females of most species nurse only their own infants (though communal nursing occurs in capuchin monkeys), and respond selectively to their own infant's distress calls. Although it might seem advantageous for primates to have an innate means of recognizing their young from the moment of birth, there is actually little need for this since young infants spend virtually all of their time in physical contact with their mothers. Thus, mothers are unlikely to confuse their own newborn with another.

Monkeys and apes seem to learn to identify their siblings and other relatives through contact with their mothers. Offspring continue to spend considerable amounts of time with their mothers even after their younger siblings are born.

Figure 8.7

(a) Even mothers must learn to recognize their own infants. Here, a female bonnet macaque peers into her infant's face. (b) Primates apparently learn who their relatives are by observing patterns of association among group members. Here, a female inspects another female's infant. (a, Photograph courtesy of Kathy West.)

(a)

(b)

Thus they have many opportunities to watch their mothers interact with their new brothers and sisters. At the same time, the newborn infant's most common companions are its mother and the mother's older offspring (Figure 8.8).

One shortcoming of this process is that it is clearly biased toward recognition of maternal kinship relationships. In nonmonogamous primate species, association patterns probably provide less reliable measures of paternal kinship relationships than maternal kinship relationships, for several reasons. First, in species that form multimale, multifemale groups, females may mate with several different males near the time of conception. Second, males do not often associate with their mates during pregnancy and lactation. These factors make it more difficult to detect paternal kin than maternal kin in many species.

Even though nonhuman primates may not discriminate kin from nonkin as humans do, and may have no conscious concept of kinship, they would be fulfilling the conditions of Hamilton's rule if they adopted the following policy: spend a lot

(a)   (b)

Figure 8.8

Monkey and ape infants grow up surrounded by various relatives. (a) These adult baboon females are mother and daughter. Both have young infants. (b) An adolescent female bonnet macaque carries her younger brother while her mother is recovering from a serious illness.

Figure 8.9
A female baboon allows her infant to share gum that she has extracted from an acacia tree.

of time near the female who nursed you, and direct altruistic behaviors principally toward those who also spend a lot of time close to her.

Food sharing occurs mainly among kin.

In a small number of primate species, adults voluntarily share food items with other group members. **Food sharing**, the unforced transfer of food from one individual to another, generally involves close relatives (Figure 8.9). At Gombe Stream National Park in Tanzania, Jane Goodall and her coworkers provided chimpanzees with bananas, a food the chimpanzees coveted but that doesn't grow in their home range (Figure 8.10). William McGrew of Miami University (Ohio) found that 86% of such exchanges involved maternal kin although these individuals constituted only 5% of the possible pairs of individuals in the group. Most exchanges involved mothers sharing bananas with their own offspring. Chimpanzees also share plant foods that they obtain outside the feeding area. One of us (J. B. S.) has found that most of these exchanges involved mothers and offspring, and mothers are particularly prone to share foods that their infants cannot obtain or process on their own. However, not all food sharing in chimpanzees involves kin. When male chimpanzees make a kill, they often share meat with other males and with unrelated adult females.

Figure 8.10
A female chimpanzee allows her infant to share bananas that she has received at the feeding station in the Gombe Stream Research Center. A veteran Tanzanian field assistant, Hilali Matama, monitors the situation.

Food sharing also occurs in tamarins and marmosets. These tiny monkeys are omnivores, feeding on fruit, gum and sap, nectar, insects, and small vertebrates. Young marmosets and tamarins have trouble catching large and mobile insect prey and manipulating large fruits. These foods are selectively shared with infants by older group members. In some cases, food items are spontaneously offered to infants and juveniles, and in other cases infants use specialized begging vocalizations to solicit food from others. Since tamarins and marmosets live in small monogamous or polyandrous groups, infants are probably closely related to most other group members.

Grooming is also more common among kin than nonkin.

Social **grooming** plays an important role in the lives of most gregarious primates (Figure 8.11). Grooming is likely to be beneficial to the participants in at least two ways. First, grooming serves hygienic functions as bits of dead skin, debris, and parasites are removed and wounds are kept clean and open. Second, grooming may provide a means for individuals to establish relaxed, **affiliative** (friendly) contact and to reinforce their social relationships with one another (Box 8.2). Groom-

(a)

(b)

(c)

(d)

Figure 8.11

Some of the many species of primates that groom are (a) capuchin monkeys, (b) blue monkeys, (c) baboons, and (d) gorillas. (Photographs courtesy of: a, Susan Perry; b, Marina Cords; d, John Mitani.)

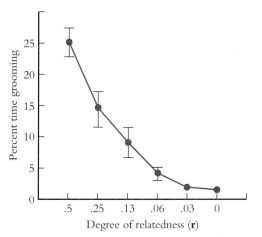

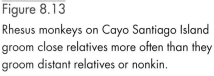

Figure 8.13

Rhesus monkeys on Cayo Santiago Island groom close relatives more often than they groom distant relatives or nonkin.

relatedness among rhesus macaques on Cayo Santiago Island. In this population, females groom close kin at higher rates than nonkin (Figure 8.13). It is interesting to note that as relatedness declined, the differences in the proportions of time spent grooming kin and nonkin were essentially eliminated. This may mean that monkeys cannot recognize more distant kin, or that the conditions of Hamilton's rule ($rb > c$) are rarely satisfied for distant kin.

Coalition formation occurs most frequently among close kin.

Most disputes in primate groups involve two individuals. Sometimes, however, several individuals jointly attack another individual, or one individual comes to the support of another individual involved in an ongoing dispute (Figure 8.14). We call these events **coalitions** or **alliances**. Support is likely to be beneficial to the individual that receives aid because it alters the balance of power among the original contestants. The beneficiary may be more likely to win the contest, or may be less likely to be injured in the confrontation. At the same time, however, intervention may be costly to the supporter who expends time and energy and risks defeat or injury when it becomes involved. Hamilton's rule predicts that support will be preferentially directed toward kin, and that the greatest costs will be expended on behalf of close relatives.

Figure 8.14

Two baboons form an alliance against an adult female.

Table 8.3   Captive pigtail macaques support close kin more frequently than distant kin. Individuals related by $r = 0.5$ include parents and offspring, and full siblings. Pairs of individuals related by $r = 0.25$ include half-siblings with each other, and grandparents and grandoffspring. Pairs of individuals related by $r = 0.125$ include cousins. (From Table 1 in A. Massey, 1977, Agonistic aids and kinship in a group of pigtail macaques, *Behavioral Ecology and Sociobiology* 2:31–40.)

| | Degrees of Relatedness | | |
	0.5	**0.25**	**0.125**
Number of individuals	38	156	48
Number of aids	173	164	13
Number of aids per pair	4.55	1.05	0.27
Number of aids per aggressive encounter	0.15	0.04	0.01

Many studies have shown that support is selectively directed toward close kin. Female pigtail macaques defend their offspring and close kin more often than they defend distant relatives or unrelated individuals (Table 8.3). Mature male bonnet macaques also intervene in support of their brothers and other male kin more often than they support unrelated males (Figure 8.15).

Females run some risk when they participate in coalitions, particularly when they are allied against higher-ranking individuals. Coalitions against high-ranking individuals are more likely to result in retaliatory attacks against the supporter than are coalitions against lower-ranking individuals. Female macaques are much more

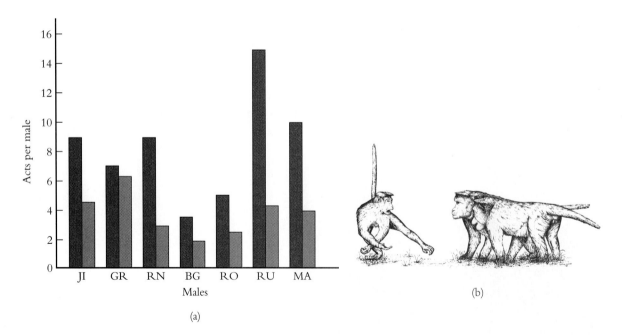

(a)

(b)

Figure 8.15

(a) In a large captive group of bonnet macaques, adult males supported related males (red bars) more often, on average, than they supported unrelated males (blue bars). Each of the seven males that had kin in the group followed this pattern. (b) Male bonnet macaques often intervene in ongoing disputes. Here, two males stand side by side and threaten a lower-ranking male, who raises his tail in submission and leaps away. (b, Drawing by Kathy West.)

likely to intervene against higher-ranking females on behalf of their own offspring than on behalf of unrelated females or juveniles. Thus, macaque females take the greatest risks on behalf of their closest kin.

Kin-based support in conflicts has far-reaching effects on the social structure of many primate groups. We have evidence of this from studying macaque, vervet, and baboon groups.

Maternal support in these species influences the outcome of aggressive interactions and dominance contests. Initially, an immature monkey is able to defeat older and larger juveniles only when its mother is nearby. Eventually, regardless of their age or size, juveniles are able to defeat everyone their mothers can defeat, even when the mother is some distance away. Maternal support contributes directly to several remarkable properties of dominance hierarchies within these species:

• Maternal rank is transferred with great fidelity to offspring, particularly daughters. For example, in a group of baboons at Gilgil, Kenya, maternal rank is an almost perfect predictor of the daughter's rank (Figure 8.16).

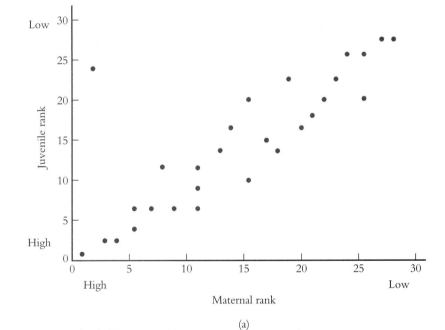

(a)

(b)

Figure 8.16
Juvenile female baboons acquire ranks very similar to their mother's rank.
(a) The anomalous point at maternal rank 2 belongs to a female whose mother died when she was an infant.
(b) Here, the dominant female of Alto's group is flanked by her two daughters, ranked 2 and 3.

- Maternal kin occupy adjacent ranks in the dominance hierarchy, and all the members of one **matrilineage** (maternal kin group) rank above or below all members of other matrilineages. A group of free-ranging baboons in Amboseli National Park, studied for many years by Jeanne and Stuart Altmann of the University of Chicago and their colleagues, provides a good example of this kind of dominance hierarchy (Figure 8.17). In 1981 the four top-ranking females were all descendants of a single female, named Alto (AL). Three of them were her daughters, and the other was her granddaughter.

- Ranking within matrilineages is often quite predictable. In most cases, mothers outrank their daughters, and younger sisters outrank their older sisters. However, there seems to be greater regularity in rank relationships within lineages in large provisioned and captive groups than in free-ranging groups like the one represented in Figure 8.17.

- Female dominance relationships are amazingly stable over time. In many groups, female dominance relationships remain the same over months and sometimes over years. In the Amboseli baboon group, there were virtually no changes in the relative ranks among females over a 10-year period, even though these females matured, gave birth, contracted illnesses, sustained injuries, and aged during this interval (Figure 8.17). The stability of dominance relationships among

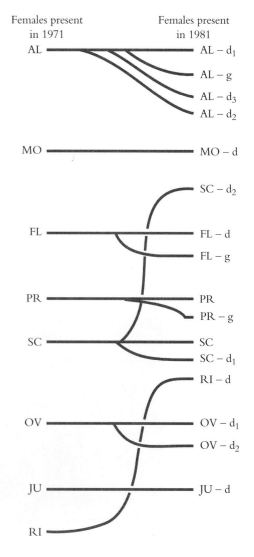

Figure 8.17

Female baboons in Amboseli National Park transfer rank to their daughters. Females are listed here in descending order of rank for two different times a decade apart. The two-letter capital codes represent particular matriarchs, the lower case letters represent the relationship (d = daughter, g = granddaughter), and the subscript numbers indicate birth order among sibling daughters. Thus AL-g is the only female granddaughter of AL, and AL-d_3 is AL's third daughter. Females that appear in 1971 but not in 1981 died during the interval, while females that appear in 1981 but not 1971 matured during the interval. Members of matrilineages generally rank together, with daughters ranking just below their mothers. There are few changes in dominance ranks over this period—only two females (SC-d_2 and RI-d) rose in rank.

females may be a result of the tendency to form alliances in support of kin. If a female knows that when she threatens another female, some of the victim's relatives will form an alliance against her, she may be reluctant to challenge females that outrank her. Even if females do not receive support in every encounter, the presence of potential allies may serve as an effective deterrent against aggression.

Reciprocal Altruism

Altruism can also evolve if altruistic acts are reciprocated.

The theory of **reciprocal altruism** relies on the basic idea that altruism among individuals can evolve if altruistic behavior is balanced between partners (pairs of interacting individuals) over time. In reciprocal relationships, individuals take turns being actor and recipient—giving and receiving the benefits of altruism (Figure 8.18). Reciprocal altruism is favored because over time the participants in reciprocal acts obtain benefits that outweigh the costs of their actions. This theory was first formulated by Robert Trivers of Rutgers University, and later amplified and formalized by others.

Three conditions occurring together favor the development of reciprocal altruism: individuals must

1. have an opportunity to interact often,
2. have the ability to keep track of support given and received, and
3. provide support only to those who help them.

Figure 8.18
Two old male chimpanzees groom each other. Reciprocity can involve taking turns or interacting simultaneously. Male chimpanzees remain in their natal communities throughout their lives and develop close bonds with one another.

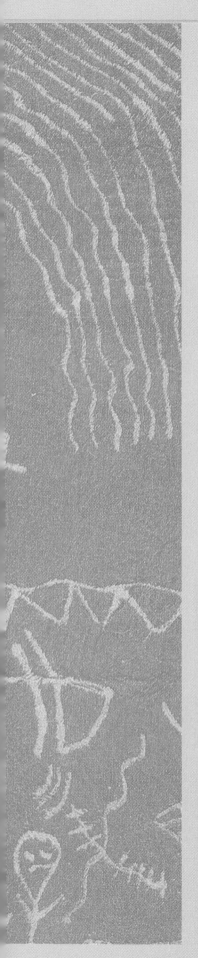

How Do We Study Primates?

HABITUATION AND RECOGNITION

At the beginning of any study the field primatologist faces a number of problems. He or she must locate subjects, habituate them to the presence of humans without disrupting their natural behavior, distinguish one group from another, and, if possible, learn to recognize animals individually. These early stages of fieldwork often represent a low point in morale and productivity. To begin with, habituating monkeys to close-range observation is extremely frustrating and dull. Little can be accomplished if your subjects vanish whenever you approach, although good work on vegetation, fecal samples, and warning cries has been conducted during this period. When the observer is finally lucky enough to get close to the subjects, attempts to recognize animals individually can create further frustration and anguish. In some studies, of course, the difficulties of individual recognition have been overcome by capturing and marking animals, while observers of species such as chimpanzees and gorillas report that their animals can be recognized individually as easily as humans, even after only one or two days' observation. For the most part, however, primatologists give thanks for cuts on the ear, misshapen noses, wounds that leave scars, and other abnormalities that facilitate recognition until, after months of frustration, individual monkeys are distinguished almost as easily as individual humans.

OBSERVATIONAL SAMPLING

Once a group of primates has been habituated and the animals can be recognized individually, the observer is ready to begin sampling their behavior. . . . Underlying such observations is a fundamental principle: one cannot record everything. Even in small groups, social behavior or feeding can occur in rapid sequences, and the observer will often want to gather carefully limited data on the durations of events, intervals between events, or sequences of events. From a large population of many behavioral acts, the observer must therefore sample a small proportion and design rules for sampling to yield data that comprise a representative, unbiased subset of all the events that have occurred.

. . . We briefly describe here perhaps the three most common methods of observational data collection.

Instantaneous Sampling. This method, also called point sampling, is used to record one or more classes of behavior at a predetermined instant in time. For example, suppose an observer wanted to determine whether individuals in a group of black-and-white colobus monkeys showed a particular

Many primate field projects benefit from the assistance of skilled observers like Mr. Mokupi Mokupi, who studies baboons in The Okavango Delta of Botswana.

pattern of feeding and resting throughout the day, perhaps feeding early in the morning, late in the afternoon, and resting at noontime. A test of this hypothesis using instantaneous sampling would proceed as follows. First, decide the time interval at which monkeys will be sampled, for example, one individual every minute. Second, produce a randomized list of the individual monkeys to be sampled. Third, beginning at the earliest light of dawn, locate the first monkey listed and record its activity at the instant that the minute changes or over a predetermined, brief interval (e.g., 5 sec) beginning on the minute. Then, repeating this procedure for the next monkey on the list, continue until all subjects have been sampled equally often for a specified length of time.

If this protocol is repeated for a number of days, and if sampling sessions are varied so that they include an equal number of sessions from dawn to dusk, the observer will find it fairly easy to compare, for each individual, the proportion of time spent in different activities at different times of day. Instantaneous sampling is often useful because it allows a large number of individuals to be sampled daily, and it minimizes the length of time between successive samples on the same individual. . . . The value of instantaneous samples is limited, however, because it provides no information on sequences of interactions or their duration.

Focal Animal Sampling. Suppose one wanted to test the hypothesis that genetically related individuals were more likely than others to groom within 1 minute after an aggressive interaction. Instantaneous sampling would clearly be inadequate to address this question, but an appropriate method could be devised if the observer simply increased the duration of time during which data were recorded. For instance, an observer might locate the first animal on the list and then follow that individual, recording all its activities, not for an instant but for some predetermined time, say 5 minutes, 10 minutes, 1 hour, or even a day. Such long-term follows of a preselected individual are an example of focal animal sampling, perhaps the most widely used method in the study of primate behavioral ecology. Focal animal sampling is an extremely versatile technique because the data it produces can be used to answer a variety of questions about frequencies (how often a behavior occurs), rates (how often a behavior occurs per unit time), sequences (how often behavior A follows behavior B), and durations (how long a behavior typically lasts). . . . One weakness of focal animal sampling, however, results from the time devoted to each sample. It is rarely possible to sample many animals on a given day, and considerable time may elapse between successive samples on the same individual. . . .

Ad libitum Sampling. A variety of conditions may conspire to make regular, ordered sampling of behavior by either instantaneous or focal animal sampling impossible. In dense vegetation, for example, it may prove extremely difficult to find "the next subject" on an observer's list, and all primatologists are familiar with the mysterious process that causes subjects to disappear from sight shortly after a focal sample on them has begun. More important, some extremely significant behaviors such as copulation may be so rare that most instances would go unrecorded even by regular observation. To overcome these limitations, a variety of ad libitum sampling techniques, each designed to supplement data gathered by more systematic methods, have been developed by different observers. Each method of ad-lib sampling has its own assets and limitations, and each reflects the different problems posed by work on different species. . . .

FIELD EXPERIMENTS

Field experiments have always played a major role in research on birds and nonprimate mammals, but until recently such experiments were rarely used by primatologists, for a number of different reasons. Most important, the same logistical factors that prevent observation at close range can discourage any thought of experimentation in the field. Accounts of the abnormal behavior that results when primates interact with humans in zoos or wild animal parks have encouraged the belief that behavior will quickly be distorted if the animals interact with their observers in any way. These considerations are certainly understandable and important, but it has also become clear during the past ten years that, with suitable precautions, well-controlled field experiments can be conducted on primates without distorting their behavior and that, as with other species, such experiments can provide new insights into behavior that cannot be obtained through observation alone.

Among primatologists, Hans Kummer and his colleagues were pioneers in the integration of observational and experimental field techniques. . . . Beginning with detailed observational data on hamadryas baboons, these investigators formulated hypotheses about the mechanisms underlying food gathering, partner preferences, and social structure. They then tested such hypotheses by experimentally manipulating food availability or social competition, either among their free-ranging subjects or in large enclosures. . . .

A second group of primatologists, following a long tradition of field experiments on bird vocalizations, has used portable tape recorders and loudspeakers to conduct playback experiments on free-ranging primates. . . . Such work was originally designed to provide new information on the use of vocalizations, and in this area it has made important contributions. . . . Perhaps more important, however, vocal playbacks offer an opportunity to test hypotheses about an animal's social or ecological knowledge because they allow an observer to mimic the presence of certain individuals or the occurrence of certain events under specified conditions.

Source (excluding photograph): From pp. 5–8 in B. B. Smuts, D. L. Cheney, R. M. Seyfarth, R. W. Wrangham, T. T. Struhsaker, 1987, The study of primate societies, pp. 1–8 in *Primate Societies,* ed. by B. B. Smuts et al., University of Chicago Press, Chicago. Copyright (1987) by The University of Chicago.

The first condition is necessary so that individuals will have the opportunity for their own altruism to be reciprocated. The second condition is important so that individuals can monitor and balance altruism given and received from particular partners. The third condition produces the nonrandom interaction necessary for the evolution of altruism. If individuals are unrelated, initial interactions will be randomly distributed to altruists and nonaltruists. However, reciprocators will quickly stop helping those who do not help in return, while they will continue to help those who do. Thus, as in the case of kin selection, reciprocal altruism can be favored by natural selection because altruists receive a disproportionate share of the benefits of altruistic acts. Note that altruistic acts do not need to be exchanged in kind; it is possible for one form of altruism (such as grooming) to be exchanged for another form of altruism (such as coalitionary support).

In primates, the conditions for the evolution of reciprocal altruism are probably often satisfied, and there is some evidence that it occurs.

Most primates live in social groups that are fairly stable, and they can recognize all of the members of their groups. We do not know whether primates have the cognitive capacity to keep track of support given and received from various partners, but we know they are very intelligent and can solve complex social problems. Thus, it seems likely that primates have both the opportunity and the capacity to practice reciprocal altruism.

In a number of species of primates, grooming is reciprocated. That is, individuals most often groom the animals from whom they receive the most grooming. Reciprocal grooming has been observed in several species of macaques, baboons, vervet monkeys, and chimpanzees. Some researchers doubt that grooming provides strong evidence for reciprocal altruism because there is little direct evidence that grooming is actually costly to individuals.

Stronger evidence of reciprocity comes from studies of coalitionary support. Researchers have reported that support in agonistic encounters is reciprocated in a variety of species, including several species of macaques and chimpanzees. A study of bonnet macaques conducted by one of us (J. B. S.) found that males often intervened in ongoing disputes among other males, and they gave the most support to the males from whom they received the most support (see Figure 8.15b). However, the number of acts of support among pairs of males was often unbalanced. In general, high-ranking males supported low-ranking partners more often than vice versa. This may be due to the fact that support was less costly to high-ranking males than to low-ranking ones. There is also evidence that grooming is sometimes exchanged for coalitionary support. Thus, male bonnet macaques most often supported the males who groomed them most frequently. Similar patterns have been observed in vervet monkeys, baboons, and chimpanzees.

These correlational studies are consistent with predictions derived from the theory of reciprocal altruism, but they do not demonstrate that altruism is contingent on reciprocation. The contingent nature of such exchanges is convincingly demonstrated by two experimental studies.

The first study was conducted by Robert Seyfarth and Dorothy Cheney of the University of Pennsylvania. To understand this experiment you need to know something about the methods that were used. The first step in their research was to tape-record vocalizations given by vervet monkeys in a broad range of situations. Then Seyfarth and Cheney assessed the relationships between the acoustic properties of the calls and the context in which the calls were given. This allowed them to identify a number of distinct calls and to develop hypotheses about the functional

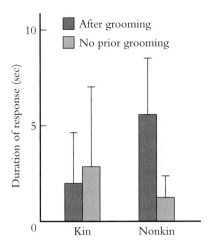

Figure 8.19
Vervet monkeys responded more strongly to recruitment calls played from a hidden speaker if the caller had previously groomed them than if the caller had not.

significance of the calls. These hypotheses were then tested in carefully designed playback experiments. In this kind of experiment, tape-recorded calls are played back to selected individuals from a hidden speaker, and the responses of the listeners are recorded on film or videotape. When they hear the tape-recorded calls, monkeys may look intently in the direction of the hidden speaker, approach the hidden speaker, or ignore the tape-recorded call. The duration of the monkeys' gaze in the direction of the hidden speaker is used to gauge the level of their interest in the tape-recorded call.

Like all monkeys, vervets spend much of their free time grooming. Vervets also form coalitions, and use specific vocalizations to recruit support. In this experiment, Seyfarth and Cheney played tape-recorded recruitment calls to individuals in two different situations. Vervet A's recruitment call was played to vervet B from a hidden speaker under two conditions: (1) after A had groomed B, and (2) after a fixed period of time in which A and B had not groomed. It was hypothesized that if grooming were associated with support in the future, then B should respond most strongly to A's recruitment call after being groomed. The vervets did just that (Figure 8.19).

The second study was designed by Frans de Waal of Emory University and focused on the relationship between grooming and food sharing in a well-established group of captive chimpanzees. At variable, prescheduled times over a three-year period, a group of chimpanzees was provided with compact bundles of leaves. Although these bundles could be monopolized by a single individual, possessors of these bundles often allowed other individuals to share some of their booty. Before each provisioning session, De Waal monitored grooming interactions among group members. He found that the possessors were much more generous to individuals who had recently groomed them than they were toward other group members (Figure 8.20). Moreover, possessors were much less likely to resist efforts to obtain food if their solicitors had recently groomed them than if they had not groomed them.

There are only a small number of well-documented examples of reciprocal altruism in nonhuman primates. However, reciprocal altruism may occur more often than it is detected by observers. Since altruism is potentially reciprocated in different currencies (grooming for support, predator defense for food, and so on), and since the actual costs and benefits associated with specific behaviors are almost impossible to quantify, it is very difficult to establish whether altruism is actually reciprocated.

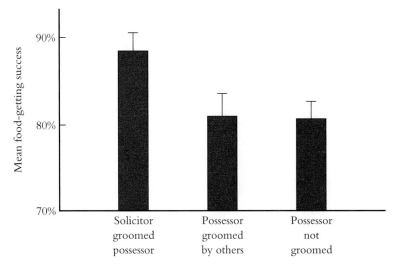

Figure 8.20

Chimpanzees were more successful in obtaining food from animals whom they had previously groomed (left bar) than from animals who had been groomed by others, or not groomed at all.

Further Reading

Gouzoules, S., and H. Gouzoules. 1986. Kinship. Pp. 299–305 in *Primate Societies,* ed. by B. B. Smuts, D. L. Cheney, R. M. Seyfarth, R. W. Wrangham, and T. T. Struhsaker. University of Chicago Press, Chicago.

Harcourt, A. H., and F. B. M. De Waal, eds. 1992. *Coalitions and Alliances in Humans and Other Animals.* Oxford University Press, Oxford.

Krebs, J. R., and N. B. Davies. 1993. *An Introduction to Behavioural Ecology.* Sinauer Associates, Sunderland, Mass., chap. 11.

Seyfarth, R. M., and D. L. Cheney. 1984. Grooming, alliances, and reciprocal altruism in vervet monkeys. *Nature* 308:541–543.

Study Questions

1. Consider the kinship diagram shown here. What is the kinship relationship (for example, mother, aunt, or cousin) and degree of relatedness (such as 0.5 or 0.25) for each pair of individuals?

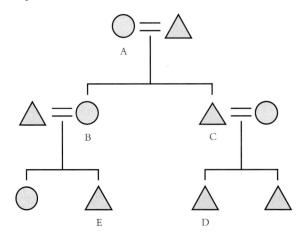

2. In biological terms, what is the difference between the following situations:
   (a) A male monkey sitting high in a tree gives alarm calls when he sees a lion

at a distance. (b) A female monkey abandons a desirable food patch when she is approached by another female.

3. In documentaries about animal behavior, animals are often said to do things "for the good of the species." For example, when low-ranking animals do not reproduce, they are said to be forgoing reproduction in order to prevent the population from becoming too numerous and exhausting their resource base. What is wrong with this line of reasoning?

4. Why is some sort of nonrandom interaction among altruists necessary for altruism to be maintained?

5. There are relatively few good examples of reciprocal altruism in nature. Why is reciprocal altruism uncommon? Why might we expect reciprocal altruism to be more common among primates than among other kinds of animals?

6. In Seyfarth and Cheney's experiment, they found that vervet monkeys tended to respond more strongly to the calls of the animals that had groomed them earlier in the day than to the calls of animals that had not groomed them. However, this effect held only for unrelated animals, not for kin. The vervets responded as strongly to the calls of grooming relatives as to those of non-grooming relatives. How might we explain kinship's influence on these results?

7. In addition to kin selection and reciprocal altruism, a third mechanism leading to nonrandom interaction of altruists has been suggested. Suppose altruists have an easily detected phenotypic trait, perhaps a red "A" on their chest or a green beard. Then altruists could use the following rule: "Do altruistic acts only for individuals who have green beards." Once the allele became common, most individuals carrying green beards would not be related to one another, so this would not be a form of kin selection. However, there is a subtle flaw in this reasoning. Assuming that the genes controlling beard color are at different genetic loci than the genes controlling altruistic behavior, explain why green beards would not evolve.

CHAPTER 9

Primate Intelligence

What Is Intelligence?

Compendiums of the names for multitudes of animals invariably include some terms that are familiar (such as "flock of sheep" and "herd of buffalo"), some that are quaint (for example, "sounder of wild hogs"), and some that are particularly apt ("gaggle of geese"). A lesser-known entry in the last category is a "shrewdness of apes." Although there may have been little empirical evidence to justify this label when it was coined, we now know that it is an accurate assessment. Primates are unusual, if not unique, in the relatively large size of their brains and the complexity of their behavior. Monkeys and apes have larger brains in relation to their body size than members of any other taxonomic group (Figure 9.1). Humans, of course, carry this evolutionary trend to an even greater extreme.

Large brains come at a substantial cost because brain tissue is extremely expensive to maintain. Our brains account for approximately 2% of our total body

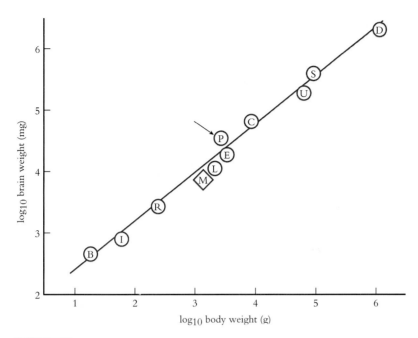

Figure 9.1

This graph plots the average value for brain and body weights for marsupials and for 10 orders of placental mammals. The red line represents the line that best fits the data for placental mammals. There is generally a uniform relationship between brain weight and body weight, with brain weight (in milligrams) being approximately equal to body weight (in grams) raised to the 0.75 power. Primates deviate from this best-fit line, which means that they have relatively large brains for their bodies. B = bats, C = carnivores, D = whales and dolphins, E = edentates, I = insectivores and tree shrews, L = rabbits and hares, M = marsupials, P = primates, R = rodents, S = seals and sea lions, U = hoofed mammals.

weight, but they consume about 20% of our metabolic energy. Natural selection does not maintain costly features unless they confer important adaptive advantages on organisms. Thus, one of the central questions of primate evolution is: why has evolution made primates so smart? Understanding the nature and causes of the cognitive abilities of nonhuman primates will help us to answer this question. We begin this chapter by considering the selective factors favoring intelligence and behavioral complexity within the order Primates, and then examine evidence regarding the extent of cognitive complexity in nonhuman primates.

Flexible problem-solving ability is a primary component of intelligence.

As you know, intelligence is notoriously difficult to define. Many researchers avoid the term entirely, referring instead to specific cognitive capacities or adaptations that humans possess. It is even more difficult to measure intelligence or evaluate the nature of cognitive adaptations in members of other species, partly because we know so little about their mental processes. Considerable ink has been spilled over the question of whether other animals have mental representations, thoughts, or consciousness.

Although we might not be able to agree on a precise definition of intelligence, most of us would probably agree that one fundamental component of intelligence

(a)

(b)

Figure 9.2

Behaviors that increase an animal's fitness are not necessarily evidence of intelligence. For example, (a) whales—like many other animals—migrate long distances to find food, and (b) a female wildebeest will defend her calf against attack. Yet neither navigational abilities nor parenting behaviors require the kind of flexibility necessary to solve novel problems—a key component of intelligence.

is the ability to solve problems in complex situations by flexible means. It is important to emphasize that intelligence and evolutionary success are not necessarily synonymous. If they were, then we would have no basis for distinguishing between a bird that recalls thousands of locations where it has stored seeds for the winter, a whale that finds its way from a beach in southern Argentina to summer fishing grounds in the Arctic Circle, a wildebeest female that manages to protect her calf from a hungry pack of hyenas, and a chimpanzee that leads his companions away from a hidden cache of prized food and returns to eat it in privacy. While the first three traits—a capacious memory, impressive navigational abilities, and effective parenting strategies—clearly contribute to the fitness of these organisms, they may be based on fixed stimuli or invariant rules that regulate behavior. Intelligence implies something more about the flexibility of the means used to solve problems. Therefore, our definition emphasizes the ability to cope with complexity and to incorporate novel solutions into existing behavioral repertoires (Figure 9.2).

Why Are Primates So Smart?

There is now considerable debate about the primary factors that favored the evolution of relatively large brains and enhanced cognitive capacities among non-human primates. Some researchers argue that ecological factors associated with locating and processing inaccessible food items are principally responsible for the evolution of intelligence among primates. Others have suggested that social demands associated with life in large, stable groups provided the primary selective force favoring cognitive complexity and intelligence among primates. We will review the rationales for each of these positions, and then consider the empirical data that bear on each.

Hypotheses Explaining Primate Intelligence

Some primatologists believe that primate intelligence evolved to solve ecological problems.

University of California, Berkeley, anthropologist Katherine Milton believes that the challenges associated with exploiting short-lived and patchily distributed foods may have favored greater cognitive abilities among primates. As we discussed in Chapter 6, fruit trees in tropical forests are patchily distributed and bear ripe fruit for relatively short periods of time. Once an animal has located a food source, it will become a dependable, but seasonal, resource over the lifetime of the individual (Figure 9.3). It would be advantageous for primates to form a detailed mental map of the sites at which fruit may be found and to be able to find their way from one food source directly to another. Milton argues that individuals who can plan routes efficiently, anticipate the availability and quality of various food resources, and keep track of changing conditions of food sources may be at a distinct advantage.

Katherine Gibson of the University of Texas, and Sue Parker of the State University of California at Sonoma, offer a second ecological hypothesis. They believe that natural selection has favored enhanced cognitive capacities among primates because many foods that primates consume require considerable skill to process. For example, some primates eat hard-shelled nuts that must be cracked open with stones or smashed against a tree trunk. Others dig up roots and tubers; extract insect larvae from bark, wood pith, or dung; and crack open pods to obtain seeds (Figure 9.4). These kinds of foods, which Gibson and Parker label **extractive foods**, are valuable because they tend to be rich sources of protein and energy. However, extractive foods require complicated, carefully coordinated techniques to process. For example, to feed on the pith of wild celery plants, mountain gorillas must first break the stem into manageable pieces, then peel the outer layers of the stalk away with their teeth, and finally pick out edible bits of the pith with their fingers. To master these skills primates must be more intelligent than other animals.

Figure 9.3

A muriqui feeds on fruit. A reliance on patchily distributed foods like fruit may place a premium on cognitive skills. (Photograph courtesy of Karen Strier.)

(a)   (b)

Figure 9.4

Primates sometimes exploit foods that are difficult to extract. Here, (a) a male chimpanzee pokes a long twig into a hole in a termite mound and extracts termites, and (b) a capuchin monkey punctures an eggshell and extracts the contents. (b, photograph courtesy of Susan Perry.)

Other scientists believe that primate intelligence evolved in order to solve social challenges.

In contrast to these ecological models, a number of primatologists believe that social challenges have provided the most important selective factor favoring the evolution of intelligence in primates (Figure 9.5). As we have seen, life in social groups means that animals face competition and experience conflict. At the same time, social life also provides opportunities for affiliation, cooperation, nepotism, and reciprocity. Nicholas Humphrey was among the first to voice these ideas. He wrote that

> . . . The life of social animals is highly problematical. In a complex society, such as those [societies] we know exist in higher primates, there are benefits to be gained for each individual member both from preserving the overall structure of the group, and at the same time from exploiting and out-maneuvering others within it. Thus, social primates are required from the very nature of the system that they create and

Figure 9.5

Living in complex social groups may have been the selective factor favoring large brains and intelligence among primates.

maintain to be calculating beings; they must be able to calculate the likely behavior of others, to calculate the balance of advantage and loss—and all this in a context where the evidence on which their calculations are based is ephemeral, ambiguous, and liable to change, not least as a consequence of their own actions. In such a situation, "social skill" goes hand in hand with intellect, and here at last the intellectual faculties required are of the highest order. The game of social plot and counter-plot cannot be played merely on the basis of accumulated knowledge, any more than can a game of chess (p. 309 in N. K. Humphrey, 1976, The social function of intellect, pp. 303–317 in *Growing Points in Ethology,* ed. by P. P. G. Bateson and R. A. Hinde, Cambridge University Press, Cambridge).

This view, that the challenges of social life have promoted the evolution of intelligence in primates, has come to be known as the **social intelligence hypothesis**.

Testing Models of the Evolution of Intelligence

Comparative tests of both the social and ecological hypotheses for the evolution of intelligence can be made by examining the relationship between the relative size of the brain's neocortex and the extent of social and ecological complexity across primate species.

Robin Dunbar, a primatologist at the University of Liverpool, pointed out that these models of the evolution of intelligence in primates generate specific predictions about the patterning of variation in intelligence or cognitive abilities among living primates. For example, Parker and Gibson's idea can be tested in theory by determining whether species that process complex, packaged foods are more intelligent than species that feed on simpler foods. However, a major obstacle to making such tests is the lack of reliable operational criteria to assess and compare cognitive abilities, ecological features, and social complexity among primate species.

The assessment of cognitive abilities is particularly problematic since we do not know exactly how the brain processes information or solves problems. Dunbar believes that the volume of the brain's neocortex provides a useful anatomical measure of cognitive capacity in primates for two reasons. First, the evolutionary changes that have occurred in the primate brain have mainly involved changes in the size and structure of the forebrain, particularly the neocortex (Figure 9.6). Second, the **neocortex** seems to be the thinking part of the brain, the part of the brain most closely associated with problem solving and behavioral flexibility. Therefore, Dunbar contends that the size of the neocortex and the **neocortex ratio**, the ratio of the size of the neocortex to the rest of the brain, can serve as a proxy for cognitive ability.

Both social and ecological hypotheses receive some support from comparative studies of the neocortex.

The ecological hypotheses for the evolution of intelligence predict that specific characteristics of the diet or the environment of particular primate species will be correlated with their cognitive abilities. For Milton's model, this means that smarter species should use more patchy food resources. Frugivores, which utilize a dispersed, patchy, and ephemeral (short-lived) food supply, would need greater cog-

Figure 9.6

These figures depict the brains of (a) a large prosimian, the indri; (b) a macaque, and (c) a human. There are three main components of the brain: the hindbrain, which contains the cerebellum and medulla (part of the brainstem); the midbrain, which contains the optic lobe (not visible here); and the forebrain, which contains the cerebrum. The cerebrum is divided into four main lobes: occipital, parietal, frontal, and temporal. The forebrain is greatly expanded in primates and other mammals, and much of the gray matter (which is made up of cell bodies and synapses) is located on the outside of the cerebrum in a layer called the cerebral cortex. The neocortex is a component of the cerebral cortex, and in mammals the neocortex covers the surface of virtually the entire forebrain.

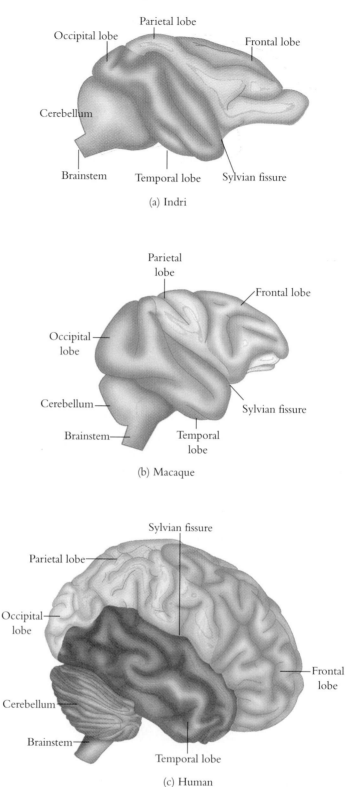

(a) Indri

(b) Macaque

(c) Human

nitive skills than folivores, whose foods are more uniformly distributed. If the hypothesis is correct, we should find that species with a high proportion of fruit in the diet will have a higher neocortex ratio. It is also possible that the size of the area contained in the mental map places the greatest demands on the brain. Frugivores typically have larger home ranges and longer day journeys than folivores.

Thus, Milton's hypothesis predicts that neocortex ratio and home-range size or day-journey length will be positively correlated. Gibson's model predicts extractive foragers will have relatively larger neocortex ratios than those who feed on more readily accessible foods.

The social intelligence hypothesis predicts that there will be a positive correlation between the complexity of social life and the neocortex ratio. Dunbar suggests that group size may be taken as a rough index of social complexity because primates in social groups need to recognize, associate, and interact with all the other members of their groups. Animals somehow keep track of their relationships with other members of their groups, particularly when they participate in social interactions that are regulated by nepotism or by reciprocity. Similarly, the decision to act aggressively or submissively, or to intervene in conflicts involving others, may depend in part on an individual's ability to remember or assess the dominance ranks of other group members. As groups grow larger, the number of pairs grows rapidly, making it considerably more difficult to keep track of social relationships. Thus, Dunbar predicts there should be a positive correlation between the size of social groups and the neocortex ratio.

Dunbar compiled data to test these predictions, gathering together data on group size, ecological parameters, and brain size from a variety of sources. He found neocortex ratio and neocortex size to be positively related to group size (Figure 9.7). That is, primates who live in large social groups tend to have larger neocortexes and greater neocortex ratios than those who live in smaller social groups. More recent work by Rob Barton of the University of Durham has extended Dunbar's result using statistical methods that control for phylogenetic affinities between species (as described in Box 4.1). Barton confirmed the relationship between group size and neocortex size. He also discovered that frugivorous monkeys and apes have larger neocortexes than other species do, and that this relationship is independent of the effects of group size. There is no evidence that home range size or extractive foraging are directly associated with neocortex enlargement.

These data provide support for the idea that social challenges and ecological parameters played a role in the evolution of cognitive capacities in nonhuman primates. However, there are two reasons to be cautious in drawing firm conclusions from these data. First, we cannot be sure that neocortex ratio or neocortex size are valid measures of cognitive ability because we know very little about how the brain processes information or shapes behavior. Second, it is always problematic to infer causal relationships from correlational data. The association between group size and neocortex ratio may be a spurious result that arises because both group size and neocortex ratio are causally related to a third variable that has not yet been identified. Thus, we now turn to observational and experimental evidence about what monkeys and apes know about their world.

KNOWLEDGE ABOUT THE ECOLOGICAL AND SOCIAL DOMAINS

If ecological pressures have favored the evolution of intelligence, then primates should be adept at solving ecological problems. Similarly, if social pressures have favored the evolution of intelligence, then primates should be adept at solving social challenges.

Certain problems arise when we attempt to measure what monkeys actually know about their world. Unlike psychologists working with human subjects, we cannot ask monkeys what they are thinking. At the same time, we cannot simply infer what they know or what they understand from what they do.

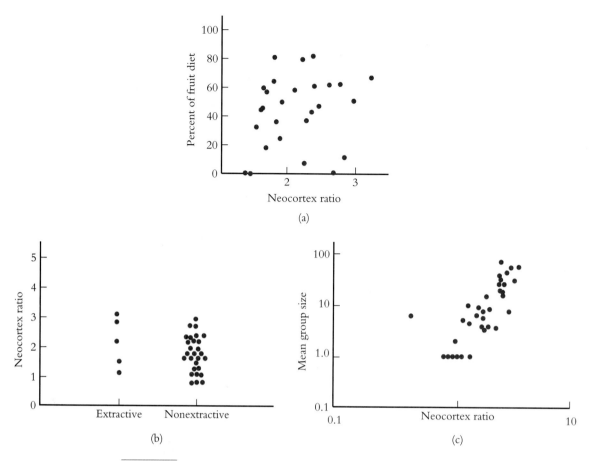

Figure 9.7

The neocortex ratio (the ratio of the volume of the neocortex to the volume of the rest of the brain) is plotted against (a) the amount of fruit in the diet, (b) extractive versus nonextractive foraging techniques, and (c) group size. Primates who live in large groups have larger neocortex ratios than those who live in smaller groups.

To see why this might be problematic, consider the following example. An observer watching a female macaque notes that she does not respond when her infant screams. How can the observer determine whether this means that the mother is (1) unaware that her infant is in distress, (2) unable to recognize her own infant's scream, or (3) reluctant to intervene against a higher-ranking female than herself? To distinguish among these three possibilities, the observer must be able to demonstrate that mothers can distinguish between calls given in different contexts, that mothers can recognize their own infants' calls and differentiate them from the calls of other infants, and that mothers can assess the nature of the risks their infants face. As daunting as these tasks seem, researchers working in field and laboratory situations have developed methods for answering these kinds of questions. As a result of this innovative experimental work, we are beginning to get some idea of what monkeys know about their world.

THE ECOLOGICAL DOMAIN

Monkeys and apes know a lot about their environments.

Monkeys are excellent naturalists. They select appropriate food items, avoid eating foods laced with toxins, know the location of food resources within their home ranges, move efficiently from one feeding site to the next, and evade predators. Although these skills are clearly related to survival and reproductive success, there is a general assumption that these kinds of ecological tasks are less cognitively demanding than social tasks. But it is possible that environmental knowledge is more complicated than we think, perhaps because we are accustomed to buying our food in grocery stores and using maps and compasses to find our way in the wilderness. Consider, for example, the tasks primates must face living in the forests of East Kalimantan, where there are more than 700 species of fruiting plants. These animals must distinguish which fruits are edible and which are not, track the status of fruit crops as they ripen, monitor the availability of fruits at different sites, make decisions about when to remain in a food patch and when to switch to a new one, and so on.

Macaques know a lot about the characteristics of the foods they eat.

Charles Menzel of Georgia State University has designed a series of experiments to find out what macaques know about the foods they eat. For example, he has compared the responses of free-ranging Japanese macaques finding indigenous foods and provisioned foods in their home ranges. When they find akebi fruits on the ground when these fruits would not normally be present, the monkeys immediately look up into trees for additional fruits. When they find chocolate on the ground, the macaques search the ground for more chocolate in the immediate vicinity, but don't look elsewhere. Apparently, these monkeys know that fruit grows on trees, but candy doesn't. When monkeys are presented with two similar unfamiliar objects, a red rubber ball with a black band and a red plastic tomato, they focus their attention on the plastic tomato, apparently orienting to generic visual features that create resemblances to familiar foods. These simple experiments suggest that monkeys have considerable, subtle knowledge about the properties of familiar food items.

Monkeys and apes construct cognitive maps of their home ranges. These maps enable them to move efficiently from one food source to another.

Perhaps the most convincing evidence of the complexity of primates' ecological knowledge comes from studies of what monkeys know about the location of food and how they move efficiently from one feeding site to the next. The mental representation of the location, availability, and quality of things in the environment is called a **cognitive map**. Cognitive maps allow animals to work out efficient routes from one site to another, saving them both time and energy. To see why cognitive maps are so useful, consider the way you navigate your own environment. You begin at home, perhaps rushing to your first class of the day. Then you might head to the library to study, or to the cafeteria for lunch. Later, you might have another class at the other end of campus. At the end of the day, you might stop at the gym to work out. When you were new to campus, you probably took the same routes every day and were afraid to stray from familiar paths for fear of getting hopelessly lost in the crowded maze of buildings. However, as time passed, you discovered new routes that shortened your walk from one place to another, and you felt more confident in your ability to get around campus. You could do this because of an ability to visualize where things were and an ability to plan efficient routes. You developed a cognitive map of your home range. Many other primates apparently have the same kinds of abilities.

Paul Garber of the University of Illinois at Urbana-Champaign has studied the foraging behavior of tamarins in the forests of Peru. Some of the plant species that tamarins feed upon are highly synchronous. This means that if one individual of a particular species is fruiting or producing gum, others of the same species are doing the same thing. Thus, when tamarins have exhausted the resources of one feeding tree, they can expect to find more food at other trees of the same species. Garber found that 70% of the time, tamarin groups moved directly from one feeding tree to the nearest tree of the same species that they had not yet depleted. These tiny monkeys apparently know the location of hundreds of food trees and can remember for days or weeks when they last fed at particular sites.

Charles Janson of the State University of New York at Stony Brook has conducted a series of field experiments to examine capuchin monkeys' knowledge of the location and quality of their food resources. Janson constructed special feeding platforms in the forest and baited the platforms with tangerines. He conducted the experiments during the winter months when virtually no other fruit was available to the monkeys. Janson varied the amount of food available on each platform, so some sites always contained more fruit than others, and no site was baited more than once per day. By following the monkeys' movements through the forest, he was able to examine their knowledge of the locations and quality of these artificial food sites. Janson found that monkeys moved to closer platforms more often than they moved to more distant ones. The capuchins preferred sites where they might expect to find more fruit over sites where they might expect to find less fruit, and they avoided visiting sites that had been depleted within the last 24 hours. These data suggest that monkeys were able to remember the location of their fruit patches, to assess the amount of fruit that would be available at each site, and to evaluate the likelihood that the crop would be renewed since their last visit.

Taken together, these data suggest that all monkeys create good cognitive maps of their home ranges and have detailed knowledge of their foods' characteristics. Advocates of the social intelligence hypothesis downplay this evidence because these skills do not necessarily distinguish primates from other kinds of animals. They point out that rodents, birds, fish, and perhaps even insects have similar skills. However, those who emphasize the complexity of ecological knowledge also contend that we don't yet understand what primates know about their environment, making it premature to conclude that other animals know as much about the nonsocial world as primates do.

SOCIAL KNOWLEDGE

One of the most striking things about primates is the interest they take in one another. Newborns are greeted and inspected with interest (Figure 9.8). Adult

Figure 9.8

In many primate species, all group members take an active interest in infants. Here a female baboon greets a newborn infant. Evidence from playback experiments, laboratory experiments, and naturalistic observations suggest that monkeys know something about the relationships among group members.

females are sniffed and visually inspected regularly during their estrous cycles. When a fight breaks out, other members of the group watch attentively. As we have seen in previous chapters, monkeys know a considerable amount about their own relationships to other group members. A growing body of evidence suggests that monkeys also have some knowledge of the nature of relationships among other individuals, or **third-party relationships**. Monkeys' knowledge of social relationships may enable them to form effective coalitions, practice deception, and manipulate other group members for political purposes.

There is evidence that monkeys and apes know something about the nature of kinship relationships among other members of their groups.

One of the first indications that monkeys understand the nature of other individuals' kinship relationships came from a playback experiment on vervet monkeys conducted by Dorothy Cheney and Robert Seyfarth in Amboseli National Park. Several female vervets were played the tape-recorded scream of a juvenile vervet from a hidden speaker. When the call was played, the mother of the juvenile stared in the direction of the speaker longer than other females did. This suggests that mothers recognized the call of their own offspring. However, even before the mother reacted, other females in the vicinity looked directly at the juvenile's mother. This suggests that other females understood who the juvenile belonged to, and that they were aware that a special relationship existed between the mother and her offspring.

Additional evidence that monkeys may understand mother-offspring relationships comes from more controlled laboratory experiments conducted by Verena Dasser working with a captive group of long-tailed macaques in Zürich, Switzerland. In these experiments, a young female was the primary subject. In the first phase, the female was shown one slide of a particular mother-offspring pair and a second slide of two monkeys who were not related as mother and infant. All of the monkeys in the slides were members of the subject's social group, and were well known to her. These trials were repeated until the female was able to select the mother-offspring pair consistently. In the second phase of the experiment, the female was again presented with two slides of group members, one depicting a mother-offspring pair and one depicting another pair of animals who were not mother and infant. However, in this phase of the experiment she was shown a different mother-offspring pair in each trial. If, *and only if,* she understood that there was a particular relationship between the mother and offspring that served as her model in training, she would select the new mother-offspring pair rather than the other pair of monkeys. The young female picked out the mother-offspring pair in nearly every trial. The monkey's performance in these experiments is particularly remarkable because the offspring in the mother-offspring pairs varied considerably in age, and in some cases the subjects had not seen the mother interacting with her child when it was an infant.

Monkeys may also have broader knowledge of kinship relationships. The evidence for this also comes from Cheney and Seyfarth's work on vervet monkeys. When monkeys are threatened or attacked, they often respond by threatening or attacking a lower-ranking individual that was not involved in the original incident, a phenomenon we call **redirected aggression**. Vervets selectively redirect aggression toward the maternal kin of the original aggressor. So, if female A threatens female B, then B threatens AA, a close relative of A. If monkeys were simply blowing off steam, or venting their aggression, they would choose a target at random. Thus, the monkeys seem to know that certain individuals are somehow related.

Several lines of evidence suggest that monkeys understand the nature of rank relation-ships among other individuals.

Since kinship and dominance rank are major organizing principles in most primate groups, it makes sense to ask whether monkeys also understand third-party rank relationships. Although information about knowledge of rank relationships is limited, the evidence suggests that monkeys may understand the relative ranks of other group members.

The most direct evidence that monkeys understand third-party rank relationships comes from playback experiments conducted on baboons in the Okavango Delta of Botswana by Cheney, Seyfarth, and one of us (J. B. S.). In this group, dominance relationships were stable, and females never responded submissively toward lower-ranking females. In this experiment, females listened to a recording of a female's grunt followed by another female's submissive fear barks. Female baboons responded more strongly when they heard a higher-ranking female responding submissively to a lower-ranking female's grunt than when they heard a lower-ranking female responding submissively to a higher-ranking female's grunt. Thus, females were more attentive when they heard a sequence of calls that did not correspond to their knowledge of dominance rank relationships among other females. Control experiments excluded the possibility that females were reacting simply to the fact that they had not heard a particular sequence of calls before. The pattern of responses suggests that females knew the relative ranks of other females in their group and were particularly interested in the anomalous sequence of calls.

Participation in coalitions may draw on sophisticated cognitive abilities.

Even the simplest coalition is quite a complex interaction. When coalitions are formed, at least three individuals are involved, and several different kinds of interactions are going on simultaneously (Figure 9.9). Consider the case in which one monkey, the aggressor, attacks another monkey, the victim. The victim then solicits support from a third party, the ally, and the ally intervenes on behalf of the victim against the aggressor. The ally behaves altruistically toward the victim, giving support to the victim at some potential cost to itself. At the same time, however, the ally behaves aggressively toward the aggressor, imposing harm or energy costs on the aggressor. Thus, the ally simultaneously has a positive effect on the victim and a negative effect on the aggressor. Under these circumstances, decisions

(a)   (a)

Figure 9.9

Primates form coalitions that are more complicated than the coalitions of most other animals. (a) In a captive bonnet macaque group, members of opposing factions confront one another. (b) Two capuchins jointly threaten a third individual out of the picture. (b, Photograph courtesy of Susan Perry.)

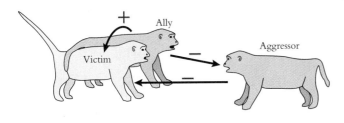

Figure 9.10

Complex fitness calculations may be involved in decisions about whether to join a coalition. By helping the victim against the aggressor, the ally increases the fitness of the victim but decreases the fitness of the aggressor.

about whether or not to intervene in a particular dispute may be quite complicated. Consider a female who witnesses a dispute between two of her offspring. Should she intervene? If so, which of her offspring should she support? When a male bonnet macaque is solicited by a higher-ranking male against a male that frequently supports him, how should he respond? In each case, the ally must balance the benefits to the victim, the costs to the opponent, and the costs to itself (Figure 9.10).

Understanding of third-party relationships may be particularly useful in managing coalitions.

Given the great complexity of even the simplest coalitions, knowledge of third-party relationships may be extremely valuable because it enables individuals to predict how others will behave in certain situations. Thus, animals who understand the nature of third-party relationships may have a good idea about who will support them and who will intervene against them in confrontations with particular opponents, and they may also be able to tell which of their potential allies are likely to be most effective in coalitions against their opponents.

Primates seem to form more complex coalitions than other animals do.

Many animals form alliances in defense of their territories and their young. However, Andrew Harcourt at the University of California, Davis, has concluded that primates use alliances in different ways from other animals. Primates seem to assess differences in competitive ability among other members of their group, and cultivate relationships with powerful individuals. For example, in many species, monkeys selectively groom higher-ranking individuals. In some of these species, grooming is exchanged for support in aggressive encounters, and individuals compete to groom high-ranking animals.

Primates may also attempt to manipulate the alliances among other group members. There are several accounts of chimpanzees attempting to prevent lower-ranking rivals from forming alliances with each other.

Deception provides further evidence of the sophistication of primate social knowledge.

Andrew Whiten and Richard Byrne of the University of St. Andrews in Scotland compiled and catalogued instances of deception in nonhuman primate species. They found evidence for a range of deceptive strategies employed by monkeys and apes (Box 9.1). Chimpanzees figure most prominently in Whiten and Byrne's compendium. Whiten and Byrne argue that these anecdotes are significant because they require considerable behavioral flexibility on the part of the actor. The acts are normal parts of the animal's behavioral repertoire, but the behaviors are used

BOX 9.1

Examples of Deception in Nonhuman Primates

The following examples of deception in nonhuman primates are adapted from A. Whiten and R. Byrne, 1988, The manipulation of attention in primate tactical deception, pp. 211–223 in *Machiavellian Intelligence,* ed. by R. Byrne and A. Whiten, Oxford University Press, Oxford.

Concealment Hans Kummer watched as an adult female hamadryas baboon spent 20 minutes inching toward a spot behind a rock. Once she reached this spot, she began to groom an adolescent male. This was an interaction not normally tolerated by the resident adult male.

Distraction Andrew Whiten and Richard Byrne observed a young male baboon attack a juvenile that screamed repeatedly. Screams are one means of recruiting support. As several adults came into view giving aggressive vocalizations, the attacker stood bipedally and stared intently into the distance as if a predator had been spotted. The adults paused and followed his gaze. Although no predator was detected, the aggressive interaction was terminated.

Creating a false image Frans de Waal, who conducted long-term observations of chimpanzees in the Arnhem Zoo (the Netherlands), described a male chimpanzee sitting with his back to a rival male. The male heard his rival give an aggressive vocalization and he grinned submissively. He used his fingers to push his retracted lips together over his teeth, altering his facial expression. He repeated this three times before the fear grin was eliminated. Then he turned to face his rival.

Manipulation of a conspecific using a social tool Robin Dunbar described one case in which an infant gelada baboon was unsuccessful in its efforts to nurse. The infant moved near the group's dominant male, and then vocalized, hit the male's back, and pulled on the long mane of hair around his shoulders. The male ignored the infant, but the infant pulled on his hair again. The male then turned and struck at the infant. In the commotion, the mother looked up. When the infant approached her again, she allowed it to nurse.

in unusual ways and contexts to achieve specific objectives that are beneficial to the actors. Thus, when a baboon suddenly looks intently toward the horizon in the midst of an aggressive contest, it uses a standard part of its behavioral repertoire. Such behavior generally means that a predator has been sighted, and is not a standard element in aggressive encounters. In this case, it has the effect of distracting the baboon's opponents and ending the conflict.

Deception among primates differs from the kinds of deception seen in other animals. Other animals feign injury to lure potential predators away from vulnerable young, they mimic the phenotypes of foul-tasting species, or they camouflage themselves to blend in with their surroundings (Figure 9.11), but there is not much flexibility in this deceptive behavior. Moreover, all of these forms of deception are principally directed at members of other species. Primates seem to be unusual in

(a)   (a)

Figure 9.11
Many animals, like this stick insect (a), camouflage themselves as a defense against predators (b). This is an effective form of deception, but does not require sophisticated cognitive abilities.

the flexible and tactical nature of their deception, and in their capacity to deceive familiar members of their own species.

A number of researchers remain skeptical about the existence of deception among nonhuman primates, and believe that each of the anecdotes catalogued by Whiten and Byrne has a simpler explanation. It is very difficult to prove that a given incident is the result of a conscious intention to deceive conspecifics. Behaviors like hiding from the dominant male while grooming a subordinate male, or distracting aggressors by feigning concern about a predator, may be random innovations that happen to work, not goal-directed strategies.

Because chimpanzees are able to manipulate other group members adeptly for strategic purposes, they may have mastered rudimentary political tactics.

In his influential book, *Chimpanzee Politics*, Frans de Waal documents how male chimpanzees achieve and maintain high-ranking positions in a captive colony at the Arnhem Zoo in The Netherlands. He vividly describes power struggles among three rival males and their efforts to contain alliances that would threaten their own positions. The central players in this drama were Yeroen, Luit, and Nikkie. Yeroen and Luit were the two oldest males in the group, while Nikkie was considerably younger. For many years, Yeroen was the undisputed leader of the group, and his relationship with Luit was stable but never entirely relaxed. Then Luit abruptly began to challenge Yeroen's position, beginning a power struggle that continued for more than a year. When Luit's challenge began, Yeroen started to associate more often with the adult females in the group who invariably supported him in contests against Luit. Luit then began to harass the females when they associated with Yeroen, but he also groomed them assiduously at other times. Gradually the females began spending less and less time with Yeroen. During this period, Luit obtained support from the third male, Nikkie. Although Nikkie did not intervene directly against Yeroen on Luit's behalf, he did turn his attentions to those females who supported Yeroen. By harassing the females when they came to Yeroen's aid, he distracted them from supporting Yeroen. This in turn enabled Luit to intimidate Yeroen. After many months, Luit successfully established himself as the most dominant male and gained the support of most of the females who had formerly supported Yeroen. But further developments continued to alter the balance of power within the group when Yeroen and Nikkie formed a coalition that ultimately undermined Luit's position. Luit initially attempted to deter his rivals from associating with one another, but he was not successful. This time, Nikkie emerged on top, and Yeroen regained some of his former power.

Polly Want a Cracker?

[Note: This reading describes the cognitive skills of Alex, a grey parrot, trained by Irene Pepperberg. Note that Alex's numerical accomplishments parallel those of apes trained in the laboratory.]

Alex is remarkable not just for the fact that he has a large vocabulary but the fact that he uses English words appropriately to get what he wants. Irene Pepperberg, who trained Alex, realized that one of the reasons that most talking birds don't seem to understand what they are saying is that people teach them to speak in ways that make it impossible for them to make the link between the sound of the word and what it stands for. For instance, how is any parrot supposed to learn that 'Hello!' is a greeting given by people when they first see each other if its main experience of the word is to have it repeated many times by people standing in front of it for minutes or even hours at a time? And how is even the most perceptive parrot supposed to realize that the phrase 'Pretty Polly' refers to the bird itself and is not just the sound that a human being makes, equivalent to a dog barking or a cow mooing?

Pepperberg reasoned that if she could devise a way of training Alex so if he used a particular word it would have important and specific consequences for him (such as the correct word for a piece of food being followed by his being given that food), he might then stand a chance of learning that the word and the result were connected. If he could do this, he might begin to show evidence of understanding the meaning of words or at least using them in the right context. The exciting thing was that not only was Pepperberg successful in teaching Alex to use words correctly, she was also able to go on and ask him questions and get answers from him, some of which, as we shall see, related directly to his understanding of numbers.

Her training method sounds bizarre, but it worked. Alex was placed on a perch near two human beings who were talking to each other and not to the parrot at all. At this stage, Alex was nothing more than a passive observer. One of the human trainers would hold up an object and say to the other 'What's this?'. The object in question would be something calculated to be of interest to a parrot such as a nut, a cork or a wooden clothes peg (parrots like things they can chew). If the second trainer correctly named the object, she would be given it and praised, watched by the parrot. But if she gave the wrong name, she would be firmly told 'No!' and the object ostentatiously taken away and hidden.

At first, such conversations and the giving or withholding of objects went on entirely between the two human trainers with Alex taking an evident interest in the proceedings. Eventually, however, he started to join in, giving the word (or, initially, an approximation to the right sound) for an object he wanted. If he gave the correct name, he, too, would be praised and given the object. From then on the trainers would direct their conversations to Alex—holding up a series of objects and either giving them to him if he named them correctly or saying 'No!' if he didn't. Alex quickly learnt the names of nine different things—paper, key, wood, peg, a piece of leather, cork, corn, nut and pasta. He could say the right name on over 80 per cent of the occasions when he was shown one of them. He could also cope with variations on the original objects, such as a piece of paper that was of a colour or shape that he had not seen before.

Alex proceeded to show his grasp of human words by spontaneously using them in entirely appropriate ways. It is disconcerting, to say the least, to watch a videotape of an unsuccessful training session in which a parrot refuses to co-operate with his trainer and then see it terminated by the bird itself moving off-camera muttering the words 'I'm going away!' His use of the word 'No!', too, had an uncanny aptness. Just as his trainers had said 'No!' if he had given the wrong word to something they had shown him, so he started saying 'No!' back to them if they gave him something he didn't like. If they gave him a cork, he might drop it and say 'No!' As his training continued, he started to replace his typical parrot speech (inadequately rendered as 'RAAAKK!') with the more and more frequent use of 'No!' to express displeasure or non-co-operation, such as when he was refusing to take part in any more naming trials. Clearly, talking to the natives in their own language was more effective than screeching at them!

It was, however, when Pepperberg moved on to teaching Alex about colour, shape and number that she opened up the possibility of giving real insights into an animal's mind. She taught Alex the names of three colours—'rose' (red), 'green' and 'blue' as well as two shapes—'three-corner' (triangle) and 'four-corner' (square). Alex could say 'four corner paper' if shown a square of paper and asked 'What shape?' and he could say 'green' if shown something green and asked 'What colour?'. He would correctly say 'green' even if the object was a shade of green he had not seen before or was itself something completely unfamiliar to him.

Pepperberg also taught Alex numbers, using the same methods that she had used before to teach him the names and colours of objects. She would put Alex on his perch and then hold a conversation with the other trainer in front of him. 'How many?', one of them would say, holding up five wooden sticks. 'Five wood', the other would reply. 'That's right', she would be told, 'FIVE wood' and be given the sticks. Eventually, Alex might intervene with 'I wood' and then later, after watching many similar conversations, he, too, would say 'Five wood' and then be praised and given the sticks to chew.

In the same way Alex was taught to give the correct response to the question 'How many?' when two, three, four or six pieces of wood or corks were held up to him. If he gave the right answer, he was given the sticks or the corks. If he said the wrong number or called the objects by their wrong name, he was told 'No!' and they would be hidden from him. The corks and sticks used in this experiment varied considerably in appearance because, as Pepperberg puts it in her report, 'of Alex's prior manipulation of these exemplars' (he chewed them to bits, in other words). But the variable physical state of the objects he was being asked about inadvertently made the experiment all the more convincing because it tested his ability to correctly identify how many things there were whatever they looked like. Two intact wooden sticks or two chewed ones or one large one and one nearly demolished one all had to be described as 'Two wood', so any simple pattern recognition of what two things looked like was quite impossible.

Once Alex had learnt to say how many corks or wooden sticks there were, he was then asked about the numbers of other objects such as keys, paper or clothes pegs. These were all things that Alex had seen before and knew the names of but he had never been formally shown collections of them and asked 'How many?'. An experiment was done to see whether he could straight away transfer his knowledge of numbers from corks and sticks to these new things. On 145 tests with the new objects, he gave a completely correct answer nearly 80 per cent of the time, saying both what they were and how many of them were being held up. Most of his incorrect answers consisted of giving the correct name but leaving out the number, such as saying 'Paper' rather than what he should have said which was 'three paper'. After further prompting ('HOW MANY paper?'), he was correct about number in 90 per cent of his 'incorrect' trials. As Hank Davis pointed out in commenting on this reluctance to give the number, it could have been because Alex had no particular interest in getting more than one of anything. What he wanted was a clothes peg to chew and it may not have mattered to him how many he was given as long as it was at least one. . . .

At this stage, however, it was not clear exactly what Alex was doing. He was clearly capable of attaching the right colour, shape and number labels to things but quite what his idea of 'number' consisted of was still vague. The objects he appeared to 'count' were very variable in size so that he couldn't just have been equating 'number' with surface area or size. They were also very different in appearance so that it seemed unlikely that he was using an estimation of number based on the pattern the objects made. But it was just still possible that he was doing one of these things (two objects do, after all, usually take up less room than four, and two things do make a different pattern to five). The only way to find out was to try and increase the novelty of the situations in which he had to apply his number labels and find out if he could still correctly apply them even to completely new objects, new arrangements and new combinations.

Pepperberg's next approach was, therefore, to try asking Alex about the numbers of totally novel objects, things that had never been used in any of his experiments before. These were such things as bottles of typewriter fluid, toy cars, washers, thimbles or antacid tablets. Obviously, Alex could not be asked what these were as he had not been taught their names, so all he had to do when asked the question 'How many?' (thimbles, cars, etc.) was to give the correct numerical answer between two and six. Even with these entirely novel objects (actually, they weren't completely novel; they had been placed where Alex could see them for a couple of days so that he wouldn't be alarmed by them, but they were novel in the sense that he had never before been asked any questions about them) even here he answered correctly on 80 per cent of trials. So three bottles of typewriter fluid would get the response 'Three' even though he had never before had to give this answer to these particular objects. Alex had clearly transferred 'Three' from a familiar to an unfamiliar situation, showing that he had an idea of what constituted 'three' that was not tied to specific situations or specific things. This is at least some evidence that he was genuinely thinking about a number.

His next task was even more complicated. Alex was shown a mixed set of objects—say two keys and two corks, making four objects altogether, and again asked, 'How many?' Again his answers were astonishingly accurate. On 70 per cent of occasions he correctly said the total number of things present, despite the inherent ambiguity of the question 'How many?' when applied to such a situation. ('How many corks?' or 'How many things altogether?'; I don't think I would be sure which was meant without a little further clarification.)

Pepperberg admits that she does not know whether Alex really does 'count' in the sense that we know it or whether he had a much cruder grasp of what it was to assign numbers to groups of objects. He certainly has the ability to give the right 'number label' in appropriate situations and to transfer it to new ones that he has not specifically encountered before. He has, it would seem, gone considerably further than the rats with their 'checking off' of tunnels in a row but quite what he thinks about is still not clear. Pepperberg herself is quite restrained in what she claims for hm and she sums up her experiments as simply providing 'evidence for certain, albeit limited, number-related abilities in the grey parrot'. Number-related abilities, certainly, and those that fall into the category of what we mean by 'thinking'— that is, having an internal representation of the external world. Whether we call it protocounting or protothinking or dispense with such prefixes altogether does not really matter. We do have evidence—real evidence—that some animals at least have the rudiments of what in ourselves we would refer to as 'thinking'.

Source: From pp. 119–25 in Marian Stamp Dawkins, 1993, *Through Our Eyes Only?*, Oxford University Press, Oxford.

It is difficult to be certain that Yeroen, Luit, and Nikkie consciously plotted their strategies or were aware of the political consequences of their behavior. De Waal believes, however, that it is a reasonable possibility and should seriously be considered. Others have reported similar though less baroque examples of manipulation of coalitions among male chimpanzees.

THEORY OF MIND

The ability to understand the mental states of other individuals is called a theory of mind.

Developmental psychologists have discovered that very young children cannot distinguish between their own knowledge and the knowledge of others. They acquire this ability only as they mature. The capacity to distinguish between the two is called a **theory of mind** and is considered an essential prerequisite for performing complex deceptions, imitating, pretending, and teaching. Recently, primatologists have begun to consider the question of whether or not nonhuman primates can attribute mental states to others.

The question of whether monkeys and apes can attribute mental states to others might seem both impossible to demonstrate conclusively (see Box 9.2) and of no real practical importance. However, a theory of mind is useful to animals living in social groups. For example, deception requires the manipulation of another individual's belief about the world. Consider a low-ranking male who unexpectedly comes upon a desirable food item. He may know from past experience that older and stronger animals routinely take such items from him. It would make sense, then, either to carry the food item away, hide it and return to it later, or to lead unsuspecting group members away from the area. But to execute this deception, the finder must first understand that his knowledge differs from the knowledge of other group members and then come up with a way to take advantage of this discrepancy effectively. A similar argument can be made for teaching. Female chimpanzees have been seen demonstrating the fundamental elements of nut cracking to their offspring. In teaching, the instructor has to realize the limits of the pupil's knowledge first. In turn, the pupil has to grasp the intent of the instructor as well as the objective of the behavior. Both imitation and teaching play an important role in the transmission of complex behavior patterns in humans and underlie the human capacity for culture.

THE GREAT APE PROBLEM

A considerable body of evidence suggests that great apes have greater cognitive sophistication than monkeys. In particular, chimpanzees seem to have a greater knowledge of the minds of others than monkeys do.

Apes seem to have more knowledge of what others are thinking than monkeys do (see Box 9.2). This capacity may permit apes to manipulate and exploit others, and it may explain why chimpanzees figure so prominently in Byrne and Whiten's compendium of deception (see Box 9.1) and de Waal's accounts of political intrigue. Knowledge of others' state of mind may underlie the capacity for empathy and may enable chimpanzees to console others after conflicts have ended. There is also experimental evidence that chimpanzees are better able to view tasks from others' perspective than monkeys are. Chimpanzees also provide the best evidence of teaching and imitation and are the most proficient tool users (Figure 9.12).

BOX 9.2

Examining Theory of Mind in Children, Monkeys, and Apes

You might be surprised, skeptical, or extremely doubtful that we would venture to draw conclusions about non-human primate ability to distinguish between their own knowledge of the world and others' knowledge of the world. How can we know this for any organism that we can't question directly? Cognitive psychologists have devised a number of extremely clever ways to address these issues that do not require language. Several experiments deal with **attribution,** the capacity to assess the knowledge or mental states of others.

To give you a concrete example, we will briefly outline the classic experiment on attribution. A young child is shown a matchbox and asked what it contains. The child normally answers that it contains matches. The child is shown that the matchbox actually contains something else, say M&M's (or the British equivalent of M&M's called Smarties). Then a newcomer enters the room, and the child is asked what the newcomer will think is in the matchbox. Children below the age of three or so invariably say that the newcomer will think the matchbox contains M&M's. Older children say that the newcomer will think the matchbox contains matches. These results suggest that young children cannot distinguish what they know about the world from what others know about the world, but that they acquire this ability as they mature.

This test cannot be applied directly to monkeys and apes, but other tests can. Cheney and Seyfarth conducted a series of experiments designed to assess attribution in monkeys. One of these experiments compared the behavior of mothers when their offspring were aware of the presence of desirable food items and when they were ignorant of the foods' proximity. In one set of trials mothers and their offspring were seated side by side, and both could see into the test area as apple slices were placed into a food bin. In the other set of trials, only the mother could see the apple slices being placed in the food bin. Then the juvenile was released into the test area. Mothers of ignorant offspring did not behave any differently from mothers of knowledgeable offspring; they did not call more often, orient more toward the food bin, or otherwise seem to communicate their knowledge to their offspring. As a result, knowledgeable offspring found the food items significantly sooner than ignorant offspring did.

These results suggest that the mothers did not differentiate between what they knew and what their infants knew, though other explanations for the mothers' behavior are possible. For example, a mother might be aware of her infant's mental state, but not use this information to alter the infant's behavior. However, this interpretation is weakened by the fact that in a parallel experiment, mothers also failed to warn ignorant offspring about the presence of a frightening and potentially dangerous situation. Certainly, there should be strong selection favoring alerting offspring to the presence of danger.

Woodruff and Premack, from the University of Pennsylvania, conducted an ingenious experiment to assess the chimpanzee's ability to attribute mental states. Sarah, a chimpanzee involved in research on language and cognition, was shown videotapes of an actor in a cage who was faced with a variety of dilemmas. In one case the actor could not

reach a bunch of fruit hanging from the top of the cage. In another he could not reach fruit just beyond the bars of the cage. Then Sarah was given a series of photographs. One of the photographs depicted the solution to the problem (such as standing on a chair), and the others showed irrelevant actions (such as reaching sideways with a stick). Sarah routinely chose the photograph that represented the appropriate solution to the problem. Premack argued that Sarah's choice of the correct solution means that she may have understood the actor's intention and desire to get the fruit. To do this, she may have attrib-

uted a state of mind to the actor. On the other hand, as Premack acknowledged, it may be that Sarah understood the problem and knew how to solve it, but did not actually attribute a state of mind to the actor.

Sarah could not solve all the attribution problems put to her, but young human children cannot solve all the attribution problems that Sarah can solve. These data do not prove that apes have a theory of mind, or that monkeys do not. But they do provide suggestive evidence regarding this question, and they do indicate that such questions are open to objective, scientific investigation.

The social intelligence hypothesis does not predict that apes will be more intelligent than monkeys.

Richard Byrne, one of the original architects of the social intelligence hypothesis, points out that great apes do not face greater social challenges than other monkeys do, but apes are nonetheless more intelligent than monkeys. The social groups of great apes are no larger and no more complex than the social groups of many monkeys. Orangutans, in fact, are largely solitary. Thus, it seems unlikely that social pressures alone are responsible for cognitive differences between monkeys and apes.

Some researchers believe that the cognitive abilities of great apes evolved in response to ecological challenges, not social challenges.

Byrne suggests that ancestral populations of great apes were once subject to strong selective pressures for more efficient feeding. Great apes have a particular need for efficient feeding techniques because they are large-bodied creatures who move relatively slowly and inefficiently and have no specialized digestive anatomy or cheek pouches to store food. The adaptive solution to this problem was the development of an ability to plan a course of action. This is turn required the ability to represent

Figure 9.12

Chimpanzees make and use tools in the wild. Here a female carefully inserts a twig into a hole in a termite mound. (Photograph by William McGrew.)

actions mentally. Byrne suggests that great apes can represent abstract problems in their minds, enabling them to simulate alternative actions and to compute potential outcomes. The ability to visualize and plan is reflected in the development of technically demanding foraging techniques, which only apes can master. Although these abilities were initially favored because they enhanced foraging efficiency, they have come to play an important role in the social lives of great apes as well.

Great apes make use of more complex foraging techniques than other primates do.

Byrne points out that great apes make use of more complicated foraging techniques than other primates do, enabling them to feed on some foods that other primates cannot process. For example, he notes that all plant foods that mountain gorillas rely upon are well-defended by spines, hard shells, hooks, and stingers. To process these kinds of foods, gorillas must perform a different sequence of steps, each structured in a particular way. Many of the foods that orangutans feed upon are also difficult to process.

Great apes also use tools to obtain access to certain foods that are not otherwise available to them. Chimpanzees poke twigs into holes of termite mounds and anthills, use leaves as sponges to mop up water from deep holes, and employ stones as hammers to break open hard-shelled nuts (see Chapter 18). Recently, Carel van Schaik of Duke University and his colleagues observed Sumatran orangutans using sticks to probe for insects and to pry seeds out of the husks of fruit. Although bonobos and gorillas have not been observed to use tools in the wild, all apes are adept tool users in captive settings.

A number of animals use tools in the wild, but chimpanzee tool use suggests that more complex cognitive processes are involved when apes use them: two or more tools may be utilized in sequence to perform a single task, one tool may be employed to make another tool, tools may be chosen before they are actually needed, tools may be modified before they are used, and tools may be manipulated in novel ways to fit new circumstances.

Although the special cognitive abilities of great apes may have evolved to solve ecological challenges, these abilities have come to play an important role in the social domain as well. The ability to represent abstract problems mentally, to simulate alternative courses of action, and to compute potential outcomes enables great apes to become adept at the "game of social plot and counter-plot," which Humphrey originally described. Apes are able to understand the minds, reactions, and likely behavior of other group members, to evaluate alternate tactics, and to manipulate and deceive others effectively.

Some researchers believe that the magnitude of the cognitive gap between monkeys and apes has been exaggerated and therefore deny that great apes pose a challenge to the social intelligence hypothesis.

Michael Tomasello and Josep Call, who recently published an encyclopedic review of studies of primate cognition, believe that great apes may not actually be smarter than monkeys. They point out that direct comparisons of monkeys' and apes' cognitive skills are scarce because relatively few cognitive experiments have been conducted on both. In the few cases in which direct comparisons can be drawn, apes don't always perform better than monkeys. Moreover, researchers don't always know what cognitive skills underlie observed behaviors, making it impossible to attribute sophisticated cognitive abilities to apes that they may not actually possess (Box 9.3). Finally, many examples of apes' superior cognitive abilities come from observing apes who have had extensive contact with humans and are not seen in apes who have been raised in more natural contexts.

Box 9.3

Walks Like a Duck, Quacks Like a Duck

One of the authors (J. B. S.) conducted research on chimpanzees at the Gombe Stream Research Center, where chimpanzees have been studied for several decades. She relates the following story:

At Gombe the terrain is steep and the undergrowth is thick. Observers typically work in pairs and try to follow individual chimpanzees for several hours at a time. The chimpanzees move effortlessly through the densest brush and up the steepest hills taking paths observers find hard to follow. As a consequence, observers may lose sight of the individual they are following or of one another. Observers developed a system for locating one another when they become separated. Taking a cue from Tarzan movies, one person would "hoot," and their partner would give an answering call.

One day I was following a large party of chimpanzees as they initiated a hunt. At Gombe chimpanzees become completely silent when they are hunting, moving very quickly. Soon a young female chimpanzee named Little Bee and I became separated from the main body of the group. Little Bee had a clubbed foot and moved more slowly than the others. I planned to let Little Bee lead me to the main party, and I fell in step behind her. Little Bee soon paused and stood motionless on the path. I heard nothing; Little Bee didn't move. Little Bee sat down and started feeding casually on some berries that grew along the path. But as the pause grew longer, I became impatient and hooted for my partner. Soon there was an answering hoot from some distance away. Little Bee cocked her head toward the sound and headed off immediately in that direction. After a

while she paused, listened, and sat down. We waited together for several minutes. This time, Little Bee looked back at me. So I hooted again to my partner. As soon as the answering call came, Little Bee headed off. This sequence repeated itself several times, until we finally reached the main party. A young male named Goblin had caught a colobus monkey and was struggling to kill it.

This episode suggests that Little Bee must have understood the meaning of the hoots. After all, if she had not understood this, she wouldn't have been able to make use of the observers' calls to find her way. And she wouldn't have waited expectantly for J. B. S. to call to her partner.

There is plenty of reason to have confidence in this assumption. Humans are very closely related to chimpanzees and are extremely similar in many ways. Thus, it seems reasonable to assume that similar behaviors are shaped by the same underlying cognitive mechanisms. Since their responses were the same, the cognitive processes that shaped those responses must be the same too. After all, if it walks like a duck and quacks like a duck, it must be a duck, right?

Maybe not. In an elegant series of experiments, Daniel Povinelli and Timothy Eddy of the New Iberia Research Center at the University of Southwestern Louisiana have demonstrated that chimpanzees and humans draw different inferences about the world from what they see. We know that these differences are not based on the ability to perceive images, because the visual systems of humans and chimpanzees are very similar. Instead, they are based on how the brain processes perceptual information, and links what animals see with what they understand.

Figure 9.13

Chimpanzees use a specific gesture to solicit (beg) for food. They hold their hand out, palm up, and reach toward the possessor.

Povinelli and Eddy worked with a group of young chimpanzees. In the testing room there was a clear partition with two holes about an arm's length apart. An experimenter sat in front of one of the holes. If a chimpanzee reached through the hole in front of the experimenter and extended his hand, he was given a piece of fruit. If he reached through the other hole, he received nothing. This experiment made use of a begging gesture, which is part of the chimpanzee's normal behavioral repertoire (Figure 9.13). The chimpanzees quickly mastered this task, receiving fruit in virtually every trial. Then, in a series of baseline trials, each chimpanzee encountered two experimenters when entering the room. One experimenter was offering a block of wood, the other a piece of fruit. The chimpanzees invariably gestured to the experimenter holding the fruit.

The baseline trials were interspersed with a series of experimental "probe" trials in which the animals encountered a novel situation when entering the testing room. In one series of probe trials, the chimpanzees encountered two experimenters, one facing them, the other with her back toward them. From the beginning of this set of trials, the chimpanzees invariably approached the experimenter who was facing them and then began begging.

This suggests that the chimpanzees understood that an experimenter could see them begging only if she were facing them. This interpretation is consistent with naturalistic observations con-ducted by Michael Tomasello and his colleagues at the Yerkes Primate Center. They found that chimpanzees use visually based gestures (e.g., begging) when individuals are oriented toward them and tactile (e.g., poking) or auditory (e.g., ground slap) gestures when recipients are oriented away from them.

However, this commonsense interpretation is apparently incorrect. In a series of other trials, the chimpanzees failed to distinguish between an experimenter who could see them and an experimenter who could not see them. Thus, when faced with two experimenters facing toward them, the chimpanzees were just as likely to beg from an experimenter who had a bucket over her head as they were to beg from an experimenter whose head was unobscured. One could argue that buckets are not natural objects, and so it's not surprising that chimpanzees failed this test or similar tests with blindfolds and screens (Table 9.1). But they also failed to distinguish between an experimenter who was standing with her back to them and looking away from them and an experimenter who was standing with her back to them and looking over her shoulder at them—a situation they must encounter frequently in nature. Apparently the chimpanzees did not understand that the experimenter's awareness of their begging gestures was predicated upon the experimenter's ability to see them. In contrast, by the age of three years, human children do quite well on these kinds of tasks, and they clearly un-

Table 9.1 An experiment showing the success rate of chimpanzees distinguishing between experimenters who could see them and experimenters who could not.

	SUCCESSFUL (%)
Baseline	0.98
Back vs. front	0.84
Bucket on head	0.58
Blindfold	0.49
Screen over face	0.54
Hands over eyes	0.53
Look over shoulder	0.50

(Adapted from Table 3 in D. J. Povinelli and T. J. Eddy, 1996, *What Young Chimpanzees Know about Seeing.* Monographs of the Society for Research in Child Development, vol. 61, no. 3, University of Chicago Press, Chicago.)

derstand the relationship between seeing and knowing.

These results are surprising and counterintuitive. It is particularly surprising because we know that chimpanzees can follow a line of sight. When they see one animal staring at something, such as a snake in the grass, they follow the animal's line of sight. Chimpanzees seem to understand that seeing is linked to knowledge of things in the real world. In humans, gaze following appears in early infancy and is linked by psychologists to the infant's emerging understanding of the mental state of attention and the theory of mind. The ability to follow a gaze suggests that chimpanzees should have the same understanding of seeing as a projection of an internal state of attention.

However, this does not seem to be the case. In another series of probe trials, chimpanzees entered the testing room and encountered two experimenters, much as they did in the other experiments. This time, one experimenter was gazing at the upper corner of the room and one experimenter was facing forward. All of the chimps turned and looked up toward the corner of the room, following the experimenter's gaze, which remained fixed on the upper corner of the room. Then, about half of the time, the chimpanzees proceeded to beg from the experimenter who was still gazing steadily at the corner of the room.

This body of experiments is instructive for two reasons. First, it provides detailed information about what chimpanzees know about seeing. Second, it provides a warning about the dangers of *assuming* that similar behaviors are based on similar cognitive underpinnings. For children, who do quite well on many of the tasks that chimps failed, visual perception is linked to understanding the relationship between seeing and knowing. But chimpanzees seem to be largely oblivious to this relationship. They simply don't seem to understand that there is such a thing as "seeing," or that others have internal visual experiences. Thus, chimpanzees may see like humans, and they may follow a gaze like humans, but they apparently don't think about seeing like humans do.

The Value of Studying Primate Behavior

As we come to the end of Part II, it may be useful to remind you of the reasons that information about primate behavior and ecology plays an integral role in the story of human evolution. First, humans are primates, and the first members of the human species were probably more similar to living nonhuman primates than to any other animals on earth. Thus, by studying living primates we can learn something about the lives of our ancestors. Second, humans are closely related to primates and similar to them in many ways. If we understand how evolution has shaped the behavior of animals so much like ourselves, we may have greater insights about the way evolution has shaped our own behavior and the behavior of our ancestors. Both of these kinds of reasoning will be apparent in Part III, which covers the history of our own lineage.

Further Reading

Byrne, R. W. 1995. *The Thinking Ape*. Oxford University Press, Oxford.

Byrne, R., and A. Whiten, eds. 1988. *Machiavellian Intelligence*. Oxford University Press, Oxford.

Cheney, D. L., and R. M. Seyfarth. 1990. *How Monkeys See the World*. University of Chicago Press, Chicago.

Russon, A. E., K. A. Bard, and S. T. Parker. 1996. *Reaching into Thought: The Minds of the Great Apes*. Cambridge University Press, Cambridge.

Whiten, A., and R. W. Byrne. 1997. *Machiavellian Intelligence II*. Cambridge University Press, Cambridge.

Study Questions

1. Diet, extractive foraging, and social challenges have all been suggested as influences on the evolution of cognitive abilities among nonhuman primates. What is the rationale for each of these models?

2. Robin Dunbar's analysis comparing primate group size and the relative size of the primate neocortex is based largely on the correlation between these two variables. What shortcomings are inherent in correlational analyses like this one?

3. Compare and contrast the forms of deception displayed by primates and by other animals.

4. What is meant by a theory of mind? Why is this topic important in primate studies?

5. Describe the patterning of alliances among primates. Explain why alliances are such complicated interactions for the participants.

6. What are the advantages and disadvantages of using experimental methods, like those developed by Cheney and Seyfarth, in studying primate social behavior? What advantages do such methods have over naturalistic observations, and vice versa?

7. What evidence do we have that monkeys have a concept of kinship?

PART THREE

The History of the Human Lineage

CHAPTER 10

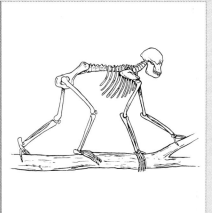

From Tree Shrew to Ape

Continental Drift and Climate Change
The Methods of Paleontology
The Evolution of the Early Primates
The First Anthropoids
The Emergence of the Hominoids

During the Permian and early Triassic periods (Table 10.1), much of the world's fauna was dominated by therapsids, a diverse group of reptiles that possessed several traits linking them to the mammals that evolved later. For example, therapsids may have been warm-blooded and covered with hair (Figure 10.1). At the end of the Triassic, most therapsid groups disappeared, and dinosaurs radiated to fill all of the niches for large, terrestrial animals. One therapsid lineage, however, evolved and diversified to become the first true mammals. These early mammals were probably mouse-sized, nocturnal creatures that fed mainly on seeds and insects. They had internal fertilization but still laid eggs. By the end of the Mesozoic era, 65 million years ago (mya), placental and marsupial mammals that bore live young had evolved. With the extinction of the dinosaurs at the beginning of the next era (the Cenozoic) came the spectacular radiation of the mammals. All of the modern descendants of this radiation—including horses, bats, whales, elephants, lions, and primates—evolved from creatures very much like a shrew (Figure 10.2).

In order to have a complete understanding of human evolution, we need to know how the transition from a shrewlike creature to modern human took place.

Table 10.1   The geological time scale.

ERA	PERIOD	EPOCH	PERIOD BEGINS (MYA)	NOTABLE EVENTS
Cenozoic	Quarternary	Recent	0.01	Origins of agriculture and complex societies
		Pleistocene	1.7	Appearance of *Homo sapiens*
	Tertiary	Pliocene	5	Dominance of land by angiosperms, mammals, birds, and insects
		Miocene	23	
		Oligocene	34	
		Eocene	54	
		Paleocene	65	
Mesozoic	Cretaceous		136	Rise of angiosperms, disappearance of dinosaurs, second great radiation of insects
	Jurassic		190	Abundance of dinosaurs, appearance of first birds
	Triassic		225	Appearance of first mammals and dinosaurs
Paleozoic	Permian		280	Great expansion of reptiles, decline of amphibians, last of trilobites
	Carboniferous		345	Age of Amphibians; first reptiles, first great insect radiation
	Devonian		395	Age of Fishes; first amphibians and insects
	Silurian		430	Land invaded by a few arthropods
	Ordovician		500	First vertebrates
	Cambrian		570	Abundance of marine invertebrates
Precambrian				Primitive marine life

Remember that according to Darwin's theory, complex adaptations are assembled gradually, in many small steps, each step favored by natural selection. Modern humans have many complex adaptations, like grasping hands, **bipedal** locomotion (walking upright on two legs), toolmaking abilities, language, and large-scale co-operation. To really understand human evolution, we have to consider each of the steps in the lengthy process that transformed a small, solitary, shrewlike insectivore scurrying through the leaf litter of a dark Cretaceous forest into a farmer leaning on her hoe after a hard day's work in the tropical sun, or into a dying soldier lying trapped in a tangle of barbed wire between hostile armies. Moreover, it is not enough to chronicle the steps in this transition. We also need to understand why each step was favored by natural selection. We want to know, for example, why a quadrupedal (walking on all fours), apelike animal should become a bipedal creature, and why complex grammatical language should replace a simpler system of vocalizations.

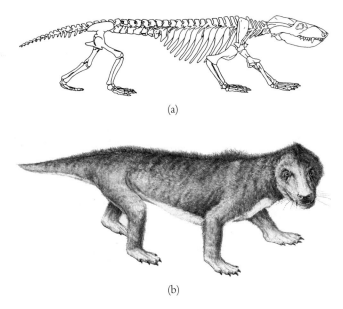

(a)

(b)

Figure 10.1

Therapsids dominated the earth about 250 mya, before dinosaurs became common. The therapsids were reptiles, but they may have been warm-blooded and had hair instead of scales. The therapsid *Thrinaxodon,* whose skeleton is shown in (a) and reconstruction shown in (b), was about 30 cm long and had teeth suited for a broad, carnivorous diet.

In this part of the text, we trace the history of the human lineage. We begin in this chapter by describing the emergence of the creatures that resemble modern lemurs and tarsiers, then document the transformation into monkeylike creatures, and finally into animals something like present-day orangutans. In later chapters, we recount the transformation from hominoid to hominid. We will introduce the first members of the human family, Hominidae, and then the first members of our own genus, *Homo,* and finally the first known representatives of our own species, *Homo sapiens.* We know something about each step in this process, although far more is known about recent periods than about periods in the most distant past. We will see, however, that there is still a great deal left to be discovered and understood.

Figure 10.2

The first mammals probably resembled the modern-day Belanger's tree shrew.

Continental Drift and Climate Change

To understand the evolution of our species, it is important to understand the geological, climatic, and biological conditions under which these evolutionary changes occurred.

When we think about the evolution of modern humans, we usually picture early humans wandering over open grasslands dotted with acacia trees—the same breathtaking scenery that we see in wildlife documentaries. As we shift the time frame forward through millions of years, the creatures are altered, but in our imagination the backdrop is unchanged. However, our mental image of this process is misleading, because the scenery has changed as well as the cast of characters (Figure 10.3).

It is important to keep this fact in mind because it alters our interpretation of the fossil record. If the environment were constant, then the kinds of evolutionary changes observed in the hominid fossil record (such as increases in brain size, bipedality, and prolonged juvenile dependence) would have to be seen as steady improvements in the perfection of human adaptations. Evolution would seem to be creating steady *progress* toward a fixed goal. But if the environment was changing, then new characteristics seen in the fossils would not necessarily represent progress in a single direction. Instead, these changes may have been adaptations to novel and impermanent environmental conditions. As we will see, the world has become much colder and drier in the last 20 million years, a change that altered the course of human evolution. If during the same period the world had become warmer rather than colder, then our human ancestors would probably have remained in the safety of the trees, and not become terrestial or bipedal. We would probably be stellar rock climbers but poor marathon runners.

The positions of the continents have changed relative to each other and to the poles.

The world has changed a lot in the last 200 million years. One of the factors that has contributed to this change is the movement of the continents, or **continental drift**. The continents are not fixed in place; instead, the enormous, relatively light plates of rock that make up the continents slowly wander around the globe, floating on the denser rock that forms the floor of the deep ocean. Around 200 mya, all of the land making up the present-day continents was joined together in a single, huge landmass called **Pangaea**. Around 125 mya, Pangaea began to

Figure 10.3

Today, East African savannas look much like this. In the past, the scenery is likely to have been quite different.

break apart into separate pieces (Figure 10.4). The northern half, called **Laurasia**, included what is now North America and Eurasia minus India, while the southern half, **Gondwanaland**, consisted of the rest. By the time the dinosaurs became extinct 65 mya, Gondwanaland had broken up into several smaller pieces. Africa and India separated, and India headed north, eventually crashing into Eurasia, while the remainder of Gondwanaland stayed in the south. Eventually, Gondwanaland separated into South America, Antarctica, and Australia, and these continents remained isolated from each other for many millions of years. South America did not drift north to join North America until about 5 mya.

Continental drift is important to the history of the human lineage for two reasons. First, oceans serve as barriers that isolate certain species from others, and so the position of the continents plays an important role in the evolution of species. Second, continental drift is one of the engines of climate change, and climate change has fundamentally influenced human evolution.

The climate has changed substantially during the last 65 million years—first becoming warmer and less variable, then cooling, and finally fluctuating widely in temperature.

The size and orientation of the continents have important effects on climate. Very large continents tend to have severe weather. This is why Chicago has much worse weather than London, even though London is much further north. Pangaea was much larger than Asia, and is likely to have had very cold weather in winter. When continents restrict the circulation of water from the tropics to the poles, world climates seem to become cooler. These changes, along with other, poorly

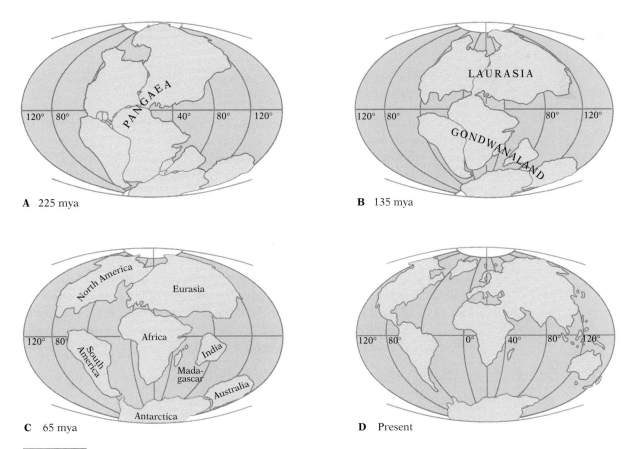

A 225 mya

B 135 mya

C 65 mya

D Present

Figure 10.4

The arrangement of the continents has changed considerably over the last 180 million years.

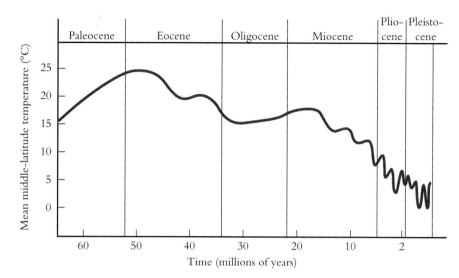

Figure 10.5

Global temperature has changed substantially in the last 65 million years. The world has cooled during most of this period. The climate has also become much more variable during the last 10 million years or so.

understood factors, have led to substantial climate change. Figure 10.5 summarizes changes in global temperature during the Cenozoic era.

To give you some idea of what these changes in temperature mean, consider that during the period of peak warmth in the early Miocene, palm trees grew as far north as what is now Alaska, rich temperate forests (like those in the eastern United States today) extended as far north as Oslo (Norway), and only the tallest peaks in Antarctica were glaciated.

The Methods of Paleontology

Much of our knowledge of the history of life comes from the study of fossils, the mineralized bones of dead organisms.

In certain kinds of geological settings, the bones of dead organisms may be preserved long enough for the organic material in the bones to be replaced by minerals (**mineralized**) from the surrounding rock. Such natural copies of bones are called **fossils**. Scientists who recover, describe, and interpret fossil remains are called **paleontologists**.

A great deal of what we know about the history of the human lineage comes from the study of fossils. Careful study of the shape of different bones tells us about what early hominids were like—how big they were, what they ate, where they lived, how they moved, and even something about how they lived. When the methods of systematics described in Chapter 4 are applied to these materials, they also can tell us something about the phylogenetic history of long-extinct creatures. The kinds of plant and animal fossils found in association with the fossils of our ancestors tells us what the environment was like—whether it was forested or open, how much it rained, and whether rainfall was seasonal.

There are several radiometric methods for estimating the age of fossils.

In order to assign a fossil to a particular position in a phylogeny, we must know how old it is. As we will see in later chapters, the date that we assign to particular specimens can profoundly influence our understanding of the evolutionary history of certain lineages or traits.

Radiometric techniques provide one of the most important ways to date fossils. To understand how radiometric techniques work, we need to review a little chemistry. All of the atoms of a particular element have the same number of protons in their nucleus. For example, all carbon atoms have six protons. However, different **isotopes** of a particular element have different numbers of neutrons in their nucleus. Carbon-12, the most common isotope of carbon, has six neutrons, and carbon-14 has eight. Radiometric methods are based on the fact that the isotopes of certain elements are unstable. This means that they change spontaneously from one isotope to another of the same element, or to an entirely different element. For example, carbon-14 changes to nitrogen-14 and potassium-40 changes spontaneously to argon-40. For any particular isotope, such changes (or, **radioactive decay**) occur at a constant, clocklike rate that can be measured with precision in the laboratory. There are several different radiometric methods:

1. **Potassium-argon dating** is used to date the age of volcanic rocks found in association with fossil material. Molten rock emerges from a volcano at a very high temperature. As a result, all of the argon gas is boiled out of the rock. After this, any argon present in the rock must be due to the decay of potassium. Since this occurs at a known and constant rate, the ratio of potassium to argon can be used to date volcanic rock. Then, if a fossil is discovered in a geological **stratum** (layer) lying under the stratum containing the volcanic rock, paleontologists can be confident that the fossil is older than the rock. A new variant of this technique, which is called **argon-argon dating** because the potassium in the sample is converted to an isotope of argon before it is measured, allows more accurate dating of single rock crystals.

2. **Carbon-14** (or **radiocarbon**) **dating** is based on an unstable isotope of carbon that living animals and plants incorporate into their cells. As long as the organism is alive, the ratio of the unstable isotope, carbon-14, to the stable isotope, carbon-12, is the same as the ratio of the two isotopes in the atmosphere. Once the animal dies, carbon-14 starts to decay into carbon-12 at a constant rate. By measuring the ratio of carbon-12 to carbon-14, paleontologists can estimate the amount of time that has passed since the organism died.

3. **Thermoluminescence dating** is based on an effect of high-energy nuclear particles traveling through rock. These particles come from the decay of radioactive material in and around the rock and from cosmic rays that bombard the earth from outer space. When they pass through rock, they dislodge electrons from atoms, so the electrons become trapped elsewhere in the rock's crystal lattice. Heating a rock relaxes the bonds holding the atoms in the crystal lattice together. All of the trapped electrons are then recaptured by their respective atoms, a process that gives off light. Researchers often find flints at archaeological sites that were burnt in ancient campfires. It is possible to estimate the number of trapped electrons in these flints by heating them in the laboratory and measuring the amount of light given off. If the density of high-energy particles currently flowing through the site is also known, scientists can estimate the length of time that has elapsed since the flint was burned.

4. **Electron-spin-resonance dating** is used to determine the age of **apatite crystals**, an inorganic component of tooth enamel, based on the presence of trapped electrons. Apatite crystals form as teeth grow, and initially they contain no trapped electrons. These crystals are preserved in fossil teeth and, like the burnt flints, are bombarded by a flow of high-energy particles that generate trapped electrons in the crystal lattice. Scientists estimate the number of trapped electrons by subjecting the teeth to a variable magnetic field, a technique called electron spin resonance. To estimate the number of years since the tooth was formed, paleontologists must once again measure the flow of radiation at the site where the tooth was found.

Different radiometric techniques are used for different time periods. Methods based on isotopes that decay very slowly, like potassium-40, work well for fossils from the distant past. However, they are not useful for more recent fossils because their "clock" doesn't run fast enough. When slow clocks are used to date recent events, large errors can result. For this reason, potassium-argon dating usually cannot be used to date samples less than about 500,000 years old. Conversely, isotopes that decay quickly, like carbon-14, are only useful for recent periods, because all of the unstable isotopes decay in a relatively short period of time. Thus, carbon-14 can only be used to date sites that are less than about 40,000 years old. The development of thermoluminescence dating and electron-spin-resonance dating is important because these methods allow us to date sites that are too old for carbon-14 dating, but too young for potassium-argon dating.

Absolute radiometric dating is supplemented by relative dating methods based on magnetic reversals and comparison with other fossil assemblages.

Radiometric dating methods are problematic for two reasons. First, a particular site may not always contain material that is appropriate for radiometric dating. Second, radiometric methods have relatively large margins for error. This has led scientists to supplement these absolute methods with other, relative methods for dating fossil sites.

One such relative method is based on the remarkable fact that every once in a while the earth's magnetic field reverses itself. This means, for example, that compasses now pointing north would at various times in the past have pointed south (if compasses had been around then, that is). The pattern of magnetic reversals is not the same throughout time, so for any given time period the pattern is unique. But the pattern for a given time *is* the same throughout the world. We know what the pattern is because when certain rocks are formed, they record the direction of the earth's magnetic field at that time. Thus, by matching up the pattern of magnetic reversals at a particular site with the well-dated sequence of reversals from the rest of the world, scientists are able to date sites.

Another approach is to make use of the fact that sometimes the fossils of interest are found in association with fossils of other organisms that existed for only a limited period of time. For example, during the last 20 million years or so there has been a sequence of distinct pig species in East Africa. Each pig species lived for a known period of time (based on securely dated sites). This means that some East African materials can be accurately dated from their association with fossilized pig teeth.

The Evolution of the Early Primates

The evolution of flowering plants created a new set of ecological niches. Primates were among the animals that evolved to fill these niches.

During the first two-thirds of the Mesozoic, the forests of the world were dominated by the **gymnosperms**, trees like contemporary redwood, pine, and fir. With the breakup of Pangaea during the Cretaceous, a revolution in the plant world occurred. Flowering plants, called **angiosperms**, appeared and spread. The evolution of the angiosperms created a new set of ecological niches for animals. Many angiosperms depend on animals to pollinate them, and they produce showy flowers with sugary nectar to attract pollinators. Some angiosperms also entice animals to disperse their seeds by providing nutritious and easily digestible fruits. Arboreal animals that could find, manipulate, chew, and digest these fruits could exploit these new niches. Primates were one of the taxonomic groups that evolved to take advantage of these opportunities. Tropical birds, bats, insects, and some small rodentlike animals probably competed with early primates for the bounty of the angiosperms.

The ancestors of modern primates were small and insectivorous, much like contemporary shrews.

To understand the evolutionary forces that shaped the early radiation of the primates (Figure 10.6), we need to consider two questions. First, what kind of

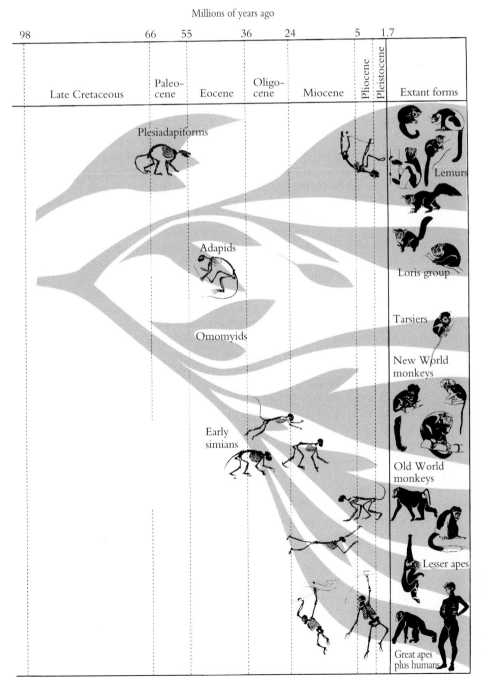

Figure 10.6
The phylogeny of modern primates and their likely links to extinct primates is shown here.

animal did natural selection have to work with? Second, what kind of animal was likely to be successful in exploiting this kind of ecological niche?

The **plesiadapiforms**, a group of fossil animals found in what is now Montana, Colorado, New Mexico, and Wyoming, give us some clue about what the earliest primates were like. These sites date from the Paleocene epoch, 65 to 55 mya, a time so warm and wet that broadleaf, evergreen forests extended to 60°N (latitude 60° north, near present-day Anchorage in Alaska). The plesiadapiforms varied from tiny, shrew-sized creatures to animals as big as a marmot, and although most species are known only from their teeth, it seems likely that they were solitary, nocturnal quadrupeds with a well-developed sense of smell. The teeth of these animals suggest that they had a range of dietary specializations, including insect-eating and seed-eating. For many years systematists classified them within the primate order because they had primatelike teeth, and their limbs and feet show adaptations to arboreal life. However, they show almost none of the special anatomical features that characterize modern primates. For example, they had claws rather than nails, and their **orbits** (eye sockets) were not fully enclosed in bone (Figure 10.7). They also have a number of derived features not shared with primates. Thus, most paleontologists now believe that the plesiadapiforms were a separate, but related, group of organisms.

Primates with modern features appear in the Eocene epoch.

The Eocene epoch (55–34 mya) was even wetter and warmer than the preceding Paleocene, with great tropical forests covering much of the globe. Primate

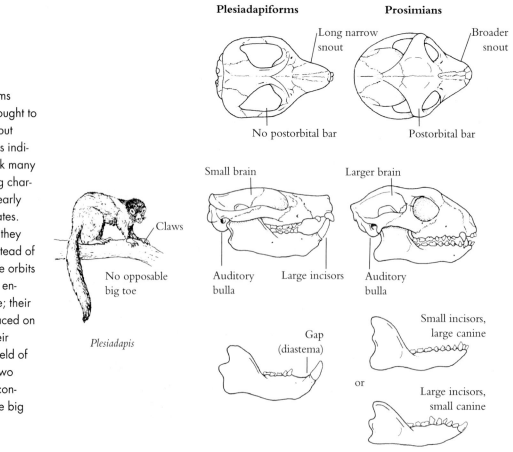

Figure 10.7

Plesiadapiforms were once thought to be primates, but closer analysis indicates they lack many of the defining characteristics of early modern primates. For example, they had claws instead of nails; their eye orbits were not fully encased in bone; their eyes were placed on the side of their head so the field of vision of the two eyes did not converge; and the big toe was not opposable.

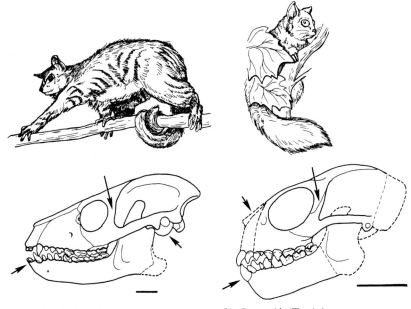

(a) Adapids (*Notharctus*)   (b) Omomyids (*Tetonius*)

Figure 10.8

Adapids were larger than omomyids, and had longer snouts and smaller orbits than the omomyids. (a) The size of the orbits suggests the adapids were active during the day, and the shape of their teeth suggests they fed on fruit or leaves. (b) Omomyids were small primates that fed mainly on insects, fruit, or gum. The large orbits suggest they may have been nocturnal. They are similar in some ways to modern tarsiers.

fossils from this period have been found in both North America and Europe. It is in these Eocene primates that we see the defining features of modern primates (see Chapter 5) for the first time. They had grasping hands and feet, with nails instead of claws, hind-limb-dominated posture, and relatively large brains. The Eocene primates are classified into two families, Omomyidae and Adapidae (Figure 10.8), both of which resembled modern prosimians. The omomyids were similar to modern tarsiers. They had huge orbits, sharp shearing teeth, and long, grasping hands—features that suggest they were arboreal insectivores who hunted at night. The absence of sexual dimorphism suggests they were solitary or lived in monogamous pairs. The adapids were more like contemporary lemurs. They had smaller orbits and more generalized dentition, suggesting they were diurnal herbivores (Figure 10.9). They were also larger than the omomyids, and their **postcranial** bones (the bones that make up the skeleton below the neck) indicate that some were active arboreal quadrupeds like modern lemurs, while others were slow quadrupeds similar to contemporary lorises. At least one species showed substantial sexual dimorphism, a feature that points to life in nonmonogamous social groups.

Anthropologists disagree about why natural selection favored the basic features of primate morphology.

There are several different theories about why the traits that are diagnostic of primates evolved in early members of the primate order. An early theory suggested that this set of traits enabled primates to occupy an arboreal niche and to make a living in the trees. According to this theory, the movement of the eyes forward in the face enhanced depth perception, an essential trait for arboreal animals leaping from branch to branch. The olfactory apparatus was reduced because an arboreal animal needed a sensitive sense of smell less than accurate vision. Grasping hands and feet were needed for climbing trees, and the need for agility and coordination led to the elaboration of areas of the brain that control movement.

However, Matt Cartmill, a Duke University anthropologist, has pointed out that many arboreal mammals lack the entire suite of adaptations that characterize

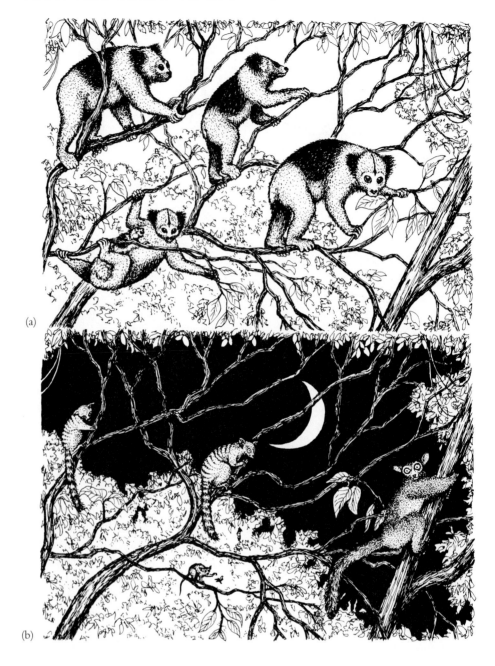

(a)

(b)

Figure 10.9

The behavior of adapids and omomyids differed.
(a) Adapids were probably diurnal; here a group forages for leaves.
(b) Omomyids were probably nocturnal; shown here are several different species.

primates. For example, squirrels scamper through trees with great aplomb, though they lack both grasping hands and stereoscopic vision. Thus, it cannot be simply life in the trees that favors the primate body plan. A closer look at the characteristics of arboreal creatures provides several clues about the function of the features that characterize modern primates. For example, small-nailed paws like those of squirrels typify animals that run on all fours along large tree branches, while grasping hands are most common among creatures, like tree shrews, that clamber among the many small branches of the dense understory. Moreover, stereoscopic vision characterizes arboreal predators like ocelots and owls, as well as primates (Figure 10.10). Cartmill argues that this evidence suggests early primates were predators who lived in the understory and located their prey by sight. The main problem with this theory is that many of the living descendants of the original primates, members of extant prosimian groups, rely heavily on sound and smell in locating

Figure 10.10

Arboreal predators, like this ocelot, typically have stereoscopic vision.

and hunting insects. They are not visually oriented predators, so we must question the notion that their ancestors were.

Robert Sussman of Washington University contends that the general characteristics of the primates reflect an initial shift from an exclusively insectivorous diet to a mixed diet. We know that the appearance of the primate order coincided with the radiation of many types of flowering plants. Increased visual acuity and other primate traits may have been advantageous for exploiting a new array of plant resources—including fruit, nectar, flowers, and gum—as well as insects. The early primates may have foraged and handled small food items in the dimness of the forest night, and this may have favored good vision, precise eye-hand coordination, and grasping hands and feet. The fact that fruit and other plant parts play an important role in the diets of many living prosimians provides some support for this view.

The First Anthropoids

During the Oligocene epoch, many parts of the world became colder and drier.

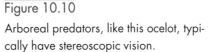

By the end of the Eocene epoch, 34 mya, the continents were more or less positioned on the globe as they are today. However, South America and North America were not yet connected by Central America; Africa and Arabia were separated from Eurasia by the Tethys Sea, a body of water that connected the Mediterranean Sea and the Persian Gulf. South America and Australia had completed their separation from Antarctica, creating deep, cold currents around Antarctica (Figure 10.4). Some climatologists believe that these cold currents reduced the transfer of heat from the equator to Antarctic regions, and may have been responsible for the major drop in global temperatures that occurred during this period. During the Oligocene epoch, 34 to 23 mya, temperatures dropped, and the range in temperature variation over the course of the year increased. Throughout North America and Europe, tropical broadleaf evergreen forests were replaced by broadleaf deciduous forests. Africa and South America remained mainly warm and tropical.

Primates similar to modern monkeys first appear in the fossil record at the Eocene-Oligocene boundary.

There are a few fragmentary fossils from sites in Algeria, Tunisia, and Oman hinting that primates similar in some ways to modern monkeys may have been present in northern Africa in the early Eocene, 54 mya. However, the earliest unambiguous anthropoid fossils were found at a site in the Fayum depression of Egypt. The Fayum deposits straddle the Eocene-Oligocene boundary, 36 to 33 mya. The Fayum is now a desert, but it was very different then. Sediments of soil recovered from the Fayum tell us that the Fayum primates lived in a warm, wet, and somewhat seasonal habitat. The plants were most like those now found in the tropical forests of Southeast Asia. The soil sediments contain the remnants of the roots of plants that grow in swampy areas, like mangroves, and the sediments suggest there were periods of standing water at the site. There are also many fossils of water birds. All this suggests that the Fayum was a swamp during the Oligocene (Figure 10.11).

Paleontologists recognize 14 genera of primates among the fossils found at the Fayum. Some of these are prosimian-like adapids and omomyids. However, there are 10 genera that share many derived features with modern monkeys and apes and are classified as anthropoids. The earliest creatures were similar to small New World monkeys. The most complete of these early fossils is *Catopithecus brownii*, which lived about 36 mya. Its teeth had sharp edges suited to shearing tough material like insect skeletons or leaves. But *Catopithecus* was about the size of a squirrel monkey (600 g or 1.3 lb). As we saw in Chapter 5, most folivores are larger than this. Thus, it is likely that *Catopithecus* subsisted mainly on insects. The molars show some adaptations for grinding, which suggests it may have supplemented insects with fruit. *Catopithecus*'s eyes were relatively small, suggesting that it was diurnal. The olfactory area was larger than the olfactory areas in any modern anthropoids except tamarins and marmosets, which rely heavily on scent marking and other forms of olfactory communication. The postcranial bones suggest that

Figure 10.11

Although the Fayum depression is now a desert, it was a swampy forest during the Oligocene.

Figure 10.12

The Fayum was home to a diverse group of primates, including propliopithecoids like *Aegyptopithecus zeuxis* (upper left) and *Propliopithecus chirobates* (upper right) and parapithecids like *Apidium phiomense* (bottom).

Catopithecus walked quadrupedally on top of branches. Other early primates from the Fayum are smaller, ranging down to the size of tamarins and marmosets (300–500 g, or 0.7–1.2 lb).

The early primates at the Fayum share many primitive features with the Eocene prosimians. However, they also share a number of derived features with contemporary anthropoids that are not seen among prosimians. For example, the orbits are fully enclosed in bone, and the chewing system is suited to generating large forces on the teeth. The dental pattern of two of the early Fayum monkeys, *Catopithecus* and a second genus, *Proteopithecus*, was the same as that of modern Old World monkeys and apes (2.1.2.3/2.1.2.3), and different than that of contemporary New World monkeys (2.1.3.3/2.1.3.3). This suggests that the lineage leading to the New World monkeys had already diverged from the lineage leading to Old World monkeys and apes.

The later Fayum fossils are divided into two groups, propliopithecoids and parapithecoids (Figure 10.12). Some researchers consider these groups infraorders, while others consider them families. The largest and most famous of the propliopithecoids is named *Aegyptopithecus zeuxis,* and it is known from several skulls and a number of postcranial bones (Figure 10.13). *A. zeuxis* was a medium-sized monkey, perhaps as big as a female howler monkey (6 kg, or 13.2 lb). It was a diurnal, arboreal quadruped with a relatively small brain. The shape and size of the teeth suggest that it ate mainly fruit. Males were much larger than females, which indicates that they probably lived in nonmonogamous social groups. Other propliopithecoids were smaller than *A. zeuxis,* but their teeth suggest that they also ate fruit, as well as seeds and perhaps gum. They were either arboreal quadrupeds or leapers. The propliopithecoids share even more derived features with modern

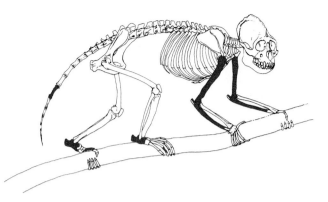

Figure 10.13

In this reconstruction of the skeleton of *Aegyptopithecus*, the postcranial bones that have been found are shown in red.

monkeys than did earlier Fayum primates, including a single fused mandible. It is possible that modern monkeys and apes are derived from members of this family. The parapithecoids were also monkeylike, but were apparently unrelated to modern anthropoids (Box 10.1).

Paleontologists are uncertain whether the modern anthropoids evolved from the omomyids or the adapids.

We do not know which of the Oligocene primate groups gave rise to modern anthropoid primates. Both morphological and molecular data suggest that living anthropoid primates are more closely related to the tarsiers than to other prosimians. The morphology of fossil omomyids clearly links them with tarsiers, and this suggests that both tarsiers and anthropoids evolved from an omomyid ancestor. However, fossil anthropoids like *Aegyptopithecus* and *Catopithecus* also share several derived features with the adapids, which lends support to the idea that adapids were the ancestors of the anthropoid primates. Given the sparseness of the fossil record, such problems are to be expected.

Primates appear in South America for the first time during the Oligocene, but it is unclear how they got there.

There are very few fossils of New World primates, and their history is very poorly understood. Fossil primates that date from the Oligocene epoch have been found in Bolivia and Argentina. The best known of these creatures is *Dolichocebus gaimanensis,* a medium-sized (3 kg, or 6.6 lb) creature that is broadly similar to the Fayum primates, except that it had three premolars as modern New World monkeys do. Its postcranial skeleton suggests that it was either an arboreal quadruped or a leaper, and its teeth suggest a diet of fruit and gum (Figure 10.15).

The origin of these primates is a mystery. The absence of Oligocene primate fossils in North America and the many similarities to Fayum primates suggest to many scientists that New World monkey ancestors came from Africa. The problem with this idea is that we don't know how they could have gotten from Africa to South America. Remember that South America separated from Africa more than 100 mya. By the late Oligocene, the two continents were separated by at least 3000 km (2000 miles) of open ocean. Some authors have suggested that primates could have rafted across the sea on islands of floating vegetation. Although there are no well-documented examples of primates rafting such distances, fossil rodents appear in South America about the same time and are so similar to those found in Africa that it seems very likely that rodents managed to raft across the Atlantic.

BOX 10.2

Missing Links

The University of Zürich anthropologist Robert Martin has pointed out that most lineages are probably older than the oldest fossils we have discovered. The extent of the discrepancy between the dates of the fossils that we have discovered and the actual origin of the lineage depends on the fraction of fossils that have been discovered. If the primate fossil record were nearly complete, then the fact that no anthropoids living more than 35 mya have been found would mean that they didn't exist that long ago. However, we have reason to believe that the primate fossil record is quite incomplete. (Figure 10.17).

Just as not all fossils of a given species are ever found, not all species are

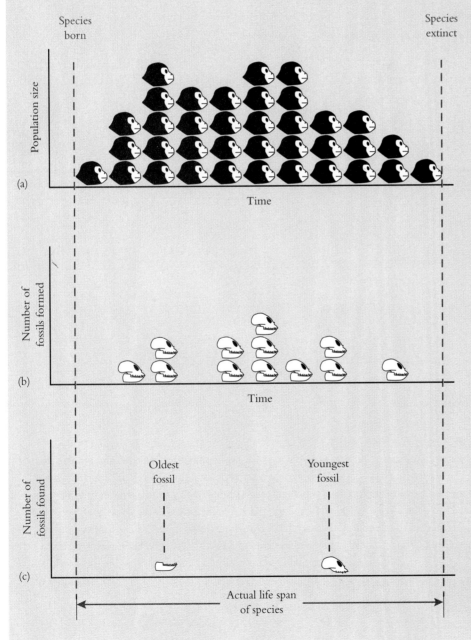

Figure 10.17

The sparseness of the fossil record virtually guarantees that the oldest fossil found of a particular species underestimates the age of the species. (a) The population size of a hypothetical primate species is plotted here against time. In this case, we imagine that the species becomes somewhat less populous as it approaches extinction. (b) The number of fossils left by this species is less than the population size. It is unlikely that the earliest and latest fossils date to the earliest and latest living individuals. (c) The number of fossils found by paleontologists is less than the number of fossils.

known to us either. How many species are missing from our data? Martin's method for answering this question is based on the assumption that the number of species has increased steadily from 65 mya, when the first primates appeared, to the present. This means that there were half as many species 32.5 mya as there are now, three-quarters as many species 16.25 mya as there are

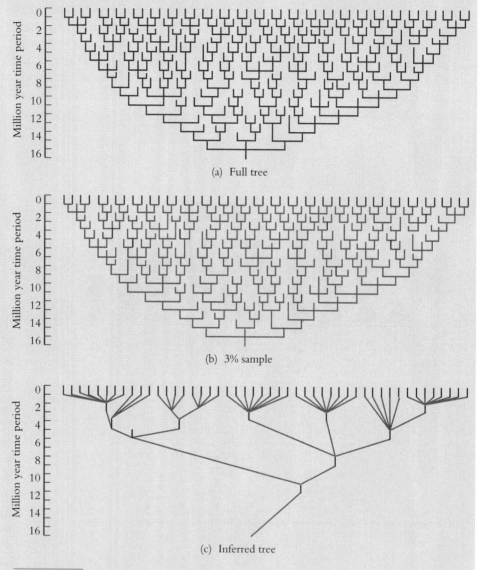

(a) Full tree

(b) 3% sample

(c) Inferred tree

Figure 10.18

The tree in (a) shows the complete phylogeny of a hypothetical lineage, and the tree in (b) shows the same phylogeny when only a small percentage of the fossil species have been discovered. The red lines show the species that have been discovered, and the gray lines show the missing species. Note that all *living* species are known, but only 3% of the fossil species have been discovered. (c) The tree we would have to infer from the incomplete data at hand differs from the actual tree in two respects: 1) it links each species to its closest known ancestor, which often means assigning inappropriate ancestors to fossil and modern species; and 2) it frequently underestimates the age of the oldest member of a clade.

now, and so on. Assuming that each species lived about 1 million years, the average life span of a mammalian species, Martin summed these figures to obtain the number of primate species that have ever lived. Then he divided the number of fossil species so far discovered by his estimate of the total number of species that have ever lived. According to these calculations, only 3% of all fossil primate species have been found so far.

Next, Martin constructed a phylogenetic tree with the same number of living species as we now find in the primate order. One such tree is shown in Figure 10.18a. He then randomly "found" 3% of the fossil species. In Figure 10.18b the gray lines give the actual pattern of descent, while the red lines show the data that would be available if we knew the characteristics of all the living species but had recovered only 3% of the fossil species. The best phylogeny possible given these data is shown in Figure 10.18c. Then Martin computed the difference between the age of the lineage based on this estimated phylogeny, and the actual age of the lineage based on the original phylogeny. The discrepancy between these values represents the error in the age of the lineage that is due to the incompleteness of the fossil record. By repeating this procedure over and over again on the computer, Martin was able to produce an estimate of the average magnitude of error, which turned out to be about 40%. Thus, if Martin is correct, living lineages are, on average, about 40% older than the age of the oldest fossil discovered.

The most intriguing hypothesis is that anthropoid primates actually appeared in Africa much earlier, when a trans-Atlantic journey would have been much easier to complete. The problem with this idea is that the fossil record from that period contains no monkey fossils. This, surprisingly enough, may not be a fatal liability for the hypothesis. There is good reason to believe that the date of the earliest fossil we have discovered usually underestimates the age of a lineage (Box 10.2). Thus, we may be nearly certain that if anthropoid primates were living in the Fayum area 36 mya, they first appeared in Africa sometime earlier. The method outlined in Box 10.2 suggests that anthropoids actually originated at least 52 mya.

The Emergence of the Hominoids

The early Miocene was warm and moist, but by the end of the period, the world had become much cooler and more arid.

The Miocene epoch began approximately 23 mya and ended 5 mya. In the early Miocene, the world became warmer, and once again Eurasian forests were dominated by broadleaf evergreens like those in the tropics today. At the end of the Miocene, the world became considerably colder and more arid. The tropical forests of Eurasia retreated southward, and there was more open, woodland habitat. India continued its slow slide into Asia, leading to the uplifting of the Himalayas. Some climatologists believe that the resulting change in atmospheric circulation was responsible for the late Miocene cooling. About 18 mya, Africa joined Eurasia, splitting the Tethys Sea and creating the Mediterranean Sea. Because the Strait of

317

Gibraltar had not yet opened, the Mediterranean Sea was isolated from the rest of the oceans. At one point, the Mediterranean Sea dried out completely leaving a searing valley thousands of feet below sea level. About the same time, the great north–south mountain ranges of the East African Rift began to appear. Because clouds drop their moisture as they rise in elevation, there is an area of reduced rainfall, called a **rain shadow**, on the lee (downwind) side of mountain ranges. The newly elevated rift mountains caused the tropical forests of East Africa to be replaced by drier woodlands and savannas.

Proconsul, the first hominoid, appeared in Africa during the late Oligocene.

The oldest hominoids are members of the genus *Proconsul* (Figure 10.19). This genus includes five species ranging from the size of a macaque (10 kg, or 22 lb) to at least the size of a bonobo (38 kg, or 84 lb). The earliest fossils, found at Losidok in northern Kenya, date to about 27 mya, and other fossils have been found at sites in Africa dated to as recent as 17 mya. Plant and animals fossils found with *Proconsul* indicate that they lived in rain-forest environments like those found in tropical Africa today. Taxonomists classify *Proconsul* as a **hominoid** because it shares a small number of features with living apes and humans. For example, *Proconsul* didn't have a tail; its eyes faced more forward; its snout was smaller than the *Aegyptopithecus*'s; and its elbow was similar to that of modern apes. *Proconsul* species also had somewhat larger brains in relationship to body size than similarly sized monkeys. Otherwise, members of the genus *Proconsul* were similar to *Aegyptopithecus* and the other Oligocene primates. Their teeth had thin enamel, which is consistent with a frugivorous diet. Compared with later hominoids, they were more lightly built and had a narrower chest and relatively short upper limbs. These postcranial features indicate that they didn't hang below branches to feed, nor did they swing from branch to branch to move through the canopy, as apes do now, but rather walked quadrupedally on top of branches like arboreal monkeys.

A number of other species of hominoids lived in Africa during the late Oligocene and early Miocene. Peter Andrews of the British Natural History Museum classifies all of these creatures together as **proconsulids**. Although they were all forest apes, and generally similar to *Proconsul,* the proconsulids show some ecological differentiation. For example, at least one seems to have been folivorous, and another may have swung from branch to branch as modern apes do. A second species was as large as a female gorilla.

Figure 10.19

Members of the genus *Proconsul* were relatively large (15 to 50 kg, or 33 to 110 lb), sexually dimorphic, and frugivorous. The skeleton of *Proconsul africanus,* reconstructed here, shows it had limb proportions much like modern-day quadrupedal monkeys.

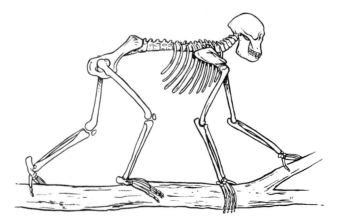

The middle Miocene epoch saw a new radiation of hominoids and the expansion of hominoids throughout much of Eurasia.

Exploration of middle Miocene (15–10 mya) deposits has produced a great abundance of new hominoid species in Africa, Europe, and Asia. Well-known examples include *Kenyapithecus* from East Africa, *Oreopithecus* (Figure 10.20), and *Dryopithecus* from Europe, and *Sivapithecus* from southern Asia. The skulls and teeth of these hominoids typically differ from those of the proconsulids in a number of ways that indicate they ate harder or more fibrous foods than their predecessors. Their molars had thick enamel for longer wear, and rounded cusps, which are better suited to grinding. Their **zygomatic arches** (cheek bones) flared farther outward to make room for larger jaw muscles. And the lower jaw was more robust, to carry the forces produced by those muscles. It seems likely that these features were a response to the climatic shift from a moist tropical environment to a drier, more seasonal environment with tougher vegetation and harder seeds. Until recently, very little was known about the postcranial skeletons of middle Miocene apes. However, in 1996 Salvador Moyà-Solà and Meike Köhler of the Institut de Paleontologia Miquel Crusafont, Spain, announced the discovery of a 9.5-million-year-old specimen of *Dryopithecus* at a site near Barcelona. This remarkable specimen includes nearly all of one hand, most of the leg and arm bones, some of the

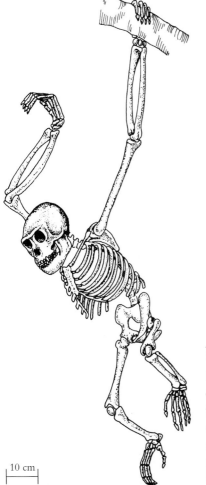

10 cm

Figure 10.20

Oreopithecus had a number of traits that are associated with suspensory locomotion, including a relatively short trunk, long arms, short legs, long and slender fingers, and great mobility in all joints. The phylogenetic affinities of this late Miocene ape from Italy are not well established.

The World of the Ancient Apes

The record of climate is constructed by unearthing fossil-rich strata on land and by piecing together long sequences from ocean cores. These archives demonstrate that over the past 50 million years, Earth has cooled, aridity has spread, and the ground has become host to a thinning vegetation. These trends in our planet's history have extended to the present and become exaggerated over the past 1 million years.

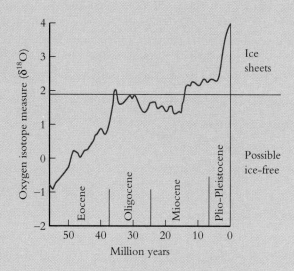

The jagged line above conveys the global schedule of ice buildup and temperature change over the past 50 million years. Its upward course depicts the chilling of the planet and the growth of glaciers.

It is extraordinary to me that such vital signs of the ancient past can be drawn. In this case, the line comes from painstaking measurements of oxygen fixed in the skeletons of single-celled organisms, the foraminifera, or *forams* for short. The remains of these microscopic beasts are abundant in long cores drilled out of the ocean depths. Frozen water, like its liquid form, contains an atom of oxygen for every two of hydrogen. But the lighter isotope of oxygen, ^{16}O, is more easily evaporated from the oceans and incorporated in glacial ice than is the heavier form, ^{18}O. As a result, the oceans fluctuate in their oxygen isotope content depending on how much ice has formed worldwide. As the lighter oxygen molecules are subtracted from the oceans and held captive by glaciers, the ocean-dwelling forams become burdened by heavy oxygen. As ice melts, the lighter isotope is returned to the oceans, and the foram skeletons indicate that this has happened.

Regardless of ice buildup, temperature also affects how easily the two kinds of oxygen are absorbed and recorded for posterity by the sensitive protozoa. And so the long, continuous burial of forams in the seafloor gives us an oxygen isotope curve, both a global thermometer and a record of ice volume. The scale on the diagram's left shows a way of measuring the oxygen isotopes in forams, measured in parts per thousand. . . .

* * *

By 20 million years ago an ape known by the name *Proconsul* had evolved in Africa. It thrived in the eastern sector and actually consisted of several different species, as we can see from fossil bones found around Lake Victoria.

The proconsuls were at home in the trees. Their bones articulated in the skeletal arrangement of an agile four-legged climber. But they possessed an odd skeleton, ill-matched to our definitive categories of ape or monkey. In one species the flexible backbone, narrow chest, and part of the pelvis most closely resemble those of a monkey the size of a baboon. Other aspects of its pelvis, however, its lack of a tail, the wide movement of its hip and shoulder, and its enlarged brain all suggest affinities with the apes. It is this essential contradiction to our present fixed boundary between monkey and ape that shakes us into recognizing that evolution has occurred. . . .

The early Miocene apes mirrored the changing world by multiplying, dividing, and spreading in the subsequent period. Between 18 and 12 million years ago, the continental plate that holds Africa and Arabia nudged farther and farther into the plate of Europe and Asia. The broad seaway that had once divided these two gigantic masses of Earth's surface became narrower and narrower. And from the exchange of species between Eurasia and Africa we see the first evidence of a land bridge between the two areas. *Proconsul*'s descendants were among the mammals who found their way into the southern zones of Europe and Asia. During this interval, at least four major groups of apes branched apart.

It was the apes' turn to toss the evolutionary dice, and they flourished in the intricate mosaic of forests and woodlands across the Old World. The significance of this acme in ape family history, 18 to 12 million years ago, cannot be overestimated. Out of this radiation of species arose a certain ancestor, the node from which the lineages of modern apes and the bipeds of Africa would branch. What caused this radiation of apes, this quickening in the labor of our origin?

In our search for clues about the environmental conditions of human origin, let us turn once again to the foram thermometer. The foram curve shows a dramatic angle at the 15- to 14-million-year mark. This emphatic shift, as we have already noted, is a hallmark of the decline, the massive rise in glacial volume, the icing of Antarctica. The rise of Afropith and Kenyapith apes in Africa, and of the Sivapith and Dryopith species in Eurasia, led to the highest diversity of apes ever recorded. This radiation of species coincided roughly with the sharp turn in the foram curve, making it tempting to link the radiation with the cooling and drying of land ecosystems. Was this trend responsible for the origin of the mid-Miocene apes?

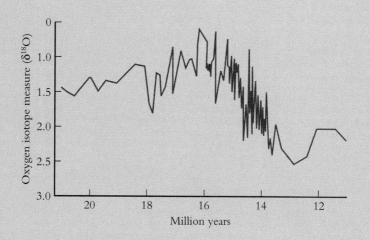

I think we must answer no. The efflorescence of Miocene apes did not correspond exactly to the global temperature decline, as a close look at the foram curve helps to show. According to recent fossil finds, the Afropith apes, which were among the first to possess a thickened cap of enamel on their teeth, arose at least 17 to 18 million years ago—about 3 million years before the frigid dive in the oxygen isotope curve. The oldest Kenyapiths also preceded this environmental episode by at least a million years.

When we consider the foram thermometer between 22 and 12 million years ago, we discover a more intriguing suggestion. Starting nearly 18 million years ago, the amount of fluctuation in the oxygen isotope curve began to increase. Between 18 and 16 million years ago, the average fluctuation was about 0.5 *parts per mil*, the units by which the ^{18}O and ^{16}O ratio is measured. This average amount of change is nearly twice that observed in the prior 4-million-year interval. It is also quite clear that this change in the degree of fluctuation was merely the beginning. It culminated 16 to 14 million years ago in an even larger average fluctuation—up to 0.68 parts per mil—coinciding with a rapid fluttering in the planet's climatic pulse. The Miocene hallmark of habitat decline, the glaciation of Antarctica, thus consisted of numerous large-scale reversals in the cooling and icing trend, not a single major change. The degree and pace of fluctuation first increased about 18 million years ago.

These observations suggest that the rise and early radiation of the mid-Miocene apes—the signal event in primate evolution of this epoch—had more to do with environmental variability than with a directional decline in habitats. According to the foram curve, fluctuation in global ice buildup was one factor, while rifting, faulting, and volcanic blanketing of the landscape helped to revamp the environmental mosaic over time and to vary its exact makeup from one region to another.

The main dental innovation of the mid-Miocene—the growth of thick enamel—equipped some of these apes to eat a wider diversity of foods, both hard and soft, than that tried by their ancestors, the thin-enameled proconsuls. Teeth have limits in their capacity to crack and crush any object placed between the jaws. The growth of thick enamel would have allowed apes the freedom to select from a broader range of foods, ushering in an era of dietary diversity.

If this interpretation is correct, thick enamel in the mouths of the Afropiths and later apes was a kind of insurance policy, enabling them to meet larger uncertainties in habitat and food than their ancestors could have survived. Oscillations over tens of thousands of years proved much more dramatic than any climatic change an ape could have experienced in its lifetime. It is in this longer time frame, as new climatic extremes arrived again and again, that the ape lineages faced their greatest tests of survival. . . .

We can also begin to discern the entire process by which the survival of species is decided as life on Earth unfolds. Every individual alive today—human or ape, ant or dandelion—belongs to a lineage whose duration is measured in geologic time. Life's vital tension thus consists of, first, the forces of stability and trend that encourage organisms to respond in a certain way according to the conditions of the moment and the predictable future; and, second, the forces of uncertainty, which favor a means of flexibility, not merely the single way that yields maximum benefit in a flinch of time, a season or a life.

Source: From pp. 50–55 in Richard B. Potts, 1996, *Humanity's Descent: The Consequences of Ecological Instability*, William Morrow, New York. By permission of William Morrow & Company, Inc.

vertebrae and ribs, part of the collarbone (or clavicle), and part of a cranium. The anatomy of these bones clearly indicates that *Dryopithecus* was an arboreal creature that moved through the canopy by swinging from one branch to another, as contemporary orangutans do, and did not walk along the top of branches like the quadrupedal proconsulids did.

There are no clear candidates for the ancestors of humans or any modern apes, except perhaps orangutans.

The evolutionary history of the apes of the Miocene is poorly understood. There were many different species, and the phylogenetic relationships among them remain largely a mystery. We have no clear candidates for the ancestors of any modern apes, except for the orangutan, who shares a number of derived skull features with *Sivapithecus* of the middle Miocene. We can establish no clear links between gorillas or chimpanzees and any of the Miocene apes. Once again, this is not too surprising given the sparseness of the paleontological record.

We can be almost certain that the earliest hominids evolved from some type of Miocene ape, but we have no idea which one it was. Nonetheless, the study of ape and hominid origins in the Miocene has been very useful for paleontologists. John Fleagle of the State University of New York at Stony Brook points out that, in the process of considering and ultimately rejecting many different candidates for the "missing link" between apes and early hominids, paleoanthropologists have come to realize that the key features distinguishing the earliest hominids from ancestral hominoids were not the features that they had originally been looking for—the big brains and small teeth that characterize modern humans. Instead, it was a suite of skeletal adaptations for bipedalism that marked the primary hominid adaptation.

During the early and middle Miocene, ape species were plentiful and monkey species were not. In the late Miocene and early Pliocene, many ape species became extinct and were replaced by monkeys.

Apes flourished during the Miocene, but all but a few genera and species eventually became extinct. Today there are only gibbons, orangutans, gorillas, and chimpanzees. We don't know why so many ape species disappeared, but it seems likely that many of them were poorly suited to the drier conditions of the late Miocene and early Pliocene.

The fossil record of Old World monkeys is quite different from that of the apes. Monkeys were relatively rare and not particularly variable in the early and middle Miocene, but the number and variety of fossil monkeys increased in the late Miocene and early Pliocene. Thus, while there are many more extinct ape species than living ones, the number of living monkey species greatly exceeds the number of extinct monkey species.

Once again, the fossil record reminds us that evolution does not proceed on a steady and relentless path toward a particular goal. Evolution and progress are not synonymous. During the Miocene, there were dozens of ape species but relatively few species of Old World monkeys. Today there are many monkey species and only a handful of apes. Despite our tendency to think of ourselves as the pinnacle of evolution, the evidence suggests that taken as a whole our lineage was poorly suited to the changing conditions of the Pliocene and Pleistocene.

Further Reading

Andrews, P., and L. Aeillo. In press. Hominoid evolution leading to human or-
        igins. In *Human Evolution: A Multidisciplinary Approach,* ed. by C. Su-
        sanne and M. D. Garralda. Academic Press, London.
Conroy, G. C. 1990. *Primate Evolution.* W. W. Norton, New York.
Fleagle, J. 1998. *Primate Adaptation and Evolution.* 2d. ed. Academic Press, San
        Diego.
Klein, R. 1999. *The Human Career.* 2d. ed. University of Chicago Press, Chicago.
Martin, R. 1990. *Primate Origins and Evolution: A Phylogenetic Reconstruction.* Prince-
        ton University Press, Princeton, N.J.
Potts, R. 1996. *Humanity's Descent: The Consequences of Ecological Instability.* William
        Morrow, New York.
Simons, E. L., and T. Rasmussen. 1994. A whole new world of ancestors: Eocene
        Anthropoideans from Africa. *Evolutionary Anthropology* 3:128–139.

Study Questions

1. Briefly describe the motions of the continents over the last 180 million years. Why are these movements important to the study of human evolution?
2. What has happened to the world's climate since the end of the age of dinosaurs? Explain the relationship between climate change and the notion that evolution leads to steady progress.
3. What are angiosperms? What do they have to do with the evolution of the primates?
4. Why are teeth so important for reconstructing the evolution of past animals? Explain how to use teeth to distinguish between an insectivore, a folivore, and a frugivore.
5. Which primate groups first appear during the Eocene? Give two explanations for the selective forces that shaped the morphologies of these groups.
6. Why is there a problem in explaining how primates arrived in the New World?
7. Why does the oldest fossil in a particular lineage underestimate the true age of the lineage? Explain how this problem is affected by the quality and completeness of the fossil record.
8. Explain how potassium-argon dating works. Why can it be used to date only volcanic rocks older than about 500,000 years?

C H A P T E R   1 1

The Earliest Hominids

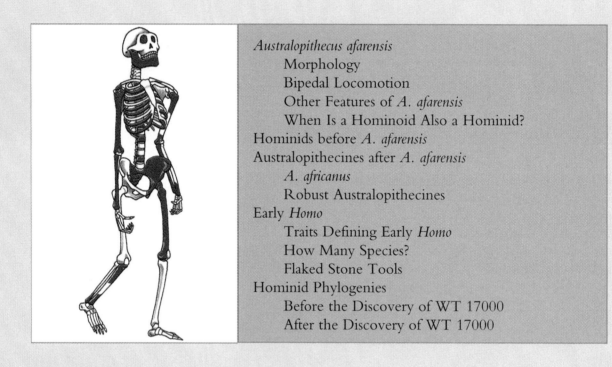

The fossil record provides little information about the evolution of the human lineage during the late Miocene, from 10 to 5 mya. Around 10 mya, several species of large-bodied hominoids that bore some resemblance to modern orangutans lived in Africa and Asia. About this time, the world began to cool, grassland and savanna habitats spread, and forests began to shrink in much of the tropics. The creatures that occupied tropical forests declined in variety and abundance, while those that lived in the open grasslands thrived. Apes were among the forest species that suffered drastic declines, particularly in Asia. We know that at least one ape species survived the environmental changes that occurred during the late Miocene because molecular genetics tells us that humans, gorillas, bonobos, and chimpanzees are all descended from a common ancestor that lived sometime between 5 and 7 mya. Unfortunately, the fossil record for the late Miocene tells us little about the creature that linked the forest apes to modern hominids.

Beginning about 5 mya, hominids begin to appear in the fossil record. These early hominids were different from any of the Miocene apes in one important way—they walked upright as we do. Otherwise, the earliest hominids were probably not much different from modern apes in their behavior or appearance.

Between 2 and 4 mya, the hominid lineage diversified, creating a community of several hominid species that ranged through eastern and southern Africa. Among the members of this community, two distinct patterns of adaptation emerged. One group of creatures, the **australopithecines**, evolved large molars that enhanced their ability to process coarse plant foods. The second group, members of our own genus, *Homo,* evolved larger brains, manufactured and used stone tools, and relied more on meat than the australopithecines did.

In this chapter we will describe the species that constituted this hominid community, and then we will discuss the role that each species played in the subsequent history of the hominid lineage.

Australopithecus afarensis

The oldest well-known hominid fossils are assigned to the species Australopithecus afarensis.

Australopithecus afarensis is best known from specimens found at Hadar, a site in the Afar depression of northeastern Ethiopia (Figure 11.1). In 1973 a French and American team headed by Maurice Taieb and Don Johanson discovered the bones of a 3-million-year-old knee joint that showed striking similarities to a modern human knee. The next year the team returned and hit paleontological pay dirt. They found a sizable fraction of the skeleton of a single individual, an amazing discovery in a discipline in which an entire species is sometimes named and classified on the basis of a single tooth. They dubbed the skeleton "Lucy," after the Beatles song "Lucy in the Sky with Diamonds" (Figures 11.2 and 11.3). Lucy, who is 3.2 million years old, was not the only remarkable find at Hadar. During the next field season, the team found the remains of 13 more individuals. This collection of fossils is formally identified as material from Afar Locality 333 (AL 333), but is often called simply the "First Family."

Figure 11.1

The Awash Basin in Ethiopia is the site of several important paleontological discoveries, including many specimens of
A. afarensis.

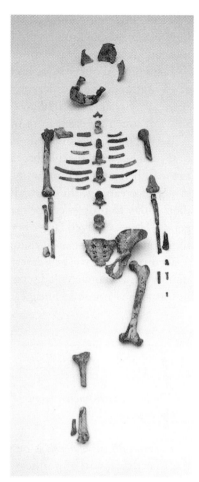

Figure 11.2

A sizable fraction of the skeleton of a single individual was recovered at Hadar. Since the skeleton is bilaterally symmetric, this means that most of the skeleton of this *A. afarensis* female, popularly known as Lucy, can be reconstructed.

Just after these fossils were discovered, the excavations were interrupted by civil war in Ethiopia, and paleontologists were unable to resume work at Hadar for more than a decade. Then a team led by Don Johanson returned to Hadar, and a separate team led by Tim White of the University of California, Berkeley, searched for fossils in the nearby Middle Awash Basin. Both groups were successful in discovering fossils from many *A. afarensis* individuals, including a nearly complete skull (labeled AL 444-2).

A. *afarensis* fossils have also been found at several sites elsewhere in Africa. During the 1970s, members of a team led by Mary Leakey discovered fossils of *A. afarensis* at Laetoli in Tanzania. The Laetoli specimens are 3.5 million years old, several hundred thousand years older than Lucy. In 1995, researchers announced interesting finds from Chad and South Africa. A French team published a description of a lower jaw from Bahr el Ghazal in Chad. This fossil, provisionally identified as *A. afarensis,* is between 3.1 and 3.4 million years old, judging by the age of associated fossils. In South Africa, Ronald Clarke and Phillip Tobias of the University of Witswatersrand found several foot bones that may also belong to *A. afarensis.* As we will see shortly, these bones are crucial for understanding how these creatures moved.

Reconstructions of the environments at these sites indicate that *A. afarensis* lived in environments ranging from woodland to dry savanna. Between 3 and 4 mya, the environment at Hadar was a mix of woodland, scrub, and grassland, while the environment at Laetoli was a dry grassland sparsely dotted with trees.

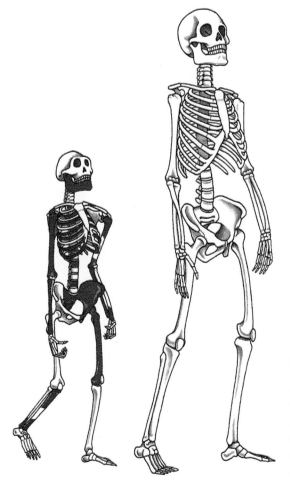

Figure 11.3

Here, Lucy's skeleton stands beside the skeleton of a modern human female. The parts of the skeleton that have been discovered are shaded. Lucy was shorter than modern females and had relatively long arms and a relatively small brain.

Morphology

The teeth and jaws of A. afarensis *share several derived features with modern humans, but the fossils also display many ancestral traits.*

The **cranium** (the skull minus the lower jaw bone) of *A. afarensis* is quite apelike. Its **endocranial volume** (the volume of its brain cavity) is 404 cc (cubic centimeters)—less than a pint. This is about the same size as the brains of modern chimpanzees. The cranium of *A. afarensis* is also primitive in a number of other details. For example, the base of the cranium is flared at the bottom and the bone is filled with air pockets, or **pneumatized;** the front of the face below the nose is pushed out, a condition known as **subnasal prognathism;** and the jaw joint is shallow (Figure 11.4). Remember that labeling a trait primitive does not imply that organisms possessing that trait are backward or poorly designed; it simply means the organisms resemble earlier forms.

The teeth and jaws of *A. afarensis* are intermediate between those of apes and humans in a number of ways. In Figure 11.5 you can see three features that illustrate this point:

- The rows of teeth in the upper and lower jaws are called the upper and lower **dental arcades**. Chimpanzees have a narrow, U-shaped dental arcade, modern

Figure 11.4

The cranium of *A. afarensis* possesses a number of primitive features, including a small brain, a shallow jaw joint, a pneumatized cranial base, and subnasal prognathism in the face. (Figure courtesy of Richard Klein.)

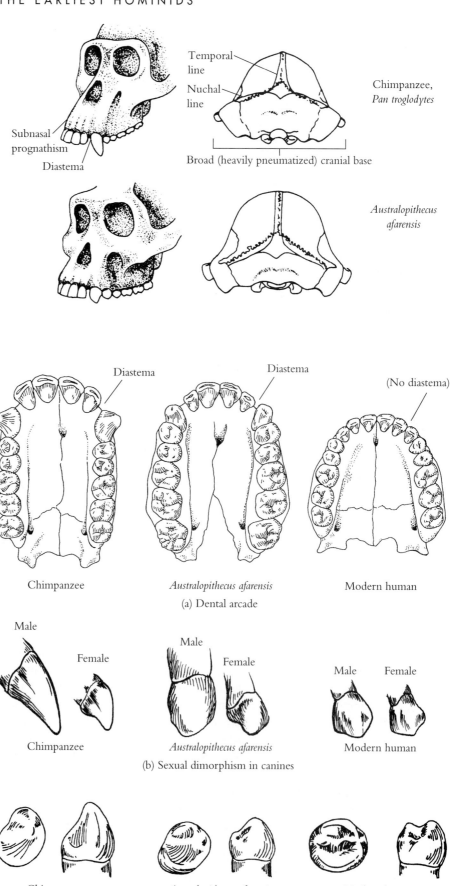

Subnasal prognathism

Diastema

Temporal line

Nuchal line

Broad (heavily pneumatized) cranial base

Chimpanzee, *Pan troglodytes*

Australopithecus afarensis

Figure 11.5

The teeth and jaws of *A. afarensis* have several features that are intermediate between those of apes and modern humans. (a) The dental arcade is less U-shaped than in chimpanzees, but less parabolic than in modern humans. (b) Chimpanzees have larger and more sexually dimorphic canines than did *A. afarensis*, which in turn had larger and more dimorphic canines than modern humans. (c) The lower third premolars of chimpanzees have only one cusp while those of modern humans have two cusps. In *A. afarensis*, the second cusp is small, but clearly present. (Figure courtesy of Richard Klein.)

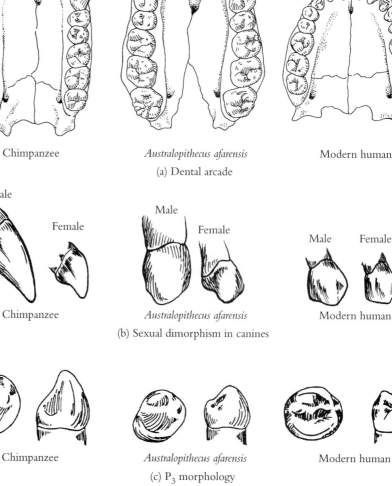

Diastema

Diastema

(No diastema)

Chimpanzee

Australopithecus afarensis

Modern human

(a) Dental arcade

Male

Female

Chimpanzee

Male

Female

Australopithecus afarensis

Male   Female

Modern human

(b) Sexual dimorphism in canines

Chimpanzee

Australopithecus afarensis

Modern human

(c) P_3 morphology

humans have a wider, parabolic dental arcade, and the dental arcade of *A. afarensis* has an intermediate V shape.

- Chimpanzees have large canine teeth. In the chimpanzee's jaw, there is a gap between the canine and the neighboring incisor called the **diastema**, which provides a space for the opposing canine when the animal's mouth is closed. Modern humans have small canines and no diastema. Here again, *A. afarensis* is intermediate between chimpanzees and modern humans in having medium-sized canines and a modest diastema.

- The first lower premolar of chimpanzees has a single cusp, while the first lower premolar in modern humans has two cusps of approximately equal size. The first lower premolar of *A. afarensis* has a small inner cusp and a larger outer cusp.

Bipedal Locomotion

Anatomical evidence indicates that A. afarensis *was bipedal.*

The transition from a forest ape to a terrestrial biped involves many new adaptations. Many of these changes are reflected in the morphology of the skeleton. Thus, by studying the shape of fossils we can make inferences about the animal's mode of locomotion. Changes in the pelvis provide a good example. The shape and orientation of the human pelvis is very different from that of forest-dwelling apes like the chimpanzee (Figure 11.6). The pelvis of *A. afarensis* is much more like a human pelvis than like a chimpanzee pelvis.

Other changes are more subtle but also diagnostic. When modern humans walk, a relatively large proportion of the time is spent balanced on only one foot. Each

Front   Top

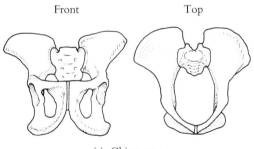

(a) Chimpanzee

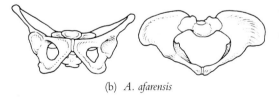

(b) *A. afarensis*

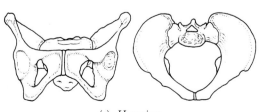

(c) *H. sapiens*

Figure 11.6

The pelvis of *A. afarensis* resembles the modern human pelvis more than the chimpanzee pelvis. Notice that the australopithecine pelvis is flattened and flared like the modern human pelvis. These features increase the efficiency of bipedal walking. The australopithecine pelvis is also much wider and narrower than that of modern humans. Some anthropologists believe these differences indicate that australopithecines did not walk the same way as modern humans do.

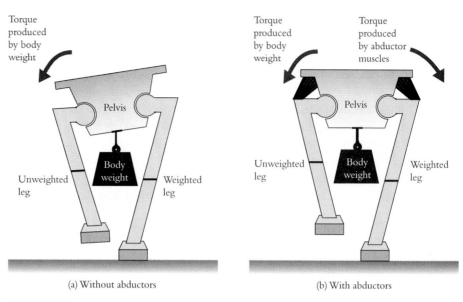

Torque produced by body weight

Torque produced by body weight

Torque produced by abductor muscles

Pelvis

Pelvis

Unweighted leg

Weighted leg

Unweighted leg

Weighted leg

Body weight

Body weight

(a) Without abductors

(b) With abductors

Figure 11.7

A schematic diagram of the lower body at the point of the stride in which all of the weight is on one leg. Body weight acts to pull down through the centerline of the pelvis. This creates a torque, or twisting force, around the hip joint of the weighted leg. (a) If this torque were unopposed, the torso would twist down and to the left. (b) During each stride the abductor muscles tighten to create a second torque that keeps the body erect.

time you take a step, your body swings over the foot that is on the ground, and all of your weight is balanced over that foot. At that moment, the weight of your body pulls down on the center of the pelvis, well inward from the hip joint (Figure 11.7). This weight creates a twisting force, or **torque**, that acts to rotate your torso down and away from the weighted leg. But your torso does not tip because the torque is opposed by **abductors**—muscles that run from the outer side of the pelvis to the femur. At the appropriate moment during each stride, these muscles tighten and keep you upright. (You can demonstrate this to yourself by walking around with your open hand on the side of your hip. You'll feel your abductors tighten as you walk, and your torso tip, if you relax these muscles. You might want to do this in private.) The abductors are attached to the **ilium** (a flaring blade of bone on the upper end of the pelvis). The widening and thickening of the ilium and the lengthening of the neck of the **femur** (the thigh bone) add to the leverage that the abductors can exert, and make bipedal walking more efficient. The existence of these morphological features in *A. afarensis* implies that it was bipedal.

The designs of the knee and foot of *A. afarensis* are also consistent with the idea that *A. afarensis* was bipedal. The modern human knee joint is quite different from the chimpanzee knee joint (Figure 11.8). Efficient bipedal locomotion requires that the knees lie close to the centerline of the body. As a result, the human femur slants down and inward, and its lower end is angled at the knee joint to make proper contact with bones of the lower leg. In contrast, the chimpanzee femur descends vertically from the pelvis, and the end of the femur at the knee joint is not slanted. The femur of *A. afarensis* has the same angled shape as that of modern humans. *A. afarensis*'s feet also show a number of derived features associated with bipedal locomotion, including a longitudinal arch and a humanlike ankle.

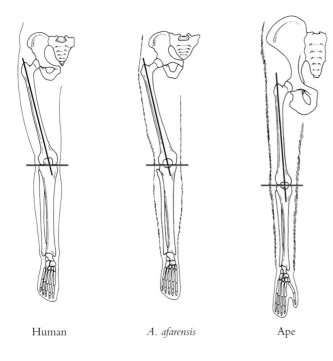

Human A. afarensis Ape

Figure 11.8

The knees of *A. afarensis* are more like the knees of modern humans than the knees of chimpanzees. Consider the lower end of the femur, where it forms one side of the knee joint. In chimpanzees, this joint forms a right angle with the long axis of the femur. In humans and australopithecines, the knee joint forms an oblique angle, causing the femur to slant inwards toward the centerline of the body. This slant causes the knee to be carried closer to the body's centerline, which increases the efficiency of bipedal walking.

It is uncertain whether A. afarensis *could walk with the same efficient, striding gait as modern humans do.*

Although there is no doubt that *A. afarensis* was bipedal, it is not entirely clear whether this hominid had the same striding gait as modern humans. Many features of the pelvis and legs of *A. afarensis* are different from modern humans, and this makes some researchers believe that Lucy and her colleagues walked with an inefficient, bent-legged gait. For example, the ilium is oriented more toward the back than in modern humans, and biomechanical calculations suggest that the abductors would be less efficient in this orientation. The legs of *A. afarensis* were also much shorter in relationship to body size than are the legs of modern humans, and this would reduce the efficiency and speed of their walking. However, other researchers have argued that *A. afarensis* individuals were efficient bipeds even though they did not walk in exactly the same way modern humans do. These researchers note, for example, that the pelvis of *A. afarensis* is much wider in relationship to its body size than in modern humans (see Figure 11.6). This extra width may have minimized vertical motion of the body during walking. The longer legs of modern humans produce the same effect.

A trail of fossil footprints proves that a striding biped lived in East Africa at the same time as A. afarensis.

The conclusion that *A. afarensis* had an efficient striding gait was dramatically strengthened by a remarkable discovery by Mary Leakey and her coworkers at Laetoli in Tanzania. They uncovered a 75-foot-long trail of footprints made by three bipedal individuals who had crossed a thick bed of wet volcanic ash about 3.5 mya (Figure 11.9). Paleontologists estimate that the tallest of the individuals that made these prints was 1.45 m (4 ft 9 in.) and the shortest was 1.24 m (4 ft 1 in.) tall. The path of these three individuals was preserved because the wet ash solidified as it dried, leaving us a tantalizing glimpse of the past. University of

Figure 11.9

Several bipedal creatures walked across a thick bed of wet volcanic ash about 3.5 mya. Their footprints were preserved when the ash solidified as it dried.

Chicago anthropologist Russell Tuttle has compared the Laetoli footprints with the footprints made by the members of a group of people called the Machigenga, who live in the tropical forests of Peru. Tuttle chose the Machigenga because their average height is close to that of the makers of the Laetoli footprints, and they do not wear shoes. Tuttle found that the Laetoli footprints are functionally indistinguishable from those made by the Machigenga, and concluded that the creatures who made the footprints walked with a fully modern striding gait.

Who made these footprints? *A. afarensis* is the likely suspect because this is the only hominid whose remains have been found at Laetoli, and *A. afarensis* is the only hominid known to have lived in East Africa at the time that the tracks were made. If *A. afarensis* did make the footprints, then those researchers who doubt that they were striding bipeds are wrong. However, some anthropologists are convinced by the anatomical evidence that *A. afarensis* was not a modern biped. If they are correct, then there must have been another, as yet undiscovered, hominid species living in East Africa 3.5 mya.

A. afarensis *probably spent a good deal of time in trees.*

All nonhuman primates, except some gorillas, spend the night perched in trees or huddled on cliffs to protect themselves from nocturnal predators like leopards. Even chimpanzees, who are about the size of Lucy, make leafy nests and sleep in trees each night (Figure 11.10). Thus, it seems likely that *A. afarensis* slept in trees too, and that their skeletons might show features related to arboreality. Anatomists such as Randall Susman of the State University of New York at Stony Brook have argued that traits of the feet, hands, wrists, and shoulder joints suggest Lucy and her friends may have been partly arboreal. For example, the bones of their fingers and toes are slender and curved like those of modern apes, and unlike modern human fingers and toes. Such hands would have been well suited to grasping and hanging onto branches. Certain distinctive features of the *A. afarensis*'s shoulder blades are also well suited to supporting the body in a hanging position.

There are several explanations for the evolution of bipedality.

Ever since Darwin, evolutionists have speculated about why bipedal locomotion was favored by natural selection. Here we examine the most popular hypotheses.

Figure 11.10

Each night, chimpanzees sleep in nests that they make by bending down branches. Sometimes, chimpanzees also rest in nests during the day, as this chimpanzee is doing.

Walking on two legs is an efficient form of locomotion on the ground. At first glance, this seems to be an unlikely possibility. After all, there are far more quadrupedal species than bipedal species, and studies of the energetics of locomotion show that terrestrial quadrupeds like horses move about more efficiently than humans do. If bipedality is a superior form of locomotion, we would expect it to be more common. However, Peter Rodman and Henry McHenry at the University of California, Davis, who studied the locomotion of humans and of quadrupedal primates on the ground, found that bipedal and quadrupedal locomotion are more or less equivalent in efficiency. This still leaves the question: if bipedal and quadrupedal locomotion are equally efficient means of traveling on the ground, why did humans become bipedal when they left the trees? Rodman and McHenry suggest that the answer has to do with the kind of quadrupedalism found in hominid ancestors. They conjecture that the ancestors of terrestrial monkeys were arboreal quadrupeds who fed on the tops of branches, while the ancestors of the hominids were suspensory feeders (hanging below branches to feed, as modern orangutans do). Selection might have favored quadrupedalism among animals descended from above-the-branch feeders and favored bipedalism among animals descended from suspensory feeders—in each case because that form of ground locomotion required fewer anatomical changes.

Erect posture allowed hominids to keep cool. Heat stress is a more serious problem in the open, sun-baked savanna than in the shade of the forest. While modern apes are restricted to forested areas, hominids apparently moved into more open areas. If an animal is active in the open during the middle of the day, it must have a way to prevent its body temperature, particularly the temperature of its brain, from becoming too high. Many savanna creatures have adaptations that allow their body temperatures to rise during the day, and then to cool off at night. They also have adaptations (such as large nasal chambers) that keep the temperature of the blood in their brain well below the temperature of the blood in the rest of their bodies. However, this option does not seem to be available to hominids. Instead, humans have evolved a system of evaporative cooling—we sweat.

Peter Wheeler of Liverpool John Moores University has pointed out that the erect posture associated with bipedal locomotion helps to reduce heat stress and lessens the amount of water necessary for evaporative cooling, in three different ways (Figure 11.11). First, the sun strikes a much smaller fraction of the body of an erect hominid than of a quadrupedal animal of the same size. An erect hominid absorbs sunlight directly only on the top of its head and shoulders, while a quad-

Figure 11.11

Bipedal locomotion helps an animal living in warm climates to keep cool by reducing the amount of sunlight that falls on the body, by increasing the animal's exposure to air movements, and by immersing it in lower-temperature air.

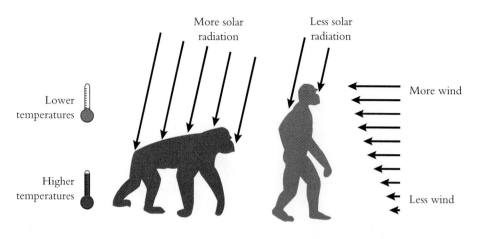

ruped absorbs the energy of the sun across its whole back and head. (Think about how much harder it is to get a tan standing up than lying down.) This difference is important because near the equator the sun is overhead during much of the middle of the day. Second, the air is warmer near the ground than it is higher up. The sun heats the soil, and the soil radiates and conducts heat back to the air, so air temperatures are highest close to the ground. Finally, wind velocities are higher 6 feet off the ground than they are 3 feet off the ground. Because moving air increases the efficiency of evaporative cooling, an erect hominid is better able to withstand the heat of the day. Using standard engineering methods, Wheeler calculated that these three factors would reduce heat stress and water consumption in bipedal hominids compared with quadrupeds. Interestingly, the same calculations show that loss of body hair provides a thermoregulatory advantage to a bipedal creature, but not to a quadruped of the same size.

Bipedal locomotion leaves the hands free to carry things. The ability to carry things is, in a word, handy. Quadrupeds can't carry things in their hands without interfering with their ability to walk and to climb. As a consequence, they must carry things in their mouths. Some Old World monkeys pack great quantities of food into their cheeks to be chewed and swallowed later. Other primates must eat their food where they find it. This can cause problems when food is located in a dangerous place, or when there is a lot of competition over food. Bipedal hominids can carry larger quantities of food in their hands and arms.

Bipedal posture allows efficient harvesting of fruit from small trees. Indiana University anthropologist Kevin Hunt has argued that the anatomy of *A. afarensis* is well suited to standing erect but not well designed for efficient bipedal walking. Hunt thinks that bipedal posture was favored because it allows efficient harvesting of fruit from the small trees that predominate in African woodlands. Two kinds of data support this hypothesis. First, Hunt found that chimpanzees rarely walk bipedally, but they spend more time standing bipedally as they harvest fruit from small trees (Figure 11.12). Using their hands for balance, they pick the fruit and slowly shuffle from depleted patches to fresh ones. Standing upright allows the chimpanzees to use both hands to gather fruit, and the slow bipedal shuffling allows them to move from one fruit patch to another without lowering and raising their body weight. Second, Hunt argues that many features of *A. afarensis* anatomy are consistent with standing upright but not with walking bipedally. He points out that the very broad pelvis of *A. afarensis* would make a stable platform for standing on both feet. Hunt also argues that the features of the shoulders, hands, and feet that many anthropologists believe are adaptations for climbing trees, are actually adaptations that enabled *A. afarensis* to hang by one hand while standing bipedally and feeding with the other hand.

Figure 11.12

Chimpanzees sometimes stand bipedally as they harvest fruit from small trees. They use one hand for balance, and feed with the other, shuffling slowly from one food patch to another.

Any or all of these hypotheses may be correct. Bipedalism might have been favored by selection because it was more efficient than knuckle walking, because it allowed early hominids to keep cool, because it enabled them to carry food or tools from place to place, and/or because it enabled them to feed more efficiently. And once bipedalism evolved, it might have facilitated other forms of behavior, like tool use.

It is true that hominids are not the only primates to make and use tools. Chimpanzees use a variety of tools in the wild. For example, they carefully prune sticks to "fish" for termites, they also use large stones to crack open hard-shelled nuts, and they sometimes brandish sticks and throw stones to intimidate their rivals (and inquisitive observers). However, we are the only primates to use tools extensively. William McGrew argues that the reason chimpanzees don't make greater use of tools is because they can't transport them easily (Figure 11.13). If a chimpanzee can't carry a tool from place to place, then it must manufacture its tools at the site where they will be used. As a consequence, it doesn't pay chimpanzees to put a lot of effort into making tools, and the tools they use are crude.

Other Features of *A. afarensis*

A. afarensis was sexually dimorphic in body size.

Anatomical evidence suggests that there was considerable variation in the body size of the adults in the Hadar *A. afarensis* population. The bigger individuals were 1.51 m (5 ft) tall and weighed about 45 kg (108 lb). The smaller ones were about

Figure 11.13

It is awkward for chimpanzees to carry things. Sometimes they carry small objects in their mouths. They can only carry large objects if they walk bipedally. Here a male chimpanzee drags a log as he performs an aggressive display. (Photograph courtesy of William C. McGrew.)

1.05 m (3 ft 6 in.) and weighed about 29 kg (70 lb). Thus, the big ones were about 1.5 times larger than the small ones. There are two possible explanations for variation in the Hadar population. The variation could represent sexual dimorphism, with the larger adults males and the smaller ones females. The difference between the Hadar males and females approaches the magnitude of sexual dimorphism in modern orangutans and gorillas and is considerably greater than that in modern humans, bonobos, and chimpanzees.

On the other hand, it is possible that large and small individuals represent two different species. This interpretation was initially supported by the fact that larger individuals were much more common at Laetoli, which was then a savanna, while the smaller ones were mainly found in what were then forested environments. Some researchers also believe that the smaller individuals were less suited to bipedal locomotion than the bigger ones were. This suggests that one species was big, fully bipedal, and adapted for savanna life. The morphology of the smaller species represented a compromise between tree climbing and walking—an adaptation for life in the forest.

The fossils recently discovered in Ethiopia by Tim White and Donald Johanson suggest that the two-species theory is wrong. The new fossils encompass the full range of variation in size at a single site and over a fairly narrow range of dates. Moreover, a large femur shows many of the same features seen in Lucy's more petite femur. The new data also suggest that both large and small forms were found in the full range of environments from forest to savanna. All of these facts suggest that this assemblage of fossils represents a single, sexually dimorphic species.

There is no evidence of stone tool manufacture or hunting in A. afarensis.

The large premolars of *A. afarensis* suggest they were frugivores, and the thick enamel on the teeth suggests they may have eaten nuts, grains, or hard fruit pits. There is reason to believe that *A. afarensis* may have eaten meat, although we don't know how important a role it played in their diets. The role of meat eating and hunting in the lives of early hominids will be discussed more fully in Chapter 12.

There are no artifacts (tools, weapons, implements) associated with *A. afarensis*, and we assume that they did not make stone tools. Of course, they may have manufactured simple tools that have not been recovered. Chimpanzees make simple tools from plant materials (Box 11.1), so it is plausible that *A. afarensis* did too.

BOX 11.1

Chimpanzee Toolmakers

In the wild, chimpanzees make and use tools to perform a variety of different tasks. Chimpanzees use long, thin branches, vines, stems, sticks, and twigs to poke into ant nests, termite mounds, and bees' nests to extract insects and honey. They sometimes scrape marrow and other tissue from the bones and braincases of mammalian prey with wooden twigs. They pound stone hammers against heavy, flat stones, exposed rocks, and roots to crack open hard-shelled nuts (Figure 11.14). They wad up leaves, dip them into pools of rainwater that have collected in hollow tree trunks, and then suck water from these leafy "sponges." They also use leaves as napkins to wipe blood, urine, semen, and feces from their bodies. Chimpanzees sometimes flail branches to defend themselves against predators and to intimidate their opponents. Finally, objects are used to communicate with other chimpanzees, to facilitate social interactions, and to enhance aggressive displays.

In many cases, chimpanzees modify natural objects for specific purposes. That is, they are toolmakers, as well as tool users. For example, chimpanzees studied at Mt. Assirik in Senegal by Caroline Tutin and William McGrew use twigs to extract termites from their mounds. The chimpanzees detach the twigs from the bush or shrub, strip off any leaves growing from the stem, peel back the bark from the stem, and then clip the twig to an appropriate length. Chimpanzees do not modify all of the materials they use as tools. The stone hammers and anvils for cracking open hard-shelled nuts are carefully selected, used repeatedly, and sometimes moved from one site in the forest to another, but they are not deliberately altered.

Approximately half of the tasks for which chimpanzees manufacture and use tools are related to food processing and food acquisition. In some cases, these tools allow chimpanzees to exploit food resources that other animals cannot use. For example, many animals are attracted to termite mounds when the winged insects emerge in massive swarms to form new colonies, but only chimpanzees can obtain access to the termites inside the mounds at other times of the year. Some of the shells of nuts that chimpanzees eat in the Taï Forest (of the Ivory Coast) are so hard that they can be opened only by repeatedly battering the shell with a stone hammer.

Figure 11.14
In West Africa, chimpanzees use stones to crack open hard-shelled nuts.

When Is a Hominoid Also a Hominid?

The shared derived characters that distinguish modern humans from other living hominoids are bipedal locomotion, a larger brain, and several features of dental morphology.

When we say that *A. afarensis* is the earliest hominid, we mean that it is the oldest fossil that is classified with humans in the family Hominidae. The rationale for including *A. afarensis* in the Hominidae is based on similarities in shared derived characters that distinguish humans from other living primates.

There are three categories of traits that separate modern humans from contemporary apes:

1. We walk bipedally as a matter of course.
2. We have much larger brains in relation to body size.
3. Our dentition and jaw musculature are different from those of apes in a number of ways. These include a wide parabolic dental arcade, thick enamel, reduced canine teeth, and larger molars in relation to the other teeth.

To be classified as a hominid, an animal must display at least some of these characteristics. *A. afarensis* is bipedal, and shares many dental features with modern humans. However, the brains of *A. afarensis* were no bigger than those of contemporary chimpanzees. As a consequence, these creatures are included in the same family as modern humans, but not in the same genus.

Hominids before *A. afarensis*

Until recently, we knew very little about the kind of creature that gave rise to *A. afarensis*. However, in 1994 and 1995 descriptions of two new hominid species were published, both older than *A. afarensis*. While the relationship between these two species to each other and to *A. afarensis* is still unclear, they vastly increase our knowledge of the human lineage before *A. afarensis*.

Australopithecus anamensis *was bipedal but had a more apelike skull than later australopithecines.*

In 1994, members of an expedition led by Meave Leakey of the National Museums of Kenya found fossils of a new hominid species at Kanapoi and Allia Bay, two sites near Lake Turkana in Kenya. These fossils are dated to between 3.9 and 4.1 mya, making them about a half a million years older than *A. afarensis* (Figure 11.15). Among these specimens are parts of the upper and lower jaw, part of a **tibia** (the larger of the two bones in the lower leg), and numerous teeth. In addition, parts of a **humerus** (the upper arm bone) found at Kanapoi in the early 1970s can now be associated with the new finds. These fossils are sufficiently different from *A. afarensis* that Leakey assigned them to a new species, *Australopithecus anamensis*. In the language of the people living around Lake Turkana, *anam* means lake.

A. anamensis shares a number of derived features with *A. afarensis* and later australopithecines. It had large molars with thick enamel, reduced canines, and the shape of the knee and ankle joints strongly indicate that it was bipedal. Thus, there is little doubt that these creatures are closely related to later australopithecines. However, *A. anamensis* lacks several derived traits seen in *A. afarensis*. For example,

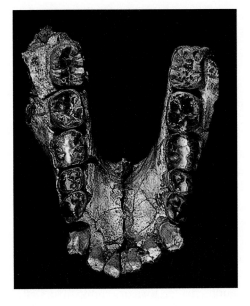

Figure 11.15

A. anamensis lived about 4 mya, half a million years before *A. afarensis*. Fragments of limb bones, the upper and lower jaws, and many teeth have been recovered.

the ear holes are small and shaped like ellipses as they are in living apes, while the ear holes of later australopithecines are larger and more rounded; the dental arcade is more like the U shape seen in chimpanzees and gorillas, and less like the V-shaped dental arcade seen in *A. afarensis;* and the chin recedes more sharply than in *A. afarensis* (Figure 11.16). This suite of more primitive characters distinguishes *A. anamensis* from *A. afarensis* and is consistent with the hypothesis that *A. anamensis* is an ancestor of *A. afarensis.*

Fossils of other animals found at Kanapoi and Allia Bay suggest that *A. anamensis* lived in a mixture of habitats including dry woodlands, gallery forests lining rivers, and more open grasslands.

Ardipithecus ramidus *may be the common ancestor of humans and chimpanzees.*

Beginning in 1992, members of an expedition led by Tim White of the University of California, Berkeley, found fossils of a still more primitive hominid at Aramis, a site in the Middle Awash Basin of Ethiopia, just south of Hadar. These fossils, dated to 4.4 mya, include teeth and jaws, the lower part of the skull, and parts of the upper arms. White and his colleagues Gen Suwa of Tokyo University and Berhane Asfaw of Addis Ababa named the species *Ardipithecus ramidus,* from the words *ardi* meaning "ground or floor," and *ramis* meaning "root" in the local Afar language. In 1994, Yohannes Haile Selasse, a member of White's team, found

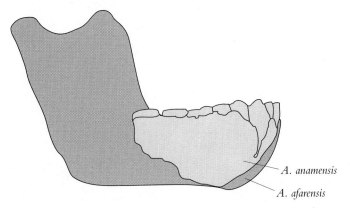

A. anamensis

A. afarensis

Figure 11.16

Australopithecus anamensis had a more receding chin than later australopithecines did, as can be seen by superposing side views of the lower jaw of *A. anamensis* (yellow) and *A. afarensis* (blue).

90 more fossils, comprising about 45% of a complete skeleton including bones of the hand, foot, pelvis, arm, leg, and skull. However, as of this writing, analysis of the fossils has only just begun, and descriptions of the new discoveries have not yet been published.

One feature clearly marks *Ardipithecus ramidus* as a hominid—the position of the opening on the bottom of the skull through which the spinal cord passes. This opening, called the **foramen magnum**, is located forward under the skull, as it is in humans and australopithecines, rather than toward the back of the skull, as it is in apes. While this not a certain indicator of habitual posture, the forward placement of the foramen magnum is associated with bipedal locomotion in *A. afarensis* and in modern humans. *Ardipithecus ramidus* shares several additional derived features with other hominids. These include smaller, more incisor-like canine teeth, which, unlike the ape canines, are not sharpened by the first premolar, and a number of features of the arm bones.

Other traits of *Ardipithecus ramidus* are much more apelike. The cheek teeth are smaller in relation to body size than in other hominids; the enamel is thinner than in other hominids; and the canines, although smaller than in chimpanzees and gorillas, are larger than in the australopithecines. In addition, the **deciduous** (baby) molars and jaw joint are very similar to those of apes, and the base of the skull is more pneumatized than in *A. afarensis*.

It is also interesting that *Ardipithecus ramidus* appears to have lived in a more forested environment than *Australopithecus afarensis* or *A. anamensis*. This conclusion is based on numerous fossils of wood and seeds that were found at the site and come from plant species that grow in woodland habitats. Animal fossils from Aramis are also mainly woodland species: most of the antelope fossils are of kudus, browsers that typically live in woodlands (Figure 11.17). Moreover, more than 30% of the mammalian fossils are of colobine monkeys, which are now found mainly in forested areas.

The apelike anatomical features of *Ardipithecus ramidus* and its apparent preference for forest environments provide tantalizing hints that this species may rep-

Figure 11.17

Ardipithecus ramidus fossils are found in association with the fossils of woodland species like this kudu. This suggests that *A. ramidus* lived in more forested environments than australopithecines did.

resent the earliest stage of hominid evolution, and may be the root of the tree linking the human lineage with that of the apes. However, we will know much more when the remaining fossil material has been described.

Australopithecines after *A. afarensis*

More than one species of hominid lived in Africa between 3 and 2.5 mya.

Between 3 and 2 mya, at least four and perhaps as many as six species of hominids lived in Africa. Moreover, two or even three species lived in the same place at the same time. Four of the species are usually assigned to the same genus as Lucy and her kin: *Australopithecus africanus, A. robustus, A. boisei,* and *A. aethiopicus.* One, or perhaps two, species are placed in the same genus as modern humans, *Homo.* The sites at which early hominid fossils have been found in Africa are located on the map in Figure 11.18.

A. africanus

A. africanus was similar to A. afarensis.

This species was first identified in 1924 by Raymond Dart, an Australian anatomist living in South Africa. Miners brought him a piece of rock from which he painstakingly extracted the skull of an immature, small-brained creature. Dart formally named this fossil *Australopithecus africanus,* which means "the southern ape of Africa," but there is now evidence that members of the same species may have been distributed throughout southern and eastern Africa. Dart's fossil is popularly

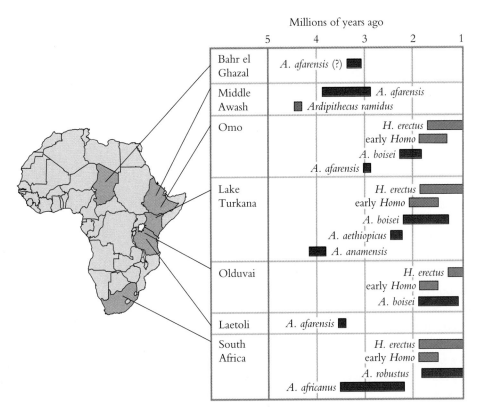

Figure 11.18

Hominid fossils have been discovered at a number of sites in eastern and southern Africa. Fossils assigned to the genus *Australopithecus* are in red, those assigned to the genus *Homo* are in blue, and *Ardipithecus ramidus* is in brown. The fossils discovered at Bahr el Ghazal in Chad, first described in late 1995, have been provisionally assigned to the species *Australopithecus afarensis* pending more detailed analysis.

known as the Taung child. Dart believed that the Taung child was bipedal because the position of the foramen magnum was more like that of modern humans than that of apes. At the same time, he observed many similarities between the Taung child and modern apes, including a relatively small brain. Thus, he argued that this newly discovered species was intermediate between apes and humans. His conclusions were roundly rejected by members of the scientific community because in those days most physical anthropologists believed that large brains evolved in the human lineage before bipedal locomotion. Controversy about the taxonomic status of *A. africanus* continued for the next 30 years.

The creature that caused all the controversy was a small biped with relatively modern dentition and postcranial skeleton. As with *A. afarensis,* there is pronounced sexual dimorphism, both in canines and body size. Males stood 1.38 m (4 ft 8 in.) tall and weighed 41 kg (90 lb), while females were 1.15 m tall (3 ft 11 in.) and tipped the scales at about 30 kg (66 lb). The teeth are more modern than those of *A. afarensis* in several ways. The brain averaged 442 cc, somewhat larger than the average for *A. afarensis.* However, the difference between *A. afarensis* and *A. africanus* is small compared with the variation among individuals and is likely due to sampling variation. The members of this species lived from 3.5 to 2.3 mya.

The postcranial skeleton is virtually identical to *A. afarensis.* Dart's notion that the Taung child was bipedal was controversial because it relied primarily on the location of the foramen magnum, and he had no postcranial bones to corroborate his conclusions. This is partly because Dart's original find came from the rubble of mining operations where dynamite was used to excavate, and the fossils were not discovered in their original location. Subsequently, many adult skulls and postcranial bones were found at two other sites in South Africa, Makapansgat and Sterkfontein. The hip bone, pelvis, ribs, and vertebrae of *A. africanus* are much like those of Lucy and the First Family. The Taung child was thus proven to be fully bipedal, just as Dart had originally claimed (Figure 11.19).

The skull of A. africanus *has a number of derived features. Only some of these are shared with modern humans.*

A. africanus is less primitive than *A. afarensis* in a number of important details. For example, the base of the skull has fewer air pockets, the face is shorter and less protruding below the nose, the canines are less dimorphic, and the base of the cranium is bent further upwards, or flexed. In all these traits *A. africanus* is more like modern humans and less like earlier primates. However, *A. africanus* exhibits a number of derived characters that it does not share with modern humans. Most

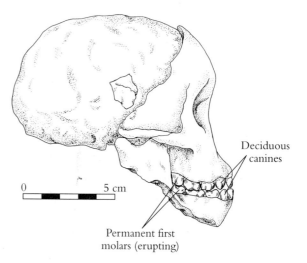

Figure 11.19

The first australopithecine specimen was identified in South Africa by Raymond Dart in 1924. Dart named it *Australopithecus africanus,* which means "the southern ape of Africa," but the specimen is often called the Taung child. Dart concluded that the Taung child was bipedal. Even though it had a very small brain, he believed it to be intermediate between humans and apes. His conclusions were not generally accepted for nearly 30 years. (Figure courtesy of Richard Klein.)

Deciduous canines

0   5 cm

Permanent first molars (erupting)

of these characters seem to have to do with heavy chewing using the molars. The molars are quite big, and the lower jaw is larger and sturdier than in modern humans.

A. africanus *may have been more arboreal than* A. afarensis.

Two recently described fossils from the South African site of Sterkfontein suggest that *Australopithecus africanus* had a more apelike physique than *A. afarensis*. The first fossil, an incomplete skeleton labeled Stw 431, indicates that the arms of *A. africanus* were significantly longer relative to its legs than were the arms of *A. afarensis* suggesting that *A. africanus* was more arboreal than its East African contemporary. This idea is supported by a second fossil also found at Sterkfontein. This fossil, dubbed Little Foot and formally labeled Stw 573, includes four bones from the foot of a single individual and dates to between 3 and 3.5 mya (Figure 11.20). The bones from the back of the foot show a number of humanlike features associated with bipedal locomotion, such as a weight-bearing heel. However, the bones at the front of the foot are not at all humanlike. In fact, Clarke and Tobias argue that the surface of the big-toe joint clearly indicates this creature could grasp things like tree branches with its foot.

It seems likely that the ancestors of the australopithecines were forest apes, with an apelike physique. Thus, we have a puzzle: how can *A. afarensis*, which is older and has a primitive skull, have a more humanlike body, while *A. africanus*, which is younger and has a more modern skull, have a more apelike body? The puzzle has two solutions. First, it could be that the southern African species, *A. africanus*, retained the apelike body of its ancestors, while the East African species evolved a more modern body with longer legs and shorter arms. In this case, if we find older hominid fossils in southern Africa, they should share the apelike body with *A. africanus*. Alternatively, it could be that the more modern bodies evolved throughout Africa, and then the trend was reversed in southern but not eastern Africa.

A. africanus *matured rapidly like chimpanzees, not slowly like humans.*

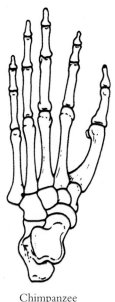

Chimpanzee   Stw 573   Modern human

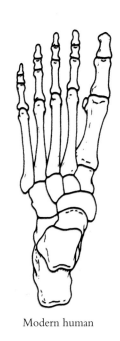

Figure 11.20
At Sterkfontein, parts of a 3- to 3.5-million-year-old foot have been discovered (Stw 573). While the back of the foot (shaded) has features associated with bipedal locomotion, the morphology of the big toe shows the creature was still able to grasp things like tree branches. If this specimen is from an australopithecine, then probably these creatures were partly arboreal.

For many years it was thought that the Taung child was about six years old when it died. This estimate was based on the assumption that the australopithecines matured at the same rate as modern humans do. Recently, we have learned that this assumption is wrong—australopithecines grew up quickly. The evidence for this conclusion is based on careful study of australopithecine teeth by Holly Smith of the University of Michigan. Smith compared the relative rates of development for different teeth in australopithecines, apes, and modern humans. She found that humans develop more slowly than chimpanzees do, and their teeth grow more slowly as well. However, there are also differences in how quickly each kind of tooth develops in each species. For example, in humans, the crown of the canine is fully developed shortly after the crown of the incisors is developed and long before the first premolar is developed. In chimpanzees, the canines develop more slowly, and canine crown development is not completed until more than a year after the incisors. Smith used this information to assess the developmental rate of the australopithecines. When she used the human developmental pattern as a guide, she got quite different estimates for the age of the Taung child using different kinds of teeth. When she assumed that australopithecines had the same developmental pattern as chimpanzees, all the teeth produced the same estimate of the child's age. This suggests that the australopithecine developmental pattern was more like that of chimpanzees than like that of modern humans (Figure 11.21).

Recently Timothy Bromage of City College of New York and M. C. Dean of the University of London have found another way to estimate the chronological age of fossils. Their method relies on the fact that as teeth grow, enamel is laid down in discrete layers that are formed at a constant rate. Bromage and Dean used an electron microscope to make a highly magnified image of the tooth of an immature australopithecine. Then, by carefully counting the layers of enamel, they calculated the age at which the creature died. Like Smith, they concluded that australopithecines developed rapidly, as chimpanzees do. So the Taung child was not six years old when it died; it was only about three.

These results are important because they suggest that australopithecine infants did not have as long a period of dependency as human children do. Many anthropologists believe that a number of the fundamental features of human foraging societies, such as the establishment of home bases, sexual division of labor, and extensive food sharing were necessitated by a long period of infant dependency (Box 11.2). If australopithecine infants matured at the same rate as modern chim-

Figure 11.21

The development of australopithecines was probably more like modern chimpanzees than modern humans. This chimpanzee is about four years old, nearly old enough to be weaned.

BOX 11.2

A Surprise from Ethiopia:
Australopithecus garhi

In 1999, just as this edition of our book was going to press, a new member of the hominid family made its scientific debut. A research team led by Berhane Asfaw (from the Rift Valley Research Service) and Tim White announced the discovery of a new species of australopithecine from the Awash valley of Ethiopia, *A. garhi*. Garhi means "surprise" in the Afar language, and this new species is aptly named because it is an unexpected creature in several ways.

Asfaw, White, and their colleagues recovered hominid remains from a site called Bouri, the ancient site of a shallow freshwater lake not far from where *A. afarensis* was first found. In 1996, the team unearthed a number of postcranial bones, including the partial skeleton of one individual. These specimens were well preserved and could be securely dated to 2.5 mya, but they didn't contain diagnostic features that could be used to assign them to a specific species. However, the next year the research team discovered a number of cranial remains about 300 meters from the site where the postcranial remains were unearthed. These pieces of the skull, maxilla (upper jaw), and teeth came from the same stratigraphic level as the postcranial remains and revealed some important clues about the taxonomic identity of the Bouri hominids.

Like *A. afarensis* and *A. africanus*, these creatures had small brains (approximately 450 cc) perched above a prognathic face. Their canines, premolars, and molars were generally larger than those of *A. afarensis* and *A. africanus* (Figure 11.22). Certain detailed features of the dentition differed from *A. afarensis* as well. *A. garhi* also had a sagittal crest, a ridge of bone that runs over the top of the skull. Asfaw and White assigned the Bouri specimens to a new species of the genus *Australopithecus* because they differed from any of the known australopithecines but lacked the derived characters associated with early *Homo*, such as a larger brain.

So far, the postcranial remains found near the type specimen of *A. garhi* cannot be assigned to a particular species, but they do reveal some interesting developments in the hominid lineage. In chimpanzees, the upper bone of the arm (the humerus), the lower bones of the arm (the radius and ulna), and the upper bone of the leg (the femur) are about

Figure 11.22

In 1999, the first descriptions of a 2.5 million-year-old hominid species from Ethiopia were published. This new species, named *A. garhi*, had a small brain and prognathic face. It also had a sagittal crest, a fin of bone that runs along the top of the skull. Compared to *A. afarensis* and *A. africanus*, *A. garhi* had broader canines and larger molars.

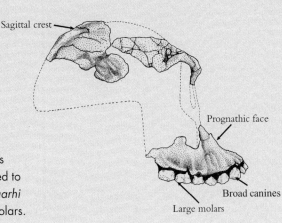

Sagittal crest

Prognathic face

Broad canines

Large molars

the same length, while in modern humans, the bones of the forearm have become shorter, and the femur has become longer than the humerus. *A. afarensis* resembles chimpanzees in the proportions of these bones. However, reconstructions of the humerus, radius, ulna, and femur of the hominids at Bouri suggest that these creatures' femurs had become somewhat elongated, while the relative proportions of the humerus and forearm bones had not changed. The postcranial remains suggest that there was considerable variation in the size and robustness of the Bouri hominids. This variation may reflect the fact that males were larger than females, but there is not yet enough evidence to be certain that this was the case.

Perhaps the most unexpected thing about the Bouri hominids is that they apparently made and used stone tools. Although we know that hominids were making and using stone tools 2.5 mya, it was generally believed that the tools were made by members of the genus *Homo*, not by australopithecines. Evidence from Bouri forces us to question this conclusion. Stone tools make characteristic marks on animal bones (see Chapter 12), and at Bouri there are a number of animal bones that carry the telltale marks of stone-tool use. It appears that the Bouri hominids used stone tools to dismember animal carcasses, cut meat from bones, and smash open bones to extract the marrow. (It is not clear whether the Bouri hominids acquired animal carcasses by hunting or scavenging.) If the Bouri hominids used tools, they must have been quite resourceful creatures. None of the debris normally associated with tool manufacturing is found near the butchery sites, and there are no raw materials suitable for tool making available in the immediate area. The Bouri hominids must therefore have traveled some distance to obtain material to make tools, which they manufactured off-site and then carried to the sites where they were used.

A. garhi's discovery raises new questions about the phylogeny of early hominid species. As Asfaw and his colleagues point out, *A. garhi* appeared at the right time and right place to be the ancestor of the first members of the genus *Homo* (Figure 11.23). Although this possibility

Figure 11.23

One possible phylogeny for *A. garhi* is depicted here. *A. garhi* derived from *A. afarensis*, which also gives rise to *A. africanus* in southern Africa and *A. aethiopicus* in eastern Africa. In this new scheme, *A. garhi* is the ancestor of early *Homo*, *A. africanus* gives rise to *A. robustus*, and *A. aethiopicus* gives rise to *A. boisei*. While this scenario is plausible, a number of other phylogenies are also consistent with the evidence.

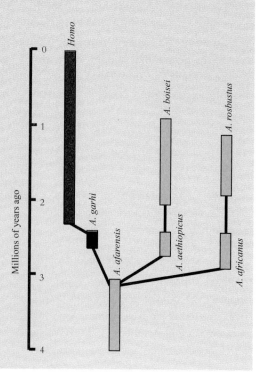

panzees do, then it seems likely that these features were not yet part of the hominid adaptation.

Robust Australopithecines

Australopithecus robustus is a more recent species than the three other australopithecines discussed so far. It had teeth and skull structures specialized for heavy chewing.

At another South African site, Kromdraai, Robert Broom, a retired Scottish physician and avid paleontologist, discovered fossils that seemed very different from the *A. africanus* specimens previously found at Sterkfontein. These creatures were much more robust (sturdier, with heavier bones) than their neighbors, and so Broom called his find *Paranthropus robustus*. This choice was questioned by many of his colleagues, who thought he had discovered another species of *Australopithecus*—most paleontologists now classify Broom's fossils as *Australopithecus robustus*. However, there are some who argue that the robust australopithecines are so different from *A. afarensis* and *A. africanus* that they should be placed in a separate genus. If so, Broom's original classification would be restored.

There is a large sample of *A. robustus* fossils from Kromdraai and also from a second site nearby, named Swartkrans. These creatures appeared about 1.8 mya and disappeared about 1 mya. Their brains averaged about 530 cc. *A. robustus* males stood about 1.32 m (4 ft 6 in.) tall and weighed about 40 kg (88 lb); females were about 1.10 m tall (3 ft 7 in.) and weighed 32 kg (70 lb). Their postcranial anatomy shows they were undoubtedly bipedal. Like *A. africanus* they share several derived features with humans. In addition, they share several specialized features for heavy chewing with *A. africanus,* but these features are much more pronounced in *A. robustus*. Major cranial differences among these australopithecines are shown in Figure 11.24. The molars are enormous, the lower jaw is very large, and the entire skull has been reorganized to support the massive chewing apparatus. For example, there is a pronounced **sagittal crest**, a fin of bone that runs along the centerline of the skull making it look a bit like a punk rocker's Mohawk haircut. The sagittal crest enlarges that surface area of bone available for attaching the **temporalis muscles**, one of the muscles that work the jaw. You can easily demonstrate for yourself the function of the sagittal crest. Put your fingertips on your temples and clench your teeth—you will feel the temporalis muscle bunch up. As you continue clenching your teeth, slowly move your fingertips upward until you can't feel the muscle any more. This is the top point of attachment for your temporalis muscle, about 1 in. above your temple. Australopithecines had much bigger teeth than we do and so needed larger temporalis muscles, and these muscles required more space for attachment to the skull. In *A. africanus* the muscles ex-

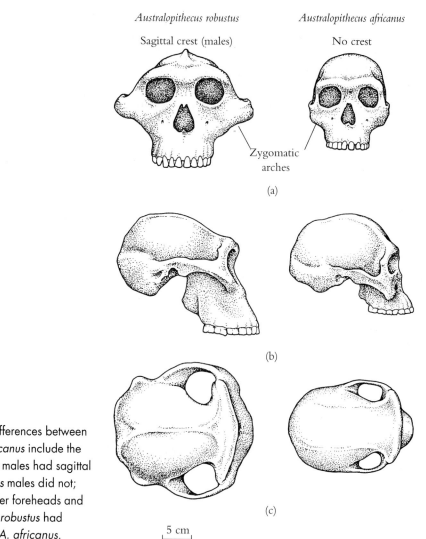

Australopithecus robustus

Sagittal crest (males)

Australopithecus africanus

No crest

Zygomatic arches

(a)

(b)

(c)

5 cm

Figure 11.24

The principal cranial differences between *A. robustus* and *A. africanus* include the fact that (a) *A. robustus* males had sagittal crests while *A. africanus* males did not; (b) *A. robustus* had flatter foreheads and flatter faces; and (c) *A. robustus* had shorter snouts than did *A. africanus.*

panded so that they almost met at the top of the skull. When teeth became even larger in *A. robustus,* even more space for muscle attachment was needed, and the sagittal crest served this function (Figure 11.25). Other distinctive features of

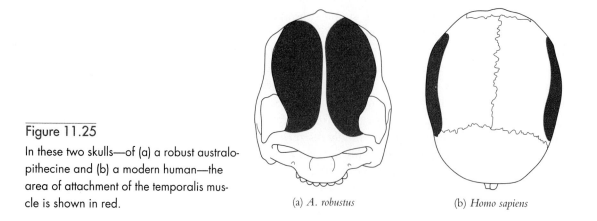

Figure 11.25

In these two skulls—of (a) a robust australopithecine and (b) a modern human—the area of attachment of the temporalis muscle is shown in red.

(a) *A. robustus*

(b) *Homo sapiens*

A. robustus were also accommodations to the enlarged teeth and jaw musculature. For example, the cheek bones (zygomatic arches) are flared outward to make room for the enlarged temporalis muscle. The flared cheek bones caused the face to be flat, or even pushed in.

It is unclear what the robust australopithecines were doing with this massive chewing apparatus. Until recently, most anthropologists believed the robust australopithecines relied more on plant materials that required heavy chewing than *A. africanus* did. In other animals, large grinding teeth are often associated with a diet of tough plant materials, while omnivorous animals typically have relatively large canines and incisors. Moreover, the wear patterns on *A. robustus* teeth suggest that they ate very hard foods like seeds or nuts. However, recent research on the chemistry of *A. robustus* teeth suggests that they probably also ate substantial amounts of meat (Box 11.3).

Hominids are larger than most other primates, probably as an adaptation to terrestrial life.

It used to be thought that the robust australopithecines were taller and considerably heavier than *A. afarensis* and *A. africanus*. However, recent estimates of body size (Box 11.4) indicate that *A. africanus* and *A. robustus* were roughly the same size.

The australopithecines were smaller than most adult humans but larger than most other present-day primates. Their size may be due to the fact that they were terrestrial—terrestrial primates are larger on average than arboreal ones (Figure 11.26), perhaps because terrestrial animals are freed from the constraints that limit the body size of their arboreal counterparts. Small animals move through the trees more easily than large ones do—arboreal life favors the lithe, slender physique of the gibbon, not the massive bulk of the gorilla. As we noted in Part Two, large body size may also confer protection against predators.

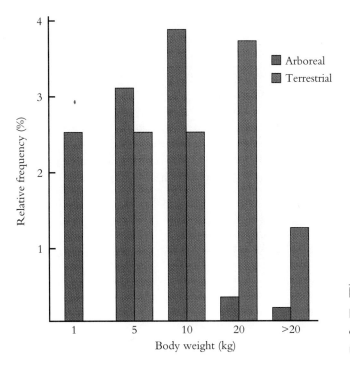

Figure 11.26

In general, modern arboreal primate species (green) are smaller and lighter than modern terrestrial primate species (brown).

BOX 11.3

Chemical Clues about the Diet of Robust Australopithecines

Two kinds of chemical analyses of the fossilized teeth of robust autralopithecines suggest that they may have eaten substantial amounts of animal matter. The first kind of evidence comes from the obscure but useful fact that the ratio of strontium (Sr) to calcium (Ca) in the bones of living animals declines as these elements are passed up the food chain: plants have higher Sr/Ca ratios than do the grazing herbivores that eat them, and grazing herbivores that feed mainly on grasses have higher Sr/Ca ratios than the carnivores who prey on them. Browsing herbivores constitute the major exception to this rule. Browsers, such as deer and kudu, mainly eat the leaves and shoots of woody plants, but they have low Sr/Ca ratios like carnivores.

To find out what *A. robustus* was eating, Andrew Sillen of the University of Cape Town measured the Sr/Ca ratios in the fossilized bones of *A. robustus* and in the fossilized bones of a number of animal species found at Swartkrans. He needed to survey a number of species to be sure that fossil specimens showed the same relative Sr/Ca ratios as living members of the same species do (Figure 11.27). He found that they do, so we can be confident that the Sr/Ca ratios were not changed by geochemical processes after the animals died. Sillen's measurements of the Sr/Ca ratio for *A. robustus* suggest these creatures were either browsers or carnivores but not grazing herbivores.

To determine whether *A. robustus* was a browser or a carnivore, we turn to another line of evidence. It turns out there are two chemically distinct kinds of photosynthesis. Trees, shrubs, and other woody plants utilize one type, called C_3 photosynthesis, and grasses use a second type, called C_4 photosynthesis. C_4 plants and the animals that eat them have higher concentrations of the heavy isotope of carbon, carbon-13, than do C_3 plants and the animals that eat C_3 plants. Julia Lee-Thorp of the University of Cape Town and her colleagues measured the amount of carbon-12 and carbon-13 in the fossilized teeth of a number of species at Swartkrans, including *A. robustus*. The levels of carbon-13 found in the teeth of *A. robustus* are higher than the levels found in browsing species like kudu. This indicates that *A. robustus* either ate grass and grass seeds or ate animals that ate grasses. Since the Sr/Ca ratios suggest that *A. robustus* did not eat grasses, these data indicate that *A. robustus* was at least partially carnivorous.

Taken together, these data suggest that *A. robustus* may have consumed some animal protein. Nonetheless, their massive teeth may have evolved for chewing tough plant material. For example, they might have specialized on tough plant material during the dry season but had a more diverse diet during the rest of the year.

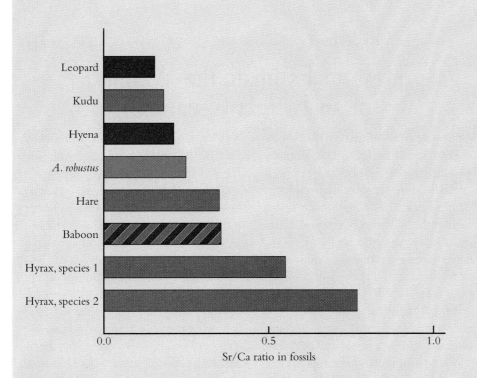

Figure 11.27

Measuring the ratio of strontium to calcium in the bones of both living and fossil species can help paleontologists deduce the kind of diets extinct creatures had. In general, the ratio of strontium (Sr) to calcium (Ca) in the bones of living animals declines as you move up the food chain from pure herbivory to carnivory. An exception is browsing herbivores (like kudu), which have low Sr/Ca ratios. Here the Sr/Ca ratios for typical herbivores (green) and carnivores (red) are plotted along with the Sr/Ca ratio for *A. robustus* (blue). Baboons, which combine herbivory and carnivory, are plotted in red and green. The low Sr/Ca ratio in *A. robustus* is consistent with carnivory or browsing herbivory, but not with general herbivory.

A. boisei was a robust robustus.

Another australopithecine with large molars was discovered by Mary Leakey at Olduvai Gorge, Tanzania, in 1959 (Figure 11.28). This specimen, officially labeled Olduvai Hominid 5 (OH 5), was first classified as *Zinjanthropus boisei*. *Zinj-* derives from an Arabic word for "East Africa," and Charles Boise was a donor who was funding Leakey's research at the time. Leakey's find was later reclassified *Australopithecus boisei* because of its affinities to the South African forms of *A. robustus*. Those who advocate reclassifying *A. robustus* as *Paranthropus robustus* would assign Leakey's find to the same genus, and call it *P. boisei*. The discovery of OH 5 was important, partly because it ended nearly 30 frustrating years of work in Olduvai Gorge in which there had been no dramatic hominid finds. It was only the first of a very remarkable set of fossil discoveries at Olduvai.

Essentially, *A. boisei* is an even more robust *A. robustus*—that is, a hyperrobust australopithecine. The molars are larger than those of *A. robustus,* even when differences in body size are taken into account, and the skulls are even more specialized for heavy chewing. *A. boisei* was also somewhat larger than *A. robustus* in body size.

353

BOX 11.4

How to Estimate the Weight of an Extinct Creature

Anthropologists estimate the body weight of fossil hominids like *A. robustus* by studying the relationship between skeletal anatomy and body weight in living species. To see how this is done, consider the following example. Anatomists know that the area of joints is related to the loads that they must bear, and we have seen that the australopithecines were bipeds like humans. This means that their knees and ankles bear their full body weight. Thus, an anthropologist might try to estimate the size of *A. robustus* using data on the area of knee joints in living humans. To do this, she would measure the area of knee joints and body sizes in a number of people and then use statistical techniques to derive a mathematical formula relating body weight to knee-joint size. She would then measure the area of the knee joint of the fossil, and use the mathematical formula to estimate the weight of the hominids that lived millions of years ago.

The key to getting valid estimates using this type of technique is to choose the right anatomical feature and the right living primate species. Some anatomical features are much better predictors of body weight than others. For example, Henry McHenry of the University of California, Davis, points out that the first robust australopithecine ever found had teeth and jaws comparable to those of a female gorilla weighing as much as 90 kg (198 lb), but had a foot the size of a 34-kg (75-lb) modern human. It is also important to choose the right species. If our anthropologist used data from chimpanzees instead of humans, she would get a very different estimate of the weight of *A. robustus* because chimpanzees and humans have very different body proportions and distribute their body weight quite differently among their limbs.

McHenry solved this problem in the following way. He derived the relationship between body weight and various anatomical features (such as the area of the knee joint, the area of the elbow joint, or limb-bone thickness) for a number of different humans and chimpanzees. Then he used these relationships to obtain weight estimates for *A. afarensis*. The results based on chimpanzee features were highly divergent. For instance, the estimate of *A. afarensis*'s body weight based on chimpanzee knees was twice the estimate based on chimpanzee elbows. In contrast, estimates of *A. afarensis*'s weight based on various human features were quite similar. This suggests that the body proportions of the australopithecines were more similar to humans than to chimpanzees, and that the best estimates of body weight for *A. afarensis* are those based on human models. Using the same approach, McHenry discovered that *A. africanus* and *A. robustus* weighed about the same.

A. boisei appears in the fossil record at about 2.2 mya in eastern Africa. It became extinct about 1.3 mya, although the exact date for its disappearance is difficult to establish.

Australopithecus aethiopicus *is another robust form that displays some of the primitive characteristics of* A. afarensis.

Figure 11.28

Olduvai Gorge in Tanzania has been the site of many important paleontological and archaeological finds. Louis and Mary Leakey worked in Olduvai for more than 30 years.

Alan Walker of Pennsylvania State University recently discovered the skull of another robust australopithecine at a site on the west side of Lake Turkana in northern Kenya. Walker called it the "Black Skull" because of its distinctive black color, but its official name is KNM-WT 17000 (Figure 11.29). (This identification system is not as obscure as it may seem. KNM stands for Kenya National Museum, and WT stands for West Turkana, the site at which the fossil was found.) This creature lived about 2.5 mya, and in some ways it is intermediate between *A. afarensis* and the robust australopithecines. For example, the hinge of the jaw in *A. afarensis* is very similar to the jaw joints of chimpanzees and gorillas. The later australopithecines and the earliest members of the genus *Homo* have a modified jaw hinge. WT 17000 has the same primitive structure as *A. afarensis*. On the other hand, WT 17000 also shows many of the hyperrobust features of *A. boisei*. Also, both species were quite sexually dimorphic, they had similar postcranial anatomy, and they were equipped with relatively small brains in relation to body size.

Leakey and Walker initially classified WT 17000 as a form of *A. boisei*, but most anthropologists prefer to place it in its own species, *A. aethiopicus* or *Paranthropus*

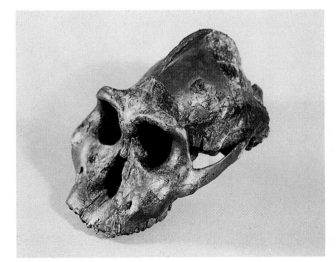

Figure 11.29

This is the Black Skull (WT 17000), which was discovered on the west side of Lake Turkana in Kenya. (Photograph courtesy of Alan Walker.)

Walking through Time

[*Note:* The cover of this book is based upon a photogrammetric image of one of the Laetoli footprints. This footprint was made by a hominid who walked across moist ash about 3.6 mya. This article describes the painstaking efforts that have been made to preserve the Laetoli footprints for future generations.]

Who has not walked barefoot on a beach of crisp sand and, bemused, examined the trail of footprints, paused, then looked back to see the tide wiping them away? So ephemeral are the traces of our passing.

Yet, astonishingly, the tracks of extinct animals have survived for aeons under unusual circumstances of preservation, recording a fleeting instance millions of years ago. Preservation of such traces occurs under conditions of deep burial whereby the sand or mud into which the prints were impressed is changed into stone, later to be exposed by erosion.

When, in 1978, fossil footprints of an extinct human ancestor were discovered during a palaeontological expedition led by Dr. Mary Leakey, scientific and public attention was immense. The prints, partly exposed through erosion, were discovered at the site of Laetoli, to the south of the famed Olduvai Gorge in Tanzania, where Louis and Mary Leakey did their pioneering work researching human evolution.

The footprints at Laetoli, dated at around 3.6 million years, resolved one of the major issues of contention in palaeoanthropology (the study of early mankind), a field characterized by fierce rivalries of discovery and interpretation. At Olduvai, Laetoli, and other sites in Africa and beyond, the search for evidence regarding human development has focused on the discovery of fossilized bones. But while fossils have been the primary means of understanding our past, they cannot yield all the answers to the great debates that have beset the study of human evolution. One debate has been over the development of the brain in relation to our ancestors' ability to walk upright. Since Darwin's time it was thought that once upright posture and bipedalism had developed, the hands were then free to evolve manipulative skills. Stone tool making, it was supposed, was the critical factor in the emergence of early man. This view, however, was not universally accepted. Some be-

Two members of the conservation team excavate footprints from the Laetoli trail. In the foreground is the fiberglass cast that was taken when the footprints were initially discovered by Mary Leakey's team.

The southern portion of the Laetoli trackway is shown here. At the far end, a Masai man stands guard over the excavations.

lieved that "the brain led the way," not erect posture. Although functional analysis of hominid bones from Africa pointed to early bipedalism, the fossils themselves could not provide the definitive answer.

The Laetoli trackway settled the issue. Excavated by Mary Leakey and her team in 1978 and 1979, the trackway consists of some 70 footprints in two parallel trails about 30 meters long, preserved in hardened volcanic ash. The best-preserved footprints are unmistakably human in appearance, and yield evidence of soft tissue anatomy that fossil bones cannot provide. It is significant that the earliest stone tools known are about 2.6 million years old, made nearly a million years *after* the footprints at Laetoli. The Laetoli hominids were therefore fully bipedal well before the advent of tool making—an event considered to define the beginning of culture—and the traces they left behind provide evidence that the feet led the way in the evolution of the modern human brain.

THE CONSERVATION PROBLEM

The footprints at Laetoli, recorded by the Leakey team using various techniques including molding, casting, and photogrammetry, were reburied in 1979 as a means of preservation. After the trackway's reburial, the site revegetated. Although its condition was not known, nor was it visited frequently because of its remoteness, it was feared that the trackway might be deteriorating because of the impact of root growth, especially from acacia trees.

Following a request to the Institute by the Tanzanian Antiquities Unit for assistance in conserving the site, a GCI-Tanzanian team undertook a preliminary investigation in mid-1992. The team opened a 3-by-3 meter trench which confirmed fears that root growth had caused damage, though the full extent could not be determined. Where root growth had not affected the tracks, preservation was excellent, validating the Leakey team's decision to rebury the site, and confirming a practice increasingly adopted by archaeologists to conserve excavated sites.

In 1993, field testing was undertaken by the project team, and a year later 69 trees on the burial mound were poisoned, the site mapped, and measures taken to prevent erosion. In addition, the original cast of the trackway, stored at Olduvai since 1979, was remolded to make a new master cast and additional casts were made with the assistance of staff from the National Museum of Tanzania. Since the trackway itself was too fragile to be remolded, the master cast provides the most accurate replica possible. The new casts also guided re-excavation of the trackway.

In the subsequent two years, campaigns were undertaken during the dry season. A joint team of specialists, including conservators, archaeologists, scientists, and photogrammetrists, re-excavated half of the trackway, recorded its condition stereophotographically, extracted stumps and roots, stabilized the surface, and reburied it using synthetic geotextile materials layered into the overburden of sand and soil. Geotextiles provide protection against root penetration, yet allow the trackway surface to "breathe"—that is, to maintain moisture equilibrium between the subsurface of the trackway and the atmosphere. For a period during the field work, the site was open to palaeoanthropologists for further study. Because of its fragility, it could only be exposed for a very limited period. Upon completion of the fieldwork a monitoring and maintenance plan for the site was designed for implementation by the Tanzanian authorities to ensure the long-term survival of the tracks. Meetings with the local community and a traditional blessing ceremony by Maasai religious leader and healer in the area—the *Loiboni*—were held to emphasize the significance of the trackway and explain the need for its protection.

Many opinions were voiced as to the best method to save the tracks, and the strategy of reburial was debated at length. Among other ideas have been proposals to uplift the entire trackway (or only the individual footprints) and move it to the National Museum in Dar es Salaam; or, to build a shelter over the site and open it to the public. The latter is impractical, at least for the moment, because of the site's remoteness and difficulty of access, and the lack of infrastructure for displaying, staffing, and securing the site. The former assumes that the tracks have scientific value only and thus overlooks their cultural significance.

The Laetoli footprints are the most ancient traces yet found of humanity's ancestors. To move the site *in toto* or, worse, to remove only the prints, would be contrary to the widely accepted ethic of conservation in context. The prints were impressed in volcanic ash in that location 3.6 million years ago, in sight of the Sadiman volcano 20 kilometers away, whose subsequent ash falls buried them under 30 meters of deposit. Over the aeons the landscape eroded; until less than a few inches of soil covered the surface at the time of discovery. Powerful arguments can be mustered to save the site

When the conservation project was completed in 1996, the preservation team convened a meeting with members of the local Masai community. The preservation of the site and the safety of the footprints will depend largely on the goodwill and cooperation of the local residents.

in its original setting. The Tanzanian authorities themselves committed to this approach.

It is indisputable that burial is an effective preservation measure, if other requirements such as vegetation control through maintenance are also met. Only if these criteria are not achievable should other options be considered. While lifting the tracks is doubtless technically possible, it would be enormously costly, require much research, and risk damage or loss. For these reasons, the decision to rebury the site has been made, and if future conditions allow the site to be opened to visitors, it will have been saved.

Because the Laetoli tracks have been reburied, a permanent display was installed at the Olduvai Museum, which overlooks the gorge where Mary and her husband, Louis S. B. Leakey, made so many of their famous discoveries. The Laetoli room displays the cast of the trackway and photographs and text (in English and Swahili) that explain why the site was reburied and how it is being protected.

TRANSCENDING TIME

The footprints at Laetoli represent an immense distance in time. While we are used to bandying terms like "a million years," we cannot really comprehend them on a human scale. We are comfortable with one or two thousand years. They are within the frame of recorded history, spanning the last few hundred human generations. The Laetoli tracks are of another dimension, taking us back perhaps more than one hundred and eighty *thousand* generations.

The question has been asked why the Getty Conservation Institute, whose work is preservation of the cultural heritage, should be involved in saving a fossil site, even one of immense significance in the study of evolution. The answer to this question will be clear to those who have trod the beach and pondered their own trail of footprints, for there can be scarcely anything so evocative as the Laetoli trail marking humanity's long, wondrous, and mysterious journey. As a nexus between cultural heritage and science, so often uncomprehended in today's world of big science, the footprints are a poignant reminder of our ancient origins. Let Mary Leakey have the last word in talking of one of the hominids who made the trail: "At one point, and you need not be an expert tracker to discern this, she stops, pauses, turns to the left to glance at some possible threat or irregularity, and then continues to the north. This motion, so intensely human, transcends time. Three million six hundred thousand years ago, a remote ancestor—just as you or I—experienced a moment of doubt."

Source: Adapted from an article by N. Agnew and M. Demas that originally appeared in the GCI Newsletter, *Conservation,* vol. X, no. 1, 1995.

aethiopicus (Figure 11.30). As we will see, the Black Skull's discovery confounded paleontologists because it contradicted many hypotheses about the phylogenetic affinities among the australopithecines. However, before describing the conundrum created by the Black Skull, we need to introduce one more member of the hominid community that lived between 3 and 2 mya.

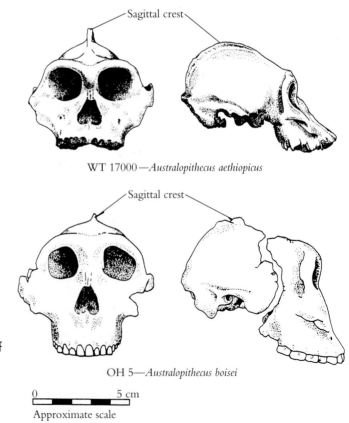

WT 17000—*Australopithecus aethiopicus*

OH 5—*Australopithecus boisei*

0   5 cm
Approximate scale

Figure 11.30

A. aethiopicus (WT 17000) shows some of the hyperrobust features of *A. boisei,* such as extremely large molars and a robust skull, but it is considerably older. (Figure courtesy of Richard Klein.)

Early *Homo*

Fossils of early Homo, *a hominid with a larger brain and more humanlike teeth, have been discovered at many sites in East Africa.*

In 1960, while working at Olduvai Gorge with his parents, Louis and Mary Leakey, Jonathan Leakey found pieces of a hominid jaw, cranium, and hand. The Leakeys assigned this specimen, labeled Olduvai Hominid 7 (OH 7), to the genus *Homo* because they believed the cranial bones indicated that it had a much larger brain than *Australopithecus* and had small, more humanlike teeth. Louis Leakey named the species *Homo habilis,* or "handy man," because he believed he had found the hominid responsible for the simple, flaked stone tools discovered nearby. Later the Leakeys found several more fossils that they assigned to *H. habilis,* including pieces of a second cranium, part of a foot, and more teeth. These fossils range in age from 1.6 to 1.9 million years old.

Many paleoanthropologists doubted that the Leakeys had found the oldest member of our genus. After all, the Leakeys' estimate of *H. habilis*'s brain size was based on only part of the cranium, and even then the complete cranium was estimated to be at most 50% larger than the brains of australopithecines. There was plenty of room for skepticism. However, in 1972, Bernard Ngeneo, a member of Richard Leakey's team, found a nearly complete skull at a site called Koobi Fora on the eastern shore of Lake Turkana (Figure 11.31). This skull proved that large brains were indeed present in a hominid living 1.9 mya. This find is usually referred to by its official collection number, KNM-ER 1470. (ER stands for East Rudolf—Lake Turkana was known as Lake Rudolf when Leakey began his work there.) The most striking thing about ER 1470 is that it had a much larger brain than any known australopithecine (Figure 11.32). Its endocranial volume is 775 cc, 75% bigger than the brains of similarly sized specimens of *A. africanus.* However, compared to OH 7, it had a relatively large, australopithecine-like face. Since the discovery of ER 1470, fossils of early *Homo* have been found at several other sites including Olduvai Gorge, the Omo Basin, Sterkfontein, and a site near Lake Malawi. We will refer to all of these fossils as early *Homo.*

Traits Defining Early *Homo*

Fossils similar to OH 7 and ER 1470 share several derived characters with Homo sapiens.

Figure 11.31

The discovery of ER 1470 at a site on Lake Turkana confirmed that there was at least one large-brained hominid in East Africa 2.5 mya. (Photograph courtesy of Alan Walker.)

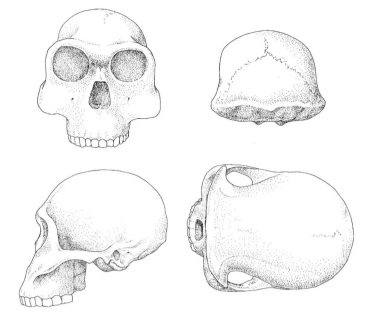

Figure 11.32

ER 1470 had a larger brain, less robust skull, and more humanlike dentition than earlier hominids. (Figure courtesy of Richard Klein.)

KNM-ER 1470

OH 7 and ER 1470 are distinctly different from any of the australopithecines and similar to us in the following ways:

- The brain size of these fossils is substantially bigger than that of any of the australopithecines.
- The teeth are smaller and have thinner enamel, and the dental arcade is more parabolic, than is found in the australopithecines.
- The skulls are more rounded, there are fewer air pockets in the bottom of the skull, the face is smaller and protrudes less, and the jaw muscles are reduced in size compared with the australopithecines.

Most authorities agree that these creatures were sufficiently similar to us that they should join modern humans in the genus *Homo*. For many years, anthropologists classified all these specimens together as members of a single species, *Homo habilis.*

How Many Species?

Fossils traditionally assigned to H. habilis *are quite variable. As a result, some anthropologists believe there were two species of early* Homo.

Since the discovery of ER 1470, many other fossils have been assigned to the species *H. habilis.* The skulls and teeth of these fossils are quite variable. Some of them, like KNM-ER 1813 (Figure 11.33), have more humanlike teeth and less-robust skulls with smaller faces and teeth than ER 1470, but they also have smaller brains than ER 1470, often around 500 cc (Figure 11.34). Other fossils are like ER 1470, with larger brains but also larger teeth and faces.

The postcranial materials assigned to early *Homo* are equally variable. As we have seen, the australopithecines have numerous features that suggest a partly arboreal life—they had small bodies, long upper arms, and curved finger and toe

Figure 11.33

ER 1813 (shown here) had more modern teeth but a smaller brain than ER 1470. This variation has led some researchers to suggest that there may have been more than one species of early *Homo*. (Photograph courtesy of Michael Day.)

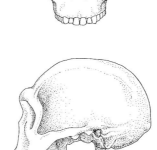

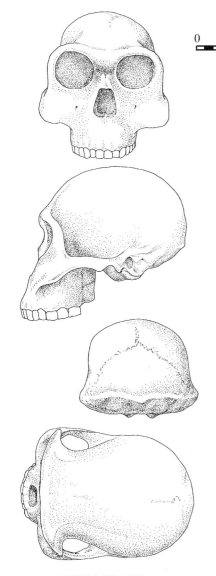

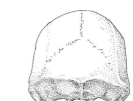

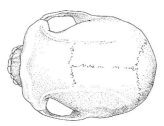

(a) KNM-ER 1470   (b) KNM-ER 1813

Figure 11.34

Fossils assigned to *Homo habilis* show a mixture of derived and ancestral traits. (a) Some, like ER 1470, have larger brains but more robust skulls and teeth. (b) Others, like ER 1813, have more humanlike teeth and a less robust skull but also small, australopithecine-sized brains. Anthropologists are divided about whether these fossils are members of the same species. (Figure courtesy of Richard Klein.)

bones. The first fragmentary postcranial material found at Olduvai Gorge suggested that *Homo habilis* was very similar to the australopithecines from the neck down. A recent discovery from Olduvai Gorge adds credence to this picture. During an expedition led by Don Johanson, Tim White discovered a specimen including an incomplete cranium and numerous postcranial bones. This specimen, labeled OH 62, had a very short stature and surprisingly long arms suggesting that the postcranial skeletons of at least some early *Homo* species were similar to Lucy 1.4 million years earlier. However, other postcranial fossils assigned to early *Homo* indicate these creatures had relatively long legs, more like modern humans than like australopithecines. Many anthropologists believe that the shorter, more australopithecine OH 62 postcranial material should be associated with the smaller-brained but more modern crania like ER 1813, and that the longer-boned, more modern postcranial bones should be associated with the larger-brained, more robust skulls like ER 1470. However, others caution that the more modern postcranial material is from a hominid that appears slightly later in the fossil record, *Homo erectus*. (You will learn more about *H. erectus* in Chapter 13.)

There is a lively debate among paleoanthropologists about how to interpret this morphological variation. Some believe that the pronounced difference in size among specimens assigned to *H. habilis* could represent sexual dimorphism within a single species. Others believe that there were two species of relatively large-brained hominids present in East Africa 2 mya. They believe that the small-brained, less-robust individuals should be classified as *H. habilis* while the large-brained, more robust ones with more modern postcrania are a second species of early *Homo* that should be named *H. rudolfensis*. Presently, not enough is known about these creatures to resolve this debate, and so we will finesse the problem by referring to all of these creatures as "early *Homo*."

Flaked Stone Tools

The earliest stone tools appear in association with early Homo.

As we mentioned earlier, *H. habilis* was given this name ("handy man") because the fossils were found in association with ancient stone artifacts. These artifacts, collectively referred to as the Oldowan tool industry, are very simple. They consist of rounded stones, like the cobbles once used to pave city streets, that have been flaked (chipped) a few times to produce an edge (Figure 11.35). Although these artifacts are very crude, it is clear that they have been deliberately modified. The Oldowan artifacts are quite variable, but this variation does not seem to be related to different functional uses or differences in manufacturing practices or styles over time. Instead, most of the variation seems to be due to differences in the raw materials used to make them. There is now evidence that the **flakes** (small, sharp chips) struck from these cobbles were at least as useful as the **cores** themselves (Box 11.5).

Until the discovery of *A. garhi* in 1999, most researchers assumed that all of the stone tools found at these sites were made by early *Homo*. They made this assumption because they did not think australopithecines were smart enough to have made stone tools. After all, the brains of the australopithecines were about the same size as the brains of other apes. While chimpanzees make and use a variety of tools and sometimes use stones to crack nuts, they don't modify stones to make tools in the wild. Thus, most researchers assumed that australopithecines did not make stone tools either. However, two recent discoveries have raised the possibility that

Summary of Plio-Pleistocene Hominids

	ARDIPITHECUS RAMIDUS	*AUSTRALOPITHECUS ANAMENSIS*	*AUSTRALOPITHECUS AFARENSIS*	*AUSTRALOPITHECUS AFRICANUS*	*AUSTRALOPITHECUS (PARANTHROPUS) AETHOPICUS*
Dates	4.4 mya	4.2–3.9 mya	3.9–2.8 mya	3.5–2.3 mya	2.7–2.3 mya
Sites	Middle Awash, Ethiopia	Lake Turkana, Kenya	Middle Awash, Omo, Ethiopia; Bahr el Ghazal, Chad; Laetoli, Tanzania	Taung, Sterkfontein, Makapansgat, South Africa	Lake Turkana, Kenya
Height (m)	??	??	Males: 1.51 Females: 1.05	Males: 1.38 Females: 1.15	??
Weight (kg)	??	50	Males: 45 Females: 29	Males: 41 Females: 30	??
Brain size (cc)	??	??	380–485	430–520	410
Skull form	Foramen magnum under cranium	Small external acoustic meatus, pneumatized cranial base	Prognathic face, pneumatized cranial base	Shorter face, less pneumatized, more flexed basiocranium	Flat basiocranium, very pneumatized cranial base, dish-shaped face
Jaws & teeth	Thin enamel, reduced canines	Large molars, thick enamel, reduced canines, U-shaped dental arcade	Large molars, thick enamel, reduced canines, less U-shaped dental arcade	Larger molars, more reduced canines	Very large molars and premolars
Physique	Very powerful arms	Bipedal but with very robust arms	Bipedal but with long arms, curved fingers, short legs, and conical rib cage	Longer arms and shorter legs than *A. afarensis*. Possibly had opposed big toe.	??

Summary of Plio-Pleistocene Hominids (*continued*)

	AUSTRALOPITHECUS (PARANTHROPUS) BOSEI	*AUSTRALOPITHECUS (PARANTHROPUS) ROBUSTUS*	*HOMO HABILIS (LARGE OR RUDOLFENSIS)*	*HOMO HABILIS (SMALL)*
Dates	2.2–1.3 mya	1.8–1.0 mya	1.9–1.6 mya	1.9–1.6 mya
Sites	Omo, Ethiopia; Lake Turkana, Kenya; Olduvai, Tanzania	Taung, Kromdraai, Swartkrans, South Africa	Omo, Ethiopia; Lake Turkana, Kenya	Olduvai, Tanzania; Sterkfontein, South Africa
Height (m)	Males: 1.37 Females: 1.24	Males: 1.32 Females: 1.10	1.5	1.0
Weight (kg)	Males: 49 Females: 34	Males: 41 Females: 30	??	??
Brain size (cc)	500–550	500–550	600–800	500–650
Skull form	Sagittal crest, flat face, extreme post-orbital constriction	Sagittal crest, flat face, extreme post-orbital constriction	More robust	Less robust
Jaws & teeth	Very large molars and premolars, deep mandibles	Very large molars and premolars, deep mandibles	Large jaw and molars	Smaller, more modern molars
Physique	Small but powerfully built; long arms, short legs	Small but powerfully built; long arms, short legs	Larger, with longer legs	Smaller, with longer arms and shorter legs

Bifacial chopper

Hammer stone

Discoid

Flake scraper

Polyhedron

Flake

0   5 cm

Heavy-duty (core) scraper

Figure 11.35

Early hominids, probably early *Homo,* made stone tools like these. Researchers are not sure how these tools were used. Some think the large cores shown here were used for various tasks, while others think the flakes were the real tools and the rounded stones are simply what was left after flakes had been struck off.

australopithecines might have been proficient toolmakers. First, animal bones found at Bouri show cut marks made by stone tools. Large mammals were disarticulated and defleshed, and their long bones were broken open. *A. garhi* is the only hominid known from this area during this time period. Second, new discoveries from West Turkana provide good evidence that hominids were proficient toolmakers by 2.3 mya. Researchers have found an extensive array of stone artifacts there, including flakes, cores, hammer stones, and related debris. Preservation conditions at this site were so good that the workers were able to fit a number of flakes back onto the cores from which they had been removed. Some of the cores were coarse-grained cobbles from which only a few flakes had been removed. But there were also a number of fine-grained lava pebbles from which up to 30 flakes had been systematically removed. What is remarkable about this site is the technical expertise of the toolmakers. The creatures took into account the quality of raw materials that were available and the natural morphology of the blocks they were working with, carefully maintaining precise flaking angles during the entire tool making sequence. It is not clear who made these tools. Both robust australopithecines and *A. garhi* were present in East Africa 2.3 mya, but none of the clearly

BOX 11.5

Ancient Toolmaking and Tool Use

Recently, Kathy Schick and Nicholas Toth of Indiana University have done many experiments with simple stone tools. They have mastered the skills needed to manufacture the kinds of artifacts found at Oldowan sites and learned how to use them effectively. Their experiments have produced several interesting results. Schick and Toth believe the Oldowan artifacts that archaeologists have painstakingly collected and described are not really tools at all. They are cores left over after striking off small, sharp flakes, and the flakes are the real tools. Schick and Toth find that the flakes can be used for an impressive variety of tasks, even for butchering large animals like elephants. In contrast, while the cores can be used for some tasks like chopping down a tree to make a digging stick or a spear, or cracking bones to extract marrow, they are generally much less useful. Schick and Toth's conclusion is supported by a microscopic analysis of the edges of a small number of Oldowan flakes, which indicates that they were used for both woodworking and butchery.

Schick and Toth have also been able to explain the function of enigmatic objects that archaeologists call spheroids. These are smooth, approximately spherical pieces of quartz about the size of a baseball. Several suggestions for their function were put forth, including processing plants and bashing bones to extract marrow. Some researchers thought the spheroids were part of a bola, a hunting tool used on the grasslands of Argentina. In a bola, three stones are connected by leather thongs and thrown so that they tangle the prey's legs. Schick and Toth have shown there is a much more plausible explanation for these stones. If a piece of quartz is used as a hammer to produce flakes, bits of the hammer stone are inadvertently knocked off. The hammer surface is no longer flat, so the hammer stone is shifted in the toolmaker's hand. Thus, the hammer gradually becomes more and more spherical. After a while, a quartz hammer becomes a spheroid.

Perhaps the most remarkable conclusion that Schick and Toth drew from their experiments is that early hominid toolmakers were right-handed. Schick and Toth found that right-handers usually hold the hammer stone in their right hand and hold the stone to be flaked in the left hand. After driving off the first flake, they rotate the stone clockwise and drive off the second flake. This sequence produces flakes in which the **cortex** (the rough, unknapped surface of the stone) is typically on the right side of the flakes (Figure 11.36). On flakes made by left-handers, the cortex is typically on the left side. With this result in hand, they studied flakes from sites at Koobi Fora, Kenya, dated 1.9 to 1.5 mya. Their results suggest that most of the individuals who made these flakes were right-handed.

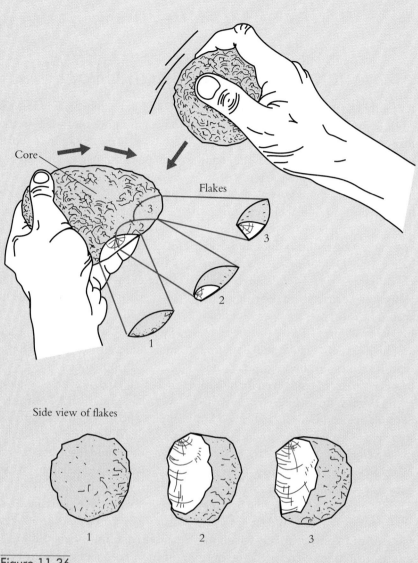

Core

Flakes

3

2

1

Side view of flakes

1   2   3

Figure 11.36

This diagram demonstrates why right-handed flint knappers make distinctive flakes. When a right-handed person makes a stone tool, she typically holds the hammer stone in her right hand and the core to be flaked in the left hand. When the hammer stone strikes the core, a flake spalls off leaving a characteristic pattern of rays and ripple marks centered around the point of impact (here shown in red). The knapper then rotates the core and strikes it again. If she rotates the core clockwise as shown, the second flake has the percussion marks from the first flake on the upper left, and part of the original rough surface of the rock, or cortex (here shown in brown), on the right. If she rotates the core counterclockwise, the cortex will be on the left and the percussion marks on the upper right. Modern right-handed flint knappers make about 56% right-handed flakes and 44% left-handed flakes, and left-handed flint knappers do just the opposite. A sample of flakes from Koobi Fora dated to between 1.5 and 1.9 mya contains 57% right-handed flakes and 43% left-handed flakes, suggesting that early hominid tool makers were right-handed.

established specimens of early *Homo* are quite that old. Thus, these tools may have been made by robust australopithecines, *A. garhi*, or by older members of the genus *Homo* who have not yet been discovered.

Hominid Phylogenies

Before the Discovery of WT 17000

Until the late 1980s there was general agreement about the basic form of the hominid phylogeny.

Remember, the goal of our enterprise is to reconstruct the evolutionary history of human beings. To do so, we must understand the phylogenetic relationships among our ancestors. This has become a rather difficult problem in recent years. Until the late 1980s there was general agreement about some important features of the phylogeny of hominids. These features are illustrated in Figure 11.37. This phylogeny embodies two important ideas:

1. Either *A. africanus* or *A. afarensis* gave rise to both the later australopithecines and the genus *Homo*.
2. The robust australopithecines formed a single lineage in which *A. boisei* de-

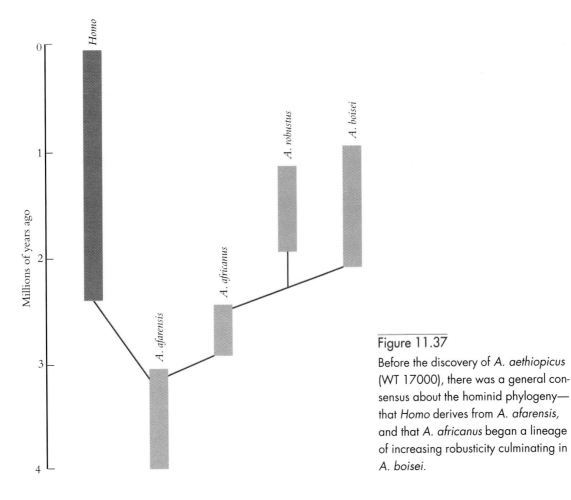

Figure 11.37

Before the discovery of *A. aethiopicus* (WT 17000), there was a general consensus about the hominid phylogeny— that *Homo* derives from *A. afarensis*, and that *A. africanus* began a lineage of increasing robusticity culminating in *A. boisei*.

scended from *A. robustus,* and *A. robustus* descended from *A. africanus.* (In some formulations, *A. robustus* descended from *A. afarensis.*)

This phylogeny reflects the fact that *A. afarensis* shows more primitive (apelike) features than the other australopithecines do, and is based on the plausible assumption that robustness is a derived character. For example, *A. africanus* has larger teeth and is otherwise more robust than *A. afarensis. A. africanus* also has several other derived features, such as the shape of the jaw joint and the degree to which the bottom of the cranium is flexed upward. The robust australopithecines share these derived features and are even more robust. Thus, the simplest phylogeny accounting for these observations has *A. afarensis* giving rise to *A. africanus,* which in turn gives rise to *A. robustus,* and then to *A. boisei.*

There was some question about whether *Homo* should be shown as deriving directly from *A. afarensis* or from *A. africanus.* The fact that *A. africanus* has larger teeth and a more robust skull than either *A. afarensis* or early *Homo* suggests that early *Homo* diverged from the australopithecine line sometime before large teeth and robust skulls began to evolve. This would mean that early *Homo* evolved directly from *A. afarensis.* However, early *Homo* also shares many derived features with *A. africanus* and the robust australopithecines but not with *A. afarensis.* This implies that early *Homo* diverged from the australopithecines after these features had evolved, and that early *Homo* evolved from *A. africanus.* Clearly, no one phylogeny can fit all of the data, and any phylogeny must assume that some similarities are due to convergent evolution. If early *Homo* is derived from *A. africanus,* then we would have to accept a pattern of morphological change in which first, small teeth got larger and skulls became more robust, and later, teeth got smaller and skulls became less robust again. Many anthropologists believe that such evolutionary reversals are unlikely to occur, and therefore favor the idea that early *Homo* evolved from *A. afarensis.*

After the Discovery of WT 17000

The discovery of A. aethiopicus *at Lake Turkana challenged the consensus that early* Homo *evolved from* A. afarensis.

As you may have noticed, all of these phylogenetic scenarios are missing one actor—*A. aethiopicus.* Remember that this species, represented by the Black Skull or WT 17000, has many of the primitive features that characterize *A. afarensis,* but also has very large teeth and a hyperrobust skull like *A. boisei.* Also remember that WT 17000 is dated to about 2.5 mya.

These facts threw the world of paleoanthropology into a tizzy because they contradicted the phylogenetic models just described in two fundamental ways. First, recall from Chapter 4 that the fossil record can be used to determine which traits are ancestral and which are derived by assuming that traits appearing earlier in the fossil record are ancestral. But remember also that this assumption can lead to errors because the fossil record is incomplete, as we saw in Chapter 10. The discovery of WT 17000 exposed just such an error. Before its discovery, paleoanthropologists had sensibly assumed that the robustness was a derived state among the australopithecines because the fossil record appeared to show that the teeth became larger and the skull became more robust over time. The fact that the skull of WT 17000 is both very robust and very old invalidates this assumption.

Second, the fact that WT 17000 combines a hyperrobust skull with primitive features of *A. afarensis* means there has been extensive convergence in the evolution of the robust australopithecines as well as in the evolution of early *Homo*. Remember that *A. africanus* shares a number of derived characters with *A. robustus* and *A. boisei,* but not with *A. afarensis* or *A. aethiopicus.* If we believe that the robust features linking WT 17000 to the robust australopithecines are homologous, then we must assume that the similarities between *A. africanus* and the robust australopithecines are due to convergence. If, on the other hand, we assume that the similarities in the lower portion of the crania of *A. africanus, A. robustus,* and *A. boisei* are homologous, then we must assume that large teeth and robust skulls evolved independently in *A. aethiopicus* and the other robust australopithecines.

Adding WT 17000 to the picture makes many alternative phylogenies equally plausible, depending on what is assumed to be homologous and what is assumed to be convergent. Recently, Randall Skelton of the University of Wyoming and Henry McHenry of the University of California, Davis, performed a phylogenetic analysis of 77 traits of the skull of early *Homo* and the five australopithecine species. They sorted the traits into groups according to their function and their location in the body. For example, one functional group of traits included features thought to be adaptations for heavy chewing, such as large teeth and a sagittal crest, and another group included all of the features of the bottom of the cranium. Then Skelton and McHenry determined the best trees to fit different combinations of traits. They found that the branching pattern of the best tree depends on which groups of traits are included. For example, the best fit based on traits in the heavy-chewing complex groups together all the robust australopithecines, and shows early *Homo* as deriving from *A. afarensis* (Figure 11.38). In contrast, when the traits in

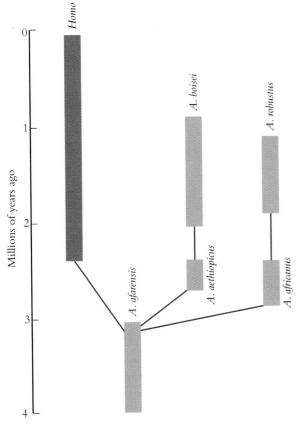

Figure 11.38

If we assume that the similarities in traits related to heavy chewing are homologous, then we obtain a phylogeny in which early *Homo* derives from *A. afarensis,* and there were two distinct lineages of robust australopithecines.

the heavy-chewing complex are eliminated, the results are consistent with the radically different tree shown in Figure 11.39. In this case, *A. aethiopicus* is unrelated to the other robust australopithecines, and early *Homo* derives from *A. africanus*.

The profound and disruptive effect of the discovery of WT 17000 suggests that hominid evolution is more complex than we had previously thought. The tree diagrams we have seen so far have only a few branches. Evolutionary theory does not give us any reason to believe that phylogenies must be simple, or even that they are likely to be simple. It is much more plausible that phylogenies have many branches, and look more like bushes than trees.

The absence of a secure phylogeny for the early hominids does not prevent us from understanding human evolution.

It is easy to be discouraged by the uncertainties in the hominid fossil record, and to doubt that we know anything concrete about our earliest ancestors. Although there are many gaps in the fossil record and much controversy about the relationship among the early hominid species, we do know several important things about them (Figure 11.40). In every plausible phylogeny, the human lineage is derived from one of the less robust australopithecines, *A. africanus* or *A. afarensis*.

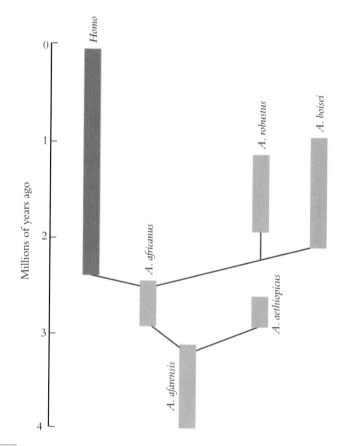

Figure 11.39

If we assume that similarities in traits other than those involved in heavy chewing are homologous, then we might obtain the phylogeny shown here, in which *A. afarensis* gives rise to *A. africanus*, which in turn gives rise to early *Homo* and later robust australopithecines.

Figure 11.40

Although many of the details of early hominid history remain elusive, we know that sometime about 5 mya, a small-brained bipedal ape moved out of the forest and into more open habitats. One of this ape's descendants developed a somewhat larger brain and the ability to make tools from stone.

The earliest australopithecines were small bipeds who were adept in trees. Their teeth and jaws were suited for a generalized diet. The males were considerably larger than the females, and their brains were the same size as those of modern apes. This is the kind of creature that linked the apes of the Miocene to the earliest members of our own genus *Homo*.

Further Reading

Conroy, G. 1997. *Reconstructing Human Origins: A Modern Synthesis*. W. W. Norton, New York.

Foley, R. 1987. *Another Unique Species*. Longman Scientific and Technical, Harlow, U.K.

Klein, R. 1999. *The Human Career*. 2d. ed. University of Chicago Press, Chicago.

Schick, K., and N. Toth. 1993. *Making Silent Stones Speak*. Simon & Schuster, New York.

Study Questions

1. Explain why *A. afarensis* is classified as a hominid.
2. Explain how anthropologists conclude that the australopithecines were bipedal.
3. Outline three reasons why natural selection may have favored bipedal locomotion in the hominid lineage. In each case, explain why hominids became bipedal but other terrestrial primates, like baboons, did not.

4. What is the evidence that australopithecines spent more time in trees than modern humans do?
5. Did the australopithecines make tools?
6. Describe two kinds of evidence suggesting that australopithecines matured rapidly like modern apes rather than more slowly like modern humans.
7. How do anthropologists estimate the weight of extinct hominids?
8. What is the evidence that early hominid toolmakers were right-handed?
9. Explain why the discovery of WT 17000 led to important changes in our ideas about hominid phylogenies.

CHAPTER 12

The Lives of Early Hominids

The history of the human lineage is more than a list of species and dates. It is the story of real creatures who foraged for food, were eaten by predators, searched for mates, formed relationships, gave birth, raised their offspring, and died. These events, set in the changing environment of tropical Africa several million years ago, transformed a small, bipedal ape into a modern human. It is not enough to know that hominid brains got bigger and hominid teeth got smaller. We want to understand *why* these changes occurred. What was it in the day-to-day life of these early hominids that caused the big-brained, small-toothed individuals to leave more offspring than their fellows did? We also want to understand the evolution of the behaviors that distinguish modern humans from other primates. Did early hominids eat meat? Did they hunt? Did they use tools? Did they share food? Did males and females perform different subsistence tasks?

From Ape to Human

We want to explain how our apelike ancestors were transformed into people whose lifestyle was similar to that of modern human food foragers.

Remember that our goal throughout this text is to understand how the human lineage evolved, step by small step, from the earliest mammals to contemporary peoples. In this chapter we will consider what the lives of the early hominids were like. We will also examine the impact an innovation in technology—the introduction of stone tools by early hominids, probably early *Homo*—had on hominid evolution. Unfortunately, there is relatively little material to work with because behavior leaves few traces in the fossil record. To reconstruct early hominid behavior, anthropologists must supplement the fossil record with information about the environmental conditions early hominids faced and comparative information about the behavior and ecology of other primates and carnivores. Anthropologists do have one concrete source of information about behavior—stone artifacts. **Archaeologists** (scientists who study the material remains of past cultures and peoples) have conducted detailed studies of dense concentrations of bones and Oldowan tools at several sites in the Olduvai Gorge, trying to decipher clues about the behavior of the creatures that used them. As mentioned earlier, most anthropologists believe these tools were manufactured by early *Homo,* but some believe that robust australopithecines may also have been toolmakers.

Many researchers have drawn on the behavior of modern apes, particularly chimpanzees, to reconstruct the behavior of early hominids, and have turned to modern foraging peoples to reconstruct the behavior of later hominids. Neither apes nor modern foraging groups provide perfect analogues for early hominids. Nonetheless, as William McGrew has observed in his book *Chimpanzee Material Culture* (1992, Cambridge University Press, pp. 21–22), these data

> . . . supply the closest living approximations for calibrating the past process of hominisation. Sometime in the Miocene, an ancestral hominoid whose closest living analogue is an African pongid [ape] was set on a course that led to humanity. Sometime in the Pleistocene an early human being emerged whose closest living analogue is a tropical hunter-gatherer. . . . However crude, these two models of behaviour give us starting and ending points for reconstructing the process of human emergence.

Studying chimpanzees helps us to understand the behavior of the australopithecines by telling us what is possible with an ape-sized brain.

Chimpanzees help us to develop ideas about the origins of human behavior because they display precursors (simpler forms) of a number of traits that are fundamental elements of traditional human societies (Figure 12.1). Chimpanzees living under natural conditions make a variety of tools from plant materials. They also hunt mammalian prey, share food, show some evidence of sex differences in subsistence behaviors, and commit lethal assaults on members of other chimpanzee communities. However, it is important to keep in mind that the behavior of our other hominoid relatives is quite different. It is currently believed, for example, that gorillas, orangutans, and bonobos do not use tools in the wild. Moreover, differences in tool use between species of great apes are not clearly linked to differences in social organization, habitat preferences, diet, arboreality, manual

Figure 12.1

The great apes provide one source of insight about the lives of early hominids.

dexterity, or cognitive abilities. Thus, we can't *assume* that chimpanzees and the earliest hominids behaved in similar ways, even if they lived under similar ecological circumstances. But we can use information derived from observations of contemporary chimpanzees to develop insights about the *likely* form of particular behaviors among early hominids. For example, it seems reasonable to assume that if early hominids hunted, their hunting strategies would have been similar to those of chimpanzees. This kind of data is particularly valuable because evidence of the earliest forms of hunting activities is not likely to have been preserved in the archaeological record.

The shared characteristics of modern foraging people tell us what kinds of behaviors are imposed by the economics of hunting and gathering.

We can also gain insights from the subsistence strategies of contemporary peoples, called **foragers** or **hunter-gatherers**, who live in small-scale societies and depend on hunting game and gathering wild plant foods. Detailed ethnographic studies have been conducted on several groups who have survived into the 20th century, such as the !Kung San of Botswana, the Aché of Paraguay, the Mbuti and Efe pygmies of the Democratic Republic of the Congo (Figure 12.2), the Hadza of Tanzania, the Agta of the Philippines, and various groups of Arctic Inuit (Eskimo) and Australian aborigines. These peoples are similar in many ways. They do not cultivate crops or raise domesticated livestock. Instead, they subsist on wild plant foods that they gather and on game that they hunt. Women mainly gather plants, capture small prey, prepare food, and care for their children, while men mainly hunt larger game. Food, particularly meat, is widely shared within the group of people who live together. Such groups typically contain several pair-bonded couples and their children, a form of social organization not found in any other primate. Both men and women invest heavily in their offspring. Unlike herding and farming peoples, foragers do not establish permanent settlements, do not have substantial economic inequality, and do not recognize any formal political leadership.

Yet it would be a mistake to think of modern foragers as living fossils who preserve lifeways handed down unchanged from the Stone Age. There is good evidence that some contemporary foragers practiced agriculture some time in the past. Nor are they cut off from the rest of the world. Many foraging groups have been pushed into marginal habitats, like the arid Kalahari Desert, by the relentless expansion of their economically and militarily more powerful neighbors. Others

Figure 12.2

An Efe man carries a duiker, a small forest antelope, that he has killed in the Ituri Forest of the Democratic Republic of the Congo. The Efe are food foragers who subsist mainly on wild plant foods and wild game. Food foragers have traditionally made use of simple technology, accumulated few material goods, and lived in egalitarian societies. (Photograph courtesy of Robert Bailey.)

have engaged in regular economic exchanges with neighboring groups for hundreds, perhaps thousands, of years. Modern foragers often trade for metal tools, cooking pots, and other modern gadgets. In short, these are fully modern people, with sophisticated languages, elaborate tools, complex ideologies, an extensive knowledge of the natural world, and a growing knowledge of other cultures.

Despite these trends, contemporary foragers help us to understand early hominids in a number of ways. For example, the fact that contemporary foraging peoples are similar to each other in many ways suggests that the economics of foraging imposes constraints on the size and complexity of their societies. The fact that food sharing, particularly the exchange of meat, plays a central role in the lives of all foraging peoples, alerts us to the importance of these features in the history of the human lineage.

Early Hominid Environments

During the late Miocene and Pliocene, eastern Africa became colder, drier, and more seasonal, causing forest habitats to contract and woodland and savanna habitats to expand.

The first step in reconstructing the lives of the earliest hominids is to figure out what the world's climate used to be like. In particular, since tropical forests today are where we find many monkey and all ape species, it makes sense to examine the climate changes in these regions before and during the evolution of the first hominids. At the beginning of the Miocene (23 mya), most of the world's tropical regions were wet and warm. The temperature did not vary much during the course of the year, and rainfall was distributed evenly. As a result, unbroken tropical rain forests covered the tropics and extended as far north as London. Sometime during the Miocene, the earth's temperature began to decline. This global cooling caused two important changes in the weather of the African tropics. First, the total amount of rain that fell each year declined. Second, the rainfall became more seasonal, meaning there were several months each year when no rain fell.

As the tropical regions of Africa became drier, moist tropical forests shrank and drier woodlands and grasslands expanded. This conclusion is partly based on comparative studies of contemporary environments, which indicate that ecological conditions change predictably as the climate becomes drier and more seasonal: dense tropical forests first are transformed into open woodlands and then, if the drying trend continues long enough, give way to tropical grasslands, or savannas (Figure 12.3).

Early hominids lived in dry, seasonal environments.

Primates, like other animals, were forced to respond to these gradual ecological challenges. Some species, including many of the Miocene apes, apparently failed to adapt, and became extinct. Others retreated to the depths of the forests and carried on their lives much as before. Changes brought about through generations of natural selection allowed a few species eventually to move down from the trees, out of the rain forests, and into the woodlands and savannas. Hominids were among these pioneering species. We know that most of the early hominid sites were in dry, seasonal environments from the types of fossilized pollen grains and animal fossils found there. This means that all of the early hominid species spent at least part of their time in savannas and dry woodlands. The transition for our hominid ancestors from a partly arboreal life in the forest to a largely terrestrial life in the savannas and woodlands was not simple or sudden. As we will explain shortly, this transition probably involved changes in body size, group size, antipredator strategies, diet, social organization, and behavior.

(a)

(b)

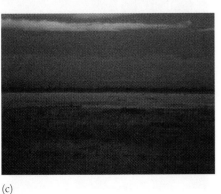

(c)

(d)

Figure 12.3

As environments become drier, (a) dense tropical forests give way to (b) woodlands, (c) open savannas, and (d) deserts. (a, Photograph courtesy of Lynne Isbell).

Early Hominid Ecology—The Emergence of Meat Eating

Contemporary foragers rely more heavily on meat than any other primates do.

Most primates subsist on fruit, leaves, and insects. Very few species regularly capture or eat mammalian prey, and no primate hunts large game. In contrast, modern human foragers are skilled hunters who kill and eat the largest animals in their environments. The dependence on meat varies widely among foraging groups. No group relies 100% on meat and fish, although the Arctic Inuit (or Eskimo) come close, eating no vegetables or fruit during the long winter months. In other modern food foraging groups, the proportion of meat in the diet varies from 70% of total calories for the Aché, who live in the forests of Paraguay (Figure 12.4), to around 30% of total calories for the !Kung and the Hadza, who live on the African savannas.

Why Meat Eating Is Important

Many anthropologists believe that meat eating played an important role in the evolution of modern human lifeways.

In contemporary foraging groups, meat eating is invariably linked to the maintenance of home bases, to extensive food sharing, and to the sexual division of labor. There are two reasons why meat eating plays such an important role in the lives of food foragers. First, meat is worth lugging around. Compared with most vegetable resources, meat provides large amounts of calories and nutrients in a relatively compact, transportable form. This makes it practical for hunters to move meat from the kill site back to their group, which in turn facilitates food sharing and the sexual division of labor, as we will see shortly.

Even more important, many anthropologists believe that a heavy dependence on meat makes food sharing necessary because without it, hunting is not feasible. And food sharing is important because it is a social behavior leading to other kinds of cooperation. Let's examine how food sharing provides insurance against the risks inherent in hunting. Hunting, especially for hunters who concentrate on large game, is a boom-or-bust activity. When a hunter makes a kill, there is a lot of very high-quality food available. However, hunters are often unlucky, and each time a hunter sets out to hunt there is a fairly high probability of returning empty-

Figure 12.4

Meat makes up about 70% of the diet of the Aché, a group of foragers from Paraguay. Here, an Aché man takes aim at a monkey. (Photograph courtesy of Kim Hill.)

handed and hungry. Food sharing greatly reduces the risks associated with hunting by averaging returns over a number of hunters.

To see why this argument has such force, consider the following simple hypothetical example. Suppose there are five hunters in a group that subsists entirely on meat. Hunters are able to hunt every day, and each hunter has a 1-in-5 (0.2) chance of making a kill and a 4-in-5 (0.8) chance of bringing back nothing. Further suppose people starve after 10 days without food. The probability of starvation for each hunter over any 10-day period is obtained by multiplying the probability of failing on the first day (0.8) by the probability of failing on the second day (0.8), and so on, to get

$$0.8 \times 0.8 \times 0.8 \times 0.8 \times 0.8 \times 0.8 \times 0.8 \times 0.8 \times 0.8 \times 0.8 \approx 0.1$$

Thus, there is about 10% chance that a hunter will starve over any 10-day period. With these odds, it is impossible for people to sustain themselves by hunting alone.

Of course, hunting would be more practical if we improved the odds. But a comparison with chimpanzee hunting provides good reason to think these are realistic values for early hominids. Craig Stanford of the University of Southern California and his colleagues have carefully analyzed records of hunting by chimpanzees at Gombe Stream National Park in Tanzania. In about half of the hunts the chimpanzees succeeded in killing at least one monkey, and sometimes they made more than one kill. The average number of monkeys killed per hunt was 0.84. However, males hunted in groups that contained on average seven males. Dividing 0.84 by 7 gives the average number of monkeys killed per male per hunt, which comes out to 0.14. Thus, on any given day, each male chimpanzee had only a 15% chance of making a kill, which is less than the 20% chance we posited for human hunters at the outset of our example.

Now let's consider how sharing food alters the probability of starvation. If each hunter has a 0.8 chance of coming back empty-handed, then the chance that all five hunters will come back on a given evening without food is

$$0.8 \times 0.8 \times 0.8 \times 0.8 \times 0.8 \approx 0.33$$

Thus, on each day there is a $\frac{1}{3}$ chance that no one will make a kill. If the kill is large enough to feed all members of the group, then no one will go hungry as long as someone succeeds. The chance that the five hunters will all face starvation during any 10-day period is

$$0.33 \times 0.33 \times 0.33 \times 0.33 \times 0.33 \times 0.33$$
$$\times 0.33 \times 0.33 \times 0.33 \times 0.33 \approx 0.000015$$

Sharing reduces the chance of starvation from 1 chance in 10 to roughly 1 chance in 10,000. Clearly, the humans could also reduce the risks associated with hunting even further if there were alternative sources of food that unsuccessful hunters might share. For example, suppose one member of the group hunted while the others foraged, and they all contributed food to a communal pot.

The fact that food sharing is mutually beneficial is not enough to make it happen. As we pointed out in Chapter 8, food sharing is an altruistic act. Each individual will be better off if he or she gets meat but does not share it. For sharing to occur among unrelated individuals, as it often does in contemporary foraging societies, those who do not share must be punished in some way, such as being excluded from future sharing, or by being forced to leave the group.

Food sharing transforms hunting from a risky enterprise to a prudent one. In all contemporary foraging groups, hunting is associated with extensive food sharing and with sexual division of labor. In nearly all foraging groups, men take primary responsibility for hunting large game and women take primary responsibility for gathering plant foods. Since food sharing and sexual division of labor are linked with hunting in modern foragers, it would be more plausible to think that early hominids relied on meat if we thought that they shared food with one another or specialized in different resource-gathering strategies. As we will see, data on wild chimpanzees suggest this is a likely possibility.

Comparison with Hunting in Chimpanzees

In the past, many anthropologists thought it was unlikely that the australopithecines ate very much meat. Hunting, they argued, requires careful planning, subtle co-ordination, and extensive reciprocity. These interactions would require cognitive and linguistic skills that the relatively small-brained australopithecines must have lacked. Two recent studies of chimpanzee hunting suggest that we must revise this view. In the Taï Forest of the Ivory Coast, Christophe and Hedwige Boesch of the University of Basel studied patterns of chimpanzee predation. Craig Stanford has also studied hunting behavior by chimpanzees at Gombe.

Chimpanzees hunt monkeys, bushpigs, or small antelope, but they rarely scavenge.

In the Taï Forest and at Gombe, chimpanzees participate in hunts approximately once every two or three days. At Gombe, hunting occurs most often during the dry season when the chimpanzees form relatively large parties. Most hunts at both Gombe and Taï focus on monkeys, particularly red colobus (Figure 12.5). Gombe chimpanzees also prey on bushpigs and bushbucks (a sheep-sized forest antelope) that they encounter on the forest floor (Figure 12.6). Unlike some other meat-eating species, chimpanzees never feed on the carcasses of dead animals they encounter opportunistically, and they rarely take fresh kills from other predator species.

(a)   (b)

Figure 12.5

Chimpanzees hunt small mammals, particularly red colobus monkeys. (a) A red colobus monkey carrying an infant leaps from one tree to another at Gombe Stream National Park in Tanzania. (b) A male chimpanzee at Gombe consumes a red colobus monkey. (Photographs courtesy of Craig Stanford.)

Chimpanzee Prey at Gombe, 1974–1991

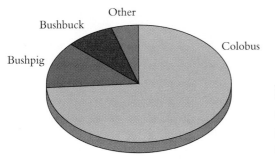

Figure 12.6

Chimpanzees at Gombe most commonly hunt red colobus monkeys, although they sometimes prey on bushpigs, bushbucks, and other animals.

Chimpanzees are effective hunters who sometimes plan cooperative hunts.

At Gombe and Taï, many red colobus hunts begin when the chimpanzees encounter a group of monkeys in the forest or detect them at a distance. They often spend some time on the ground gazing up into the canopy before they begin to hunt. Then one or more chimpanzees climb into the trees and give chase. Others watch from the ground. While most hunts at Gombe seem to arise out of opportunistic encounters with groups of monkeys, about half of the hunts at Taï begin before there is any sign that monkeys are nearby. In these cases, the chimpanzees fall completely silent, stay very close to one another, frequently change direction, and stop often to search for monkeys in the canopy.

The chimpanzees in these two study areas are effective hunters, making kills in approximately half of their hunts. The probability of making a kill depends, in part, on the size of the hunting party. Larger parties are more successful in capturing prey than smaller parties or single individuals are (Figure 12.7). At both Gombe and Taï, most hunts take place when the chimpanzees are in groups. This suggests that hunts may be cooperative endeavors, but there is some dispute over the extent to which the participants actually coordinate their actions in hunting. At Gombe, hunters sometimes pursue different prey during a single hunt, and most observers have concluded that the hunters are acting independently rather than cooperatively. At Taï, the majority of hunts appear to be collaborative efforts in which the hunters perform different but complementary actions in pursuit of a single victim. Some chimpanzees may drive the prey toward waiting hunters, some hunters may chase prey while others block escape routes, or several hunters may jointly encircle the prey.

These observations invalidate the argument that a small-brained hominid must have lacked the necessary cognitive and linguistic skills for hunting. However,

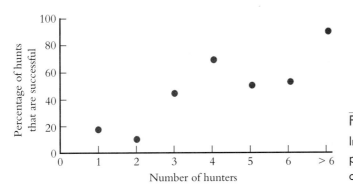

Figure 12.7

In the Taï Forest, large chimpanzee hunting parties are more likely to be successful in capturing prey than smaller parties are.

(a)   (b)

Figure 12.8

In seasonal habitats, the landscape changes dramatically over the course of the year.
Here you see the same area in Amboseli National Park (a) just after the seasonal rains
have ended and (b) at the end of the long dry season.

they do not tell us how likely it is that the australopithecines actually hunted. To
answer this question, we need to turn to other lines of evidence.

Seasonality and Meat Eating

Increased seasonality in rainfall favors a greater dependence on certain food types.

Herbivorous animals that live in seasonal environments are faced with a quite
variable food supply. When the rain falls, grasses sprout, flowers blossom, fruit
ripens, and there are many pools of fresh water (Figure 12.8a). Food is abundant
and diverse. Since food is plentiful, there is little competition among species over
access to food. During the dry season, grasses go to seed, flowers die, and all but
a few permanent water holes dry up (Figure 12.8b). Both the abundance and
diversity of food are greatly reduced (Figure 12.9). There is very little to eat for
most herbivorous animals, which leads to intense competition among certain species. Many species respond to this problem by focusing their foraging efforts on a
particular type of food during the dry season. For example, in Amboseli National

(a)   (b)

Figure 12.9

In seasonal habitats, resource availability is strongly affected by rainfall. (a) After the
rains, lush growth sprouts, and baboon foods are abundant and diverse. (b) By the end
of the dry season, the ground is parched, new growth has withered, and less food is
available to baboons.

Figure 12.10

A male baboon consumes a newborn bush-buck in the Okavango Delta of Botswana.

Park, baboons spend much of their time during the dry season digging up grass corms (the bulbous storage organ in grass stems). Another way to cope with shortages of plant foods is to become more omnivorous (Figure 12.10). Baboons at many different sites increase their meat intake during the dry season (Figure 12.11).

It is plausible that the less-robust australopithecines, from which modern humans probably evolved, relied on hunting during the dry season.

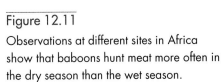

Cambridge University anthropologist Rob Foley has suggested that the hominid lineage may have adapted to increased seasonality in two different ways. The robust australopithecines with their huge jaws and chewing muscles may have specialized in processing tough plant materials in much the same way modern baboons specialize on corms. The less robust species, on the other hand, may have shifted to a more omnivorous diet and relied more heavily on meat.

Archaeological Evidence in Early Hominids

Around 2.4 mya, flaked stone tools appear at a number of sites in Africa. These tools allow us to make direct inferences about meat eating by the creatures that made them.

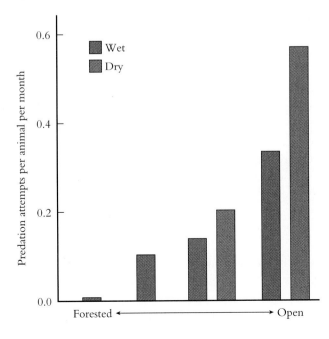

Figure 12.11

Observations at different sites in Africa show that baboons hunt meat more often in the dry season than the wet season.

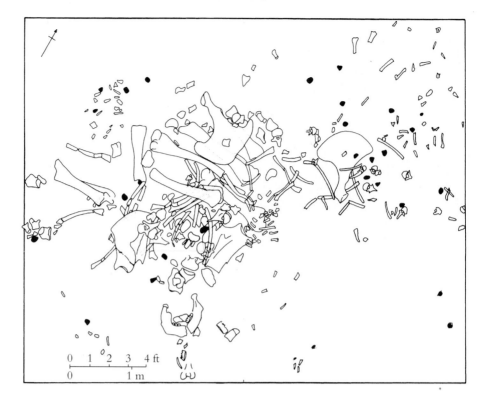

Figure 12.12

In this site map for one level of Bed I at Olduvai Gorge, most of the bones are from an elephant and tools are shown in black.

 At several archaeological sites in East Africa, Oldowan tools have been found along with dense concentrations of animal bones.

Archaeological sites with early stone tools occur at the Olduvai Gorge of Tanzania, Koobi Fora in Kenya, and a number of sites in Ethiopia. But the sites in Bed I of the Olduvai Gorge excavated by Mary Leakey have been analyzed most extensively. These sites, which are dated at 2 to 1.5 mya, measure only 10 to 20 m in diameter. They are littered with fossilized animal bones (Figure 12.12). The densities of animal bones in the archaeological sites are hundreds of times higher than in the surrounding areas or in modern savannas. The bones belong to a wide range of animal species, including bovids (like present-day antelope and wildebeest), pigs, equids (horses), elephants, hippopotamuses, rhinoceros, and a variety of carnivores (Figure 12.13).

Along with these bones, Mary Leakey found several kinds of stone artifacts: cores, stone flakes removed from the cores, battered rocks that may have been used as hammers or anvils, and some stones that show no signs of human modification or use. Recall from Chapter 11 that researchers originally believed hominids used the cores and discarded the flakes, but recent work suggests that both the cores and flakes were used as tools. The artifacts were manufactured from rocks that came from a number of different spots in the local area, and the nearest source of some of the raw material was several kilometers away (Figure 12.14).

 The association of hominid tools and animal bones does not necessarily mean that early hominids were responsible for these bone accumulations.

It is easy to jump to the conclusion that the association of hominid tools and animal bones means that the same creatures were responsible for both making the tools and accumulating the bones. Perhaps these were camps much like the camps of modern human foragers, or perhaps they were sites where hominids butchered

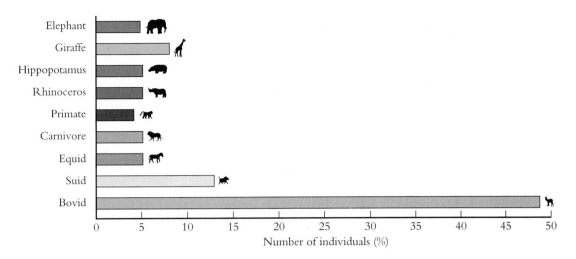

Figure 12.13

The bones of many different mammals were found at one archaeological site at Olduvai. Bovids (which include antelopes, gazelles, sheep, goats, and cattle) clearly outnumber all other taxa.

their prey, leaving their tools behind when they finished their work. However, it may also be a coincidence that the animal bones and stone tools have been found together. Bones may accumulate at particular spots for several different reasons. They may have been deposited there by moving water, many animals may have died there of natural causes, or the bones may have been collected there by non-hominid carnivores such as lions or hyenas. Sometime after the bones accumulated, perhaps hundreds of years later, hominids might have visited these sites and left their tools behind.

Archaeologists have resolved some of the uncertainty about these sites by studying how contemporary kill sites are formed.

One way to resolve the ambiguity of these sites is to draw on **taphonomy**, which is the study of the processes that affect an animal's remains from the time

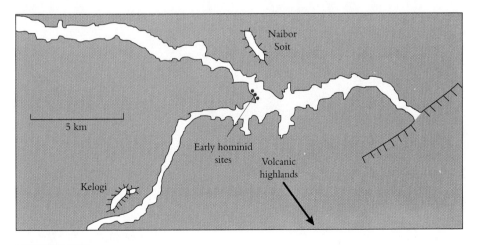

Figure 12.14

This map of Olduvai Gorge shows the major Bed I sites. Some of the tools were made of quartzite from Naibor Soit, others were made from gneiss from Kelogi, and some used lava cobbles that came from streams in the volcanic highlands to the south.

it dies to the time it becomes a fossil. Taphonomists study the characteristics of contemporary kill sites—spots where animals have been killed, processed, and eaten by various predators, including humans. Taphonomists observe how each type of predator consumes its prey—noting such things as whether limb bones are cracked open for marrow, which bones are carried away from the site, how stone tools are used, and how bones are distributed by predators at the site. These data enable archaeologists to develop a profile of the characteristics of kill sites created by different types of predators. Taphonomists also assess the characteristics of archaeological sites and compare these with the profiles of modern kill sites. In some cases, they are able to determine whether an archaeological site is a kill site and if so, what type of predator was present.

Taphonomic analyses of the Bed I sites at Olduvai Gorge suggest that hominids processed bones at these sites.

Studies of the Olduvai sites tell us, first of all, that the bones were not accumulated by natural processes.

The bones at the Olduvai sites were not deposited by moving water. Animals sometimes drown as they try to cross a swollen river or when they are swept away in flash floods (Figure 12.15). The bodies are carried downstream, and sometimes the carcasses accumulate in a sinkhole or on a sandbar. As the bodies decompose, the bones are exposed to the surrounding elements. The study of modern sites shows that sediments deposited by rapidly moving water have many distinctive characteristics. For example, water-borne sediment particles sort by size—they separate out from the water at different rates, which is evident by the size gradations in the resulting beds of rock. The sediments surrounding Olduvai sites do not show any of the features characteristic of sediments deposited by rapidly moving water.

Figure 12.15

Sometimes large numbers of wildebeest drown when trying to cross swollen rivers.

The concentrations of bones are not due to the death of a large number of animals at one spot. Sometimes many animals die in the same place. For example, in severe droughts, large numbers of animals may perish near water holes. Mass deaths usually involve members of a single species. Moreover, there is typically little mixing of bones from different carcasses, as the bodies decompose where they have fallen. The bones at the Olduvai sites come from a number of different species, and bones from the carcasses are jumbled together.

Early hominids were responsible for some of the concentrations of bones but not all of them. Nonhuman carnivores often create dense bone concentrations. Spotted hyenas, for example, often carry and drag carcasses from kill sites to their dens so that they can feed their young and avoid competition with other carnivores. It is possible that hominids dropped their tools near carnivore kill sites but had nothing to do with the bones themselves. However, a close look at the bones suggests otherwise. Many processes leave marks on bones, including the teeth of carnivores gnawing on the meat and the stone tools used by early hominids to butcher prey. Taphonomists have shown that stone tools leave a distinctive mark on bones. Flaked stone tools have microscopic serrations on the edges and make very fine parallel grooves when they are used to scrape meat away from bones (Figure 12.16). Some of the animal bones at Olduvai show extensive cut marks, indicating that early hominids must have processed meat at these sites.

There are additional reasons to suspect that early hominids were at least partly responsible for the bone collections at the Olduvai sites. Remember from Figure 12.14 that some of the stone used to make the tools found at the Olduvai sites was transported there from some distance away. If these collections of bones were created exclusively by nonhominid carnivores, then we must imagine that Oldowan hominids regularly carried stone over long distances and then left them at spots where they were likely to encounter large, dangerous carnivores. This seems implausible.

Taphonomic analyses suggest that hominids were not active at all of the Olduvai sites. At one site in particular the pattern of bone accumulation is very similar to the pattern of bone accumulations near modern hyena dens. Moreover, there are no stone tools, and the animal bones show no cut marks.

Hunters or Scavengers?

There has been much controversy about whether the Oldowan hominids were hunters or scavengers.

The archaeological evidence indicates that Oldowan hominids ate meat. They processed the carcasses of large animals, and we assume they ate the meat they cut from the bones. Meat-eating animals can acquire meat in several different ways: by stealing kills made by other animals, by opportunistically exploiting the carcasses of animals that die naturally, or by hunting and capturing prey themselves.

There has been considerable dispute among anthropologists about how early hominids acquired meat. Some have argued that hunting, division of labor, use of home bases, and food sharing emerged very early in hominid history. Others think the Oldowan hominids would have been unable to capture large mammals because they were too small and too poorly armed. These researchers view these hominids as scavengers who occasionally appropriated kills from other predators or fed on animals that died of natural causes. There would have been no food sharing or division of labor because there was no need—meat was only a small part of their

Tooth   Stone tool

Bone

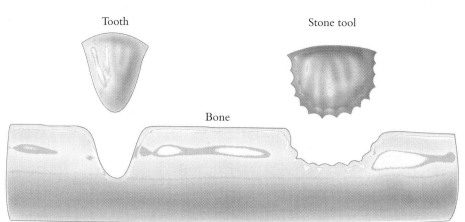

(a)

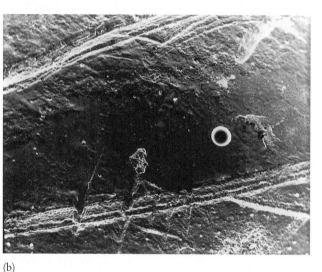

(b)

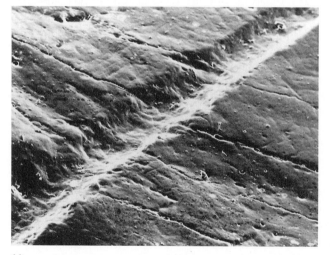

(c)

Figure 12.16

The marks made on bone by teeth differ from the marks made by stone tools. (a) The smooth surfaces of teeth leave broad, smooth grooves on bones, while the edges of stone tools have many tiny, sharp points that leave fine parallel grooves. Cut marks made by (b) stone tools and (c) carnivore teeth can be distinguished when they are examined with a scanning electron microscope. These are scanning electron micrographs of 1.8-million-year-old fossil bones from Olduvai Gorge.

An Oldowan Challenge

A recurrent pattern at some of the world's oldest archaeological sites is the presence of carcasses of "megafauna" (elephants, hippopotamuses) along with simple stone artifacts. For instance, two archaeological sites from Olduvai Gorge in Tanzania have yielded elephant skeletons (an *Elephas recki* in Bed I and a *Deinotherium*, an extinct form with downward curving, digging tusks, in lower Bed II). At each of these sites, dated between 1.7 and 1.5 million years ago, an array of Oldowan artifacts was found. At Koobi Fora in northern Kenya as well, parts of a hippopotamus carcass dating to 1.9 million years ago were found with simple stone tools.

Two major questions have emanated from such evidence. First of all, is there a *causal* relationship between these giant animal bones and these crude stone artifacts, or are these coincidental associations, places where humans discarded artifacts and these large mammals died independently, perhaps near the edge of a drinking spot along a watercourse or under the cool shade of a stand of trees? Second, if there could be cause or meaning connecting these stone artifacts and these pachyderms, what possible role could the stones have played? How could the world's simplest stone tools be used to process meat from animals weighing many thousands of pounds, with skins that can be over an inch thick?

We have had two opportunities to put stone artifacts to the ultimate test—to butcher elephants (which had died of natural causes). Somewhat

Butchering the world's largest terrestrial mammal with the world's simplest flaked stone technology: using a stone flake, Kathy Schick and Ray Dezzani cut through the one-inch-thick hide of an African elephant that died of natural causes. Most scavengers cannot or will not attack a fresh elephant carcass, so early Stone Age hominids might have moved into an "open niche" through the exploitation of such resources. To get a sense of the task required, imagine cutting through a car tire with a razor blade.

daunted, we approached our task equipped with simple lava and flint flakes and cores, which looked more and more paltry as we got closer to the impressive body. Initially, the sight of a twelve-thousand-pound animal carcass the size of a Winnebago can be quite intimidating—where do you start? We had never seen a field manual on pachyderm butchery, and they aren't like smaller animals: you cannot move the body around (for instance, flip it over to get a better vantage) without heavy power machinery. You have to play the carcass where it lies. In fact, once the upper side has been filleted, most of the lower half remains untouched and almost inaccessible unless you dismantle the skeleton of the animal, a positively arduous task.

Despite the success of our tools in dozens of other butcheries, we were not really sure they were up to this task. We were amazed, however, as a small lava flake sliced through the steel gray skin, about one inch thick, exposing enormous quantities of rich, red elephant meat inside. After breaching this critical barrier, removing flesh proved to be reasonably simple, although the enormous bones and muscles of these animals have very tough, thick tendons and ligaments, another challenge met successfully by our stone tools.

Throughout these and many other butcheries, our tools soon became strewn around the carcass, as we used one for slitting here, another for filleting over there, and another for hacking at a tough muscle attachment. It was always simple enough to grab another tool or knock another flake off a nearby core. It was easy to see how artifacts would become lost and engulfed in the task, to be left behind with any animal remains ultimately abandoned.

Although we feel that the butchery of such pachyderms was probably a rare event in the early Stone Age, and probably resulted from scavenging rather than hunting, such experiments demonstrate that the simplest stone

No beast too large. The use of early types of stone tools in the experimental butchery of an elephant that died of natural causes. Such carcasses could have provided occasional huge "bonanzas" of meat to hominid groups that possessed flaked stone cutting edges. Nicholas Toth defleshes the carcass while Jack Fisher, now at the University of Montana, loads meat into canisters for weighing.

technologies can be used to process even the largest terrestrial mammals. Cut marks from stone tools have been found on some *Elephas* bones at Olduvai, which indicate that stone tools had some causal relationship with the carcass. Since modern scavengers normally do not eat a dead elephant until it has decomposed for several days, such carcasses may have provided occasional bonanzas for early Stone Age hominids, at least until chased away by larger scavenging social carnivores.

SIMPLE FLAKES, POWERFUL TOOLS

In many considerations of Oldowan tools, the flake has been treated simply as waste, something removed in order to make the presumed core tool. But our experiments in butchery drive home an important realization: with a simple flake a hominid could open up a whole world of possibilities.

Source: From pp. 166–169 in K. D. Schick and N. Toth, 1993, *Making Silent Stones Speak: Human Evolution and the Dawn of Technology,* Simon & Schuster, New York. Reprinted by permission.

(a)   (b)

Figure 12.17

Competition among predators at kill sites is frequently intense. Here (a) lions have scavenged prey from a pack of hyenas, and (b) the hyenas fight to get it back.

diet. However, as we will see shortly, neither view is likely to be entirely correct. Data on contemporary carnivores and chimpanzees suggest that early hominids were unlikely to have relied entirely on scavenged kills, but the appearance of hunting behavior does not necessarily mean that the use of home bases, division of labor, or food sharing had yet developed.

 For most contemporary carnivores, scavenging is as difficult and dangerous as hunting.

To resolve the debate about whether early hominids were hunters or scavengers, we must first rethink popular conceptions of scavengers. Scavenging is not an occupation for the cowardly or lazy. Scavengers must be brave enough to snatch kills from the jaws of hungry competitors, shrewd enough to hang back in the shadows until the kill is momentarily left unguarded, or patient enough to follow herds and take advantage of natural mortality. Studies of contemporary carnivores show that the great majority of scavenged meat is acquired by taking a kill away from another predator. Most predators respond aggressively to competition from scavengers. For example, lions jealously guard their prey from persistent scavengers that try to steal bits of meat or to drag away parts of the carcass. These contests can be quite dangerous.

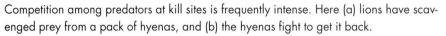 *Most large mammalian carnivores practice both hunting and scavenging.*

We also tend to think that some carnivores, like lions and leopards, only hunt, while others, like hyenas and jackals, only scavenge. The simple dichotomy between scavengers and hunters collapses when we review the data on the behavior of the five largest of Africa's mammalian carnivores (lion, hyena, cheetah, leopard, and wild dog). The fractions of meat obtained by scavenging vary from none for the cheetah, to one-third for hyenas, with the others ranging somewhere in between. Contrary to the usual stereotypes, the noble lion is not above taking prey from smaller competitors, including female members of his own pride, and hyenas are accomplished hunters (Figure 12.17). For most large carnivores in eastern Africa, hunting and scavenging are complementary activities.

Notice that, while there are some mammalian carnivores that never scavenge, none subsists entirely by scavenging. It would be difficult for any large mammal

Figure 12.18

Wildebeest migrate in vast herds that once numbered millions of animals.

to do so. For one thing, many prey species are movable feasts, migrating over long distances. For example, in the vast herds of wildebeest in the Serengeti Plain (Figure 12.18), many animals die from natural causes and their carcasses do provide ready meals for scavengers. However, for much of the year, mammalian carnivores cannot follow these migrating herds very far because they have dependent young that cannot travel long distances (Figure 12.19). Only avian scavengers that can soar over great distances, like the griffon vulture, rely entirely on scavenging (Figure 12.20). So, during most of the year, mammalian carnivores rely on nonmigratory prey, such as waterbuck and impala. During this time, acquiring meat solely by scavenging is not possible: natural mortality among resident species is not high enough to satisfy the caloric demands of carnivores, so they must hunt and kill their own prey, scavenging when the opportunity arises (Figure 12.21).

Scavenging might be more practical if carnivores switched from big game to other forms of food when migratory herds were not present. This is a plausible option for early hominids. Most groups of contemporary foraging people rely

Figure 12.19

As dusk falls, a trio of cheetah cubs wait for their mother to return from hunting.

Figure 12.20

Vultures, which soar on thermals and have enormous ranges, are the only carnivores that rely entirely on scavenging.

heavily on gathered foods—including tubers, seeds, fruit, eggs, and various invertebrates—in addition to meat. A few, such as the Hadza, obtain some of their meat from scavenging as well as from hunting. Early hominids might have relied mainly on gathered foods and scavenged meat only opportunistically.

Taphonomic evidence suggests that early hominids may have acquired meat both by scavenging and by hunting.

As we saw earlier, predators often face stiff competition for their kills. An animal that tries to defend its kill risks losing it to scavengers. Thus, leopards drag their kills into trees and eat the meat in safety. Other predators, like hyenas, sometimes rip off meaty parts of the carcass, such as the hindquarters, and drag their booty away to eat in peace. This means that limb bones usually disappear from a kill site first, while less meaty bones, like the vertebrae and skull, disappear later or remain at the kill site (Figure 12.22). If hominids obtained most of their meat from scavenging, we would expect to find cut marks mainly on bones left at kill sites by predators, like vertebrae. If hominids obtained most of their meat from their own kills, we would expect to find cut marks mainly on large bones, like limb bones.

Figure 12.21

Male lions sometimes take kills from smaller carnivores and from female lions.

Figure 12.22

After other predators have left, vultures consume what remains at the kill site.

However, at Olduvai Gorge, cut marks appear on both kinds of bones—those usually left to scavengers, and those normally monopolized by hunters (Figure 12.23).

Taken together, the evidence suggests that early hominids acquired meat by a mix of hunting and scavenging. The fact that chimpanzees regularly hunt but never scavenge suggests that hunting played some role in early hominid subsistence. The fact that most large mammalian carnivores in eastern Africa both hunt and scavenge further supports the idea that early hominids also did both. Finally, the evidence from tool marks on bones indicates that humans sometimes acquired meaty bones before, and sometimes after, other predators had gnawed on them.

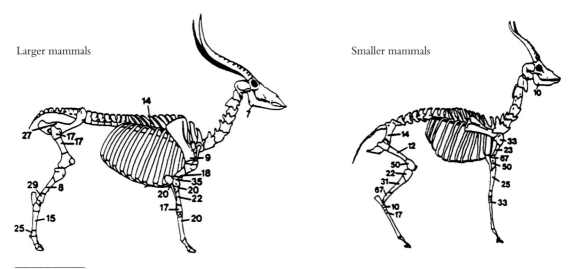

Figure 12.23

The numbers of cut marks on different kinds of bones vary at one site in the Olduvai Gorge. There are more than twice as many pieces of bones from large mammals (more than 250 lb) as there are from small mammals (less than 250 lb). The numbers in the figure are the percentages of all bones of a particular type that bear cut marks. For example, 14% of all rib bones of larger mammals showed cut marks. In general, cut marks are concentrated on bones that have the most meat.

Early Hominid Social Organization

In this section, we turn from the question of how early hominids (australopithecines and early *Homo*) made a living, to consider the social context in which these creatures lived. Did they live alone? What kinds of relationships did males and females form? Did males and females eat the same foods and perform the same tasks? Did they establish permanent home bases? Again, researchers draw on comparative evidence from other primates, particularly chimpanzees, and ethnographic data from studies of contemporary foragers to shed light on these questions.

Terrestrial life may have favored the formation of large multimale, multifemale groups among early hominids.

Comparative analyses indicate that terrestrial primates usually live in bigger groups, on average, than arboreal primates do. Baboons often live in groups of 60 individuals, while arboreal howler monkeys usually live in groups that number less than a dozen. Large group size may be a form of defense against terrestrial predators since animals living in large groups are less vulnerable to predators than animals living in smaller groups are (see Chapter 7).

Large groups tend to contain several adult males. As we have seen in Chapter 7, it is difficult for a single male to maintain exclusive access to a large number of females, so large groups tend to contain several adult males. In such groups there are no permanent bonds between males and females, and males invest little in offspring. In nonhuman primates, terrestrial life is generally associated with multimale, multifemale social organization (Table 12.1).

The fossils themselves give us another reason to believe that hominids lived in large groups. The australopithecines and early hominids showed pronounced dimorphism in body size, as males were substantially larger than females. Sexual dimorphism in body size is consistent with the formation of multimale or one-male groups.

Sexual Division of Labor

The diets of male and female chimpanzees differ, and may be a model for the evolution of the sexual division of labor in humans.

In all modern foraging peoples, there are differences in the kinds of tasks performed by men and by women. Men generally hunt large game, while women

Table 12.1   Comparative data of nonhuman primates suggest that terrestrial species are more likely to form multimale, multifemale groups than one-male, multifemale groups, while the reverse is true among arboreal species. (From figure 7.8h in R. Foley, 1987, *Another Unique Species*, Longman Scientific and Technical, Harlow, U.K.)

	NUMBER OF SPECIES	
FORM OF SOCIAL ORGANIZATION	TERRESTRIAL	ARBOREAL
Multimale, multifemale	13	6
One-male, multifemale	4	10

Figure 12.24

A !Kung San woman carries her young child on her back and digs for roots and tubers in the Kalahari Desert of Botswana. (Photograph courtesy of Nicholas Blurton Jones.)

spend most of their time gathering plant foods, preparing food, and caring for their children. When and how did the sexual division of labor first arise in hominids? There is no hard evidence to answer these questions. Taphonomists can determine whether the makers of stone tools were right-handed or left-handed, but they can't tell whether the toolmakers were males or females. In many reconstructions of the behavior of early humans, sexual division of labor is linked to the enlargement of the brain in hominids and to an increase in the dependence of infants at birth. These trends in turn restricted mothers' abilities to care for themselves and their offspring (Figure 12.24). As a result, pair-bonding, male investment in young, and economic cooperation between men and women became necessary. Is this the most likely scenario for the origins of the sexual division of labor? Again we turn to our nearest contemporary relatives, the chimpanzees, to give us some clues.

Among chimpanzees, hunting mammalian prey is mainly a male activity, while searching for insects is primarily a female activity. Chimpanzees frequently feed on various types of insects, including termites and ants. Sex differences in insectivory are reported at a number of sites, and in every case, females are more active insect predators than males are. At Gombe Stream National Park, for example, females are more likely to have traces of termites and weaver ants in their feces than males are, while the opposite pattern holds for vertebrate remains (Figure 12.25).

At both Taï and Gombe, male chimpanzees are more likely to be found in hunting parties than females are, and they make more kills than females do. This sex difference in hunting is apparently not due to an inability of the females to catch or subdue prey on their own. At Taï, for example, females who join the hunts with males are as successful at capturing prey as the males are, but females do not join hunting parties often.

It is not clear what leads to different foraging behaviors in male and female chimpanzees.

Since female chimpanzees can hunt monkeys and other prey species, and males can capture termites and ants, it is not entirely clear why males and females have different foraging strategies. William McGrew has compiled a list of several factors

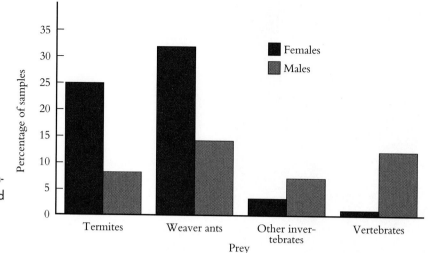

Figure 12.25

Analyses of fecal samples show that female chimpanzees eat more termites and ants than males do, while males eat more vertebrate prey than females do.

that may favor differences in foraging strategies among male and female chimpanzees:

- *Males are larger.* Sexual selection, which has favored larger body size, greater strength, and more highly developed weaponry among males than among females, may make it less costly for males to hunt monkeys and other large prey than for females to do so. Their physical size and strength may also place males at less risk from the fierce counterattacks of their prey.
- *Insects are less costly to exploit.* For females, who must meet the energy demands of pregnancy and lactation, it is important to minimize the costs of acquiring resources. Termite mounds, and to a lesser extent ant nests, are a reliable and localized resource. Females make regular circuits of known mounds, moving directly from one site to the next. At Gombe, chimpanzees have "fished" at some of the same termite mounds for more than 20 years.
- *Males range more widely than females do.* Males travel farther on most days than females do. At Gombe, males travel on average 4.9 km each day, while females travel 3.0 km per day. As a result, males may encounter prey species more often than females do.
- *Males are less likely to have prey stolen.* When females capture prey, they are likely to have their kills stolen by larger and more dominant males. Although females may also be deprived of their food resources, as when they get supplanted from productive holes in termite mounds by other chimpanzees, it is a relatively simple matter to begin fishing again in another spot.
- *Males are unencumbered by infants.* The need to carry and protect offspring may also hinder female ability to hunt fast-moving, arboreal prey effectively. In contrast, while females fish for termites, their offspring can play, suckle, rest, watch, or perfect their own termite fishing skills (Figure 12.26).

Although these data do not tell us when or why sexual division of labor arose in the human lineage, they inform us that sex differences in subsistence strategies are not necessarily linked to increases in brain size, extended infant dependence, and economic cooperation among males and females. It is possible that early hominid males and females practiced slightly different subsistence strategies, much as modern chimpanzees do.

Figure 12.26

A female chimpanzee relaxes near a termite mound, and her infant probes the side of the mound with its finger. (Photograph courtesy of William C. McGrew.)

Food Sharing

Chimpanzee food sharing provides a useful model for the same behavior in early hominids.

Within all modern foraging groups, hunting and food sharing are closely linked (Figure 12.27). As we explained earlier in this chapter, hunting makes sharing necessary, and sharing makes hunting feasible. Thus, if meat eating played an important role in the lives of early hominids, then reciprocal exchanges of food must have been important as well. Again, we look to chimpanzees for insights about the origins of this behavior.

Food sharing is uncommon among nonhuman primates, and when it does occur, it is generally limited to exchanges of plant foods among parents and their infants. Chimpanzees, however, constitute an exception to this general rule. Food is frequently shared by mothers with their infants, and even adults share food under certain circumstances. Sharing between mothers and infants generally involves plant foods, while sharing among adults generally involves meat.

The patterns of food sharing between mothers and their infants have been most carefully analyzed at Gombe. There, mothers are most likely to share foods that are difficult for the infants to obtain or to process independently (see Chapter 8).

Figure 12.27

Efe men butcher a duiker and distribute the meat. In human foraging communities, meat is always shared. (Photograph courtesy of Robert Bailey.)

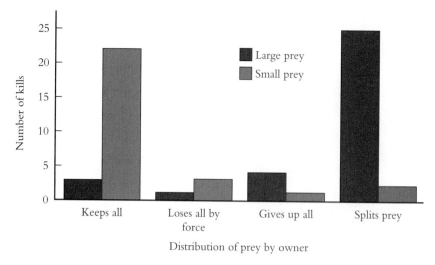

Figure 12.28

In the Taï Forest, chimpanzees sometimes share their kills. Infant and juvenile monkeys (small prey) are generally consumed by the captor, but adult monkey carcasses are generally divided by the captor and shared with other chimpanzees.

For example, infants have a hard time opening hard-shelled fruits and nuts and extracting seeds from sticky pods. A mother will often allow an infant to take bits of these items from her own food or sometimes she will spontaneously offer them to her infant.

When vertebrate prey are captured by chimpanzees, they are dismembered, divided, and sometimes redistributed among members of the foraging party. At Taï Forest, small prey are generally retained by the captor, while larger prey are typically divided among several individuals (Figure 12.28). The distribution of meat spans a continuum from outright coercion to apparently voluntary donations. High-ranking males sometimes take kills away from lower-ranking males, and adult males sometimes take kills away from females. At Gombe, about one-third of all kills are appropriated by higher-ranking individuals. More often, however, kills are retained by the captor and shared with others that cluster closely around. Males, who are generally the ones who control the kills, share food with other males, adult females, juveniles and infants (Figure 12.29).

Archaeological Evidence

Thus far we have seen that chimpanzees hunt, share food, and show sex differences in foraging strategies, but there is still a sizable gap between their behavior and that of modern human foragers. Chimpanzees hunt often, but meat represents a relatively small fraction of their diet. Chimpanzees share food, but they do not

Figure 12.29

Male chimpanzees sometimes share their kills with other males, sexually receptive females, and immatures. (Photograph courtesy of Craig Stanford.)

depend on one another for food and they obtain relatively little of their food from reciprocal exchanges. Male and female chimpanzees differ in the frequency with which they hunt mammalian prey and insects, but both sexes spend much of their time foraging for the same foods, and there is considerable overlap in their diets. Modern human foragers rely on meat to a great extent, share food widely among members of the group, and show much more pronounced sexual division of labor than chimpanzees do. As William McGrew notes, chimpanzees and modern humans represent the ends of a continuum. Where did the australopithecines and early *Homo* fall along this continuum? To answer this question, we return to the archaeological sites at Olduvai, which have played a key role in anthropologists' understanding of this issue.

Some archaeologists believe that artifacts and animal remains associated with early hominid fossils are the remains of home bases.

Remember that some archaeological sites at Olduvai are densely littered with fossil animal bones and stone tools. Some archaeologists, particularly the late Glyn Isaac, have suggested that these sites were hominid home bases. These researchers have speculated that early hominids acquired meat by hunting or scavenging, and then brought pieces of the carcass back to the home base, where it could be shared. The dense collections of bones and artifacts were thought to be the result of prolonged occupation of the home base (and sloppy housekeeping). At one Olduvai Bed I site there is even a circle of stones (Figure 12.30), dated to 1.9 mya,

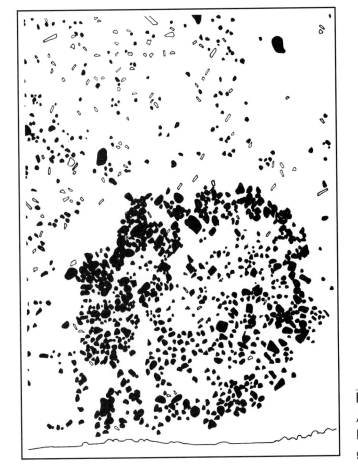

Figure 12.30

A circle of stones at one Olduvai Bed I site has been interpreted by some archaeologists as the remains of a simple shelter.

that is similar to the circles of stones anchoring the walls of simple huts constructed by some foraging peoples in dry environments today. The stones are associated with many Oldowan tools and bone fragments from a variety of prey species.

Isaac envisioned the hominid home bases to have much the same form and function as the camps of modern foragers. Nearly all contemporary foraging peoples establish a temporary camp, where food is shared, processed, cooked, and eaten. The temporary camp is also the place where people weave nets, fletch (add feathers to) arrows, sharpen digging sticks, string bows, make plans, resolve disputes, tell stories, and sing songs (Figure 12.31). Since foragers often shift their home bases from one location to another, their camps are simple. Generally, they consist of modest huts or shelters built around a series of hearths.

Several lines of evidence suggest that these sites were not early hominid home bases.

The possibility that hominids established home bases nearly 2 million years ago is an arresting notion. It would mean that many of the features that define modern foraging societies, such as food sharing and division of labor, were already well established. The importance of this issue has prompted a number of researchers to give close scrutiny to Isaac's ideas. Making use of modern taphonomic data, these researchers have made the following observations:

Both hominids and nonhominid carnivores were active at the Olduvai sites. Many of the bones at the Olduvai sites were gnawed by nonhominid carnivores. Sometimes the same bones show both tooth marks and cut marks, but some show only the marks of nonhominid carnivores.

Hominids and nonhominid carnivores apparently competed over kills. The bones of nonhominid carnivores are overrepresented at the Olduvai sites. This means they occur more often than expected based on their occurrence in other fossil assemblages or based on modern carnivore densities. Perhaps the carnivores were killed (and eaten) when attempting to scavenge hominid kills or when hominids attempted to scavenge their kills. Hominids may not have always won such contests—some fossilized hominid bones show the tooth marks of other carnivores.

Modern kill sites are often the scene of violent conflict among carnivores. This conflict occurs among members of different species as well as the same species. It is especially common when a small predator, like a cheetah, makes a kill. The kill attracts many other animals, most of which are able to displace the cheetah.

The bones accumulated at the Olduvai sites are weathered. When bones lie on the surface of the ground, they crack and peel in various ways. The longer they remain exposed on the surface, the greater the extent of this weathering. By carefully watching bones in modern habitats, taphonomists can calibrate the weathering process. Then, by measuring the amount of weathering on fossilized bones, they can determine how long the bones lay on the ground before being buried. Some of the bones at the Olduvai sites were exposed to the elements for at least four to six years (Figure 12.32).

The Olduvai sites do not show evidence of intensive bone processing. The bones at these sites show cut marks and tooth marks, and many bones were apparently smashed with stone hammers to remove marrow. However, the bones were not processed as intensively as bones by modern hunters are.

For several reasons these data are hard to reconcile with the idea that the Olduvai sites were home bases—places where people eat, sleep, tell stories, and care for their children. First, foragers today use dogs, fire, and thorn fences to prevent carnivores from visiting their camps. It is hard to imagine that early hominids could have occupied these sites if lions, hyenas, and saber-toothed cats were regular, albeit unwelcome, guests. Second, bones at the Olduvai sites appear to have ac-

Figure 12.31
The Efe build temporary camps in the forest and shift camps frequently. (Photograph courtesy of Robert Bailey.)

cumulated over a period of years. Contemporary foragers usually abandon their home bases permanently after a few months because the accumulating garbage attracts insects and other vermin. Even though they revisit the same areas regularly, they don't often reoccupy the old sites. Finally, the fossilized bones found at Olduvai were not fully processed. Modern foragers typically process meat and bone to get every last bit of meat, marrow, and sinew from them. All of the soft tissue is removed from the bone and eaten, and bones are shattered and boiled to extract the marrow.

Hominids may have brought carcasses to these sites, and processed the carcasses with flakes made from previously cached stones.

If these sites were not home bases, then what were they? Richard Potts, an anthropologist at the Smithsonian Institution, suggested that they were places where hominids worked but did not live. He believes that hominids brought their

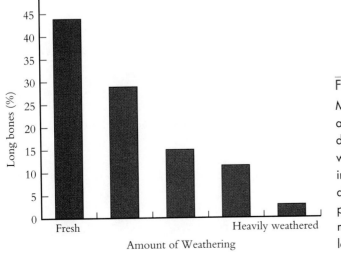

Figure 12.32
Many of the bones at Bed I sites at Olduvai are heavily weathered, suggesting that they were deposited and exposed to the elements over a fairly long period of time.

kills to these sites and dismembered their carcasses there. Some of the carcasses were scavenged by hominids from other carnivores, and some of the hominids' kills were lost to scavengers. If these were butchery sites, then we can explain why bones accumulated over such a long time, why bones of nonhominid carnivores were present, and why bones were not completely processed.

But why did hominids need to transport their kills to these sites? Potts thinks that hominids needed tools to process the meat. They couldn't always be sure of finding appropriate rocks for toolmaking at the site of their kills. They couldn't leave their kills unguarded while they went off to fetch their tools, lest some hungry creature steal their supper. This means that they had to carry the meat to where their tools were kept, or keep their tools with them all of the time. Potts suggests that the best strategy would have been to keep caches of tools scattered around the landscape, and then to carry carcasses that they acquired to the nearest cache. This is the most efficient solution because it minimizes the amount of energy expended while traveling with stone tools and meat.

To use stone caches, hominids must have been able to remember the locations of caches and navigate to them from kill sites—behavior seen in modern chimpanzees.

If hominids did cache their tools, and returned to these sites repeatedly, they must have been able to remember where they left their tools and to plot a route that would take them from the site of a kill to the location of the nearest tool cache. Observations of chimpanzee behavior suggests that hominids may have had such skills. Chimpanzees in some parts of West Africa use special hammer stones to crack open nuts against heavy stone anvils, roots, or emergent rocks (Figure 12.33). In the Taï forest, hammer stones vary in weight (approximately 0.5 to 18 kg, or 1 to 40 lb) and hardness. There are relatively few stones that make adequate hammers in the forest, so chimpanzees carry hammer stones from one place to another. (Although it might seem sensible to carry the nuts to the stones rather than vice versa, this scenario may present two difficulties. First, chimpanzees would have trouble carrying many nuts in their arms. Second, they might have trouble defending a pile of nuts between trips to the hammer stone.) Since some of the stones are quite heavy, and the chimpanzees have no knapsack or basket to carry them in, it is sometimes quite awkward for them to move the stones. When the chimpanzees choose a stone for cracking the hardest nuts, they seem to take into account the weight of the stone and its distance to the tree to which they need to carry it. They carry heavy stones over long distances only to crack the hardest

Figure 12.33

A young chimpanzee in Assirik, Senegal, uses a hammer stone to crack a hard nut. (Photograph courtesy of William C. McGrew.)

nuts. They seem to know where the stones are, how far away they are, whether one stone is closer or farther away than another, and how to get from any stone to any nut tree. If chimpanzees can make assessments like these and draw on memory in this way, then it is likely that early hominids could also.

Further Reading

Conroy, G. 1997. *Reconstructing Human Origins: A Modern Synthesis*. W. W. Norton, New York.

Foley, R. 1987. *Another Unique Species*. Longman Scientific and Technical, Harlow, U.K.

McGrew, W. C. 1992. *Chimpanzee Material Culture*. Cambridge University Press, Cambridge, U.K.

Potts, R. 1984. Home bases and early hominds. *American Scientist* 72:338–347.

Study Questions

1. Why are studies of chimpanzee behavior relevant to our understanding of early hominids?
2. How did the climate of Africa change at the end of the Miocene? Discuss the ways these changes may have affected the evolution of early hominids.
3. Why does a reliance on meat eating favor food sharing?
4. Describe the evidence suggesting that early toolmakers were partially responsible for the bone accumulations at the Olduvai Bed I sites.
5. How do scavengers like hyenas usually acquire their meat?
6. Why is it unlikely that the Olduvai Bed I sites are the remains of hominid home bases?

The Ancients

Between 1.7 and 1.8 mya, a new kind of hominid appeared in Africa. These creatures, called *Homo erectus* by some paleoanthropologists and *Homo ergaster* by others, were fully committed to life on the ground. They invented a new kind of tool technology and may have learned to master fire and to hunt large game. In this chapter, we trace the path of *H. erectus/ergaster* as they migrated from Africa and spread throughout temperate Asia and perhaps Europe. Over the next 1.7 million years, these creatures gradually evolved larger brains and more sophisticated technology than the hominids that came before them, and these evolutionary changes led to the emergence of our own species, *Homo sapiens*, around 100,000 years ago. We will draw on evidence from the fossil and archaeological records to reconstruct the transformation from early *Homo* to the earliest members of our own species, *Homo sapiens*.

We will see that anthropologists are sharply divided about how this transition occurred. Some believe that a single hominid species was gradually transformed from *Homo erectus/ergaster* into a larger brained hominid, "archaic *Homo sapiens*," and then into modern *Homo sapiens*. Other anthropologists believe that *Homo erectus/ergaster* diverged into several distinct species as it spread from Africa through the tropical and temperate parts of the Old World. According to this view, African and Eurasian lineages independently evolved larger brains and more modern morphology, but modern humans are all descended from the African species alone.

Homo erectus

Fossils of the new hominid that appeared on the African scene between 1.7 and 1.8 mya have been found at Olduvai Gorge (Tanzania), Lake Turkana (Kenya), Konso-Gardula (Ethiopia), Swartkrans (South Africa), and a number of other sites in Africa. Figure 13.1 shows a very well preserved skull (labeled KNM-ER 3733) from Lake Turkana that was found in 1976 by a team led by Richard Leakey. Until recently, most paleontologists assigned these fossils to the species *Homo erectus*, based on their similarities to fossils that had previously been discovered in Indonesia. However, a growing number of paleontologists have come to believe that the African fossils belong to a different species, and they prefer the name *Homo ergaster* for the African forms.

As we will see, this is not just a dispute about names—important questions about the history of the human species are at stake. But before this dispute can make sense, you need to know something about the morphology of these creatures and what kinds of tools they used. We need names to attach to our descriptions, even though any labels that we adopt are potentially contentious. For the present, we will use *Homo erectus* for any statement that applies to both Asian and African fossils, and specify Asian *H. erectus* or African *H. erectus* when we are talking about fossils from a specific region.

Skulls of H. erectus *differ from those of both* H. habilis *and modern humans.*

Skulls of *Homo erectus* retain many of the characteristics of earlier hominids (Figure 13.2):

• browridges
• a skull that narrows markedly behind the eyes

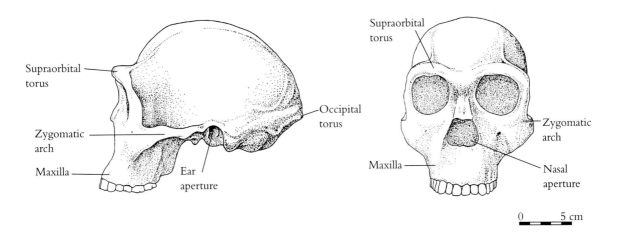

Figure 13.1

Homo erectus skulls, like the skull of KNM-ER 3733 illustrated here, show a mix of primitive and derived features. (Figure courtesy of Richard Klein.)

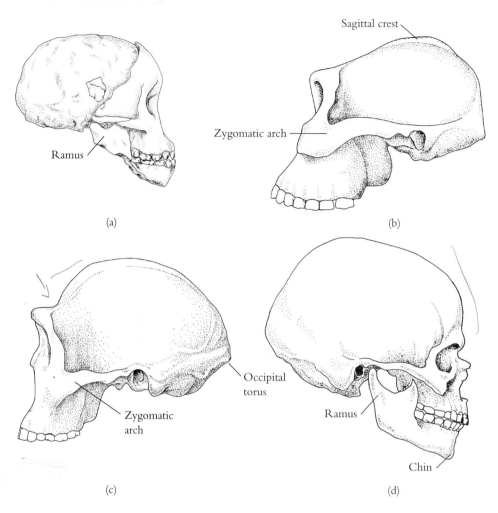

Figure 13.2

Shown here are the skulls of (a) *A. africanus,* (b) *A. robustus,* (c) *H. erectus,* and (d) modern *H. sapiens.* (Figures courtesy of Richard Klein.)

- a receding forehead
- a flattened skull relative to modern humans
- a broad, flat face
- no chin

H. erectus also shows several derived features shared by modern humans:

- a smaller, less prognathic face
- a higher skull than earlier hominid species
- a smaller and shorter ramus of the mandible. (The ramus is the part of the lower jaw that swings upward and forms a hinge with the upper jaw.)
- smaller teeth than australopithecines
- a vestibular (inner ear) system similar to that of modern humans

It is tempting to think of evolution as a unidirectional process in which earlier hominids were smoothly transformed into modern humans. However, African *H. erectus* confounds this notion, as it also shows some unique derived features—features not shared with either earlier hominids or modern humans:

- larger browridges
- a longitudinal ridge, called a **sagittal keel**, running along the top of the skull in some individuals
- a horizontal ridge at the back of the skull called the **occipital torus**

The sagittal keel is shaped like a shallow upside-down "V," much like the bottom of a row boat. The sagittal keel does not expand the area of attachment of the temporalis muscles, as the much sharper sagittal crest seen in robust australopithecines does. The function of the sagittal keel in *H. erectus* is unclear. In Asian *H. erectus* these features are more exaggerated, and the bones of the top of the skull are enormously thick.

Many of the derived features of the skull of *H. erectus* probably reflect a greater emphasis on tearing and biting with canines and incisors, and a reduced emphasis on heavy chewing with the molars. All their teeth are smaller than the australopithecines', but the molars are reduced relatively more than the incisors. The enlargement of the browridges and the point at the back of the skull are structural changes necessary to buttress the skull against novel stresses created by an increased emphasis on tearing and biting.

H. erectus had substantially larger brains than earlier hominids. The average brain volume for these creatures was about 1000 cc (1000 cc is 1 liter, just a little more than a quart), compared with an average brain size of 630 cc for *H. habilis*. The earliest African *H. erectus* brains measured around 900 cc. The brains of later Asian *H. erectus* were around 1100 cc, still considerably smaller than modern human brains, which average around 1400 cc. As we will see, however, *H. erectus* had much larger bodies than *H. habilis*, and it seems likely that their brains were not actually larger than the brains of earlier hominids in proportion to their body size.

H. erectus is the first hominid to have a projecting nose like that of modern humans. In *H. erectus*, the hole in the skull behind the fleshy nose, called the **nasal aperture**, projects forward and downward, resembling modern humans more than earlier hominids or other apes. To paleontologists this feature indicates that *H. erectus* had a nose that stuck out from the face with nostrils pointing downward. The purpose of such a nose is probably to conserve water: when moist air cools inside the nose, some of the water condenses and remains in the body rather than being exhaled. Russell Tuttle of the University of Chicago suggests that the projecting nose might also provide cooling for the brain.

The postcranial skeleton of African H. erectus *is much more similar to modern humans than to earlier hominds, but it still differs from ours in interesting ways.*

Kimoya Kimeu, the leader of the Koobi Fora paleontological team, found a fossilized skeleton of a 12-year-old African *H. erectus* male in 1984 (Figure 13.3). The find, labeled KNM-WT 15000, was made on the west side of Lake Turkana,

Figure 13.3

Kimoya Kimeu is an accomplished field researcher. He has made many important finds, including WT 15000.

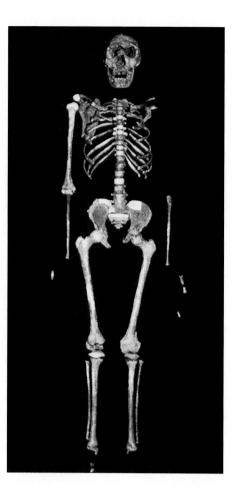

Figure 13.4

The fossil of the *H. erectus* boy, WT 15000, found at West Turkana, is amazingly complete. (Photograph courtesy of Alan Walker.)

the same region where WT 17000 (discussed in Chapter 11) was found. The boy's skeleton provides us with a remarkably complete picture of the *H. erectus* body—even delicate ribs and vertebrae are preserved (Figure 13.4). An extensive analysis of this skeleton, coordinated by Alan Walker of the Pennsylvania State University, tells us much about this youngster's body and his way of life.

Remember that the australopithecines, and probably early *Homo*, were bipeds but had long arms, short legs, and other features suggesting that they still spent a considerable amount of time in trees. In contrast, WT 15000 had the same body proportions as people who live on tropical savannas today: long legs and narrow hips and shoulders. WT 15000 also had short arms relative to australopithecines. Taken together, these features suggest that African *H. erectus* was as fully committed to life on the ground as modern humans are.

WT 15000 tells us several other interesting things about African *H. erectus*:

1. They were quite tall. WT 15000 stood about 5 ft 4 in. (1.625 m) in height. If *H. erectus* growth patterns were comparable to those of modern humans, he would have been about 6 feet (1.9 m) tall as an adult, several inches taller than the average American adult male. However, he was also heavily muscled. Think of him as a young shooting guard or small forward on a basketball team.
2. The openings in some of the vertebrae for the spinal cord were significantly smaller than in modern humans. This suggests that *H. erectus* had much less precise control of its breathing than modern humans do. As we will see in Chapter 15, this may indicate that *H. erectus* did not have language.
3. The pelvis indicates that the ratio of the size of the mother's birth canal to the size of the newborn head was the same in *H. erectus* as it is for modern humans.

The Boy from Nariokotome

The boy was a tall individual who was not unusual for his population. Our stature estimate of five feet three inches as a boy or six feet one inch as an adult was based on information about Nilohamitic tribesmen, people like the Dinka of Sudan who live close to where the boy's fossilized bones were discovered. We had used the same dataset to estimate the boy's body weight at about 106 pounds at the time of death and about 150 pounds if he had grown to adulthood. He was, by anybody's reckoning, a long, skinny kid who would have grown up to be a beanpole of a man. We did similar calculations for five other *Homo erectus* individuals from East Africa whose bones had been found previously. Adding them to the boy's estimate, we found that *Homo erectus* in Africa had an average stature of five feet seven inches (ranging from five feet two inches to six feet one inch) and an average weight of 128 pounds. I calculated that, if these individuals were alive today and if by some bizarre chance they were all males, this *erectus* "population" would rank among the tallest 17 percent of human populations worldwide. But it is far more likely that some of these individuals were females (for that is what their anatomy indicates to me). In that case, the *erectus* population would rank among an even smaller percentage of the world's populations in terms of height. Their size is truly astounding.

We were not grounding our estimates on the undemonstrated assumption that people who live in that region of eastern Africa are somehow more closely related to the ancient hominids who lived there 1.5 million years earlier. After all, 15K was an entirely different species from the modern humans of the southern Sudan or northern Kenya today. And Chris's careful forays through the available data on leg bone length and height had demonstrated to us that sharing African descent was not enough to make a particular population a reliable reference sample. There is, for example, a great deal of information in the literature about the size, shape, and proportions of the bodies of Americans, both male and female. But when Chris compared data on African Americans, Americans of Caucasian ancestry, and African blacks, he found remarkable discrepancies. In terms of the relationship between tibial length and stature, Chris has concluded that the U.S. blacks are as different from African blacks as they are from U.S. whites. Obviously, body shape and proportion are both labile and variable. They seem to respond to environment, or more specifically, ambient temperature and the demands of thermoregulation. Therefore, we chose the reference sample for our stature estimate to incorporate one of the most fascinating things we had discovered about the boy: he showed exactly the same physiological adaptations to climate 1.5 million years ago that some modern humans do.

Like other mammals, humans adapt to the environment in which they live in various physical ways. One of the most obvious is through body proportions, following what is known as Allen's rule, which states that people (and other mammals) adapt to the ambient temperature of their environment

through variations in limb length. Hotter climates pose problems of heat dissipation. Where it is dry and hot, natural selection favors those with long, slender limbs that increase the body's surface area and thus help dissipate heat through sweating. (In hot, wet climates, like rain forests, sweating is less effective. People adapted to those environments often have the same body width as those in hot, dry areas but are short, like Pygmies, to keep their body mass low.) Conversely, cold climates make heat retention difficult, conditions that favor individuals with short, stubby limbs and a low ratio of surface area to volume. The classic contrast in anthropological textbooks is side-by-side photographs of an Arctic Eskimo (short arms, legs, fingers, and toes) and an equatorial Dinka (elongated arms, legs, fingers, and toes). Their bodies reflect the inescapable demands of thermoregulation.

The measurements most often used to demonstrate the working of Allen's rule are the brachial index (the ratio of the length of the upper arm to that of the forearm) and the crural index (the ratio of the length of the thigh to that of the lower leg). High values in these ratios mean that an individual is cold-adapted; low values indicate heat adaptations, because it is the more distal, or peripheral, parts of the body that lengthen. Fortunately, we had the right body parts to calculate both indices. We found that, compared to living African (tropical climate) and living European (temperate climate) populations, the boy was extraordinary. He was not simply tropical; he was hypertropical, with ratios below the range of ratios recorded for living Africans.

The antiquity of this bodily adaptation to heat stress told us something else too: the boy was probably running around in hot, open country and *sweating*. Although it is indirect evidence, the boy's body build suggests that he, and all *Homo erectuses*, had lost whatever body fur or hair our more ancient ancestors probably possessed. If he had been hairy or furry, then panting (and avoiding activity during the hottest hours of the day) would have served as his main mechanism for heat loss, as it does for the other animals of the African savanna. Because he had no furry protection from the harsh equatorial sun, the boy was probably also very darkly pigmented. Another fascinating feature reflects another aspect of the boy's adaptation to functioning in such a hot and arid environment; he had a nose, a real nose. Bob Franciscus and Erik Trinkaus of the University of New Mexico documented the fact that *erectus* was the earliest species to have a projecting, human-type nose. In contrast, australopithecines and *Homo habilis* had flat, apelike noses, so that (prior to *erectus*) the nostrils leading to the nasal aperture—the opening of the respiratory tract through the nostrils—were sunken into the surface of the face rather than being part of an external nose. The analysis conducted by Bob and Erik suggested that this new shape of the nose allowed a greater volume of incoming air to be moisturized before it reached the lungs and yet also permitted moisture to be conserved and reclaimed as air was exhaled. Most mammals, presumably including australopithecines, possess effective physiological and behavioral mechanisms for conserving water, such as resting more and moving less when temperatures are high or keeping to the shade whenever possible. You have to ask why *Homo erectus* would need to add another mechanism: What had changed? The answer must be related to activity patterns rather than climate or habitat, since *Homo erectus* lived in many of the same places and with the same animals as *Australopithecus* or *Homo habilis*. Retaining the humidity of the nasal and respiratory mucosa would be an important improvement that would enhance the boy's ability to be active—running, walking hard, digging, carrying things, climbing trees—even during the hottest part of the day. Like his long legs, narrow pelvis, and energetically efficient body build, the boy's nose was yet another signature that he was a member of a species that (like later Englishmen) went out in the midday sun.

Another, more generalized way of adapting to hot climates is through relative body breadth. Chris models body breadth crudely as a cylinder, with hip width providing the diameter of the cylinder and body height determining its length. In the Eskimo-Dinka comparison, you cannot help seeing that the body shapes of these two groups vary as well as the limb lengths: Eskimos tend to be short and stocky (a short, squat cylinder); Dinka tend to be long and lean (an elongated, narrow cylinder). In previous research Chris had shown that the dimensions of the body cylinder are closely correlated with the mean temperature at which different groups live. The boy was slimhipped and tall, so it was no surprise to find that, once again, his body shape looked most like that of modern Africans living on the equator. Using a data base developed by Erik Trinkaus, we could even retrodict (predict backward) where such an individual "ought" to live. The answer was that he was adapted to a mean ambient temperature of about 85° to 87°F (or 29.2° to 30.8°C), just about what is recorded for the Lake Turkana region today. It was a satisfying confirmation of our interpretation. If the retrodiction had been for an ambient temperature of, say, 50°F (10°C), we would have had to go back and rethink the relationship between body build and temperature. For fun we also retrodicted the mean temperature of a Neandertal, a specimen of the type of fossil humans who inhabited Europe and the Near East during glacial times (about 100,000 to 35,000 years ago). If the boy was hypertropical, then the Neandertal we used fell at the other extreme: he was hyperarctic. The mean annual temperature for Neandertals came out to be 30°F (−1°C), which today would place them inside the Arctic Circle.

We had ample evidence that, from head to toe, the boy was a thoroughly tropically adapted individual. It seemed impossible that he would compromise these adaptations as he grew up, which meant that, had he lived, the boy would have been likely to follow the same rules of growth as do other people with similar body proportions and tropical adaptations. This was why we used information from Nilohamitic tribesmen in reconstructing the boy's stature from the length of his femur or tibia. It was not a matter of believing they shared a closer *genetic* relationship with *Homo erectus*; rather we knew that these modern populations shared a common *environment* with *Homo erectus* to which they had adapted similarly.

We also found out that the boy was incredibly strong. One of Chris's areas of expertise is in the application of engineering beam theory to human bones. Simply put, beam theory demonstrates that the way to make a beam stronger (more resistant to breaking) is to arrange the substance of the object so that most of the material is aligned along the axis where the beam is subjected to the greatest stresses. A beam subjected to equal stresses from all directions ought to be cylindrical, with a circular cross section. But, for example, a horizontal beam that is most likely to be bent along a vertical or superior-inferior axis should be shaped differently to prevent bending. Thus, this beam's cross section ought to be thicker from top to bottom (on the superior-inferior axis) than side to side.

Now substitute a long bone like a femur for the beam, change the beam's orientation from horizontal to vertical, and make the predominant stress anterior-posterior rather than superior-inferior. Although a femur or any other long bone is basically cylindrical in shape, the femur has remodeled to increase its mechanical strength, so that its cross section is elliptical rather than circular. The placement of the extra bone tissue—the material that thickens the bone in an anterior-posterior axis—tells you the axis in which the greatest, habitual stress was produced by the muscular actions of the owner of the femur. The main difference between beams and bones is that beams, once manufactured, do not change their dimensions, whereas bones respond to the stresses caused by movements and usage by remodeling (adding or subtracting bone tissue) throughout an individual's life. Take away those stresses, or change them, and the bones will remodel once again.

With Trinkaus and Clark Larsen, two colleagues who are experts on Neandertals, Chris and I surveyed the mechanical strength of the femurs of a chronological sequence of fossils from *Homo habilis* to *Homo erectus* to archaic *Homo sapiens* (Neandertals) to modern humans. The earlier members of our genus, including the Nariokotome boy, had incredibly robust bones compared to ours. In fact, we could document a steady decline in the robustness of the femur at midshaft (measured as the area of cortical bone in a cross section) through our sequence. This change is not just a downward trend; it is an exponential decline in cortical area (standardized for body weight) over a period from 1.89 million years to today, which implies an exponential decline in activity or mechanical stress on the femur through that period. Even as an adolescent, 15K apparently maintained remarkably high activity levels, for he resembles adults of *Homo erectus* in this regard. As a result, he attained exceptional strength relative to modern humans.

I couldn't help but smile inwardly at the image I was developing of this boy: all long legs, jutting elbows and knees, and probably a klutz—superbly strong but tripping over his own feet.

Source: From pp. 194–200 in Alan Walker and P. Shipman, 1996, *The Wisdom of the Bones: In Search of Human Origins*, Alfred A. Knopf, Inc., New York. Reprinted by permission of Alfred A. Knopf, Inc.

This relationship suggests that, like modern humans and unlike most other mammals, babies had to be born before brain development could be completed. (Had they stayed in the uterus the extra months necessary to finish brain development, their heads would have been too large to pass through the birth canal.) If true, then young *H. erectus* would have matured slowly and been dependent on their mothers for an extended period of time, much as modern children are.

Prolonged dependence of infants and the reduction of sexual dimorphism may be linked. Females may have had difficulty providing food for themselves and their dependent young. If *H. erectus* hunted regularly, males might have been able to provide high-quality food for their mates and offspring. Monogamy would have increased the males' confidence in paternity and favored paternal investment. Females might have shared plant foods with their mates as a means of exchange. This scenario fits with the fact that *H. erectus* was much less sexually dimorphic than previous hominids were. *H. erectus* males were only 20% to 30% larger than females. Reduced sexual dimorphism in turn suggests there was less competition among males for access to females, perhaps as a result of a shift toward a monogamous mating system with substantial paternal investment in offspring.

Acheulean Tools

African H. erectus *made fancier tools than earlier hominids did.*

African *H. erectus* made a variety of stone tools, some simple, some complex. Like *H. habilis*, *H. erectus* continued the tradition of making simple Oldowan choppers, perhaps when they needed a serviceable tool in a hurry—as foraging people do today. But other tools made by African *H. erectus* were more sophisticated than the tools of early *Homo*. These were a totally new kind of stone tool called a **biface**. To make a biface, the toolmaker (or **knapper**) strikes a large piece of rock from a boulder to make a core, and then flakes this core on all sides to create a flattened form with an edge along its entire circumference. The most common type of biface, called a **hand ax**, is shaped like a teardrop and has a sharp point at the narrow end (Figure 13.5). A **cleaver** is a lozenge-shaped biface with a flat, sharp edge on one end, while a **pick** is a thicker, more triangular biface. Bifaces are larger than Oldowan tools, averaging about 15 cm (6 in.) in length, and sometimes reaching 30 cm (12 in.). Paleoanthropologists call this collection of tools the

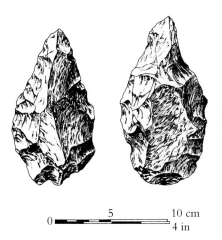

Figure 13.5

Acheulean hand axes, like the two shown here, were teardrop-shaped tools created by removing flakes from a core. The smallest ones would fit in the palm of your hand, and the largest ones are more than 0.6 m (2 ft) long.

0   5   10 cm
             4 in

Acheulean industry after the French town of St. Acheul, where hand axes were first discovered. The oldest Acheulean tools found at West Turkana date to about 1.6 million years ago, about the same time that *H. erectus* first appeared in Africa. The Acheulean period lasted from about 1.6 mya to about 300,000 years ago.

The standardized form of hand axes and other tools in the Acheulean industry suggests that these hominids had a specific design in mind when they made each tool. Oldowan tools have a haphazard appearance, and no two are alike. This lack of standardization suggests that early *Homo* picked up a core and created flakes without trying to create a tool with a particular shape, perhaps because the small flakes were the actual tools. It is easy to see how a biface might have evolved from an Oldowan chopper by extending the flaking around the periphery of the tool. However, Acheulean tools are not just Oldowan tools with longer edges—they are designed according to a uniform plan. Hand axes have regular proportions: the ratio of height to width to thickness is remarkably constant from one ax to another. African *H. erectus* must have started with an irregularly shaped piece of rock and whittled it down by striking flakes from both sides until it had the desired shape. Clearly, all the hand axes would not have come out the same if their makers hadn't shared an idea for the design.

Hand axes were probably used to butcher large animals.

If hand axes were designed, what were they designed for? The answer to this question is not obvious because hand axes are not much like the tools made by later peoples. There have been a number of proposals:

1. Hand axes were used for butchering large animals. *H. erectus* acquired the carcasses of animals like zebra or buffalo either by hunting or scavenging, and then used a hand ax as a modern butcher would use a cleaver, to dismember the carcass and cut it into useful pieces.
2. Hand axes were used for digging. Contemporary foragers in savanna environments spend a lot of their time digging up edible tubers. Although modern peoples generally use sharpened sticks for digging, some archaeologists have suggested that hand axes may have been used for this purpose. Digging tools would also have been useful for capturing burrowing animals like warthogs and porcupines, and for making wells to acquire water.
3. Hand axes were used to strip bark from trees to get at the nutritious cambium layer underneath.
4. Hand axes were used in hunting and hurled at prey animals.
5. Hand axes weren't tools at all. Instead they were "flake dispensers" from which *H. erectus* struck flakes to be used for many day-to-day purposes.

While we are not certain how hand axes were used, two kinds of evidence support the first hypothesis—that they were heavy-duty butchery tools. Kathy Schick and Nicholas Toth, whose investigations of the function of Oldowan tools we discussed in Chapters 11 and 12, have also done experiments using Acheulean hand axes for each of the tasks listed above. From these experiments, Schick and Toth conclude that hand axes are best suited to butchery. The sharp end of the hand ax easily cuts through meat and separates joints, while the rounded end provides a secure handle. The large size is useful because it provides a long cutting edge as well as the cutting weight necessary for a tool without a handle to be effective. Schick and Toth's results are supported by the work of University of Illinois paleoanthropologist Lawrence Keely, who performed microscopic analysis of wear patterns on a small number of hand axes. He concluded that the pattern of wear is consistent with animal butchery.

Evidence from Olorgesaillie, a site in Kenya at which enormous numbers of hand axes have been found, indicates that these axes may also have served as flake dispensers. A team led by Rick Potts has discovered most of the fossilized skeleton of an elephant in association with numerous small stone flakes dated to about 1 mya. The flakes are chipped in a way that is consistent with their being used to butcher the elephant, and the elephant bones show stone-tool cut marks. Careful examination of the flakes reveals that they were struck from an already-flaked core, such as a hand ax, not from an unflaked cobble. Moreover, the flakes were removed from the hand axes in ways that did not sharpen the hand-ax edge. Taken together, this evidence indicates that *H. erectus* struck flakes from hand axes and used them as butchery tools.

The Acheulean industry remained remarkably unchanged for almost 1 million years.

Acheulean tools first appear in the fossil record at about 1.6 mya. There is relatively little change in the Acheulean "tool kit" from its first appearance until *Homo sapiens* emerged around 300,000 years ago. Tools that were made half a million years apart are no more variable than tools made at different sites during the same time period.

The persistence of hand axes is amazing. They were made with little modification for more than 1 million years. Most anthropologists assume that the knowledge necessary to make a proper hand ax was passed from one generation of *H. erectus* to the next by teaching and imitation. It is a remarkable notion that this form of knowledge might have been faithfully transmitted and preserved for so long in a small population of hominids spread across great distances.

H. erectus Peoples the World

H. erectus *appeared in Eurasia at least 500,000 years ago and perhaps as early as 1.8 mya.*

Until recently, most anthropologists believed that *H. erectus* first appeared in Asia long after *H. erectus* appeared in eastern Africa. However, the dating of Asian *H. erectus* materials has been quite problematic. Until the mid-1990s the oldest Asian specimens of *H. erectus* were thought to be those unearthed by Eugene Dubois near the Solo River in Java during the 19th century (Figure 13.6). Dubois made the first finds of *H. erectus* anywhere in the world and named the species. Dubois's fossils were generally believed to be about 500,000 years old, though this date was not very well established. In 1994, Carl Swisher and Garness Curtis, both from the Berkeley Geochronology Center, reported new research indicating that two *H. erectus* sites in Java were actually 1.6 to 1.8 million years old. Swisher and Curtis used the newly developed argon-argon dating technique, which allowed them to date small crystals of rock from the sites at which the fossils had been found many years earlier. Some paleontologists are skeptical of these new dates because they are inconsistent with the paleomagnetic dates and other fauna found at the same sites. These skeptics argue that the rock crystals dated by Swisher and Curtis may have been transported to the sites by water or may have eroded out from older sediments. However, some support for this early date comes from the recent discovery of a *H. erectus* fossil nearly as old at Dmanisi (in the Caucusus

Early Middle Pleistocene: 1mya–500 kya

● *Homo erectus*

● Archaic *Homo sapiens*

Figure 13.6

By the beginning of the Middle Pleistocene, *H. erectus* had spread throughout temperate Eurasia. The earliest specimens of archaic *H. sapiens* also appeared in Africa and Europe sometime during the early Middle Pleistocene. The fossils found at these sites are discussed in the text.

Mountains of the Republic of Georgia). There, a lower jaw was recovered with Oldowan tools in sediments whose paleomagnetic polarization is consistent with a date of either 1.1 or 1.8 mya.

There is also evidence for *H. erectus* elsewhere in eastern Asia. In the 1920s and 1930s, several different investigators found *H. erectus* fossils at Zhoukoudian cave (also called Choukoutien) near Beijing. The Zhoukoudian site, which is somewhere between 250,000 and 500,000 years old, is remarkable for providing large numbers of fossils, tools, and other artifacts that shed light on the way of life of *H. erectus*.

It is not clear whether H. erectus *ever extended its range to Europe.*

There is no doubt that hominids reached Europe by 500,000 years ago because there are many hominid fossils from European sites that date to between 250,000 and 500,000 years ago, when *H. erectus* was present throughout East Asia. However, it is not clear whether the hominids in Europe and East Asia during this period were members of the same species. As we will discuss in more detail below, most paleoanthropologists classify the European fossils from this period as archaic *Homo sapiens*, not *H. erectus*.

There are only two sites in Europe with hominid fossils that are dated to over 700,000 years ago. One of these sites is in Crepano, Italy, where researchers have found a fossil skull that is very similar to other *H. erectus* fossils. The other site is at Trinchera Dolina in Spain and is about 100,000 years older than Crepano. In Trinchera Dolina the fossils, while fragmentary, seem to resemble archaic *Homo sapiens* rather than *H. erectus*. In both cases, there are reasons to suspect that the fossils might be significantly younger than the dates given, so it is possible that hominids were rare or absent in Europe before 500,000 years ago. The idea that humans did not arrive in Europe until at least 500,000 years ago is supported by the fact that palaeontologists have excavated many sites in Europe from before 500,000 years ago and have found thousands of fossil animals, but no trace of hominids or their tools.

There was extensive geographic variability among populations of H. erectus.

Specimens of *H. erectus* from eastern Asia differ morphologically from African specimens. In the Asian fossils the derived features that characterize *H. erectus* are more exaggerated. The skull is thicker, the browridges are more pronounced, the sides of the skull slope more steeply, the sagittal keel is more exaggerated, and the occipital torus (the point at the back of the skull) is more pronounced (Figure 13.7). Moreover, the eastern Asian forms do not show the increase in cranial capacity that characterizes contemporary populations in Africa and western Eurasia. It is important to notice that some of these eastern Asian forms may have been about 500,000 years old, but they are less like modern humans than are the African forms of *H. erectus*, which are one million years older.

Interestingly, the tools used by the Asian forms also differ from the tools used by African *H. erectus*. While Acheulean tools appear by 1.6 mya in Africa, they never appear in eastern Asia. Instead, east Asian *H. erectus* specimens are associated with the simpler Oldowan tool industry. Some paleoanthropologists think that the differences in tool technology between Asia and Africa represent cognitive differences between the two types—that African forms of *H. erectus* made fancier tools because they were smarter. Other researchers argue environmental differences between the two regions may have caused hominids to choose different tools and tell us nothing about the relative intelligence of African and Asian *Homo erectus*. They point out that during the Middle Pleistocene much of eastern Asia was covered with dense bamboo forests. Bamboo is unique among woods because it forms sharp, hard tools suitable for butchering game, perhaps making hand axes less necessary to hunters in Asia.

Some anthropologists believe that African and East Asian populations were distinct species.

The differences between African and Asian *H. erectus* lead some paleontologists, including Chris Stringer of the British Museum and Richard Klein of Stanford, to

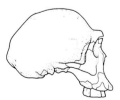

African *Homo erectus*
(KNM-ER 3733)

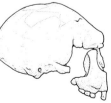

Indonesian *Homo erectus*
(Sangiran 17)

Figure 13.7

African and Indonesian *H. erectus* differed in a number of ways. Most notably, Indonesian *H. erectus* had a receding forehead and a smaller brain than its African counterparts.

conclude that the eastern Asian hominids represent a separate species. They hypothesize that the vast distances across Eurasia reduced gene flow between African and Asian populations, allowing them to diverge and form new species. Other authorities, such as G. Phillip Rightmire of Binghamton University and Milford Wolpoff of the University of Michigan, argue that *H. erectus* was a single species whose morphology varied from east to west. While the distance across Asia reduced gene flow, there was not time for *H. erectus* to split into reproductively isolated species. This dispute is difficult to resolve because to be certain whether African and Asian forms were really independently evolving lineages, we would have to know whether there was gene flow between eastern and western populations—something that we can't easily learn from the fossil record. Paleontologists have to answer this question using morphological data alone. They must ask whether Asian and African forms are as different from each other as are typical species of extant primate genera.

This question has important implications for understanding how modern humans evolved. If African and Asian *H. erectus* were different species, then modern humans can only be descended from one of them, and, assuming there are no living *H. erectus* hidden away in a Himalayan lost world, the other species is extinct. On the other hand, if African and Asian *H. erectus* were one species, *H. erectus* from all parts of the world may have contributed genes to modern populations. We will return to this controversy in Chapter 14.

For the present, we will have to be satisfied with explaining why, if there are two species, the younger Asian fossils retain the name *H. erectus* while the older African fossils have to be assigned a new name. The explanation is based upon the rules of biological nomenclature. Dubois's find is the **type specimen** for the species because it is the first fossil for the species to be discovered and described. Therefore, if systematists finally agree that eastern and western populations of *H. erectus* are distinct species, then the eastern Asian form will keep the name *H. erectus* because it is defined by the type specimen. The western form will be defined by a different type specimen and given another name; supporters of the two-species view have proposed calling it *Homo ergaster*. The name means "work man" and was first applied to a specimen from East Turkana.

Cultural Adaptations of *H. erectus*

H. erectus *was the first primate to extend its range well out of the tropics.*

The map in Figure 13.6 shows that *H. erectus* made its way out of the tropics and settled in places like northern China that are sometimes very cold today. These places were probably also cold when *H. erectus* lived there.

During the last 700,000 years there have been many long, cold glacial periods punctuated by short, warmer interglacial periods. Figure 13.8 shows estimates of global temperatures for this time period, and Box 13.1 describes how these estimates were obtained from cores taken from deep-sea sediments. Notice that the world climate has fluctuated wildly in the last 700,000 years. From geological evidence, we know that during the cold periods glaciers covered North America and Europe, and Arctic conditions were prevalent. These cold periods were interspersed with shorter warm periods in which the glaciers receded and the forests returned. Fossil pollen found at Zhoukoudian indicates that *H. erectus* only lived there during the warmer interglacial periods. It is important to keep in mind that

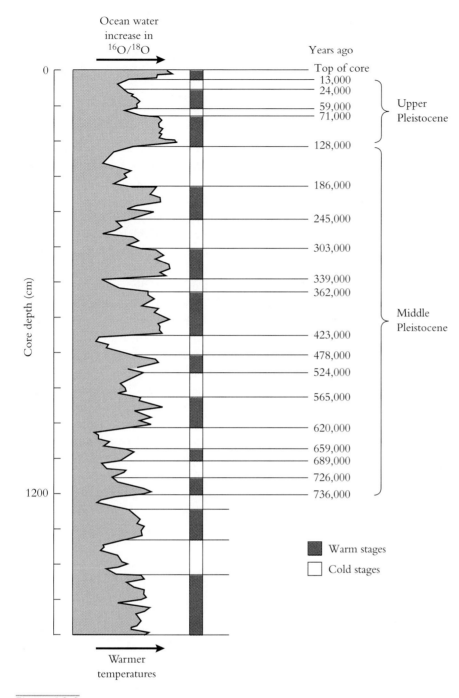

Figure 13.8

The ratio of ^{16}O to ^{18}O in seawater over the last 700,000 years indicates that there have
been wide swings in world temperature, but on average, the world has been colder than
it is now. Larger ocean values of $^{16}O/^{18}O$ signify higher global temperatures.

the interglacials were warmer than frigid glacial periods but still much colder than
the tropical climates in which African *H. erectus* evolved. We are now living in an
interglacial period, and today Zhoukoudian has a climate similar to Chicago. To
survive the rigors of cold, temperate winters outside the tropics, *H. erectus* must
have had some new ways of getting along in the world.

BOX 13.1

Reconstructing Ancient Climates Using Deep-Sea Cores

Beginning about 35 years ago, oceanographers began a program of extracting long cores from the sediments that lie on the floor of the deep sea (about 6000 m below the surface.) Data from these cores have allowed scientists to make much more detailed and accurate reconstructions of ancient climates. Figure 13.8 shows the ratio of two isotopes of oxygen, $^{16}O/^{18}O$, derived from different layers of a deep-sea core. Since different layers of the core were deposited at different times over the last 700,000 years and have remained nearly undisturbed ever since, each layer gives us a snapshot of the relative amounts of ^{16}O and ^{18}O in the sea when the layer was deposited on the ocean floor.

These ratios allow us to estimate ocean temperatures in the past. Water molecules containing the lighter isotope of oxygen, ^{16}O, evaporate more readily than molecules containing the heavier isotope, ^{18}O. Snow and rain have a higher concentration of ^{16}O than the sea does because the water in clouds evaporates from the sea. When the world is warm enough that few glaciers form at high latitudes, the precipitation that falls on the land returns to the sea, and the ratio of the two isotopes of oxygen remains unchanged (Figure 13.9). When the world is colder, however, much of the snow falling at high latitudes is stored in immense continental glaciers like those now covering Antarctica. Because the water locked in glaciers contains more ^{16}O than the ocean does, the proportion of ^{18}O in the ocean becomes greater. Therefore, the concentration of ^{18}O in seawater increases when the world is cold and decreases when it is warm. This means that scientists can estimate the temperature of the oceans over the last 700,000 years by measuring the ratio of ^{16}O to ^{18}O in different layers of deep-sea cores.

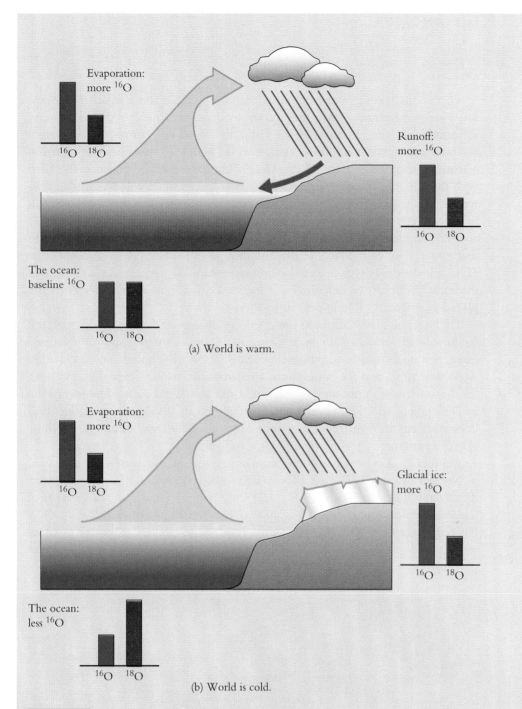

(a) World is warm.

(b) World is cold.

Figure 13.9

Water evaporating from the sea is enriched in ^{16}O, and therefore so is precipitation.
(a) When the world is warm, this water returns rapidly to the sea, and the concentration
of ^{16}O in seawater is unchanged. (b) When the world is cold, much precipitation remains
on land as glacial ice, and so the sea becomes depleted in ^{16}O.

It is uncertain whether H. erectus *hunted large mammals, although circumstantial evidence suggests they regularly ate meat.*

There is fierce controversy among paleoanthropologists about whether *H. erectus* was a big-game hunter. The issues and the evidence are similar to the controversy surrounding hunting by early *Homo.* Until the last 10 years or so, the association of bones and Acheulean tools at many sites led people to conclude that *H. erectus* was an accomplished hunter (Figure 13.10). The cave at Zhoukoudian is also filled with the bones of large game, particularly two species of deer. It is clear that they were brought to the cave by a predator. The bones of *H. erectus* and their tools have also been found in the cave. However, there are also fossilized hyena bones and feces in these caves. Many of the animal bones show evidence of hyena tooth marks, and most of the faces of the hominids are absent. Thus, it is possible that *H. erectus* was the prey at Zhoukoudian rather than the predator. As in the case of the early hominids we described in Chapter 12, it is unclear whether *H. erectus* played an important role in processing these bones.

The more recently excavated site at Olorgesaillie provides better (although still circumstantial) evidence for meat eating in *H. erectus.* At this site the fossilized bones of a single, nearly complete elephant skeleton are found in close association with stone tools, and the bones show stone-tool cut marks. There is no evidence for the activity of nonhominid predators at this site. Unlike Zhoukoudian or the earlier Olduvai sites, the association of a single isolated skeleton with stone tools does not seem to be the result of repeated use of a shade tree, water hole, or some other environmental feature by hominids and animals.

Another kind of evidence comes from a unique fossil discovered at Koobi Fora in Northern Kenya. This fossil (KNM-ER 1808) is from a *H. erectus* woman who died about 1.6 mya (Figure 13.11). Her long bones are covered with a layer of abnormal, amorphous bone that is up to 7 mm (0.028 in.) thick (Figure 13.12). Paleontologist Alan Walker of the Pennsylvania State University wanted to know what caused this anomaly, so he consulted with medical school colleagues at Johns

Figure 13.10

Many anthropologists believe that *H. erectus* hunted.

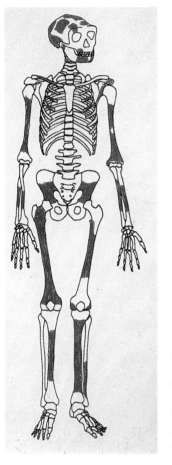

Figure 13.11

Much of the skeleton of a *H. erectus* female (KNM-ER 1808) was discovered at Koobi Fora. The bones marked in red are those that were recovered. (Diagram by Alan Walker.)

Hopkins University. The answer was surprising—ER 1808 suffered from vitamin A poisoning (Figure 13.13). One way a hunter-gatherer can acquire enough vitamin A to poison herself is by eating the liver of a large predator, like a lion or a leopard. The same symptoms have been reported for Arctic explorers who ate the livers of polar bears and seals. If this woman died from eating a predator's liver, then we must assume that *H. erectus* ate meat. Of course, we don't know how this

Figure 13.12

The long bones of KNM-ER 1808 are covered in a layer of thick, amorphous bone. A fragment of the tibia of KNM-ER 1808 (bottom) is compared with the tibia of a normal *H. erectus* (top). (Photograph courtesy of Alan Walker.)

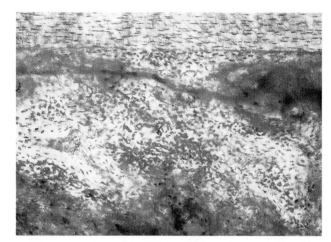

Figure 13.13

This is a microscopic view of the bone structure of KNM-ER 1808. Note that the small band of normal bone at the top looks very different from the puffy, irregular diseased bone in the rest of the picture. (Photograph courtesy of Alan Walker.)

woman acquired the liver. It is possible that she scavenged it from a dead animal, or that the predator was killed in a contest over a carcass.

Several lines of circumstantial evidence suggest that *H. erectus* relied on meat, at least during part of the year. First, hand axes seem to be well suited to the butchery of large animals. Second, the teeth of *H. erectus* are well suited for biting and tearing, and less well suited for chewing tough plant foods. As we pointed out in Chapter 12, meat, seeds, and fibrous plant materials are the main alternative food sources for primates during the dry season. If *H. erectus* was not eating tough plant foods during the dry season, then it might have relied on meat. Third, *H. erectus* was able to live outside of the tropics. Other primates rely heavily on fruit and other tender parts of plants for survival. To survive a temperate winter when such things are not available, *H. erectus* might have relied on meat. Other than humans, very few primates live in places with severe winters, and these are only small, peripheral populations of species that are usually more tropical.

H. erectus *probably controlled fire.*

Imagine living in northern Europe, China, or the American Midwest without shoes, parkas, scarves, hats, mittens—or, for that matter, without any clothes at all. Imagine living through a cold winter without a roof, windows, or solid walls to protect you from the wind, rain, and snow. Worse yet, imagine living through the winter without a source of heat—no heating pad, no radiator, and no fire. This seems like a difficult and perhaps impossible challenge for an animal that evolved in the tropics and moved to temperate climates. It is hard to believe that *H. erectus* could have survived the winter in these environments unless they knew how to control fire.

There are many sites at which fossils or tools of *H. erectus* are associated with the remains of fire. The earliest example comes from Koobi Fora near Lake Turkana, where archaeologists found baked earth beside stone tools at a site dated to about 1.6 mya. Better evidence that *H. erectus* used fire comes from the caves at Zhoukoudian. Here, huge amounts of ash and charcoal and numerous charred animal bones are associated with the bones and tools of *H. erectus*. While much of this ash may have been the result of natural fires, some layers contain thin lens-shaped deposits of ash that seem more like the remains of hearths. There are also small depressions that measure about two feet across filled with charred bones at Vertesszöllös, a site in Hungary, that have been dated to several hundred thousand years ago.

In each of these cases, archaeologists were skeptical that these sites really contained the remains of campfires: the charred bones and ash could have been produced by natural fires that swept through sites where hominids had lived, perhaps many years after hominids had occupied them. Without clear evidence of the remains of stone hearths (which do not appear in the fossil record for certain until about 100,000 years ago), there seemed to be no way to be sure that hominids used fire.

However, archaeologist Randy Bellomo of the University of South Florida has recently conducted a series of experiments strongly indicating that the baked earth at Koobi Fora is the remains of an ancient campfire. Bellomo made campfires and set fire to grass and tree stumps in several habitats and carefully examined the remains. He found that the soil under campfires reached much higher temperatures than the soil under grass fires or the soil surrounding smoldering tree stumps. The higher temperatures produced two distinctive characteristics:

1. There was a bowl-shaped layer of highly oxidized soil directly under the fire.
2. The soil under the fire became highly magnetized compared with the surrounding soil, and this magnetization was quite stable.

The soil beneath stump fires and grass fires did not have either of these features (Figure 13.14). He found that stump fires left a distinctive hole that was filled with unconsolidated sediments. With these experimental results in hand, Bellomo studied the so-called campfire sites at Koobi Fora. One of them clearly shows the diagnostic features of a campfire: it contains a bowl of oxidized soil that has been magnetized to the same degree as his experimental samples. Moreover, the magnetization is in the same direction as the earth's magnetic field would have been 1.6 mya, and not in the present direction. Thus, it seems likely that hominids have been controlling fire for more than a million and a half years.

Recent excavations at Swartkrans Cave in South Africa uncovered more evidence of fire use by *H. erectus*. There anthropologists found numerous fragmented fossil bones of antelope, zebra, warthog, baboon, and *Australopithecus robustus* that all show evidence of having been burned. Such burnt fossils were found at 20 different levels dated from 1.5 to 1 mya. Oldowan tools and *H. erectus* fossils were also found on some of these levels. To determine whether these bones had been burnt in campfires, C. K. Brain of the Transvaal Museum and Andrew Sillen of the University of Capetown compared the fossils to modern antelope bones burned at a range of temperatures. They found that at the high temperatures characteristic of long-burning campfires (300° to 400°C, or 508° to 688°F) the microscopic structure of the bone underwent clear, observable changes—changes also found in the burnt fossils at Swartkrans. Since Bellomo's experiments suggest that wildfires rarely produce high temperatures, the presence of bones burnt at high temperature is good evidence that these bones were burned by hominids in campfires.

Figure 13.14

Experimental analyses provide diagnostic cues for distinguishing between naturally occurring fires, like this one in the Okavango Delta, and manmade fires. The results strongly suggest that *H. erectus* controlled fire.

Middle Pleistocene Hominids: Archaic *Homo sapiens*

Sometime during the first half of the Middle Pleistocene (0.9 to 0.13 mya) hominids with larger brains and more modern skulls appeared in Africa and western Eurasia.

Geologists subdivide the Pleistocene into three parts: the **Lower**, **Middle**, and **Upper Pleistocene**. The Lower Pleistocene begins about 1.64 mya, a date that coincides with a sharp cooling of the world's climate. The beginning of the Middle Pleistocene is marked by the appearance of immense continental glaciers that covered northern Europe for the first time 900,000 years ago, and its end is defined by the termination of the penultimate glacial period, about 130,000 years ago. The Upper Pleistocene ends with the beginning of the present warm, interglacial phase of the world climate about 12,000 years ago.

Sometime during the first half of the Middle Pleistocene, hominids with larger brains, higher and more rounded skulls, and more rounded browridges appear in the fossil record. Figure 13.15 shows two nearly complete crania from this period, the Petralona cranium found in Greece and the Kabwe (or Broken Hill) cranium from Zambia. These individuals, who both died approximately 400,000 years ago, had substantially larger brains than *H. erectus*—between 1200 and 1300 cc. Also notice that the skulls share a number of derived features with modern humans, including more vertical sides, higher foreheads, and a more rounded back than *H. erectus*. These creatures retained many primitive features, including a long, low skull; very thick cranial bones; a large, prognathic face; no chin; and very large browridges. Their bodies remained much more robust than modern humans. Fossils with similar characteristics have been found at other sites in Africa (e.g., Ndutu in Tanzania, Bodo in Ethiopia) and western Eurasia (e.g., Maurer in Germany, Boxgrove in England) that date to about the same time (Figure 13.16). There is no sign of these kinds of fossils in eastern Asia during this period, although they do appear later.

Traditionally, paleoanthropologists have referred to the larger brained, more modern-looking hominids of the Middle Pleistocene as **archaic *Homo sapiens***, and we will continue that usage. However, as we will explain below, there is considerable controversy about whether these creatures belonged to one species or several, and, among those anthropologists who believe that there were several species, there are fierce debates about which specimens belong to which species.

Scientists are uncertain about when archaic *H. sapiens* first appears. The earliest evidence for archaic *H. sapiens* may come from Trinchera Dolina in the Sierra de Atapuerca in northern Spain, which we mentioned earlier. This site has yielded a number of hominid fossils, including part of an adolescent's lower jaw and most of an adult's face. While the Trinchera Dolina fossils are too fragmentary to provide an estimate of endocranial volume, they exhibit a number of facial features that are seen in later archaic *H. sapiens*. These fossils have been dated to about 800,000 years ago using paleomagnetic methods, but rodent fossils found at the same site indicate a more recent date, perhaps 500,000 years ago. Similar uncertainty afflicts our estimates of the ages of the other early archaic *H. sapiens* fossils, so the best that we can do is to bracket the first appearance of these creatures between 800,000 and 500,000 years ago.

During the second half of the Middle Pleistocene, archaic H. sapiens *spread to eastern Asia where they may have coexisted with* H. erectus.

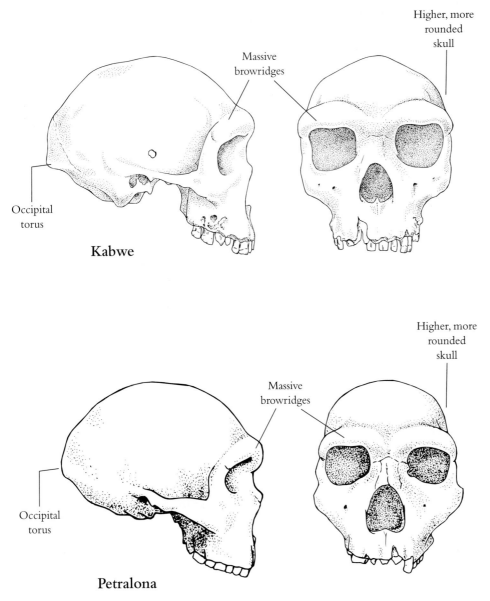

Kabwe

Higher, more rounded skull

Massive browridges

Occipital torus

Petralona

Higher, more rounded skull

Massive browridges

Occipital torus

Figure 13.15

Sometime between 800 kya and 500 kya, hominids with higher, more rounded crania and larger brains first appear in the fossil record, and by 400 kya such archaic *Homo sapiens* were common in Africa and western Eurasia. This figure shows two archaic *Homo sapiens* fossils, one from Kabwe (sometimes called Broken Hill) in Zambia, and the second from Petralona in Greece. Both are approximately 400,000 years old. (Figure courtesy of Richard Klein.)

Hominids with larger brains and more rounded skulls have been found at several sites in China that date to approximately 200,000 years ago. The most complete and most securely dated fossil is from Yinkou (also called Jinniushan) in northern China. This specimen (Figure 13.17), which consists of a cranium and associated post-cranial bones, is similar to early archaic *H. sapiens* fossils from Africa and Europe. Like the crania found at Kabwe and Petralona, it shares a larger brain case (about 1300 cc) and more rounded skull with modern humans, but it also has massive browridges and other primitive features. Other archaic *H. sapiens* fossils have been found at Dali in North China, and Maba in southern China, and are probably somewhat younger than the Yinkou specimen.

It is possible that archaic *H. sapiens* may have coexisted with *H. erectus* in East Asia during this period. Fossils of *H. erectus* that are between 200,000 and 300,000 years old have been found both at Zhoukoudian (Skull V) and at Hexian (in southern China). The fossils of other animals found in association with *H. erectus* fossils at two sites in Java (Ngandong and Sambungmachan) are consistent with an age of between 300,000 and 250,000 years old. However, the teeth of bovids

Later Middle Pleistocene: 500 kya–137 kya

Figure 13.16

During the later Middle Pleistocene, archaic *H. sapiens* spread throughout Africa and western Eurasia. During most of this period, *H. erectus* remained common in East Asia. The fossils found at these sites are discussed in the text.

(buffalo and antelope) associated with the hominid fossils at these sites have been dated to about only 27,000 years ago using ESR (electron spin resonance) techniques.

During the later Middle Pleistocene, the morphology of archaic H. sapiens *in Europe diverged from those in Africa and Asia.*

Figure 13.17

Hominids with the characteristics of archaic *Homo sapiens* appeared in East Asia later than in western Eurasia. This figure shows an archaic *Homo sapiens* from Yinkou in northern China. It is approximately 200,000 years old.

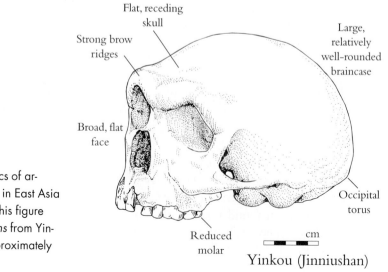

A large sample of fossils from a recently excavated site in Spain provides evidence that archaic *H. sapiens* in Europe had begun to diverge from other archaic populations during the Middle Pleistocene. This site, Sima de los Huesos ("Pit of the Bones") in the Sierra de Atapuerca only a few kilometers from Trinchera Dolina, is a small cave 13 meters below the surface in which over 2,000 bones from at least 24 different individuals have been found. These bones, which date to about 300,000 years ago, include several nearly complete crania as well as many bones from other parts of the body. Figure 13.18 shows one of the crania, labelled SH 5. Like other archaic *H. sapiens*, these skulls mix derived features of modern humans and primitive features associated with *H. erectus*. However, the crania from Sima de los Huesos also share a number of derived characters not seen in archaic *H. sapiens* living at the same time in Africa. Their faces bulge out in the middle and have doubled-arched browridges, and the backs of some of the SH skulls are rounded. The fossils from Sima de los Huesos also have relatively large cranial capacities, averaging around 1390 cc, close to the average value for modern humans. This complex of characters is significant because they are also shared by the **Neanderthals**, the hominid that dominates the European fossil record from 127,000 to 30,000 years ago. This connection suggests that the lineage leading to the Neanderthals had begun to diverge from other archaic populations by about 300,000 years ago.

Not all of the Sima de los Huesos crania express all the Neanderthal features to the same degree, and this variability has helped to settle an important question about hominid evolution in Europe. Other fossils from later Middle Pleistocene Europe also share some derived traits with Neanderthals, but the extent of their similarity to Neanderthals varies from site to site. As a result, paleoanthropologists used to debate whether these differences reflected variability within a single population of hominids or whether there was more than one type of hominid present in Europe during the Middle Pleistocene. The large and variable sample of individuals at Sima de los Huesos has put this debate to rest. These individuals, who likely belonged to a single population, vary in their expression of Neanderthal

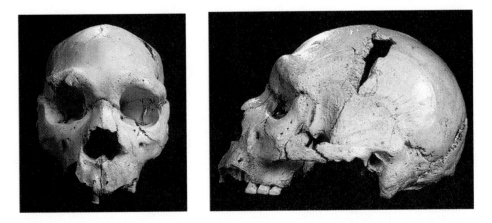

Figure 13.18

The many hominid fossils found at Sima de los Huesos in Spain provide evidence that hominids in Europe began to evolve a distinctive cranial morphology at least 300 kya. The features, which include rounded browridges, a large pushed-out face, a skull with a rounded back, and a large brain, are important because they are shared with the Neanderthals, the hominids that dominate the European fossil record during the Upper Pleistocene.

traits and this variation mirrors the kind of variation seen at different sites in Europe. This suggests that the features defining the Neanderthal had begun to evolve in European populations by 300,000 years ago, but these traits evolved independently, not as a single functional complex.

A different sequence of events seems to have occurred in Africa during the later part of the Middle Pleistocene. There, hominids show no trace of Neanderthal features. Instead, these hominids seem to be more similar to modern humans than their predecessors. For example, the cranium from Djebel Irhoud in Morocco (Figure 13.19) has a smaller face tucked under the cranium and a larger, rounder cranial vault.

About 250,000 years ago, archaic H. sapiens *shifted to a new stone tool kit in Africa, western Asia, and Europe, but not in eastern Asia.*

The tools used by early archaic *H. sapiens* are similar to those used by *H. erectus*. At most sites, tool kits are still dominated by Acheulean hand axes and other core tools, but the hand axes are better made in some cases. However, beginning about 250,000 years ago, hand axes became much less common and were replaced by tools that were manufactured by producing sizable flakes, which were then further shaped or "retouched." We will describe this kind of tool kit in more detail below. Tools of this kind are not found in eastern Asia, where sites continue to be dominated by simple, Oldowan-like tool kits.

Archaeologists do not know what caused this shift in stone-tool technology. It is not associated with any obvious morphological change in the fossil record. Moreover, the new tools appear more or less simultaneously in Europe and Africa, even though hominids in these areas were apparently becoming morphologically distinct.

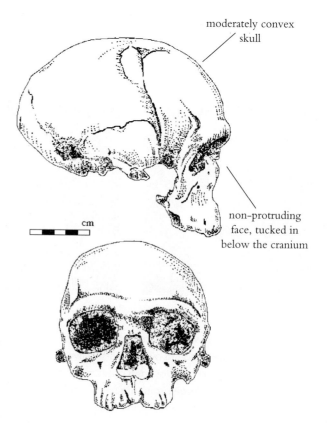

moderately convex
skull

non-protruding
face, tucked in
below the cranium

cm

Figure 13.19

This cranium found at Jebel Irhoud in Morocco has been dated to between 90 kya and 190 kya using ESR methods and faunal associations. Like other African hominid fossils from the Late Middle Pleistocene, it lacks the diagnostic features of Neanderthal crania. Instead, the Jebel Irhoud cranium shares features with modern humans, including a smaller, flatter face tucked under the cranium as well as a larger, rounder cranial vault.

Jebel Irhoud 1

There is good evidence that archaic H. sapiens *hunted big game.*

The first solid evidence for hunting big game also comes from this period. On the island of Jersey, the remains of a large number of fossilized bones of mammoths and woolly rhinoceros have been found at the base of a cliff. Some of the bones were of adults who were certainly large enough to be immune to most predators. Moreover, the carcasses have clearly been butchered with stone tools. In some instances, the skull cavity has been opened, presumably to extract the brain tissue. In some places, bones have been sorted by body parts (heads here, limbs there, and so on). All this suggests that early humans drove the animals over the headland, butchered the carcasses, and ate them.

More evidence for hunting comes from three wooden throwing-spears found in an open-pit coal mine in Schöningen, Germany. This site is dated to about 400,000 years ago. Anthropologists are confident that these were throwing-spears because they closely resemble modern javelins. They are about 2 m (6 ft) long and, like modern throwing spears, are thickest and heaviest near the pointed end, gradually tapering to the other end. Such spears are used for hunting by some modern people, and it seems plausible that they were used in the same way by archaic *H. sapiens*. This hypothesis is strengthened by the fact that along with the spears, anthropologists found the bones of hundreds of horses, many of which show signs of having been processed using stone tools.

Anthropologists disagree about how to classify Middle Pleistocene hominids.

This disagreement about classification stems from different ideas about how modern humans evolved. One school of thought, most closely associated with University of Michigan anthropologist Milford Wolpoff, holds that hominids in Africa and Eurasia formed a single interbreeding population throughout the Pleistocene. The size of Africa and Asia limited gene flow and allowed regional differences to evolve, but there was always enough interbreeding to guarantee that all hominids belonged to single species during this entire period. Thus, Wolpoff and others prefer to include all hominids who lived during the Middle Pleistocene in a single species, *Homo sapiens*, as shown in Figure 13.20a.

Other anthropologists believe that as hominids spread out of Africa they split into several new species. Within this school of thought there are differences about the details of the phylogenetic history. The diagrams in Figure 13.20b and 13.20c illustrate two recent hypotheses. According to G. Phillip Rightmire at Binghamton University, African and Asian *H. erectus* are members of the same species. However, he draws a distinction between early Middle Pleistocene archaic *H. sapiens* and those that came later. He classifies the early archaic forms as *Homo heidelbergensis* (named after a fossil found near the German city of Heidelberg) and groups the upper Middle Pleistocene archaic *H. sapiens* in Europe and the later Neanderthals into one species, *Homo neanderthalensis*. Stanford anthropologist Richard Klein believes that hominid populations in Africa, western Eurasia, and eastern Eurasia were genetically isolated from each other during most of the Pleistocene and represent three distinct species. In Africa, *Homo ergaster* gradually evolved into *Homo sapiens* around 400,000 years ago. Isolated in Asia, *Homo ergaster* evolved into *Homo erectus* early in the Pleistocene. Klein considers the development of larger brains and more modern looking skulls in the Asian fossils, like those found at Yinkou, to be the result of convergent evolution and includes them within *H. erectus*. Similarly, once hominids reached Europe about 500,000 years ago, they became isolated from African and east Asian populations, and diverged to become *Homo neanderthalensis*.

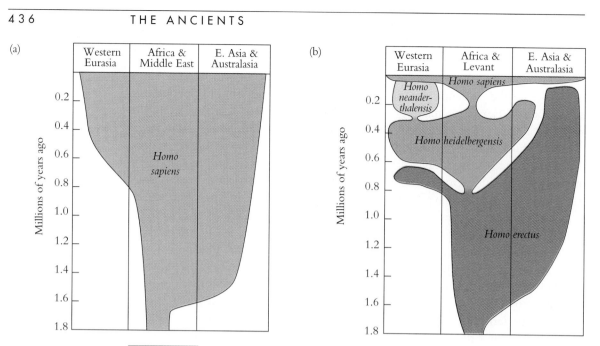

Figure 13.20

Numerous phylogenies have been proposed to account for the temporal and geographical patterns in hominid evolution during the Pleistocene. These figures illustrate three recent proposals. [a] Milford Wolpoff holds that beginning about 1.8 mya there was only one human species. Because there have been no speciation events since that time, all human fossils should be classified as *H. sapiens*. [b] G. Phillip Rightmire believes that both African and Asian specimens of *H. erectus* should be classified as a single species. Approximately 800 kya a larger brained species, *H. heidelbergensis*, evolved in Africa and eventually spread to Europe and perhaps east Asia. In Europe, *H. heidelbergensis* gave rise to the Neanderthals, while in Africa it gave rise to *H. sapiens*. [c] Richard Klein argues that *H. ergaster* evolved in Africa about 1.8 mya and soon spread to Asia, where it differentiated to become a second species *H. erectus*. About 500 kya, *H. ergaster* spread to western Eurasia, where it evolved into *H. neanderthalensis*. In Africa, *H. ergaster* evolved into *H. sapiens* about the same time.

Upper Pleistocene Hominids: Neanderthals and Their Contemporaries

We now turn to perhaps the most famous ancient human population, the Neanderthals who lived in Europe and western Asia from about 127,000 to 30,000 years ago. We will see in this chapter and the next that some anthropologists believe modern humans are descendants of the Neanderthals, while many others believe that Neanderthals are a side branch of the human family tree that became extinct and that all modern peoples are descended from African populations living about the same time. Nonetheless, we will devote a lot of space to what the fossil and archaeological records tell us about the Neanderthals and relatively little space to their African and Asian contemporaries. The reason for this bias is simple: we know vastly more about the Neanderthals than we do about their contemporaries in other parts of the world. This is in part because Europe is better studied paleontologically than Africa, and in part because Africa was very dry during this

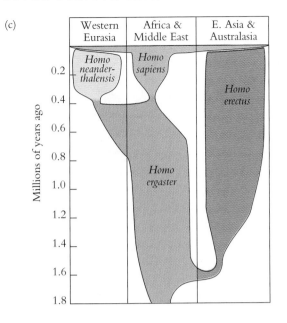

(c)

period, and hominid population densities were probably very low. What we do know about the African and West Asian Upper Pleistocene hominids suggests that they were behaviorally very similar to the Neanderthals, and thus by studying the Neanderthals we may derive some insights about other contemporary hominids.

The last warm interglacial period lasted from about 130,000 to about 75,000 years ago. For most of the time since then the global climate has been colder, sometimes much colder.

The data in Figure 13.21 give a quite detailed picture of global temperatures over the last 100,000 years. These data are based on the $^{16}O/^{18}O$ ratios of different layers of cores as described earlier in the chapter, except that these cores were taken from deep inside the Greenland ice cap. Since snow accumulates at a higher rate than sediments on the ocean bottom, ice cores provide a more detailed picture of past climates than ocean cores.

We are currently living in a typical, warm interglacial period. The last such period began about 130,000 years ago and lasted approximately 50,000 years (see Figure 13.21). We hedge on the ending date because interglacial periods have very distinct beginnings but rather indefinite ends. During the last interglacial, the world was generally much warmer than it is today. We know this from several lines of evidence. Plankton species currently living in subtropical waters (like those off the coast of Florida) extended as far as the North Sea during the last interglacial. The kinds of fossil fauna found in Europe are restricted to more tropical areas today; for example, the remains of a hippopotamus have been found under Trafalgar Square, which lies in the center of London. The range of plant life differed as well: in Africa, the rain forest extended beyond its present boundaries, and in temperate areas, broadleaf deciduous forests extended farther north than they do today.

As the last glacial period began, sometime between 100,000 and 75,000 years ago, the world slowly cooled. In Europe temperate forests shrank and grasslands expanded. The glaciers grew, and the world became colder and colder—not steadily, but with wide oscillations from cold to warm. Finally, when the glaciation was

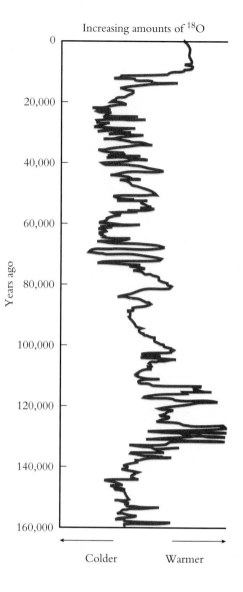

Increasing amounts of ^{18}O

Years ago

Colder   Warmer

Figure 13.21

Fluctuations in the ratio of ^{18}O to ^{16}O over the last 160,000 years (from a core drilled in the deepest part of the Greenland ice cap) indicate that about 100,000 years ago the world's climate got colder and less stable. In ice cores, more ^{18}O relative to ^{16}O indicates higher average global temperatures. (In the deep-sea cores, in contrast, more ^{18}O relative to ^{16}O indicates *lower* temperatures.)

at its greatest depth (about 30,000 years ago), gigantic continental glaciers covered most of Canada and much of northern Europe. Sea levels dropped so low that the outlines of the continents were often quite different than they are today: Asia and North America were connected by a land bridge that spanned the Bering Sea; the islands of Indonesia joined Southeast Asia in a landmass called Sundaland; and Tasmania, New Guinea, and Australia joined into a single continent called Sahul. Eurasia south of the glaciers was a vast, frigid grassland, punctuated by dunes of **loess** (fine dust produced by glaciers) and teeming with animals—woolly mammoths, woolly rhinoceros, reindeer, aurochs (the giant wild ox that is ancestral to modern cattle), musk oxen, and horses.

Sometime during the last interglacial period, a morphologically distinct group of hominids, the Neanderthals, came to dominate Europe and the Near East.

In 1856, workers at a quarry in the Neander Valley in western Germany found some unusual fossil bones. The bones found their way to a noted German anatomist, Hermann Schaafhausen, who declared them to be the remains of a race of humans who lived in Europe before the Celts. Many experts examined these curious finds and drew different conclusions. T. H. Huxley, one of Darwin's staunchest supporters, suggested that they belonged to a primitive, extinct kind of human, while the Prussian pathologist Rudolf Virchow proclaimed them to be the bones of a modern person suffering from a serious disease that had distorted the skeleton. Initially, Virchow's view held sway, but as more fossils were discovered with the same features, researchers became convinced that the remains belonged to a distinctive, extinct kind of human. The Germans called this extinct group of people the Neanderthalers, meaning "people of the Neander Valley" (*thal*, now spelled *tal*, is German for "valley"). We call them the Neanderthals, and they were characterized by several distinctive, derived features.

1. *Neanderthals had large brains.* The Neanderthal braincase is much larger than that of archaic *H. sapiens*, ranging from 1245 to 1740 cc, with an average size of about 1520 cc. In fact, Neanderthals had larger brains than modern humans, whose brains average about 1400 cc. It is unclear why the brains of Neanderthals were so large. Some anthropologists point out that the Neanderthals' bodies were much more robust and heavily muscled than those of modern humans and suggest that the large brains of Neanderthals reflect the fact that larger animals usually have bigger brains than smaller ones.

2. *Neanderthals had more rounded crania than* H. erectus. The Neanderthal skull is long and low, much like the skulls of archaic *H. sapiens*, but relatively thin walled. The back of the skull has a characteristic rounded bulge or bun and does not come to a point at the back like a *H. erectus* skull. There are also detailed differences in the back of the cranium.

3. *Neanderthals had big faces.* Like *H. erectus* and archaic *H. sapiens*, the skulls of Neanderthals have large browridges, but they are larger, rounder, and stuck out less to the sides. Moreover, the browridges of *H. erectus* are mainly solid bone, while those of Neanderthals are lightened with many air spaces. The function of these massive browridges is not clear. The face, and particularly the nose, is enormous—every Neanderthal was a Cyrano, or perhaps a Jimmy Durante (Figure 13.22).

4. *Neanderthals had small back teeth and large, heavily worn front teeth.* Neanderthal molars are smaller than those of *H. erectus*. Neanderthal molars had distinctive **taurodont roots** (Figure 13.23) in which the pulp cavity expanded so that the roots merged, partially or completely, to form a single broad root. Neanderthal

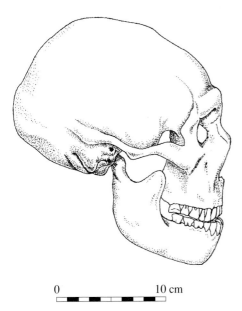

0   10 cm

Figure 13.22

The skulls of Neanderthals, like this one from Shanidar Cave, Iraq, are large and long, with large browridges and a massive face. (Figure courtesy of Richard Klein.)

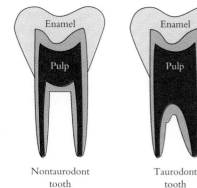

Figure 13.23

In Neanderthal molars the roots often fuse together partially or completely to form a single massive taurodont root. The third root is not shown.

Nontaurodont tooth

Taurodont tooth

incisors are relatively large and show very heavy wear. Careful study of the wear patterns indicates that Neanderthals may have pulled meat or hides through their clenched front teeth. Also there are often microscopic, unidirectional scratches on the front of the incisors as if Neanderthals had held meat in their teeth while cutting it with a stone tool. Interestingly, the direction of the scratches suggests that most Neanderthals were right-handed, just as the Oldowan toolmakers were.

5. *Neanderthals had robust, heavily muscled bodies.* Like *H. erectus* and archaic *H. sapiens*, Neanderthals were extremely robust, heavily muscled people (Figure 13.24). Their leg bones are much thicker than ours; the load-bearing joints (the knees and hips) were larger; the **scapulae** (shoulder blades) show more extensive muscular attachments; and the rib cage is larger and more barrel-shaped. All these skeletal features indicate that Neanderthals were very sturdy and

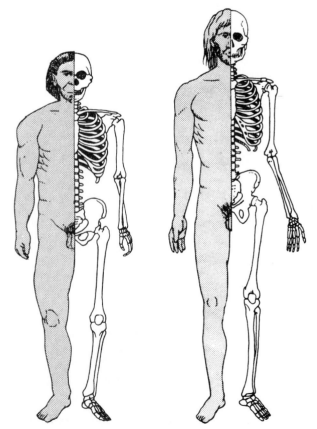

Figure 13.24

The Neanderthals were very robust people. Contrast the bones of this skeleton (left) with the bones of a modern human (right).

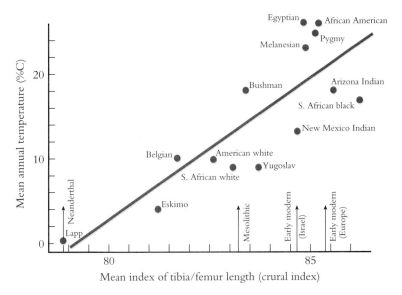

Figure 13.25

People have proportionally longer arms and legs in warm climates than in cold climates. Local temperature is plotted on the vertical axis, while crural index is plotted on the horizontal axis. Smaller values of the crural index are associated with shorter limbs relative to body size. Populations in warm climates tend to have high crural index values, and vice versa. Neanderthals had a crural index similar to the Lapps, who live above the Arctic Circle. Fossil hominids are marked with arrows.

strong, weighing about 30% more than contemporary humans of the same height. A comparison with data on Olympic athletes suggests that Neanderthals most resembled the throwers—hammer, javelin, and discus throwers, and shot putters. They were a few inches shorter, on average, than modern Europeans and had relatively larger torsos and shorter arms and legs.

This distinctive Neanderthal body shape may have been an adaptation to conserve heat in their glacial environment. In cold climates animals tend to be larger and have shorter and thicker limbs than do individuals of the same species in warmer environments. This is because the rate of heat loss for any body is proportional to its surface area, and so any change that reduces the amount of surface area for a given volume will conserve heat. Animals can reduce their surface area volume ratio in two ways: by increasing their overall body size or by reducing the size of their limbs. In contemporary human populations there is a consistent relationship between climate and body proportions. One way to compare body proportions is by calculating the **crural index**, which gives the ratio of the length of the shin bone (tibia) to the length of the thigh bone (femur). As Figure 13.25 shows, people in warm climates tend to have relatively long limbs in proportion to their height. Note that Neanderthals resembled modern peoples who live above the Arctic Circle.

Contemporaries of the Neanderthals living in Africa did not display the same distinctive morphological features.

The fossil record in Africa indicates fairly clearly that African humans of this period were not like the Neanderthals, although just what they were like is unclear because the fossil record for this time in Africa is variable and difficult to interpret. Some fossils, like those from Border Cave in South Africa and the Omo-Kibish Formation in southern Ethiopia, are similar to modern humans (Figure 13.26). They have reduced browridges, small faces, and higher, more rounded crania. However, in both cases there are reasons to suspect that the fossils actually come from much more recent populations. At Border Cave, for example, the fossils were recovered from the rubbish pit of a farmer who excavated the cave for fertilizer, so there is doubt about which level they came from. At Omo-Kibish, the very modern Omo 1 fossil was found at the same level as another specimen, Omo 2,

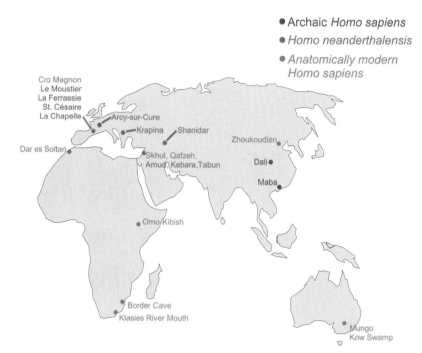

- Archaic *Homo sapiens*
- *Homo neanderthalensis*
- *Anatomically modern Homo sapiens*

Figure 13.26

At the beginning of the Upper Pleistocene (137 kya–10 kya), Neanderthals became common in western Eurasia, and shortly thereafter anatomically modern humans appeared in Africa and the Middle East. Archaic *H. sapiens* remained common in East Asia. Around 40,000 years ago, anatomically modern humans came to predominate throughout the Old World. The fossils found at these sites are discussed in the text.

which has much more primitive characteristics, suggesting that one of the two fossils intruded into the stratum in which it was found. Like Omo 2, fossils from Laetoli (Tanzania) are robust skulls that mix modern features with features characteristic of *H. erectus*. Another interesting skull was found at the Dar es Soltan Cave in Morocco. It is more than 40,000 years old and shows none of the distinctive Neanderthal features, except for prominent browridges (Figure 13.27).

Figure 13.27

The fossil skull found at Dar es Soltan in Morocco shows that African contemporaries of the Neanderthals did not have the distinctive features of Neanderthals: the face does not protrude, and the braincase is higher and probably shorter than that of Neanderthals. (Figure courtesy of Richard Klein.)

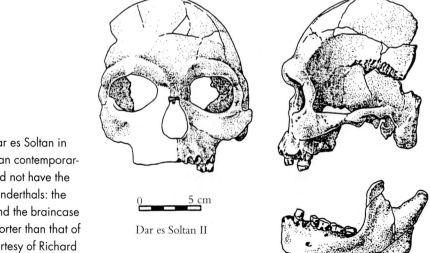

Dar es Soltan II

Despite the difficulties in interpreting the African record, three things are clear:

1. Like the Neanderthals, African hominids of the period show large cranial volumes, ranging from 1370 to 1510 cc.
2. None of the African fossils show the complex of features that are diagnostic of the European Neanderthals.
3. While the fossils are variable and some are quite robust, they are all more like modern humans than the Neanderthals are.

Mousterian and Middle Stone Age Tools

Neanderthals and their contemporaries in Africa are associated with Mousterian tools, a fancier and more diverse tool kit than the Acheulean industry of African H. erectus *and archaic* H. sapiens.

Archaeological sites that date to the Neanderthal period are dominated by flake tools. This means that Neanderthals and their contemporaries struck flakes from cores and then used the flakes as their tools. In contrast, as we saw earlier in the chapter, Acheulean sites are dominated by large hand axes and choppers made from cores. Hand axes are almost completely absent from most Neanderthal sites. During this period the stone-tool industry found in Europe is called the **Mousterian** (or sometimes the Middle Paleolithic), and the very similar industry found in Africa and in southern and eastern Asia is called the **Middle Stone Age (MSA)**. You may recall from Box 11.5, p. 367, that some archaeologists believe Oldowan hominids used mainly flake tools as well. However, unlike the small, irregular Oldowan flakes, the Neanderthals and their contemporaries produced quite symmetric, regular flakes using sophisticated techniques. One method, called the **Levallois technique** (after the Parisian suburb where such tools were first identified), involves three steps. First the knapper prepares a core with one precisely shaped convex surface. Then the knapper makes a striking platform at one end of the core. Finally, the knapper hits the striking platform, knocking off a flake whose shape is determined by the original shape of the core (Figure 13.28). A skilled knapper could produce a variety of tools by modifying the shape of the original core.

Mousterian and MSA tools are more variable than Acheulean tools. One of the most influential students of Mousterian tools, the French archaeologist François Bordes, classified tools into a large number of distinct types based on their shape and inferred functions: points, **side scrapers** (flake tools bearing a smooth edge on one or both sides), **denticulates** (tooth-shaped tools with a serrated edge), and so on. Instead of finding a random mix of tool types at each site, Bordes found that all sites fell into one of four categories that he called **Mousterian variants**, and that each variant had a different mix of tool types. Bordes thought that different sites from the same period showed different Mousterian variants, and that different levels of the same site were also characterized by different variants. He concluded that these sites were the remains of four wandering groups of Neanderthals, each preserving a distinct tool tradition over time and structured much like modern ethnic groups.

Recent studies give reason to doubt Bordes interpretation. The use of better dating techniques seems to indicate that the four Mousterian variants tended to follow one another chronologically; that is, they differed over time but not by geographic area. Moreover, tool assemblages from some sites excavated after Bordes analysis was completed do not match any of the recognized variants.

Flake the margin of the core:

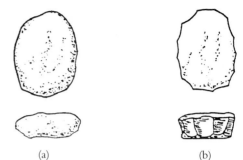

(a)   (b)

Figure 13.28

This figure shows how to make a Levallois tool. (a) The knapper chooses an appropriate stone to use as a core. The side and top views of the unflaked core are shown. (b) Flakes are removed from the periphery of the core. (c) Flakes are removed radially from the surface of the core, using the flake scars on the periphery as striking platforms. Each of the red arrows represents one blow of the hammer stone. (d) The knapper continues to remove radial flakes until the entire surface of the core has been flaked. (e) Finally, a blow is struck (large red arrow) that frees one large flake (outlined in red). This flake will be used as a tool. (f) At the end, the knapper is left with the remains of the core (left) and the tool (right). (Figure courtesy of Richard Klein.)

Prepare the surface of the core:

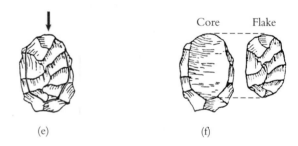

(c)   (d)

Remove Levallois flake:

Core   Flake

(e)   (f)

Many archaeologists believe that the variation among sites results from differences in the kinds of activity performed at each locale. For example, Lewis Binford (of Southern Methodist University) and Sally Binford argued that differences in tool types depend on the nature of the site and the nature of the work performed. Some sites may have been base camps where people lived, while others may have been work camps at which people performed subsistence tasks. Different tools may have been used at different sites for woodworking, hide preparation, or butchering prey. To test this idea, the Binfords assigned functions to the tools before doing their analysis, and then showed that different sites are characterized by functionally related clusters of tools. Thus, several tools that seemed to be well suited for butchery were found at the same site. More recently, however, microscopic studies of wear patterns on Mousterian tools suggest that the majority of the tools were used only for woodworking. As a result, there seems to be no association between a tool type (such as a point or side scraper) and the task for which it was used.

A related explanation for the variability in Mousterian tools focuses on the availability of raw materials for toolmaking. When suitable raw materials are close by and easily available, people don't spend much time repairing a particular tool—they simply make a new one. When raw materials are difficult to acquire, perhaps because they come from a distant source, people mend and retouch tools over and

Table 13.1 Tools at Aquitaine in southern France are generally made of materials that can be found less than 5 km (3 miles) from the site. Raw materials rarely come from more than 30 km (19 miles) away. (Adapted from Table 70 in C. Stringer and C. Gamble, 1993, *In Search of the Neanderthals*, Thames and Hudson, New York.)

DISTANCE FROM SITE (KM)	PERCENT MATERIAL FOUND ON SITE
<5	55–95
5–20	2–20
30–80	<5

over again. University of Pennsylvania archaeologist Harold Dibble has argued that many of the differences in tool types recognized by Bordes and others are really just different stages in the life histories of individual tools. For example, a toolmaker might start out by making a scraper with a single sharp edge. When the sharp edge was dulled, she might sharpen the other side, and when both sides were worn, she might retouch both sides again. Archaeologists like Bordes might classify each stage as a different tool type. If so, they would inadvertently label tools from sites where tools had been extensively retouched as different than those at sites where tools were used briefly and discarded.

Stone-tipped spears and stone axes are made by **hafting** (attaching the tool to a handle). Hafting is an extremely important innovation because it greatly increases the efficiency with which humans can apply force to stone tools. (Imagine using a hammer without a handle.) Microscopic analyses of the wear patterns on Mousterian and MSA tools suggest that some tools were hafted, probably to make spears. The discovery of a 3.4-m-long (10 ft) wooden spear in the rib cage of an elephant indicates that Neanderthals also used these weapons. The Neanderthals made virtually no use of bone, and they did not decorate their tools in any way that has survived in the archaeological record.

Neanderthals usually made tools from stone acquired locally. When archaeologists excavate a Mousterian site, they typically find that the stone used to make the tools can be found within a few kilometers of the site (Table 13.1). As we will see in Chapter 14, this pattern is altered considerably when modern humans arrive on the scene.

Neanderthal Lifeways

There is much evidence that Neanderthals and their contemporaries hunted large game.

Neanderthal sites contain the bones of many animals alongside Mousterian tools. European sites are rich in the bones of red deer (called elk in North America), fallow deer, bison, aurochs, wild sheep, wild goat, and horse, while eland (a large antelope that looks like a Brahman cow with corkscrew horns), wildebeest, zebra, and bushpig are found at African MSA sites (Figure 13.29). Archaeologists find few bones of very large animals such as hippopotamus, rhinoceros, and elephant, even though they were plentiful in Africa and Europe.

By itself this evidence does not mean that the Neanderthals killed these animals, and this has provoked as much debate as similar ambiguity has for earlier hominids. Some archaeologists, like Lewis Binford, believe that Neanderthals and their con-

(a)   (b)

Figure 13.29

Neanderthals and their contemporaries hunted large and dangerous game, like (a) red deer and (b) eland.

temporaries in Africa never hunted anything larger than small antelope, and even these prey were acquired opportunistically, not as a result of planned hunts. Any bones of larger animals were acquired by scavenging. As evidence in support of this view, Binford points to the fact that skulls and foot bones of eland predominate at Klasies River Mouth, an MSA site in South Africa, and these body parts are commonly available to scavengers. Binford believes that the hominids of this period did not have the cognitive skills necessary to plan and organize the cooperative hunts necessary to bring down large prey.

Others, such as Stanford archaeologist Richard Klein, believe that Neanderthals were proficient hunters who regularly killed large animals. He points out that animal remains at sites that date from this period are often made up almost exclusively of the bones of one or two prey species. For example, at the Klasies River Mouth site, the vast majority of bones are from eland. Similarly at Mauran, a site in the French Pyrenees, over 90% of the assemblage is from bison and aurochs. The same pattern occurs at sites across Europe. The prey animals are large creatures, and they are heavily overrepresented at these sites compared with their occurrence in the local ecosystem. It is hard to see how an opportunistic scavenger would acquire such a nonrandom sample of the local fauna. Moreover, the bones of prime-aged adults are well represented at such sites, a pattern not associated with contemporary African scavengers like hyenas, who prey on old or young animals more often than prime-aged adults (Figure 13.30). Instead, a large quantity of remains are from mature, healthy individuals—the pattern associated with catastrophic events in which whole herds of animals are killed. Several sites consist of huge jumbles of bones and tools at the bottoms of cliffs, suggesting that Neanderthals and their contemporaries hunted large game by driving them over cliffs. It also turns out that the pattern at Klasies River Mouth is far from universal. There are several sites, like Combe Grenal in France, where the bones from the meatiest parts of prey animals are overrepresented. Moreover, the cut marks on these bones suggest that Neanderthals stripped off the fresh flesh.

Figure 13.30

Hyenas frequently scavenge kills and prey mainly on very young and very old individuals.

There is little evidence for shelters or even organized campsites at Neanderthal sites.

There are many Neanderthal sites, some very well preserved, where archaeologists have found concentrations of tools, abundant evidence of toolmaking, many animal remains, and concentrations of ash. Most of these Mousterian sites occur in caves or in **rock shelters**, places protected by overhanging cliffs. This does not necessarily mean that Neanderthals were all cavemen or cavewomen, however. Cave sites are less likely to have been destroyed by erosion, and they are also relatively easy to locate because the openings to many caves from this period are still visible. Most archaeologists believe that cave sites represent home bases—semipermanent encampments from which Neanderthals sallied out to hunt and to forage.

Most sites at which Mousterian tools or Neanderthal fossils are found lack postholes or other features that archaeologists interpret as the remains of hearths or shelters. The few exceptions to this rule occur near the end of the Mousterian period. For example, evidence of hearths occurs at Vilas Ruivas, a site in Portugal dated to about 60,000 years ago.

Neanderthals probably buried their dead.

The wealth of complete Neanderthal skeletons suggests that, unlike their hominid predecessors, Neanderthals frequently buried their dead. Burial protects the corpse from dismemberment by scavengers and thus preserves the skeleton intact. Detailed study of the geological context at sites like La Chapelle-aux-Saints, Le Moustier, and La Ferrassie in southern France also supports the conclusion that Neanderthal burials were common.

It is not clear whether these burials had a religious nature, or if Neanderthals buried their dead just to dispose of the decaying bodies. Anthropologists used to interpret some sites as ceremonial burials in which Neanderthals were interred along with symbolic materials. In recent years, skeptics have cast serious doubt on such interpretations. For example, anthropologists used to think that the presence of fossilized pollen in a Neanderthal grave in Shanidar Cave, Iraq, was evidence that the individual had been buried with a garland of flowers. However, more recent analyses revealed that the grave had been disturbed by burrowing rodents, and it is quite plausible that they brought the pollen into the grave.

Neanderthals seem to have lived short, difficult lives.

Detailed study of Neanderthal skeletons indicates that Neanderthals didn't live very long. The human skeleton changes throughout the life cycle in characteristic ways that can be used to estimate the age at which fossil hominids died. For example, human skulls are made up of separate bones that fit together in a three-dimensional jigsaw puzzle. When children are first born, these bones are still separate, but later they fuse together forming tight, wavy joints called **sutures**. As people age, these sutures are slowly obliterated by bone growth. By assessing the degree to which the sutures of fossil hominids have been obliterated, anthropologists can estimate how old the individual was at death. Several other skeletal features can be used in the same way. All of these features tell the same story—Neanderthals died young. None lived beyond the age of 40 to 45 years.

Many of the older Neanderthals suffered disabling disease or injury. For example, the skeleton of a Neanderthal man from La Chapelle-aux-Saints shows symptoms of severe arthritis that probably affected his jaw, back, and hip. By the time this fellow died, around the age of 45, he had also lost most of his teeth to gum disease. Another individual, Shanidar 1 (from the Shanidar site in Iraq), had suffered a blow to his left temple that crushed the orbit (Figure 13.31). Anthropologists believe that his head injury had led to partial paralysis of the right side of his body, which in turn had caused his right arm to wither and his right ankle to become arthritic. Other Neanderthal specimens display bone fractures, stab wounds, gum disease, withered limbs, lesions, and deformities.

In some cases, Neanderthals survived for extended periods after injury or sickness. For example, Shanidar 1 lived long enough for the bone surrounding his injury to heal. Some anthropologists have proposed that these Neanderthals would have been unable to survive their physical impairments—to provide themselves with food or to keep up with the group—had they not received care from others. Some researchers argue further that these fossils are evidence of the origins of caretaking and compassion in our lineage.

Katherine Dettwyler, an anthropologist at Texas A & M University, is skeptical of this claim. The existence of individuals who recovered from severe injuries and lived with serious physical impairment does not necessarily imply that they had received care from others, or that other members of their communities showed compassion for them. In some contemporary societies, disabled individuals are sometimes able to support themselves, and they do not necessarily receive compassionate treatment from others. In addition, nonhuman primates sometimes sur-

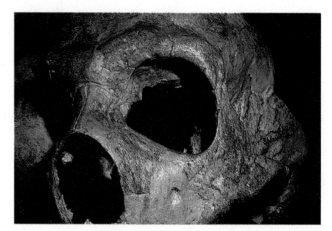

Figure 13.31

Many Neanderthal skeletons show signs of injury or illness. The orbit of the left eye of this individual was crushed, the right arm was withered, and the right ankle was arthritic.

Summary of Pleistocene Hominids

	HOMO HABILIS (LARGE OR RUDOLFENSIS)	*HOMO HABILIS* (SMALL)	AFRICAN *HOMO ERECTUS* (*ERGASTER*)	ASIAN *HOMO ERECTUS*
Dates	1.9–1.6 mya	1.9–1.6 mya	1.6 mya–500 kya	1.1 mya–300 kya
Sites	East and South Africa	East and South Africa	East and South Africa	China and Indonesia
Height (m)	1.5	1.0	1.3–1.9	??
Weight (kg)	??	??	56–64	56–64
Brain size (cc)	600–800	500–650	900–1000	1000–1250
Skull form	More robust	Less robust	Flat, thick skull, large browridge, pointed occipital	Thicker skull, larger browridge, sagittal keel
Jaws & teeth	Large jaw and molars	Smaller, more modern molars	Robust jaw, smaller teeth than *H. habilis*	Similar to African *H. erectus*
Physique	Larger, with longer legs	Smaller, with longer arms and shorter legs	Very robust, but with human proportions	Very robust, but with human proportions

Summary of Pleistocene Hominids (*continued*)

	ARCHAIC *HOMO SAPIENS*	NEANDERTHALS	ANATOMICALLY MODERN *HOMO SAPIENS*
Dates	800 kya–400 kya	300 kya–30 kya	100 kya–
Sites	Africa and temperate Eurasia	Europe and western Eurasia	Worldwide
Height (m)	??	1.5–1.7	1.6–1.85
Weight (kg)	60–72	74–78	58–68
Brain size (cc)	1100–1400	1245–1790	1200–1700
Skull form	Higher, more round skull, less protruding face	Long skull, reduced browridge, large nose	Small or no browridge, high skull, small face
Jaws & teeth	Similar to African *H. erectus*	Smaller back teeth, larger incisors	Smaller teeth, pointed chin
Physique	Very robust, but with human proportions	Very robust, shorter arms and legs	Modern skeleton

Figure 13.32

At Gombe, a male chimpanzee named Faben contracted polio, which left one arm completely paralyzed. Faben received no day-to-day help from other group members but was able to survive for many years. Here, Faben climbs a tree one-handed.

vive despite permanent disabilities. At Gombe Stream National Park, a male chimpanzee named Faben contracted polio, which left him completely paralyzed in one arm. He managed to feed himself, traverse the steep slopes of the community's home range, keep up with his companions, and even climb trees (Figure 13.32). We might also question the premise that caretaking originates with the Neanderthals. There are several examples of solicitous treatment of blind and physically impaired infant monkeys and adoption of orphaned monkeys and apes. If these primates are capable of caring for one another, it seems likely that early hominids were as well.

Further Reading

Conroy, Glenn C. 1997. *Reconstructing Human Origins: A Modern Synthesis*. W. W. Norton, New York.

Klein, R. 1999. *The Human Career*. 2nd ed. University of Chicago Press, Chicago.

Rightmire, G. P. 1992. *Homo erectus*: ancestor or evolutionary side branch? *Evolutionary Anthropology* 1:43–49.

Rightmire, G. P. 1998. Human evolution in the Middle Pleistocene: the role of *Homo heidelbergenis*. *Evolutionary Anthropology* 6:218–227.

Stringer, C., and C. Gamble. 1993. *In Search of the Neanderthals*. Thames and Hudson, New York.

Walker, A., and Shipman, P. 1996. *The Wisdom of the Bones. In Search of Human Origins*. Alfred A. Knopf, New York.

Study Questions

1. Which derived features are shared by modern humans and *Homo erectus*? Which derived features are unique to *H. erectus*?
2. What does the specimen WT 15000 tell us about *H. erectus*?

3. What is a hand ax, and what did *H. erectus* use hand axes for? How do we know?

4. What is the evidence that *H. erectus* controlled fire?

5. What led some anthropologists to propose that the fossils assigned to *H. erectus* are actually the remains of individuals belonging to two species of humans?

6. Describe the variation in hominid morphology during the Middle Pleistocene. Why do some anthropologists think there was more than one species of hominid present during this period?

7. What is the crural index? What does it measure? How does the crural index of the Neanderthals differ from modern Melanesians?

8. How did the Neanderthals differ from their contemporaries in Africa and eastern Asia?

C H A P T E R   1 4

The Moderns

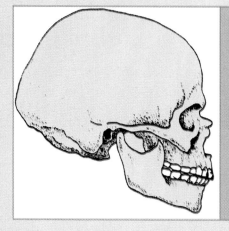

The Emergence of Anatomically Modern People
The Upper Paleolithic—the Human Revolution
    Technology and Culture
    Symbolic Behavior
    Subsistence and Social Organization
Models for the Origin and Spread of Anatomically
   Modern Humans
    Evidence from Fossils and Tool Kits
    Genetic Data

The Emergence of Anatomically Modern People

About 100,000 years ago, a new kind of human appeared. These people were more robust than any modern human population, with large bones and browridges and other primitive features. However, they shared a number of important derived features with contemporary humans that were not present in the Neanderthals or other archaic *H. sapiens*. These people are classified by anthropologists as **anatomically modern *Homo sapiens***. The qualifier "anatomically modern" emphasizes that their bodies were very similar to modern humans, but they had not yet developed the cultural traditions, symbolic behaviors, and complex technologies that we see among later peoples.

The derived features of the anatomically modern humans include (Figure 14.1):

1. *Small face with protruding chin*. The faces of anatomically modern people are much smaller than those of the Neanderthals and their contemporaries. The

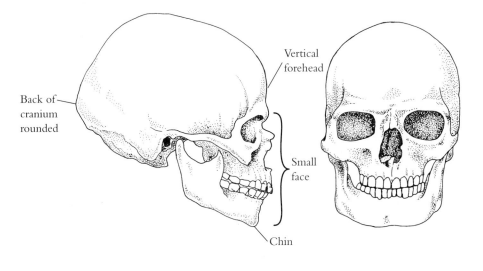

Figure 14.1
The skulls of anatomically modern humans have higher, rounder crania and smaller faces than earlier hominids, as illustrated by this skull from a man who lived about 25,000 years ago near the Don River in Russia. (Figure courtesy of Richard Klein.)

teeth and jaws of the anatomically modern *H. sapiens* were much smaller than those found in earlier hominid species, and the lower jaw bore a jutting chin for the first time (Figure 14.2). Some anthropologists believe that the smaller face and teeth were favored by natural selection because the moderns did not use their teeth as tools to the same degree that earlier people did. There has been considerable dispute about the functional significance of the chin. Some authorities contend that the chin has no function and that it arises simply as a side effect of having a smaller face. However, studies of mechanical stresses generated by chewing suggest that the protuberant chin strengthens the front of the lower jaw. In apes and early hominids, a flange on the inside of the jaw serves the same function. Some authors have argued that the reduction of the jaw in modern humans, and perhaps the mechanics of speech production, made

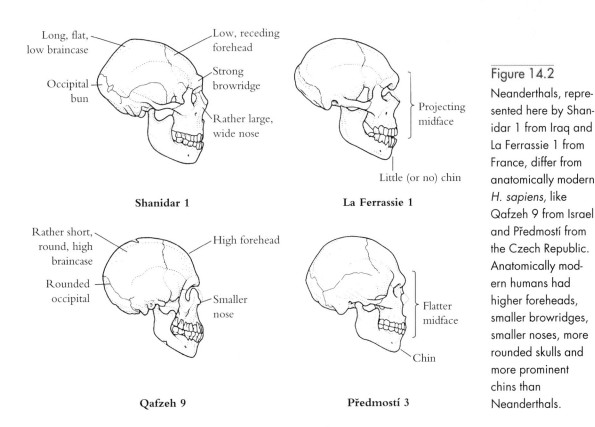

Figure 14.2
Neanderthals, represented here by Shanidar 1 from Iraq and La Ferrassie 1 from France, differ from anatomically modern *H. sapiens*, like Qafzeh 9 from Israel and Předmostí from the Czech Republic. Anatomically modern humans had higher foreheads, smaller browridges, smaller noses, more rounded skulls and more prominent chins than Neanderthals.

it advantageous to eliminate the flange inside the jaw and replace it with the chin.

2. *Rounded skull.* Modern human skulls have a high forehead, a distinctive rounded back of the cranium, and greatly reduced browridges (Figure 14.2). As we will see, anatomically modern people have left evidence of much more complex behavior than the Neanderthals did, and some anthropologists believe that these changes in the skull reflect changes in the structure of the brain that made these behaviors possible.

3. *Less robust postcranial skeleton.* The skeleton of anatomically modern humans was much less robust than Neanderthal skeletons. Anatomically modern humans had longer limbs with thinner-walled bones; longer, more lightly built hands; shorter, thicker pubic bones; and distinctive shoulder blades. Eric Trinkaus of Washington University, St. Louis, argues that these people relied less on bodily strength and more on elaborate tools and other technological innovations to do their work. Selection might have favored a more slender physique because lighter people require fewer resources to sustain themselves and to reproduce successfully.

The oldest anatomically modern humans have been found in the Middle East, but similar people may also have been living in Africa at around the same time.

The earliest anatomically modern fossils have been found at Qafzeh and Skhul caves in Israel. These fossils have been dated to 90,000 years ago using thermoluminescence and electron-spin-resonance dating techniques. There is some evidence that anatomically modern humans also lived in Africa around the same time, but the fossils are too fragmentary to be conclusive. The most extensive finds come from the Klasies River Mouth in South Africa. Excavations at this site have yielded five lower jaws, one upper jaw, a part of the forehead, and numerous smaller skeletal fragments. This site has been dated to between 134,000 and 74,000 years ago using electron spin resonance. While one of the lower jaws clearly has a modern jutting chin, and the forehead has modern-looking browridges, the fragmentary nature of the fossils makes it difficult to be sure that these are anatomically modern humans. There is also a complete anatomically modern skull from Omo-Kibish in Ethiopia, which is dated to about 130,000 years ago. As you may recall from Chapter 13, the age of this skull is uncertain because paleontologists have also found a much more primitive skull in the same soil formation only a short distance away. This has led some anthropologists to suspect that the anatomically modern skull is not 130,000 years old but much younger.

Neanderthals and anatomically modern humans seem to have coexisted in the Middle East for thousands of years.

Neanderthal fossils also have been found in Israel, at three sites named Kebara, Tabun, and Amud, which are located quite close to Qafzeh and Skhul. For many years there were no reliable absolute dates for any of these sites because they are too young for potassium-argon dating and too old for carbon-14 dating. However, most anthropologists assumed that the Neanderthals came first, and were succeeded by the anatomically modern hominids at Qafzeh and Skhul. Recently, the sites were securely dated using thermoluminescence and electron-spin-resonance methods. The results were a big surprise. The Neanderthals occupied Tabun around 110,000 years ago, some 10,000 to 20,000 years before anatomically moderns humans lived nearby. However, the Neanderthals at Kebara and Amud lived

55,000 to 60,000 years ago, which is about 30,000 years *after* anatomically modern peoples inhabited Skhul and Qafzeh.

These new dates lead to the surprising conclusion that Neanderthals and anatomically modern people coexisted for a very long time in the Middle East. They didn't necessarily live "cheek by jowl" during that time. It is more plausible that the Middle East was a boundary between the geographic ranges of the Neanderthals and the anatomically modern populations. As the climate grew colder and the ecology changed, the residents of these sites alternated. At certain times Neanderthals may have lived there, while at other times anatomically modern humans did.

The earliest anatomically modern humans seem to have had a way of life similar to that of the Neanderthals.

The earliest anatomically modern humans in the Middle East are associated with a Mousterian tool kit similar to the one used by the Neanderthals in the same region. Like the Neanderthals they buried their dead, and they subsisted by hunting the same animals. In Africa, the MSA tool kit is found at sites from this time period.

The Upper Paleolithic—the Human Revolution

Sometime between 35,000 and 45,000 years ago there was an abrupt change in tools, subsistence patterns, and symbolic expression throughout North Africa, Europe, northern Asia, and Australia.

During the Upper Paleolithic, which began 45,000 to 35,000 years ago, humans began to do a number of things that they had not done before:

- They assembled a much more elaborate tool kit.
- They made tools from a wider variety of materials, including stone, antler, ivory, and bone.
- They transported large amounts of raw material for toolmaking over long distances.
- They constructed elaborate shelters.
- They created art and ornamentation, performed ritual burials, and practiced other forms of symbolic expression.

Cambridge University archaeologist Paul Mellars has labeled this sudden, striking change in human behavior during the Upper Paleolithic "the human revolution."

The first Upper Paleolithic tools appear in the Middle East at a site in the Negev Desert of Israel called Boker Tatchit. There are four levels at Boker Tatchit, which range from 47,000 to 38,000 years old. The earliest levels contain Mousterian tools, but more recent levels contain more sophisticated tools, culminating in an Upper Paleolithic tool kit. Upper Paleolithic tools gradually spread from the Middle East to the Balkans, and arrived in France about 34,000 years ago. Upper Paleolithic industries appear in southern Asia and in Australia around this time as well. Southern Asia had long been populated by more primitive hominids, but anatomically modern humans were the first people to reach Australia.

In Europe the transition from the Mousterian to the Upper Paleolithic is associated with the appearance of anatomically modern humans and the disappearance of the Neanderthals.

The first fossils of anatomically modern humans in Europe were found in 1868 by railroad workers in southwestern France at a site known as the Cro Magnon rock shelter. This site eventually yielded bones from at least five different individuals who lived about 30,000 years ago. Since 1868, many other Upper Paleolithic sites have been discovered in Europe. The oldest sites, dating to about 40,000 years ago, are located in central Europe and northern Spain.

The Neanderthals disappear from the fossil record in western Europe around the same time that anatomically modern people appear there. We know that Neanderthals persisted at least this long because a fossil showing the classic Neanderthal features was found at St. Césaire in France. This fossil has been dated via thermoluminescence techniques to 36,000 years ago.

For a long time, anthropologists thought that the human revolution coincided with the appearance of anatomically modern humans. This is based largely on the well-documented archaeological record of the Upper Paleolithic in Europe, where it appears that a radical technical and social transformation occurred at the same time that anatomically modern humans arrived. However, this consensus was disrupted when it was learned that the anatomically modern fossils at Skhul and Qafzeh caves in Israel were 90,000 years old. This means that in the Middle East, anatomically modern peoples apparently existed for 50,000 years with no sign of the striking innovations that characterize the human revolution.

It is useful to distinguish between anatomically modern peoples who use Mousterian tools and anatomically modern peoples who practiced Upper Paleolithic industries. We will refer to the latter as Upper Paleolithic peoples.

Anatomically modern humans first entered Australia around 40,000 years ago, bringing technology similar to that of the Upper Paleolithic people in Europe.

The fossil and archaeological record in Australia is also quite good. Numerous well-dated archaeological sites indicate that anatomically modern *H. sapiens* entered Australia at least 40,000 years ago. There are three sites that may be older, including one that may be 176,000 years old, but in each case the dates are controversial. What is certain is that people occupied the entire continent by 30,000 years ago. While the fossil record for the earliest sites is poor, there are good data for sites dated to 30,000 years ago. The most complete early site is in southeastern Australia at Lake Mungo. The finds here include three hominid skulls, hearths, and ovens that have been dated to 32,000 years ago using carbon-14 dating. The skull specimens are fully modern and well within the range of variation of contemporary Australian aborigines.

The earliest Australians seem to have practiced some of the same kinds of cultural behaviors as the Upper Paleolithic Europeans did. Some of the tools found at these early sites were made from bone, there are cave paintings dated to 17,000 years ago, and there is evidence of both ceremonial burials and cremation. Around 15,000 years ago, Australians seem to have been the first people to use polished stone tools, which are made by grinding rather than flaking. In other parts of the world polished tools do not appear in the record until agriculture is introduced, about 7000 years ago.

The fact that people got to Australia at all is evidence of their technological sophistication. Much of the world's water was tied up in continental glaciers 40,000 years ago, so New Guinea, Australia, and Tasmania formed a single continent that geologists call Sahul. However, there were at least 100 km of open ocean lying between Asia and Sahul, a gap wide enough to prevent nearly all movement of terrestrial mammals across it and thus to preserve the unique, largely marsupial fauna of Australia and New Guinea. The earliest immigrants to Sahul must have known how to construct rafts or boats sturdy enough to navigate long distances in rough waters. The colonization of Sahul cannot be dismissed as a single lucky event because people also crossed another 100 km of ocean to reach the nearby islands of New Britain and New Ireland at about the same time. Thus, these peoples must have been competent seafarers.

There is little evidence for Upper Paleolithic technology in southern Asia until about 12,000 years ago.

Crude chopping tools predominated in southern Asia until recent times, and there is little sign of the elements that marked the human revolution in Europe and Australia. There is only one site in southern Asia that contains evidence of tools made from bone or antler. This site, the Batadomba Iena Cave in Sri Lanka, is dated to about 28,000 years ago, and it has yielded anatomically modern human remains, elaborate stone tools, and tools made from bone. There is no evidence of symbolic activity or ritual burials.

Negative evidence of this kind is hard to interpret. It could be that the human revolution arrived late in southern Asia. On the other hand, this part of Asia is much less well explored than Europe and Australia, and there may be many sites that have not been discovered. Since humans must have reached Sahul from southern Asia, it seems reasonable to assume that Upper Paleolithic sites lie hidden.

Technology and Culture

Upper Paleolithic peoples manufactured blade tools, which made efficient use of stone resources.

During this period, humans shifted from the manufacture of round flakes to the manufacture of blade tools. **Blades** are stone flakes that look like modern knife blades—they are long, thin, and flat and have a sharp edge. Blades have a much longer cutting edge than flakes do, so blade technology made more efficient use of raw materials than older tool technologies did. Archaeologists measure efficiency in terms of the length of the cutting edge per pound of raw materials used. The values in Table 14.1 indicate that the blade technology of Upper Paleolithic *H. sapiens* was much more efficient than that of preceding tool technologies. However, there was a cost: although blades made more efficient use of materials, they also took more time to manufacture, requiring more preparation and more finishing strokes.

The Upper Paleolithic tool kit includes a large number of distinctive, standardized tool types.

Upper Paleolithic peoples made many more kinds of tools than earlier hominids did. Chisels, various types of scrapers, a number of different kinds of points, knives, burins (pointed tools used for engraving), drills, borers, and throwing sticks are just some of the items from this tool kit.

Even more striking, the different tool types have distinctive, stereotyped shapes. It is as if the Upper Paleolithic toolmakers had a sheaf of engineering drawings on

Table 14.1 Upper Paleolithic blade technology made more efficient use of raw materials than earlier tool technologies did because blades had longer cutting edges per pound of raw material.

PERIOD	DATE OF APPEARANCE (YEARS AGO)	EFFICIENCY (IN./LB)
Upper Paleolithic	40,000	120
Mousterian	130,000	40
Acheulean	1,400,000	8
Oldowan	2,400,000	2

which they recorded their plans for various tools. When the toolmaker needed a new 4-in. burin, for example, she would consult the plan and produce one just like all the other 4-in. burins. Of course, toolmakers didn't really use drawings like modern engineers do, but the fact that we find standardized tools suggests that they carried these plans in their minds. The final shape of Upper Paleolithic tools was not determined by the shape of the raw material; instead, the toolmakers seem to have had a mental model of what the tool was supposed to look like, and imposed that form on the stone by careful flaking. This reached an apogee in the Solutrean tool tradition, which predominated in southern France between 21,000 and 16,500 years ago (Figure 14.3). Solutrean toolmakers crafted exquisite blades shaped like laurel leaves that were sometimes a foot long (28 cm), half an inch thick (1 cm), and perfectly symmetrical.

Upper Paleolithic people were also the first to shape tools from bone, antlers, and teeth. It seems clear that earlier hominids used bone for some purposes, but Upper Paleolithic people transformed bone, ivory, and antler into barbed spear points, awls, sewing needles, and beads.

Upper Paleolithic industries also varied in time and space. The first Upper Paleolithic industry, the Aurignacian, was widespread in Europe by 35,000 years ago (Figure 14.4). It is characterized by certain types of large blades, burins, and bone points. About 27,000 years ago the Aurignacian was replaced in southern France by a new tool kit, the Gravettian, in which small, parallel-sided blades predominated, and bone points were replaced by bone awls. Then about 21,000

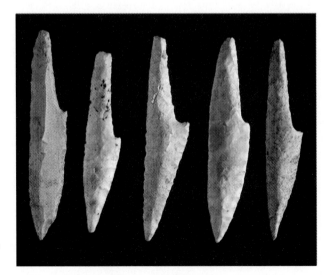

Figure 14.3

Long, thin, and delicate blades characterized the Solutrean tool tradition.

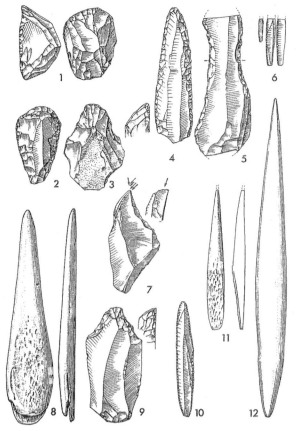

Figure 14.4

The Aurignacian tool kit, the earliest of several tool industries in the Upper Paleolithic of Europe, contains a wide variety of standardized tool types: 1, 2, 3, 9, scrapers; 4, 5, edged, retouched blades; 6, 10, bladelets; 7, a burin (a tool now used for engraving); 8, 11, bone points.

years ago, the Solutrean, with its beautiful leaf-shaped points, developed in the region. And 16,500 years ago, the Solutrean gave way to the Magdalenian, a tool kit dominated by carved, decorated bone and antler points. Other parts of Europe are characterized by different sequences of tool complexes so that, after the Aurignacian, each region typically has a distinctive material culture. Over the 23,000 years of the Upper Paleolithic in Europe there were dozens of distinctive tool kits. Contrast this variation with the Acheulean toolkit, which remained unchanged for over a million years through more than half of the Old World.

Stones and other raw materials for toolmaking were often transported hundreds of kilometers from their place of origin.

At Bacho Kiro, a 40,000 year old Aurignacian site in Bulgaria, more than half of the flint used to make blades was brought from a source 120 km (78 miles) away. Distinctive, high-quality flint quarried in Poland has been found in archaeological sites more than 400 km (248 miles) away. Seashells, ivory, soapstone, and amber used in ornaments were especially likely to be transported long distances. By comparison, as we saw in Chapter 13, most of the stone used at a Mousterian site in France was transported less than 5 km (3 miles). The long-distance movements of these resources may mean that Upper Paleolithic peoples ranged over long distances, or that they traded for these materials.

During the Upper Paleolithic, the atlatl and later the bow and arrow were developed.

Upper Paleolithic people invented new and more effective weapons. The **atlatl** (spear-thrower) is a device that greatly increases the range of a thrown spear (Figure

Figure 14.5

The spear-thrower, or *atlatl*, length-ens the arm and permits the spear to be thrown with more force over a longer distance. (Figure courtesy of Richard Klein.)

14.5). Modern demonstrations show that a spear-thrower can increase the range of a short spear from about 60 m (197 ft), when thrown by hand, to more than 150 m (492 ft). The oldest atlatls are dated to around 14,000 years ago. The first conclusive evidence for the use of the bow and arrow comes from an Upper Paleolithic site in northern Germany where 100 exquisitely crafted, split pine arrows were found and dated to about 10,000 years ago. However, since both bows and arrows are made of materials that do not preserve very well, it seems possible that these weapons were developed much earlier. This possibility is supported by indirect evidence. Small stone points like those used as arrow points by later peoples became widespread through most of the Old World around 20,000 years ago. In addition, short bone shafts that are very similar to the foreshafts of arrows used by later peoples appeared in southern Africa around the same time.

Symbolic Behavior

There is good evidence for ritual burials during the Upper Paleolithic period.

Like the Neanderthals, Upper Paleolithic peoples frequently buried their dead; unlike the Neanderthals, Upper Paleolithic burials appear to have been accompanied by ritual. Upper Paleolithic sites provide the first unambiguous evidence of both multiple burials and burials outside of caves. Upper Paleolithic people frequently buried their dead with tools, ornaments, and other objects that suggest they had some concept of life after death. Figure 14.6 diagrams the grave of a child who died about 15,000 years ago at the Siberian site of Mal'ta. Several kinds of items were buried with the child, including a necklace, a crown (diadem), a figurine of a bird, a bone point, and a number of stone tools.

Upper Paleolithic peoples were skilled artisans, sculpting statues of animals and humans, and creating sophisticated cave paintings.

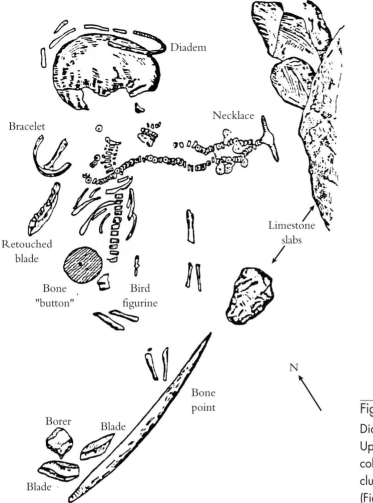

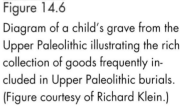

Figure 14.6

Diagram of a child's grave from the Upper Paleolithic illustrating the rich collection of goods frequently included in Upper Paleolithic burials. (Figure courtesy of Richard Klein.)

It is their art that distinguishes Upper Paleolithic peoples most dramatically from the hominids that preceded them. They engraved decorations on their bone and antler tools and weapons, and sculpted statues of animals and female figures (Figure 14.7). The female statues are generally believed to be fertility figures because they usually emphasize female sexual characteristics. Upper Paleolithic peoples also adorned themselves with beads, necklaces, pendants, and bracelets, and may have decorated their clothing with beads.

Although these artistic efforts are remarkable, it is their cave art that seems most amazing now. Upper Paleolithic peoples painted, sculpted, and engraved the walls of caves with a variety of animal and human figures. They used natural substances that included red and yellow ochre, iron oxides, and manganese to create a variety of colors for paint. They used horsehair, sticks, and their fingers to apply the paint. Their cave paintings frequently depict animals that they must have hunted: reindeer, mammoths, horses, and bison. Some figures are half human and half animal. Sometimes the creators incorporated the natural contours of the cave walls in their work, and sometimes they drew one set of figures on top of others. Some of their work is spectacular—the perspective is accurate, the characterization of behavior is lifelike, and the scenes are complex. We don't know precisely why or under what circumstances these cave paintings were made, but they represent a remarkable cultural achievement that sets the people of this period apart from earlier hominids.

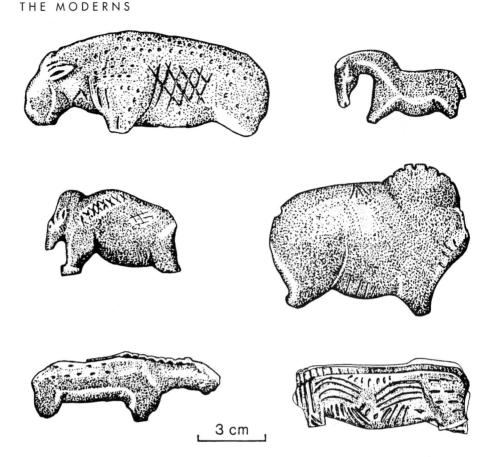

Figure 14.7
Small animal figures were carved from mammoth tusks during the Aurignacian period.

3 cm

Recently developed techniques for dating pigments have allowed scientists to determine the age of the Upper Paleolithic cave paintings for the first time. Many of the famous caves, such as Lascaux, are dated to the very end of the Upper Paleolithic, around 12,000 years ago, but spectacular paintings in the Le Chauvet Cave in France are about 30,000 years old (Figure 14.8). Most of the famous carved figurines are also less than 20,000 years old, but there are good examples of representational sculpture that date to the earliest Aurignacian. For example, Figure 14.9 shows an ivory figure with a human body and a lion's head that was found at a site in Germany dated to 32,000 years ago. There is also abundant evidence of the manufacture of beads, pendants, and other body ornaments at Aurignacian sites in France. The beads are standardized and often manufactured from materials that had been transported hundreds of kilometers. Archaeologists have found a well-preserved musical instrument that looks like a flute at an Aurignacian site in southwestern France. It is 10 cm (4 in.) long and has four holes on one side and two on the other. When this instrument was played by a professional flutist, it produced musical sounds.

Subsistence and Social Organization

During the Upper Paleolithic, Europe was a cold, dry grassland that supported a diverse population of herbivores.

The Upper Paleolithic spanned the depths of the last ice age. The area in Europe that contains the richest archaeological sites was a cold, dry grassland—something like the contemporary Arctic but without the long periods of winter darkness that

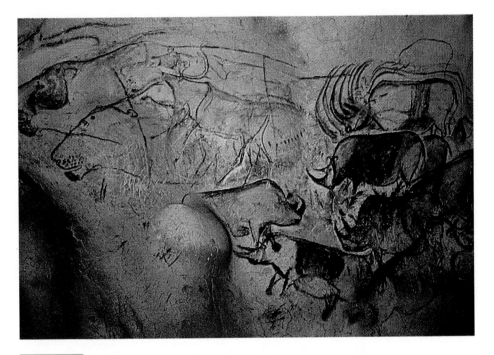

Figure 14.8
This image from Le Chauvet in France was created about 30,000 years ago. At the upper left, several lions are depicted. On the other side, there are a number of rhinoceros. At the top right, the multiple outlines suggest a rhinoceros in motion.

Figure 14.9
This ivory figure depicts a human body with a lion's head. It comes from an Aurignacian site in southern Germany and is 32,000 years old.

Why Paleolithic Art?

Pigments of the Imagination Of all the artifacts left behind by prehistoric people, painted images on rock shelters and cave walls are among the most evocative. They speak to the Western mind of religion and ritual, something deeply important to an ancient mythology. Yet the images evade ready interpretation. In recent years chemistry and physics have come to the archeologists' aid in establishing some of the context of the paintings.

The first direct dating of European cave paintings, for instance, using the accelerator mass spectrometry radiocarbon technique on charcoal particles in the pigment, was reported by a French research team in 1990. Since then some of the most important cave art sites have been examined using the same approach, although it will be many years before a significant proportion of the paintings has been dated. Precise dating of painted images may establish a chronology of Upper Paleolithic people and their work; moreover, it will eventually allow an insight into stylistic variability, which may be related to cultural expression.

Until direct dating of images was possible, indirect methods were used, such as, where feasible, the dating of material in occupation layers thought to be associated with the paintings. The French prehistorians Breuil and Leroi-Gourhan each developed chronologies of the painted caves based on their interpretation of the change of painting style through time. The uncertainty of this approach was revealed in 1992 when a team of French and Spanish scientists published radiocarbon dates on stylistically similar images of bison from the caves of Altamira and El Castillo, in Spain; and for a stylistically different bison in the cave of Niaux, in the French Pyrenees. The dates were, respectively, 14,000, 12,990, and 12,890 years. Clearly, age and style do not always coincide.

The analysis of pigments has gone beyond dating. Jean Clottes, the French government's director of prehistoric research in the Pyrenees, in collaboration with Michel Menu and Philippe Walter of the research laboratory at the Louvre Museum, Paris, gained some insight into the technology of Ice Age painting. Among their discoveries is that while the technology of paint production may change through time, style may persist.

The cave most studied so far is Niaux, in the limestone massif of the Vicdessos Valley, which is counted among the half-dozen finest examples of Ice Age art in Europe. The painted images in the virtually horizontal galleries, which extend more than 2 kilometers, were discovered in 1906 and have since been the subject of considerable scrutiny. The most decorated section of the cave system, the Salon Noir, is a side chamber—a soaring vault far from daylight where mainly black images of bison, horses, deer, and ibexes decorate the walls, arranged in panels and giving the impression of foresight and deliberation in their execution.

These bison images in the Salon Noir of Niaux Cave, Ariège, France, were done by Late Magdalenian artists, about 12,000 years ago.

The black pigment used at Niaux has long been thought to include a mixture of charcoal and manganese dioxide. Clottes and his colleagues recently applied modern analytical techniques to 59 minute samples of black and red pigment from many images throughout the cave, and from the adjoining cave of Réseau Clastre. Using a scanning electron microscope attached to an X-ray detector (to identify specific minerals) and proton-induced X-ray emission (to determine the complete elemental composition), the researchers obtained a detailed profile of the chemical components of the paint. Although all are naturally occurring and local, their combination in the paints is clearly man-made.

The painting materials—pigments and mineral extenders—were carefully selected by Upper Paleolithic people and ground to within 5 to 10 micrometers to produce a specific mix. The black pigment, as had been suspected, was charcoal and manganese dioxide. But the real interest was in the extenders, of which there seemed to be four distinct recipes, which the researchers number one through four. Extenders help to bring out the color of the pigment and, as their name implies, add bulk to the paint without diluting the color. The four recipes for extenders used at Niaux were talc; a mixture of baryte and potassium feldspar; potassium feldspar alone; and potassium feldspar mixed with an excess of biotite. Clottes and his colleagues experimented with some of these extenders and found them to be extremely effective.

In the Salon Noir most of the paintings were made with a mixture of charcoal and manganese dioxide, but it was unlike any other such combination in the cave. The particles of charcoal were unusually large and were consistent with a charcoal stick being used to sketch an outline. On top of this initial sketch was a paint composed of manganese dioxide and one type of extender, potassium feldspar and biotite (recipe four). By examining the chemical profile of minor constituents in the paint, the researchers discovered that in one panel that was examined in some detail, several different "paint pots" must have been used. While the recipe for the extender was identical throughout the panel, which consists of three bison, two different paint mixes had been employed. Clottes and his colleagues speculate that the panel may have been painted in more than one session, or perhaps worked on by different painters who followed the same recipe but obtained their components from different locations.

The overwhelming majority of paintings in the Salon Noir included recipe four, but the other three types of extender are present, too. Clottes believes that the fact that most of the paintings used recipe four implies a coherence in their execution, possibly in time but certainly in cultural context. He suggests that the ritual that undoubtedly attended the production and use of images deep underground may have extended to the manufacture of the paints themselves.

Preliminary dating of some of the images throughout the cave shows that different extenders were used at different times: those used between 14,000 and 13,000 years ago included feldspar, while later recipes, up to 12,000 years ago, favored large quantities of biotite. Whether practical issues, such as ready availability, or ritual concerns governed this shift is unknown. It is surely significant, however, that the style of the paintings done through that time remains remarkably constant.

The principal conclusion concerning the Salon Noir is that it should be regarded as an important sanctuary: the outline sketch of the panels, followed by their full execution, speaks of a premeditation absent from most of the other images, often deeper in the cave. Such premeditation is consistent with a ritual ceremony of considerable importance, suggests Clottes. Paintings made without a preliminary sketch could have been done very quickly, as the French prehistorian Michel Lorblanchet has demonstrated by experimental replication of some paintings. Produced during a single visit to a deep part of a cave, perhaps never to be revisited, such images may nevertheless have played an important role in Upper Paleolithic mythology. Chemical analysis may help define the parameters of how and when images were painted, but even these powerful techniques fail to answer the most interesting question of all: what did the images mean to Upper Paleolithic people?

occur at high latitudes. Contemporary tundra is dominated by unproductive mosses that are adapted to poorly drained Arctic soils. In contrast, the cold, dry grasslands of Europe supported large populations of a diverse assemblage of large herbivores including reindeer (called caribou in North America), horse, mammoth, bison, woolly rhinoceros, and a variety of predators, such as cave bears and wolves.

Modern humans exploited a wider range of prey species than the Neanderthals did, but the subsistence economy of the two populations was similar.

Bones found at Upper Paleolithic sites indicate that large herbivores played an important role in Upper Paleolithic peoples' diets. Like the Mousterian peoples, Upper Paleolithic humans were proficient hunters of large game. In some places they concentrated on a single species—reindeer in France, red deer in Spain (Figure 14.10), bison in southern Russia, mammoths farther north and east, and eland or buffalo in southern Africa. There are also places where they harvested several different kinds of animals. But everywhere, Upper Paleolithic peoples seem to have hunted herbivores that lived in large herds, fished, and hunted birds. In some areas like the southern coast of France, rich salmon runs may have been an important source of food.

A comparison of Middle Stone Age (MSA) and **Later Stone Age (LSA)** deposits at the Klasies River Mouth in South Africa provides another example of these contrasts. (Remember that MSA tools found in Africa are similar to Mousterian tools found in Europe, and LSA tools from Africa are similar to Upper Paleolithic tools from Europe.) The MSA people who occupied this area between 130,000 and 115,000 years ago hunted large game, particularly eland. They also consumed other species, mainly buffalo, bushpigs, and Cape fur seals. As we saw in Chapter 13, the best evidence suggests that these contemporaries of the Neanderthals were proficient hunters. The LSA people who occupied the same area around 20,000 years ago also hunted large game, although they concentrated more

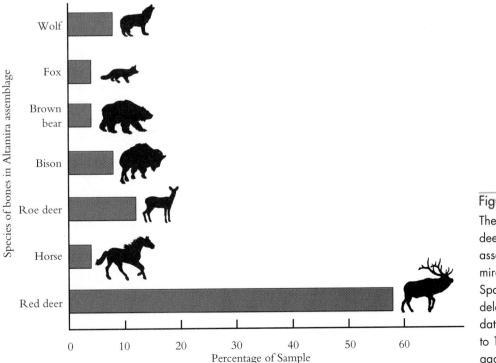

Figure 14.10
The bones of red deer dominate the assemblage at Altamira in northern Spain. This Magdelenian site is dated from 15,000 to 13,000 years ago.

on buffalo and bushpig, species that are more common but also more dangerous than eland. LSA humans hunted fur seals, but unlike their MSA predecessors, who hunted seals throughout the year, they mainly killed young seal pups. This fact suggests that LSA peoples hunted seals only during the period when the mothers and pups were on the beach and the pups were just about to be weaned. The presence of artifacts that archaeologists interpret as sinkers and fish gorges suggests that LSA peoples also fished.

The technology and culture of Upper Paleolithic peoples seem to have enabled them to cope with their environment better than the Neanderthals did.

The richness of the fossil and archaeological record in Europe allows a detailed comparison of Neanderthal and Upper Paleolithic peoples. These data suggest that Upper Paleolithic people were better able to cope with their environment than the Neanderthals were. Three kinds of data support this conclusion:

1. *Upper Paleolithic peoples lived at higher population densities in Europe.* Archaeologists estimate the relative population size of vanished peoples by comparing the density of archaeological sites per unit time. Thus, if one group of people occupied a particular valley for 10,000 years and left 10 sites, while a second group occupied the same valley for 1000 years and left 5 sites, then archaeologists would estimate that the second group had five times as many people. (In using this method it is important to be sure that the sites were occupied for approximately the same lengths of time.) By these criteria, Upper Paleolithic peoples had far higher population densities in Europe than the Neanderthals did.

2. *Upper Paleolithic peoples lived longer than Neanderthals.* Anthropologists have estimated that Late Pleistocene men sometimes reached 60 years of age, but women rarely reached 40, and childhood mortality was very high. Nonetheless, the life expectancy of these people was substantially greater than that of the Neanderthals, who rarely reached 40.

3. *Upper Paleolithic peoples were less likely to suffer serious injury or disease than were Neanderthals.* In sharp contrast to the Neanderthals, the skeletons of Upper Paleolithic people rarely show evidence of injury or disease. The rare injuries include a child buried with a stone projectile point imbedded in its spine, and a young man with a projectile point in his abdomen and a healed bone fracture on his right forearm. There are a few more instances of disease than injury among the remains of Upper Paleolithic peoples, including a young woman who probably died as the result of an abscessed tooth and a child whose skull seems to have been deformed by hydrocephaly (a condition in which fluid accumulates in the cranial cavity and the brain atrophies). Nonetheless, there is still less evidence for disease than exists among Neanderthal remains.

The peoples of the Upper Paleolithic developed more complex forms of shelter and clothing than Neanderthals did.

In what is now western and central Europe, Russia, and the Ukraine, the remains of small villages have been found. Living on a frigid, treeless plain, Upper Paleolithic peoples either hunted or scavenged mammoth and used the hairy beasts for food, shelter, and warmth. At the site of Předmostí in the Czech Republic, the remains of almost 100,000 mammoths have been discovered. Huts were constructed by arranging mammoth bones in an interlocking pattern and then draping them with skins. (Temperatures of −50°F outside provide a strong incentive for chinking the cracks!) The huge quantities of bone ash that have been found at these sites indicate that they also used the mammoth bones for fuel. A site about 470 km (214 miles) southeast of Moscow contains the remains of even larger shelters. They were built around a pit about a meter deep and covered with hides supported by mammoth bones. Some of these huts had many hearths, suggesting a number of families may have lived together (Figure 14.11).

Several lines of evidence indicate that Upper Paleolithic peoples living in glacial Europe manufactured fur clothing. First, when modern hunters skin a fur-bearing animal, they usually leave the feet attached to the pelt and discard the rest of the carcass. There are numerous skeletons of foxes and wolves complete except for their feet at several Upper Paleolithic sites in Russia and the Ukraine, suggesting that early modern humans kept warm in sumptuous fur coats. Second, bone awls and bone needles are common at Upper Paleolithic sites, so sewing may have been a common activity. Finally, three individuals at a burial site in Russia seem to have been buried in caps, shirts, pants, and shoes decorated with beads.

Some anthropologists believe that ethnic differentiation among humans began during the Upper Paleolithic.

The defining feature of an ethnic group is that people identify themselves as members of the group based on culturally defined features or markers. Some of these markers are symbolic characters, like linguistic differences, clothing styles, or distinctive patterns of ornamentation on pottery or other objects. Other markers are basic values, moral beliefs, or lifeways. For example, Fredrik Barth, a Norwegian anthropologist, studied three neighboring ethnic groups in the mountains

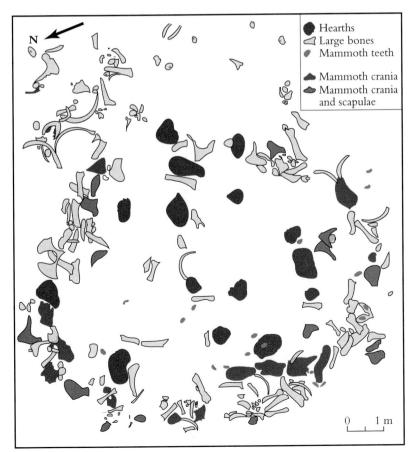

Figure 14.11

On the Russian plain, humans had to cope with harsh climatic conditions. They may have used mammoth bones to construct shelters. Mammoth bones and teeth litter this Upper Paleolithic site of Moldova. The presence of several hearths suggests this site may have been occupied by several families.

of Pakistan. One occupied the low-altitude plain and did intensive farming; a second group lived in the higher river valleys and practiced a mix of less intensive agriculture and seasonal herding; the third group lived as nomadic herders all year, moving from high to low altitude depending on the season.

Several aspects of the human revolution have led some anthropologists to hypothesize that ethnic groups first appeared during the Upper Paleolithic. First, the rapid change in tool industries and the coexistence of distinctive industries in different places suggest that, for the first time, different traditions were maintained among genetically similar people. Second, Upper Paleolithic sites show stylistic differences in utilitarian objects like stone tools. Finally, the increased degree of ecological specialization during the Upper Paleolithic is also consistent with more cultural subdivision among people living in close geographical proximity.

The social organization of contemporary foragers suggests that the Upper Paleolithic hunters lived in egalitarian kin-based bands in which food was shared and there was no formal political organization.

In the Upper Paleolithic there is little doubt that people were fully modern in their morphology. This does not necessarily mean that the behavior of Upper Paleolithic peoples was the same as that of modern peoples, if only because the ecological conditions of the last glacial period have no modern parallel. On the other hand, the behavior of modern foragers provides our best guess about what the social organization of Upper Paleolithic peoples might have resembled.

As we described in Chapter 12, the societies of modern food-foraging peoples share several general features:

1. People live in small- to medium-sized kin-based bands with their parents, siblings, cousins, and in-laws.
2. There is extensive sharing and reciprocity, particularly of food.
3. Men and women do different subsistence tasks—typically, men hunt and women gather.
4. There is little formal political organization. There are no hereditary chiefs, and there is little formal leadership, although some people may be particularly influential because of their skill or persuasiveness.

The human revolution may have been the result of either biological or cultural change.

The human revolution presents anthropologists with a fascinating conundrum. The human revolution is one of the most profound transformations of human behavior that has ever occurred. Yet it does not seem to be associated with any observable biological change in the human species. Anatomically modern *H. sapiens* manufactured both Mousterian and Upper Paleolithic tools.

There are two possible solutions to this puzzle. First, the human revolution may have been the result of some genetically based change in anatomically modern humans. For example, it seems likely that the human capacity for language is based on special-purpose structures in the brain that facilitate specific aspects of language processing. These structures can't be detected in the fossil record. If they evolved after anatomically modern peoples appeared, then this might explain why changes in skeletal morphology and behavior were not coupled.

The second possibility is that the human revolution resulted from cultural, not genetic, changes, and that it took anatomically modern humans 55,000 years to assemble the elements of the human revolution. We know that something like this happened later in human history. Around 7000 years ago an equally profound transformation was associated with the adoption of agriculture. Agriculture led to sedentary villages, social inequality, large-scale societies, monumental architecture, writing, and many other innovations. No one suggests that the agricultural revolution was the result of genetic changes among humans. It could be that the human revolution 40,000 years ago was the first cultural revolution.

Models for the Origin and Spread of Anatomically Modern Humans

The origin of anatomically modern humans is controversial. Most views fall on a continuum between two extreme hypotheses: the replacement model and the multiregional model.

Most paleoanthropologists agree that a dramatic shift in hominid morphology occurred during that last glacial epoch. About 100,000 years ago the world was inhabited by a morphologically heterogeneous collection of hominids: Neanderthals in Europe; less robust archaic *H. sapiens* in East Asia; and somewhat more modern humans in the Middle East and perhaps in Africa. By 30,000 years ago, much of this diversity had disappeared—anatomically modern humans occupied all of the Old World.

In order to understand how this transition occurred, we need to answer two questions:

1. Did the genes that give rise to modern human morphology arise in one region, or in many different parts of the globe?
2. Did these genes spread from one part of the world to another by gene flow, or through the movement and replacement of one group of people by another?

Unfortunately, genes don't fossilize, and we cannot study the genetic composition of ancient hominid populations directly. However, there is a considerable amount of evidence that we can bring to bear on these questions.

The discussion of modern human origins is sometimes framed in terms of two competing hypotheses: the replacement model and the multiregional model (Figure 14.12). According to the **replacement model**, hominid populations were [*single origin model*] genetically isolated from each other during the middle Pleistocene. As a result, different populations of *H. erectus* and archaic *H. sapiens* evolved independently, perhaps forming several hominid species. Then, between 100,000 and 200,000 years ago, anatomically modern people arose someplace in Africa and spread outward, replacing other archaic *H. sapiens* including the Neanderthals. The replacement model does not specify how anatomically modern *H. sapiens* usurped local populations of archaic *H. sapiens* in Europe, the Middle East, and Asia. However, the model posits that there was little or no gene flow as anatomically modern *H. sapiens* replaced local populations of Neanderthals and archaic *H. sapiens* in those areas.

According to the **multiregional model**, human populations throughout the world were linked by gene flow, causing anatomically modern humans to evolve as a single species from *H. erectus* into modern *H. sapiens* all over the Old World. However, since these populations were separated by great distances and experi-

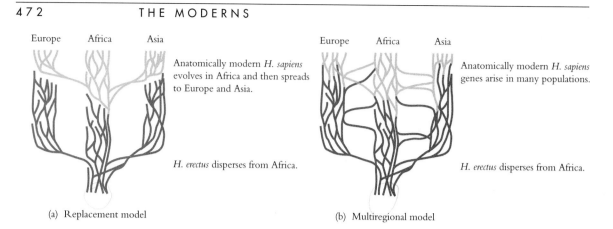

Europe   Africa   Asia

Anatomically modern *H. sapiens* evolves in Africa and then spreads to Europe and Asia.

H. erectus disperses from Africa.

(a) Replacement model

Europe   Africa   Asia

Anatomically modern *H. sapiens* genes arise in many populations.

H. erectus disperses from Africa.

(b) Multiregional model

Figure 14.12

The two main hypotheses about the origin of modern humans are diagrammed here. Both models assume that each region had several populations linked by gene flow. According to (a) the replacement hypothesis, there was no gene flow between regions, while according to (b) the multiregional model there was sufficient gene flow between regions to allow beneficial genes to spread throughout the human range but not enough to eliminate regional differences. These two models represent the opposite poles of opinion; several intermediate hypotheses have also been proposed.

enced different kinds of environmental conditions, there was considerable regional variation in morphology among them. Because a continuous line of descent can be traced from anatomically modern humans back through the transitional forms to *H. erectus,* modern humans in each region resemble the archaic *H. sapiens* from the same region.

The replacement hypothesis says that all the genes necessary for the modern human phenotype were assembled in Africa, and spread as people carrying those genes migrated out of Africa and replaced local peoples. This model is sometimes referred to as the **Out of Africa** hypothesis. The multiregional hypothesis contends that the genes creating the modern human phenotype arose in all parts of the world and were "mingled together" as people from different regions mated, thus exchanging genetic material.

Although these two hypotheses dominate the debate over the origins of modern humans, there are also several intermediate models. For example, according to one hypothesis, hominid populations in Asia and Africa were linked by gene flow, but Neanderthals were isolated in Europe. When modern humans entered Europe, they replaced Neanderthals with little gene flow. Another widely supported hypothesis holds that anatomically modern morphology evolved in Africa and spread throughout the world as anatomically modern populations moved out of Africa. However, there was extensive gene flow between this expanding population and the existing populations of archaic *H. sapiens* in each region. As a consequence, some of the genes of Neanderthals and archaic *H. sapiens* may still exist in modern populations. A third hypothesis holds that anatomically modern humans evolved in Africa, and then their genes diffused to the rest of the world by gene flow, not by the migration of anatomically modern humans and replacement of local peoples.

We will try to sketch the evidence that bears on the debate over modern human origins. We say "try" because this debate is extremely contentious, and virtually every relevant piece of evidence is challenged by advocates of one view or the other. As a result, it is nearly impossible to make an uncontroversial statement about any aspect of this question.

Evidence from Fossils and Tool Kits

A considerable body of evidence is inconsistent with the multiregional hypothesis for Europe and the Middle East.

The long period of overlap between Neanderthals and anatomically modern humans in the Middle East directly contradicts the multiregional hypothesis for this part of the world. If anatomically modern people appeared 60,000 years before Neanderthals disappeared, then it can hardly be true that all Neanderthals evolved into anatomically modern humans in that area. In fact, the foremost champion of the multiregional hypothesis, University of Michigan anthropologist Milford Wolpoff, denies that the Skhul and Qafzeh fossils are really anatomically modern humans. Instead, he believes that they and the more robust fossils from Kebara, Amud, and Tabun are part of a variable population that is unrelated either to European Neanderthals or to anatomically modern human populations. However, most anthropologists believe that Skhul and Qafzeh fossils are anatomically modern because the skulls from these sites have relatively high, rounded crania and small faces, while a cranium found at Tabun is very similar to that of European Neanderthals. Similarly, postcranial fossils from Qafzeh are completely modern, while the pelvis and femur from Tabun show all the features of classical Neanderthals.

The speed of the shift from Neanderthals to anatomically modern humans in western Europe is also difficult to reconcile with the multiregional hypothesis. The most recent Neanderthal yet found is the St. Césaire fossil, dated to about 36,000 years ago, while the earliest anatomically modern human in western Europe is dated to about 30,000 years ago. Thus, if the multiregional hypothesis were correct for western Europe, then Neanderthals would have to have evolved into anatomically modern humans in less than 6000 years!

Additional support for the replacement model comes from Cambridge University archaeologist Paul Mellars, who argues that the archaeological data indicate that Neanderthals and anatomically modern people overlapped in Europe, as well as in the Middle East. Recall that the earliest anatomically modern fossils in Europe are associated with the Aurignacian tool kit. Moreover, no Aurignacian tools have been found with Neanderthal fossils. Aurignacian tools appear in Europe about 40,000 years ago—4000 years before the Neanderthals disappeared. This brief overlap suggests that the anatomically modern humans using Aurignacian tools appeared and rapidly replaced Neanderthals. This transition seems too rapid to be the result of genetic change.

However, the situation is not quite so simple. In southern France, the transition between the Mousterian and the Aurignacian is marked by the presence of a third industry, called the **Châtelperronian**, which is intermediate between the Mousterian and the Aurignacian. Proponents of the multiregional hypothesis interpret this tool kit as evidence for the gradual transition from Neanderthal to anatomically modern human, while defenders of the replacement hypothesis argue that the Châtelperronian is the result of Neanderthals borrowing ideas and technology from the moderns. Mellars argues that three kinds of evidence support the latter view. First, if the Châtelperronian is associated with the biological transformation of the hominids living in Europe, then it should be associated with morphologically intermediate fossils. However, the association of Neanderthal fossils with Châtelperronian tools at St. Césaire and Arcy-sur-Cure in France contradicts this prediction (Box 14.1). Second, archaeological data indicate that the Châtelperronian and Aurignacian industries coexisted in southern France for hundreds of years.

BOX 14.1

Arcy–sur–Cure and the Causes of the Human Revolution

The Grotte du Renne, a site at the French town of Arcy-sur-Cure, provides important clues about whether the human revolution resulted from biological or cultural change in human populations. When the site was excavated between 1947 and 1963, archaeologists uncovered many layers that showed evidence of human activity. The lowest and therefore oldest levels yielded many Mousterian tools, and the highest and most recent levels contained Upper Paleolithic artifacts. In between, archaeologists found Châtelperronian tools and hominid fossils, including a small part of a cranium. More significantly, they also unearthed a large number of worked bone tools and 36 personal ornaments, artifacts typically associated with Upper Paleolithic peoples (Figure 14.13). However, it was not clear whether the hominids should be classified as Neanderthal or as anatomically modern humans. Thus, it was not clear what kind of hominid made the ornaments and bone tools.

Recently, two clever bits of detective work have shown that the tools and ornaments were made by Neanderthals. The first piece of the puzzle was solved by Fred Spoor and his colleagues at the University of Liverpool who were able to show that the hominids associated with the artifacts were Neanderthals. They used **computed tomography** (or **CT** scan) to show that the shape of the semicircular canals of the inner ear differed between Neanderthals and

Figure 14.13

These personal ornaments found at Arcy-sur-Cure were likely manufactured by Neanderthals living at the site about 33,000 years ago. The fact that Neanderthals could learn to make and use personal ornaments suggests that the absence of ornaments at most Mousterian sites is not due to some biologically transmitted, cognitive deficit among Neanderthals.

modern humans. Using this diagnostic feature, they were able to establish that the Grotte du Renne cranium was from a young Neanderthal child. This means that bone tools and ornaments were used at a site *occupied* by Neanderthals. It was still possible, however, that the tools and ornaments were manufactured by anatomically modern humans and subsequently acquired by Neanderthals, perhaps by trade or by scrounging. The second piece of the puzzle was provided by a French team who completed a detailed reanalysis of the Grotte du Renne record. They discovered evidence indicating that the tools and ornaments were manufactured at the site while Neanderthals were living there. One piece of bone was clearly the product of an unsuccessful effort to make an awl (a sharp tool used to punch holes in leather), the most common bone tool found at the site. There were also several ornaments made from the ends of swan bones along with the matching piece of bone from which the ornament was cut. Thus, we are now reasonably certain that both Neanderthals and anatomically modern humans carved ornaments and tools out of bone.

This conclusion alters our thinking about the causes of the human revolution. Some believe that Neanderthals and other Middle Paleolithic peoples lacked the biological capacity necessary to create art and other forms of symbolic expression. For example (as we will see in Chapter 15), there is evidence that the ability to learn and use language has

a biological basis. Perhaps, as some anthropologists have argued, Neanderthals lacked the abilities necessary for language and therefore were unable to transmit a rich symbolic culture. Others believe that the human revolution was likely the product of cultural processes, not biological changes in the hominid lineage. For example, it is possible that a technological innovation like the atlatl might have allowed for larger social groups and greater economic surplus, factors that potentiated the development of art and ornamentation. The fact that the Neanderthals who lived at Arcy-sur-Cure made and used both bone artifacts and personal adornments supports the latter position. It seems unlikely that Neanderthals would have been able to borrow ideas from their neighbors if they themselves lacked symbolic abilities. On the other hand, it is plausible that they might have adopted more sophisticated technology from neighboring peoples. The ethnographic record tells us that technological accomplishments do not provide a good measure of social or cognitive complexity. There are many cases in the historical record of technologically more advanced peoples coming in contact with technologically less advanced peoples. In such cases, it is common for the technologically less advanced to adopt ideas and techniques from the more advanced without completely abandoning their own way of life. It seems plausible that Neanderthals at Arcy-sur-Cure and other sites did just that.

Finally, other transitional tool kits have been found. As shown in Figure 14.14, each of these transitional industries is quite localized, the Châtelperronian in southern France, the Uluzzian in northern Italy, and the Szeletian in central Europe. Mellars points out that each of these transitional industries is quite distinctive, and argues that it is not likely that the very widespread Aurignacian evolved from several distinctive localized transitional industries.

Taken together, these observations suggest that many of the genes that gave rise to modern morphology arrived in Europe through the movement of people, not the flow of genes. However, they do not necessarily mean that Neanderthals contributed no genes to modern human populations. The Skhul and Qafzeh fossils

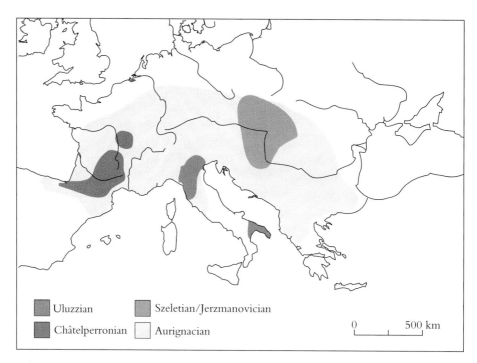

Figure 14.14

The earliest Upper Paleolithic industry, the Aurignacian, is found throughout Europe, while three industries intermediate between the Mousterian and the Aurignacian are found in localized areas. Because each of these transitional industries is distinctive, some scientists believe it unlikely that the Aurignacian evolved from one of them.

- Uluzzian
- Châtelperronian
- Szeletian/Jerzmanovician
- Aurignacian

0   500 km

are more robust than contemporary people, and their physique could be the result of gene flow between robust Neanderthal populations in Europe and less robust populations in Africa.

Fossils from eastern Asia provide the best evidence for the multiregional model.

Although the multiregional hypothesis is difficult to defend for Europe and the Middle East, it may fit data from eastern Asia. Advocates of the multiregional view, like Wolpoff, argue that there is regional similarity in the morphology of archaic and anatomically modern *H. sapiens* in both Southeast and northeastern Asia. In Southeast Asia, for example, Wolpoff argues that there are morphological similarities that link *H. erectus* living in Java half a million years ago, to archaic *H. sapiens* found at sites in Indonesia, and to contemporary Australian aborigines. These similarities include:

- thickened bones on the top of the cranium
- continuous browridges that extend unbroken over both eyes
- a large, projecting face
- massive, rounded cheekbones
- large molars and premolars
- details like a "rolled edge" on the lower part of the orbits and a distinctive ridge on the cheek

Wolpoff believes this combination of traits is unique to the ancient peoples of the Southeast Asian archipelagos and modern Australian aborigines, and that the best explanation for the similarities among these peoples is that there is genetic continuity among them. Some critics point out that most of these features are ancestral in the hominid lineage, and should not be used to infer genealogical patterns for the same reason that ancestral features should not be used in systematics (see Chapter 4). Others, like Cambridge University anthropologist Marta Lahr, point to quantitative studies showing that Australasian peoples are not clearly distinct from peoples in other parts of the world. Presently, however, most anthropologists regard the regional continuity as a viable but unproved hypothesis for eastern Asia.

The anatomy of fossil hominids from China also supports the multiregional model in eastern Asia. Some derived features of the modern human face seem to appear first in archaic *H. sapiens* fossils of eastern Asia. For example, two recently discovered skulls show that more modern looking, horizontally oriented cheekbones appeared in Asia in the Middle Pleistocene, long before these features show up in Africa or Europe. Unless they are the result of convergent evolution in archaic populations of *H. sapiens* in Asia and anatomically modern populations in Africa and Europe, they indicate that genes flowed from east to west, as well as from west to east.

These observations, if correct, suggest that genes from archaic hominid populations are present in modern humans. However, it is still possible that most of the genes that gave rise to our modern morphology arose outside of Asia, and that these genes arrived in eastern Asia through the long-distance movement of people, rather than through gene flow.

Genetic Data

Contemporary patterns of genetic variation, particularly in genes carried on mitochondria, provide information about the origin of modern humans.

By studying the patterns of genetic variation within the human species, and comparing these patterns to the predictions of population geneticists, biologists and anthropologists have been able to tease out interesting facts about the history of human populations. Much of the data used in such studies comes from genes that are carried on mitochondria. Eukaryotic cells, including those of humans, contain small cellular organelles called **mitochondria**, which are responsible for the basic energy processing that goes on in the cells. Mitochondria contain small amounts of DNA (about 0.0005% of the DNA contained in chromosomes), called, appropriately, **mitochondrial DNA (mtDNA)**. In humans and other primates, mtDNA codes for 13 proteins used in the mitochondria, for some of the RNA that makes up the structure of ribosomes, and for several kinds of transfer RNA. There are also two major noncoding regions on mtDNA.

Mitochondrial genes have several properties that make them especially useful for reconstructing the recent evolutionary history of human populations. First, unlike genes carried on chromosomes, mitochondrial genes are inherited maternally: both males and females get their mitochondria from their mother's egg cell; the mitochondria present in sperm do not get transferred to the fertilized ovum. Second, there is no recombination in mtDNA, so unless her mitochondria carry a novel mutation, a child has exactly the same mitochondrial genes as its mother. Thus, the mitochondria of every person living today is a copy, modified only by rare mutations, of the mitochondria of a single woman who lived some time in the past (see Box 14.2). Third, mitochondria accumulate mutations much faster than genes carried on chromosomes do; consequently, the mitochondrial genetic clock allows for more-accurate dating of events in the last few million years than clocks based on nuclear genes. The mitochondrial clock is like a stopwatch that keeps track of hundredths of seconds, while the genes on chromosomes are like a clock with only a minute hand. The mitochondrial genetic clock is useful for dating recent events, for the same reason that a stopwatch is useful for timing Ato Boldon and Maurice Greene in the 100-m dash—the clock changes rapidly enough to distinguish events that aren't very far apart in time.

BOX 14.2

Mitochondrial Eve

The fact that mitochondria are inherited maternally means that the mitochondria of every living person are
copies of the mitochondrion of a single
woman who lived in ages past. This
seems incredible, but it is simply a logical consequence of asexual inheritance.
To see why this is so, consider Figure
14.15. Each colored circle represents
one woman, and each row of circles
represents a population of 12 women
during a given generation. The circles
in row 1 represent the women in gen-

eration 1, the circles in the second row
are women in generation 2—all of
whom are daughters of women in generation 1—and so on. (We ignore men
because their mitochondria are not
transmitted to their children.) To keep
things simple we assume that population size is constant, so women must
produce on average just one daughter.
However, the number of daughters
born to any particular woman varies—
here we will assume that some have no
daughters, some have one, and some

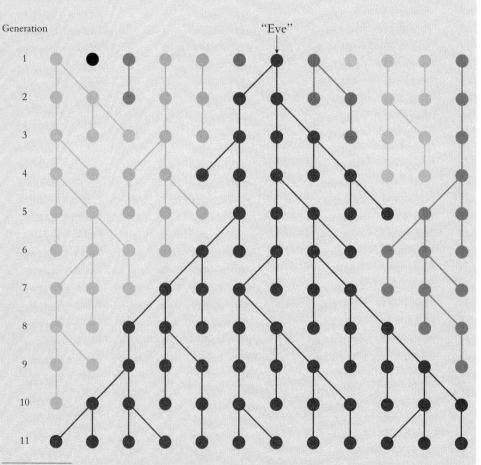

Generation

"Eve"

1

2

3

4

5

6

7

8

9

10

11

Figure 14.15

This mtDNA tree for a hypothetical population of 12 women shows why all the mitochondria in any population are descended from a single woman's mitochondria sometime in
the past.

have two. For the moment, we will also assume that there is no mutation.

To indicate that daughters inherit their mother's mitochondria, they are assigned the same color. Thus, in any generation, all of the women with the same color carry copies of the mitochondrion of the same woman in generation 1. (This color scheme does not imply that the mitochondria carried by different women have different mtDNA sequences—they could all be genetically identical or all unique.)

In generation 1 there are three women (black, dark gray, and light gray) who have no daughters. As a result, after one generation, only nine of the original 12 women and their mitochondria have descendants. However, three of the mitochondria in the first generation have two descendants in generation 2. During the second generation, three women have no daughters. However, since one of the women without a daughter has a sister who produced a daughter, copies of that woman's mitochondria are not lost. This means that the number of women's mitochondria present in the population is reduced from 12 to 7. The final outcome should be evident: as long as there are copies of mitochondria from more than one of the original women present, there is a chance that one of them will be lost because all the descendants of one of the women will have no daughters. Eventually, the mitochondria of all but one of the original women will disappear. At this point, all members of the population will be descendants of one of the original females.

We have run this argument forward in time, but it also can be run backwards (with more difficulty). That is to say, we can pick any generation and then move backward through time to find the single mitochondrial ancestor of every individual in that generation.

In this example, there are only 12 women in the population, but exactly the same thing will happen in all populations. In larger populations the process takes longer, but the logic is the same. Eventually, everyone will carry copies of the mitochondrion of a single woman—who is sometimes given the colorful but misleading name "Eve." It is important to see that this Eve is not the mother of us all, only the mother of our mitochondria. Because genes carried on chromosomes recombine, it is likely that we have ordinary chromosomal genes from many of the women in the population containing Eve. And it is important to realize that the mitochondrial Eve is not the first human woman. She may have been an anatomically modern human, or she may have been an earlier form of hominid. Eve was an otherwise unremarkable member of a (perhaps large) population whose only claim to fame is that her mitochondrion happened to survive while the mitochondria of other women did not.

In real life, the mitochondria of all living people are not identical (even though they are all replicas of Eve's mitochondrion) because mutation occasionally introduces new mitochondrial variants. Because there is no recombination, once a mutation occurs in a particular woman, it is carried by all of her descendants. The existence of mutation has two very important consequences. First, at least in principle, anthropologists can use the methods of phylogenetic reconstruction discussed in Chapter 4 to reconstruct a tree of descent among mitochondria based on derived similarities. Such reconstructions, sometimes called **gene trees**, are practical only if mutations occur frequently enough for most branches to contain mutations that allow the branches to be distinguished. It turns out that mitochondria are particularly well suited to this sort of genetic reconstruction. Second, mutation makes it possible to estimate the time that has passed since our common mitochondrial mother lived using genetic distance measures (also discussed in Chapter 4).

In practice, gene trees based on mtDNA have proven difficult to construct and to interpret. Rebecca Cann at the University of Hawaii and the late Allan Wilson collected mitochondrial DNA from 186 people, and used phylogenetic methods to reconstruct a gene tree for part of the mtDNA. They believed their results provided strong support for the replacement model, in two ways. First, all of the oldest branches of the tree they constructed terminated in the mtDNA of modern Africans, while more recent branches terminated in people from every continent, including Africa. This suggested Eve was an African. Second, the genetic distance data indicated that Eve lived between 100,000 and 400,000 years ago. Cann and Wilson interpreted these results to mean that the human species originated in Africa about 200,000 years ago, and then spread out across the world, perhaps around 100,000 years ago.

Recently, these conclusions have been roundly criticized. The most glaring problem is that Cann and Wilson accidentally misused the computer program that they chose to construct the gene tree. Other investigators have shown that the Cann-Wilson tree does not fit the data best. There are, in fact, thousands of trees that fit better, and many of them do not have early African branches. A second problem is that the connection between any speciation event and the time at which Eve lived is unclear. It could be she lived in a small founding population that gave rise to anatomically modern humans, but it is also possible she was an archaic *H. sapiens* living long before the origin of modern humans, or that she was an anatomically modern human living long after our species was born.

The human species is less genetically variable than are other species.

One measure of genetic variation is the average number of genetic differences per base-pair (nucleotide) in a particular bit of DNA within a population. To compute this measure for mtDNA, geneticists collect tissue samples from members of a population and assess part of the DNA sequence of each individual's mitochondria. Then they compute the number of positions on the mtDNA at which each pair of individuals in the sample differ, and compute the average number of differences per nucleotide over all possible pairs of individuals. As populations become more diverse genetically, the average number of genetic differences per nucleotide increases.

By this measure, humans are less genetically variable than chimpanzees, in two quite different ways. Figure 14.16 plots the amount of genetic variation within and between three geographically separated human populations—Africans, Asians, and Europeans—and the amount of genetic variation within and between chimpanzee populations in eastern, central, and western Africa. The bars on the diagonal (red) show that the amount of variation within each human population is much lower than within any of the chimpanzee populations. These data tell us that any two humans taken at random from a single population are much more similar to one another than any two chimpanzees taken at random from a single population. The off-diagonal bars (blue) plot the average differences between human populations and between chimpanzee populations. The differences between chimpanzee populations are also much greater than the differences between human populations.

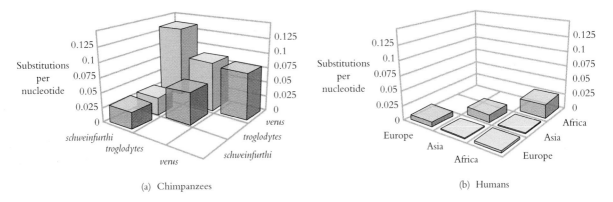

(a) Chimpanzees (b) Humans

Figure 14.16

Humans are much less genetically variable than chimpanzees. (a) Plotted here is the mean number of differences per nucleotide in one region of mtDNA for three geographically separated chimpanzee populations: *Pan troglodytes schweinfurthi* (from East Africa), *P. t. troglodytes* (from central Africa), and *P. t. verus* (from West Africa). (b) Here the same data are plotted for three geographically separated human populations (in Asia, Europe, and Africa). In both graphs, each red bar at the intersection of a row and a column gives data for comparing the population(s) identified at the ends of that row and column. Thus, the height of each red bar represents the mean number of differences between pairs of individuals drawn from the same population. For example, each pair of chimpanzees belonging to the *verus* population differ at about 7.5% of the nucleotides sampled. Each pair of individuals drawn from the most variable human population, Africans, differ at only about 2.5% of the nucleotides assessed. The height of the blue bars represents the mean number of differences between pairs of populations. On average, an individual drawn from the *verus* population differs from an individual drawn from the *schweinfurthi* population at about 13% of the nucleotides, while Africans and Europeans, the most different pair of human populations, differ at less than 0.3% of the nucleotides. Thus, the two graphs show that there is more variation within each of the chimpanzee populations than within any of the human populations, and human populations are more similar on the average to one another than chimpanzee populations are.

The most likely explanation for the small amount of genetic variation observed within the human species is that we have undergone a population explosion.

The data on genetic distance suggest that much of the human genome evolved mainly under the influence of drift and mutation. Remember that mutation introduces new genes at a very low rate, and that genetic drift eliminates genetic variation at a rate that depends on the size of the population. In large populations, drift removes genetic variants slowly, while in small populations drift eliminates variation more quickly. The effects of mutation and genetic drift will eventually balance (Figure 14.17), as the amount of variation, which we label m, arrives at an equilibrium. The value of m at equilibrium will depend on the size of the population. For a given mutation rate, bigger populations will have larger amounts of genetic variation at equilibrium. Population genetics theory says that $m = 2Nu$, where N is the number of females and u is the rate at which mutations occur.

We can substitute the measured values of m and u into the formula for m, to calculate N. Using an estimate for the mutation rate of mtDNA of $u \approx 0.0015$, and the average of the values for m given in Figure 14.17, we get $N \approx 5000$. This means that the existing amount of genetic variation within human populations

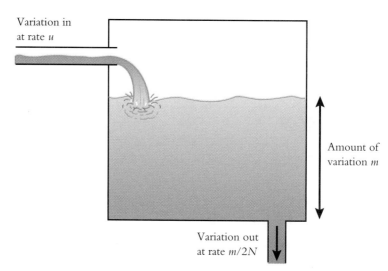

Variation in
at rate u

Amount of
variation m

Variation out
at rate $m/2N$

Figure 14.17

A physical analogy for the mutation-drift equilibrium can help to explain why large populations have more genetic variation at equilibrium than small ones do. Mutation introduces new genetic variants at a constant rate u, much like a stream of water entering a
tank. Genetic drift removes variants, like a drain at the bottom of the tank. The amount of
variation is analogous to the volume of water in the tank: As the level of water in the tank
gets higher, the pressure increases, and the water drains out more rapidly. Eventually the
rate of outflow equals the rate of inflow, and the depth of the water remains constant. In
the same way, as the amount of variation in a population increases, drift removes variation faster; eventually a steady state is reached at which the amount of variation is constant. To calculate the equilibrium amount of variation, set the rate of inflow (u) equal to
the rate of outflow ($m/2N$), and solve for m.

today is consistent with the drift–mutation equilibrium value for a population of
5000 women. Obviously this estimate of N is far lower than the size of human
populations today or even in the recent past. How can that be?

The simplest answer is that a small human population must have expanded very
rapidly sometime in the past. When a population grows in size, the equilibrium
amount of genetic variation also grows, because genetic drift removes variation
more slowly in large populations than it does in smaller ones. However, if the
population expands rapidly, it may take a long time to reach a new equilibrium
level because mutation adds variation very slowly. Thus, the low levels of genetic
variation in contemporary human populations may be the result of a rapid expansion from a small population sometime in the past.

The data do not allow us to say whether the population before the expansion
was always small, or whether it was a large population that suffered a temporary
decrease in population size, called a **bottleneck**. If a population suddenly decreases
in size, the amount of genetic variation present will be well above the equilibrium
level. Because drift is relatively rapid in small populations, the excess genetic variation is rapidly removed, and the population will then reach a new equilibrium
value relatively quickly. Thus, if a bottleneck lasts long enough for the population
to reach equilibrium, we cannot tell whether the population suffered a bottleneck,
or had always been small. In this way, population bottlenecks erase genetic evidence of previous population history.

Natural selection provides an alternative explanation for the lack of genetic
diversity in mitochondrial genes. If a strongly beneficial mutation arose in the

mitochondrial genome, it would rapidly sweep through the human population. Since there is no recombination, the particular mitochondrial genotype in which the beneficial mutation occurred would also spread, and all other mitochondrial genotypes would disappear. In this way, natural selection could cause a bottleneck in the number of distinct mitochondrial genotypes without there having been any reduction in the number of people. However, a growing number of studies of nuclear DNA tell the same story as the mtDNA and suggest that there was a bottleneck between 100,000 and 200,000 years ago. Natural selection could account for these data only if a number of separate beneficial mutations occurred throughout the nuclear genome at about the same time.

Recent research suggests that the population expansion occurred about 100,000 years ago and from a base population of about 5000 women.

A method for estimating the date of expansion and size of the initial population has been developed by anthropologists Alan Rogers and Henry Harpending of the University of Utah. They started with a sample of mtDNA from a modern population and computed the number of mutations by which each pair of people differ: some pairs have exactly the same mtDNA, some pairs differ by one mutation, some by two, and so on. Then, they created what they call a **mismatch distribution**, which plots the distribution of these differences. These distributions typically have the form shown in Figure 14.18. Using population genetics theory, Rogers and Harpending showed that the shape of these distributions changes over time. Immediately after a rapid population expansion, mutations accumulate, and the peak of the distribution shifts to the right. Thus, the current position and width of the peak can be used to estimate the time when the population expansion occurred.

Based on mtDNA data from contemporary groups, Rogers and Harpending drew two conclusions. First, the size of the human population before expansion was about 10,000 to 50,000 people. Second, there was a major expansion of human populations beginning about 100,000 years ago in Africa and about 50,000 years ago elsewhere in the world.

This scenario does not fit very well with the multiregional model because a population of 50,000 seems too small to have been connected by gene flow across the vast expanse occupied by *H. erectus*. It does fit with the replacement hypothesis because it places the initial expansion of modern humans approximately 100,000 years ago, the same time anatomically modern people appeared in the fossil record. The estimate for the rapid expansion of the human population worldwide coincides with the onset of the human cultural revolution, which archaeologists think began about 50,000 years ago (Box 14.3).

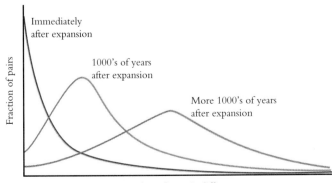

Immediately after expansion

1000's of years after expansion

More 1000's of years after expansion

Fraction of pairs

Number of genetic differences

Figure 14.18

After a population expansion, the peak of the population's mismatch distribution gradually broadens and moves to the right.

BOX 14.3

Pleistocene Park

The plot of the movie *Jurassic Park* depends on what seems like an implausible conceit: scientists recover dinosaur DNA from insects that have been fossilized in amber since the Mesozoic. Although no one has managed to clone a dinosaur, another feat almost as amazing has shed important new light on modern human origins.

A research group headed by Svante Pääbo, now at the Max Planck Institute for Evolutionary Anthropology in Leipzig, Germany, has extracted and sequenced a segment of mitochondrial DNA from a Neanderthal who died more than 30,000 years ago. Pääbo and his colleagues removed a tiny section from the humerus of the original Neanderthal fossil found in the Neander Valley of Germany in 1856. After careful testing suggested that there was enough mtDNA remaining in the fossil, they extracted mtDNA and then made many copies of the ancient mtDNA using a technique called **polymerase chain reaction (PCR)**. Since the Neanderthal fossil contained very small amounts of mtDNA, contamination with modern human mtDNA was a serious risk. However, Pääbo and his colleagues working independently in two different laboratories at the University of Munich and the Pennsylvania State University performed several experiments that convinced them that the mtDNA actually came from the cells of the long-dead Neanderthal.

Next, the Munich group determined the sequence of bases in the Neanderthal's mtDNA and compared it to mtDNA drawn from 994 modern humans from all over the world and to mtDNA drawn from 333 chimpanzees. The results are shown in Figure 14.19, which plots mismatch distributions for the three samples. The red histogram plots the fraction of pairs in the modern human sample that differ at a given number of bases. On average, modern people differ at eight positions with a range from 1 to 24 differences. The blue histogram was created by computing the number of differences between the Neanderthal mtDNA and each of the modern humans. The mean number of differences between the Neanderthal and humans is 27.2, with a range from 22 to 36. Finally, the green histogram is the result of comparing modern humans and modern chimpanzees. It shows that humans and chimpanzees differ at 55 positions on average (range = 46 to 67).

These data indicate the mtDNA of this Neanderthal was quite different from the mtDNA of modern humans. The mtDNA of this Neanderthal is well outside of the range of modern human mtDNA variation—only 0.002% of the human pairs were more different than the smallest difference between human and Neanderthal. Put another way, humans are much more different from Neanderthals than modern humans are from one another. The magnitude of the difference indicates that Neanderthal and modern humans are not closely related. Using the techniques described elsewhere in this chapter, Pääbo and his colleagues estimate that the last common ancestor of the mtDNA in the Neanderthal individual that they sequenced and modern humans lived between 550 and 690 kya. Using the same techniques, they also estimate that the last common ancestor of the mtDNA in their modern human sample lived between 120 and 150 kya.

These data are very bad news for the hypothesis that modern humans are descended from Neanderthals because they suggest that the lineage leading to Neanderthals diverged from the lineage

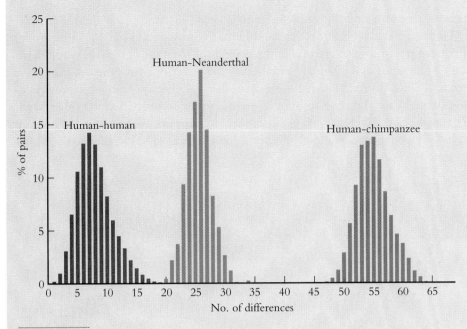

Figure 14.19

The mtDNA recovered from the Neanderthal is very different from the mtDNA of modern humans and of modern chimpanzees. Each bar in the blue histogram plots the percent of a sample of 994 humans whose mtDNA differs from that of the Neanderthal by a given number of mutations. The modern humans most similar to the Neanderthal differed by 22 mutations. The red bars give the percent of pairs drawn from the modern human sample who differ by a given number of mutations (the mismatch distribution). Because the two graphs barely overlap, there is a very small possibility (0.002%) that the Neanderthal mtDNA differs from modern human mtDNA only by chance. Put another way, if we picked one of the modern humans at random, the odds are only 2 in 100,000 that it would be as different from other humans as the Neanderthal is. The green histogram compares human and chimpanzee mtDNA. Each bar represents the percent of all possible pairs, one human and one chimpanzee, who differ by a given number of mutations. Chimpanzees are even more different from humans than Neanderthals are.

leading to modern humans long before the modern human phenotype had evolved. The situation is made even worse by the fact that the Neanderthal mtDNA is no more similar to the mtDNA of modern Europeans than it is to the mtDNA of people from other regions.

Modern African populations are genetically more variable than other populations. This fact is consistent with both the replacement and multiregional hypotheses.

Notice in Figure 14.16 that the amount of mitochondrial genetic variation within contemporary African populations is greater than the variation within European or Asian populations. Data from studies of ordinary nuclear genes tell the same story. Africans are more genetically variable than other peoples.

At first glance, this fact seems to support the Out of Africa hypothesis because we might expect African populations to be more variable than European and Asian populations, since African populations are older. However, the Out of Africa hypothesis could be true even if modern Asian, European, and African populations were equally variable. To understand this, consider the following scenarios. First, suppose that anatomically modern humans arose in Africa, and then after, say, 50,000 years of evolution on that continent, they emigrated to Asia and Europe. If African populations were unified by gene flow, then the emigrants would have carried almost all of the genetic variation present in Africa with them during their exodus. Thus, from a genetic perspective, Asian, European, and African populations would have the same amount of genetic variation today. On the other hand, if populations in different parts of Africa were isolated from each other, they would have diverged genetically (see Figure 14.20). Then, if emigrants were drawn

100 kya: Anatomically modern humans evolve and disperse throughout Africa.

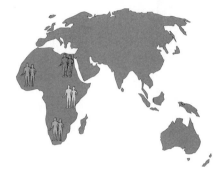

100–50 kya: Dispersed human populations diverge within Africa.

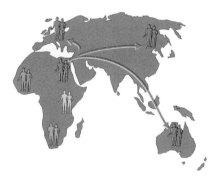

50 kya: People from one African population migrate to Eurasia and Australasia.

50 kya–present: Dispersed human populations diverge.

Figure 14.20

The Out of Africa model can explain why contemporary African populations have greater genetic variation than modern human populations in other parts of the world. This explanation requires that the dispersal of anatomically modern humans occurred in several steps:

1. Anatomically modern humans evolved and dispersed throughout Africa approximately 100 kya.
2. Over the next 50,000 years, dispersed anatomically modern populations were isolated from each other by distance, and these populations became genetically different.
3. Around 50 kya, people from one of the local African populations dispersed throughout Eurasia. Because these emigrants were from one local population, they were less variable than African populations taken as a whole.
4. Over the next 50,000 years, dispersed populations in Africa and Asia continued to diverge. Modern Eurasian populations are less variable than African populations today because Eurasian populations are descended from a single localized African population that migrated northward 50 kya. Modern African populations are more variable because they are the descendants of a number of local African populations that were spread across most of the continent.

from one local African population, they would carry with them only a fraction of the total genetic variation present in Africa. The European and Asian descendants of this regional African population would be less variable than modern populations across Africa. Thus, the amount of variability in modern African populations doesn't necessarily tell us whether the Out of Africa hypothesis is correct.

The greater amount of genetic variation among Africans could also be consistent with the multiregional hypothesis. Remember that the amount of genetic variation present in a population today depends on both the size of the most recent bottleneck and the length of time that has passed since the bottleneck occurred. Thus, greater variation among African populations is consistent with the multiregional hypothesis as long as African populations were consistently larger than Asian and European populations during the Upper Pleistocene.

The resolution of the controversy about how modern humans originated is important because it affects our view of contemporary human variation.

The outcome of the debate between the supporters of the replacement and multiregional models is important because it affects our perspective about the magnitude and significance of genetic differences among contemporary human groups. If the replacement model is correct, all humans on earth right now are descended from a lineage that radiated around 200,000 years ago. This means that genetic differences between contemporary groups originated relatively recently, and we should expect them to be relatively minor. In contrast, if the multiregional model is correct, the differences between contemporary groups represent more than 1 million years of evolution, so we might expect the differences among them to be substantially larger. We will return to this topic in Chapter 16.

Further Reading

Bahn, P. 1998. Neanderthals emancipated. *Nature* 394:719–721.

Klein, R. 1999. *The Human Career.* 2d. ed. University of Chicago Press, Chicago.

Mellars, P., and C. Stringer. 1989. *The Human Revolution.* Princeton University Press, Princeton, N.J.

Stringer, C., and C. Gamble. 1993. *In Search of the Neanderthals.* Thames and Hudson, New York.

Stringer, C. 1997. *African exodus: The Origins of Modern Humanity.* Henry Holt, New York.

Thorne, A., and M. Wolpoff. 1992. The multiregional evolution of humans. *Scientific American* (April):76–84.

Study Questions

1. What derived anatomical features distinguish anatomically modern humans from other hominids?
2. What is the human revolution? Why does it present anthropologists with a conundrum?
3. Describe the main differences between the tools of Upper Paleolithic peoples and those of their predecessors.

4. What evidence suggests that Upper Paleolithic peoples were better able to cope with their environments?

5. What is an ethnic group? What is the evidence that *H. sapiens* first became divided into ethnic groups after the human revolution?

6. Describe the two extreme models of the origin of anatomically modern people. What is the evidence for and against each model?

7. Explain why patterns of variation in mtDNA suggest that human populations have undergone a recent expansion.

C H A P T E R   1 5

The Evolution of Language

Language Is an Adaptation
    Speech Production and Perception
    Grammar
    Language Capacities Are Derived
How Language Evolved
When Language Evolved
    Did Language Arise Early?
    Did Language Arise Late?

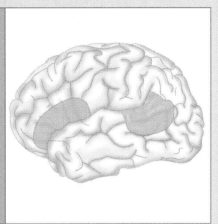

So far, our history has omitted a crucial component of the modern human phenotype—language. Sometime during the last several million years, humans evolved the ability to communicate much more complex and detailed information than any other creature. A vervet monkey can inform members of its group that there is a large, terrestrial predator nearby. A person can tell the members of her group that a pride of six lions killed a wildebeest near the big baobab yesterday just after it rained. She can also explain that the red and purple caterpillars that are found under the bark of dead mopane trees make good arrow poison, and she can confide that she thinks her husband wants to make Thag his second wife. The ability to communicate detailed information about nature, technology, and social relationships is a key part of the human adaptation. The goal of this chapter is to explore how language evolved.

We have postponed discussion of language until now because there is no consensus about *when* language evolved. Some authorities believe the australopithecines had significant language ability 4 mya, while others believe language appeared only with the modern human revolution, 40,000 years ago. We cannot reconstruct the evolutionary history of language as we reconstruct the history of bipedalism because the ability to use language leaves no clear traces in the fossil record. However, as Massachusetts Institute of Technology psycholinguist Steve Pinker has demonstrated in his book *The Language Instinct,* we can learn a great deal about *why* language evolved by studying the details of modern language. We begin by summarizing Pinker's argument that language is an adaptation, a complex structure assembled by natural selection for the purpose of communication. Then, we explain how some facts about modern language help us to understand why language was favored by natural selection. Finally, we consider the evidence about when modern language arose in the course of human evolution.

Language Is an Adaptation

Language allows speakers to modify the thoughts of their listeners.

Language is so much a part of our lives that it seems unremarkable. People living in most societies spend the majority of their waking hours immersed in a sea of words. We converse, watch TV, read books and newspapers, and listen to lectures, taking the fact of language very much for granted. But language allows us to do something amazing: modify the minds of other people. By making certain sequences of noises we cause other people to think thoughts that they otherwise might not think. You, for example, would probably not be thinking about the power of language if you were not reading these words now.

The cognitive capacities that give rise to language are products of natural selection.

Languages are clearly part of human culture. People learn the vocabulary and grammar of their native language as part of growing up in their culture. Americans learn English, Iranians learn Farsi, and the Batswana learn Setswana. Knowledge of English is transmitted from generation to generation in roughly the same way as knowledge of the rules of baseball, understanding of the principles of the internal combustion engine, or beliefs about the propriety of premarital sex are transmitted. In each case, people in different cultures learn different things from members of the previous generation.

Some anthropologists believe that biology and culture are separate domains. In their view, evolutionary explanations are appropriate for biological phenomena like bipedal locomotion but have no relevance to cultural phenomena like language. Cultural differences, the argument runs, are caused by social and cultural processes, not biology. In contrast, we believe that evolutionary principles provide useful insights about human culture and human cultural differences. The key is to focus on the evolution of psychological capacities that cause some cultural practices to persist and others to disappear. We will return to this interesting and controversial topic in Chapter 18.

Here we apply this principle to human language. There are universal cognitive capacities that allow people to learn and to use language, and these capacities cause languages to have universal properties. Thus, while human languages are part of culture, the capacities that make language possible and give human language its structure are not. They are evolved components of the human phenotype, like large molars and bipedal locomotion. As a result, language is different from other widespread cultural adaptations like metalworking, music, food preparation methods, and writing. To see why, consider the following comparisons between language and writing.

- *All human societies have language.* In contrast, the vast majority of societies did not have writing until recently. Writing systems appear to have been invented in only a few places and were then spread to other groups.
- *Groups with rudimentary technology or social organization do not have simple languages.* In fact, the languages of people with simple technology are frequently more complex and expressive than European languages. For example, Cherokee speakers have different pronouns for "me and you," "me, you, and one other person," "me, you, and several other people," "me and one other person," and "me and several other people." English speakers use "we" or "us" in all of these cases.
- *Most individuals within a society are competent users of language.* You don't have to be an English professor or a millionaire to be a fully competent language user. In contrast, in most human societies, until comparably recently, only the well-to-do and specialists like teachers, priests, and scribes were literate.
- *People do not have to be taught how to speak.* Many parents remember their child's first word, and they spend hours coaxing their infants to talk. There is not really any need to do so. In some cultures, parents rarely speak to children who are too young to talk. They ask, "Why speak to someone who can't understand what you are saying?" The children in these societies learn language from overhearing the conversations of adults and older children. As any grade school teacher will tell you, teaching six year olds to read and write is hard work. But toddlers learn to speak without help.

Language has been shaped by natural selection.

Many products of organic evolution are not adaptations. For example, in the modern world, the fact that we find electric shock unpleasant is probably useful because it allows us to learn to avoid electric fences and to keep our fingers out of light sockets. However, this aversion is not an adaptation because it has not been shaped and maintained by natural selection. Instead, it is a side effect of the basic design of our nervous systems. Some people believe that language is simply a side effect of other adaptive changes. For example, Harvard paleontologist Stephen J. Gould has argued that language is simply a byproduct of the fact that modern humans have larger brains and greater intelligence than other creatures. In the following sections, we present evidence that Gould is wrong. Language is based on a large number of morphological and mental characteristics that arose and have been maintained by the action of natural selection.

Speech Production and Perception

The morphology of the human throat differs from that of other primates.

Have you ever noticed that dogs don't have an Adam's apple? Neither do chimpanzees or any other nonhuman primate. In fact, human infants don't have an Adam's apple either. The lump that we call the Adam's apple is created by the thyroid cartilage, which surrounds the **larynx**, the sound-making organ in the throat. In most mammals, the larynx sits so high in the throat that it cannot be seen. The human larynx is placed much lower in the throat because this position enhances our ability to speak. To understand why, you have to know a little about how humans produce speech sounds.

Speech begins when air from the lungs flows through the larynx. The larynx is a valve consisting of an opening called the glottis that is bounded by two retractable flaps of flesh, the vocal folds or cords (Figure 15.1). In some animals, this valve prevents food and water from entering the lungs, and allows the animal to stiffen its torso like an inflatable boat. The larynx plays a crucial role in human speech. Air passing between partially open vocal folds causes them to open and close rapidly, generating periodic impulses of higher pressure air that we perceive as sound. The more rapidly the vocal folds open and close, the higher the pitch of the sound. By altering the shape of the vocal folds, we can change the pitch of our speech.

The rich mix of tones that emerges from the larynx is filtered and shaped by the throat and mouth. As the sound waves reverberate inside the throat, some of the tones interfere with themselves and are filtered out, while other tones reinforce themselves, creating the sounds that emerge from the mouth. The size and shape of the chambers inside the mouth and throat determine which tones are reinforced, or resonate, and which do not resonate and are filtered away. Humans form different sounds by changing the shape of these chambers. A trombone works the same way. Air passing through the lips inside the confines of the mouthpiece generates sound waves with a wide range of frequencies. By moving the slide, the trombonist alters the shape of the horn, which in turn changes the frequency at which the horn resonates. Instead of moving a slide, humans move their tongue, lips, and jaw.

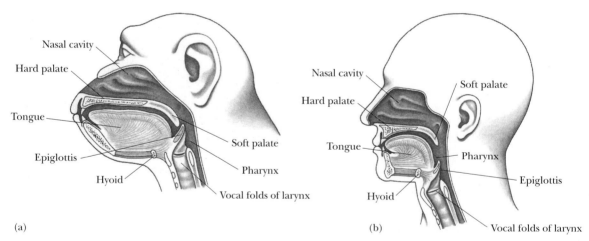

(a)
(b)

Figure 15.1

These are cutaway views of the vocal tracts of (a) a chimpanzee, and (b) an adult human. The glottis is the slit-like opening between the vocal folds. In chimpanzees and other mammals, the larynx is positioned high in the throat. When a chimpanzee swallows, the larynx rises to form a sealed passage from the nasal cavity to the lungs. This allows liquids to pass from the mouth around both sides of the air passage and into the esophagus. In adult humans, the larynx is positioned too low in the throat for such a seal to be possible. As a result, when adult humans swallow, the passage to the lungs is sealed and breathing is impossible. The position of the larynx in human infants is similar to that of chimpanzees. Note that the human tongue is shorter and thicker than the chimpanzee's tongue, which enables us to make a wider range of vowel sounds than apes can.

These derived features of the human throat are adaptations for speech production.

The production of vowel sounds depends crucially on the position of the tongue. Moving the tongue up and down, or back and forth, changes the size of the mouth and throat cavity. Different combinations of these maneuvers generate different vowel sounds by determining which harmonics are reinforced. You can prove this to yourself by pronouncing over and over the vowels "e" as in *bet* and "uh" as in *but*. Do you feel your tongue moving back and forth? This movement changes the shape of your mouth and throat causing different frequencies to resonate. Now do the same with "a" as in *bat* and "ee" as in *beet*. Do you feel your tongue moving up and down? Once again the shape of the throat and mouth changes, causing different frequencies to be filtered out.

Now we can understand why the geometry of the human throat and mouth are adaptations for speech. The human tongue is short and deep, and the back of the tongue is mainly in the throat cavity (Figure 15.1b). This geometry allows humans to produce a wide range of vowels. By moving up and down and back and forth, the tongue can simultaneously change the shape of the mouth and throat cavities. In contrast, the chimpanzee's tongue is longer and more shallow, and sits completely within the mouth (Figure 15.1a). The chimpanzee's larynx is positioned much higher in the throat. An animal with a vocal tract like that of a chimpanzee would find it very difficult to produce the range of vowels used in human languages. Thus, it is plausible that natural selection has caused the larynx to drop in the throat and the tongue to expand downward because these changes facilitate language.

This morphological change has not been without costs. The larynx sits high in the throat of other animals so that they can breathe and eat at the same time. When most animals swallow, the larynx rises up and forms a watertight seal with the entrance to the nasal cavity. Air can pass directly from the nose to the windpipe, while at the same time food and water can go around the larynx and into the stomach. This configuration is found in human infants and enables them to nurse and breathe at the same time. If you watch a baby suckle or guzzle milk from a bottle, you will be amazed at how long it can drink without coming up for air. Try it yourself and you won't last more than a minute or two. Because the mature human larynx is so much lower, it cannot form a seal with the nose cavity, and after infancy, we must alternate between breathing and drinking. Even worse, as Darwin noted in *The Origin of Species* (1859), "every particle of food and drink which we swallow has to pass over the orifice of the trachea, with some risk of falling into our lungs." As a consequence, thousands of people choke to death every year. It seems likely that in most animals the geometry of the throat has the optimal design for permitting eating and breathing at the same time, while the human design represents a compromise between the demands of eating, breathing, and speaking.

Humans can perceive speech sounds at a much more rapid rate than other sounds.

Phonemes are the basic unit of speech perception, the smallest bits of sound that we recognize as meaningful elements of language. For example, English speakers hear the difference between "l" and "r" sounds, while Swahili and Japanese speakers lump them together in a single phoneme. We have no trouble understanding normal speech at the rate of about 10 phonemes per second, and if we concentrate, we can understand people who bombard us with 25 phonemes per second. By contrast, the best Morse code operators cannot keep up once the transmission rate exceeds about 5 clicks per second. In fact, if the rate of transmission of a simple sound like a Morse code click exceeds 20 sounds per second, we hear a continuous buzz rather than a sequence of discrete sounds.

The advantage of rapid communication can be easily imagined. The early human who could decode the hurried shout, "Look out, the lion is behind you," would have a substantial edge over the poor fellow who could only say, "Speak more slowly, I can't . . ."

We perceive speech as a sequence of discrete words, a fact that greatly expands the range of thoughts we can communicate.

We perceive speech the same way we write it down—as a sequence of discrete words separated by brief silences. However, this is an auditory illusion. There are no periods of silence in ordinary speech. To prove this to yourself, listen to people speaking a language you don't understand. You won't be able to tell when one word ends and another begins. The reason that you can't tell is that there are no gaps between words. When researchers use an instrument that plots sound level against time, they note that sound level does not drop to zero between spoken words, and in fact, distinct words do not appear on the plot.

By making different combinations of words, people can generate an endless variety of meaningful signals. As far as we know, other primates use particular calls in specific contexts as if each vocalization communicated a separate thought. As we will see, vervet monkeys have a number of different alarm calls. One means roughly,

There's an eagle in the air, take care

while a second means

There's a python about, watch out

Such a system is very limited. Modern humans can remember between 45,000 and 90,000 words. Thus, if we used the vervet system, and could remember this many calls, we could communicate something like 100,000 thoughts. This sounds like a lot, but if you think about it, you will see that this limit would be very constraining. There might be one vocalization for

Thag is sleeping with my husband

a second for

Thag is sleeping with your husband

a third for

Thag wants to sleep with my husband

a fourth for

Thag used to sleep with my husband

and so on. Just keeping track of the affairs in Thag's hunter-gatherer band (or one floor of a college dorm) could easily use up all the available vocalizations. By recombining the thousands of words in any language, we can communicate a virtually endless number of thoughts.

Grammar

Grammatical rules allow us to interpret sequences of words.

Basing communication on words that can be recombined solves one problem but creates a second: a collection of words, by themselves, does not communicate meaning. Consider the sentences

> *Org killed your brother*
> *Your brother killed Org*

The words are the same, but the meanings produced in your brain, and therefore the course of action you might take, are very different. One situation calls for revenge, and the other calls for keeping a close watch on Org's relatives. We hear speech as a sequence of words, like cars coming down a one-lane off-ramp. In order to create meaning, our brains must interpret this information using rules, called **grammar**, that specify meaning. In this case, grammatical rules tell us who is the killer and who is the victim. These are not the consciously known rules that modern Western people learn in school, but the unconscious rules that allow almost every person to effortlessly speak and understand spoken language.

Grammatical rules allow people to make useful distinctions.

The world's languages share grammatical rules that allow people to use sequences of words to express useful ideas. Some of these rules were listed in a paper written by Pinker and Paul Bloom of the University of Arizona:

- Words must belong to categories such as "noun," "adjective," and "verb," which correspond to basic features of the world that people need to talk about—in this case, things, the qualities of things, and actions, respectively.
- There are rules that allow the listener to figure out what the speaker intended to say about the relationship between things, qualities, and actions. Such rules allow us to determine that Org is a killer in the first sentence above and a corpse in the second. In English these rules are based on the order of the words—the actor comes before the verb and the recipient of the act comes after the verb. In other languages these distinctions are made by modifying the nouns to indicate, for example, which person is the killer and which is the victim. Linguists refer to this part of grammar as **syntax**.
- Words are grouped into phrases that also belong to categories such as "noun phrase" and "verb phrase." By combining words in this way, language can express a wide range of thoughts with a small number of words. The noun *brother* by itself is not very informative. However, by combining *brother* with other

words in noun phrases, you can refer to *my brother, your brother, your older brother, your husband's brother,* and so on.

- There are rules that allow a listener to determine which words the speaker meant to group together in phrases. These rules allow us to distinguish between the sentences

Your brother killed my father
My brother killed your father

by telling us that *your* is to be grouped with *brother* in the first sentence, and with *father* in the second sentence.

- Verbs are modified to indicate when events occurred. We can, for example, distinguish between the statements

Org killed your brother
Org will kill your brother

Verbs are also modified to indicate the relative timing of two events. Thus,

Org had killed your brother when Thag showed up

tells the listener that Thag did not witness the killing.

There are many other grammatical rules for forming questions, for indicating states of mind, and for reducing the amount of memory necessary to process a sentence.

This system is not perfect. Even with all the grammatical rules in action and with help from context, there are many sentences that allow for more than one meaning. As examples, Pinker offered the TV blurb "On tonight's program Dr. Ruth will discuss sex with Dick Cavett," and the newspaper headline "Stud tires out." Nonetheless, by combining discrete words using grammatical rules, humans can communicate an incredible range of useful information.

The psychological mechanisms that structure language are different from other cognitive abilities.

Many researchers see language as a manifestation of a general-purpose learning mechanism. According to this view, we humans learn language the same way we learn to play the piano or to solve differential equations. We learn many complex, rule-bound skills that are peculiar to life in a particular culture. No one would argue that our brains evolved in response to learning to play the piano or solving differential equations. Therefore, we must have a powerful and *very general* learning mechanism that allows us to acquire diverse skills. Language is not an adaptation, the argument runs; it is just another complex skill acquired by the same general mechanism. According to this view, there is no special-purpose complexity to explain.

However, several lines of evidence suggest that people acquire the grammatical rules that underlie language using special-purpose cognitive mechanisms specifi-

Happily Blinkered Humans

There is one more reason we should stand in awe of the simple act of learning a word. The logician W. V. O. Quine asks us to imagine a linguist studying a newly discovered tribe. A rabbit scurries by, and a native shouts, "Gavagai!" What does *gavagai* mean? Logically speaking, it needn't be "rabbit." It could refer to that particular rabbit (Flopsy, for example). It could mean any furry thing, any mammal, or any member of that species of rabbit (say, *Oryctolagus cuniculus*), or any member of that variety of that species (say, chinchilla rabbit). It could mean scurrying rabbit, scurrying thing, rabbit plus the ground it scurries upon, or scurrying in general. It could mean footprint-maker, or habitat for rabbit-fleas. It could mean the top half of a rabbit, or rabbit-meat-on-the-hoof, or possessor of at least one rabbit's foot. It could mean anything that is either a rabbit or a Buick. It could mean collection of undetached rabbit parts, or "Lo! Rabbithood again!," or "It rabbiteth," analogous to "It raineth."

The problem is the same when the child is the linguist and the parents are the natives. Somehow a baby must intuit the correct meaning of a word and avoid the mind-boggling number of logically impeccable alternatives. It is an example of a more general problem that Quine calls "the scandal of induction," which applies to scientists and children alike: how can they be so successful at observing a finite set of events and making some correct generalization about all future events of that sort, rejecting an infinite number of false generalizations that are also consistent with the original observations?

We all get away with induction because we are not open-minded logicians but happily blinkered humans, innately constrained to make only certain kinds of guesses—the probably correct kinds—about how the world and its occupants work. Let's say the word-learning baby has a brain that carves the world into discrete, bounded, cohesive objects and into the actions they undergo, and that the baby forms mental categories that lump together objects that are of the same kind. Let's also say that babies are designed to expect a language to contain words for kinds of objects and words for kinds of actions—nouns and verbs, more or less. Then the undetached rabbit parts, rabbit-trod ground, intermittent rabbiting, and other accurate descriptions of the scene will, fortunately, not occur to them as possible meanings of *gavagai*.

But could there really be a preordained harmony between the child's mind and the parent's? Many thinkers, from the woolliest mystics to the sharpest logicians, united only in their assault on common sense, have claimed that the distinction between an object and an action is not in the world or even in our minds, initially, but is imposed on us by our language's distinction between nouns and verbs. And if it is the word that delineates the thing and the act, it cannot be the concepts of thing and act that allow for the learning of the word.

I think common sense wins this one. In an important sense, there really are things and kinds of things and actions out there in the world, and our mind is designed to find them and to label them with words. That important sense is Darwin's. It's a jungle out there, and the organism designed to make successful predictions about what is going to happen next will leave behind more babies designed just like it. Slicing space-time into objects and actions is an eminently sensible way to make predictions given the way the world is put together. Conceiving of an extent of solid matter as a thing—that is, giving a single mentalese name to all of its parts—invites the prediction that those parts will continue to occupy some region of space and will move as a unit. And for many portions of the world, that prediction is correct. Look away, and the rabbit still exists; lift the rabbit by the scruff of the neck, and the rabbit's foot and the rabbit ears come along for the ride.

What about kinds of things, or categories? Isn't it true that no two individuals are exactly alike? Yes, but they are not arbitrary collections of properties, either. Things that have long furry ears and tails like pom-poms also tend to eat carrots, scurry into burrows, and breed like, well, rabbits. Lumping objects into categories—giving them a category label in mentalese—allows one, when viewing an entity, to infer some of the properties one cannot directly observe, using the properties one *can* observe. If Flopsy has long furry ears, he is a "rabbit"; if he is a rabbit, he might scurry into a burrow and quickly make more rabbits.

Moreover, it pays to give objects several labels in mentalese, designating different-sized categories like "cottontail rabbit," "rabbit," "mammal," "animal," and "living thing." There is a tradeoff involved in choosing one category over another. It takes less effort to determine that Peter Cottontail is an animal than that he is a cottontail (for example, an animallike motion will suffice for us to recognize that he is an animal, leaving it open whether or not he is a cottontail). But we can predict more new things about Peter if we know he is a cottontail than if we merely know he is an animal. If he is a cottontail, he likes carrots and inhabits open country or woodland clearings; if he is merely an animal, he could eat anything and live anywhere, for all one knows. The middle-sized or "basic-level" category "rabbit" represents a compromise between how easy it is to label something and how much good the label does you.

Source: From pp. 153–155 in S. Pinker, 1994, *The Language Instinct,* William Morrow, New York. Copyright 1994 by Steven Pinker. By permission of William Morrow and Company, Inc.

cally shaped by natural selection to allow people to learn and to use language. We shall consider some of this evidence.

Some intelligent people have serious linguistic deficits, and some glib people aren't very smart.

If the abilities to learn and to use language were based solely on general cognitive ability, then there should be a direct relationship between cognitive abilities and linguistic skills. However, this is not the case. Some stroke victims lose nearly all of their linguistic ability, but their intelligence is not diminished. These people are not fluent language users, and they have particular difficulty with syntax, but they do not show any other cognitive deficiencies. There are also people who speak grammatically, but are otherwise mentally impaired. For example, many children suffering from spina bifida are severely retarded, and they cannot learn to read, write, or do simple arithmetic. Nonetheless, some of them speak fluently and at length, using grammatically correct sentences. However, the content of their speech is nonsensical.

Children know some universal grammatical principles without having to learn them.

Most children begin to learn the grammar of their native language when they are about two years old, and by the time they are three, most children make relatively few grammatical errors. Many linguists believe that while children have to learn the surface grammar of their own language—how to speak properly in English, Spanish, or Farsi—they do not need to learn the deep, universal grammar that underlies all languages. This knowledge is innate. In fact, these linguists think that without some innate knowledge of the structure of language, children could not learn language at all. The following example shows that children know about a rule that governs the way words are put together (the part of grammar known to linguists as **morphology**) without ever having heard an example of the rule.

In English, compound words can be formed from irregular plural nouns but not regular plural nouns. For example, in Chapter 12 we discussed the important information that can be gleaned from the study of teeth marks and tool marks on ancient bones. Surely, it must be that marks were often made by more than one tool, so if we can say *teeth marks,* why can't we say *tools marks?* Similarly, why is a house *mice-infested* but not *rats-infested?* Linguists believe that this results from an innate aspect of language processing. Namely, when your brain is mentally putting compound words together, it retrieves the root words from a mental dictionary first, then makes the compound, and finally applies the rule for forming regular plurals. When we want to talk about bones that have been marked by several tools, we first assemble the compound, *tool-mark* and then use that as input to the rule that forms regular plurals. *Teeth marks* is okay because the plural *teeth* is irregular, and must be memorized and stored in the mental dictionary along with *tool.*

Research by psycholinguist Peter Gordon of the University of Pittsburgh shows that kids know this rule without ever having heard a compound containing a plural noun. Gordon showed three- to five-year-old children a toy monster, and said, "Here is a monster that likes to eat mud. What do you call him?" And, then to get the ball rolling, he told them the answer, "A mud-eating monster." He then asked, "What would you call a monster that eats mice?" and they sensibly answered, "A mice-eating monster." But when he asked, "What do you call a monster that eats rats?" they did not reply "A rats-eating monster," as logic would seem to dictate. Instead, they gave the grammatically correct answer, "a rat-eating monster." Gordon also collected a large sample of naturalistic speech, which shows that plurals containing compounds are extremely rare, so rare in fact that it is likely that none of the children had ever heard an example. Nonetheless, they followed the rule with great fidelity.

Children can create a grammatically complete language by themselves in a single generation starting only with a vocabulary.

Further evidence that at least some grammar rules are innately understood comes from two unusual situations in which a generation of children were surrounded by adults who spoke a simplified, grammatically incomplete language. In both cases, the children invented a complete grammar to go along with the vocabulary that they learned from adults.

During the last few centuries, plantation owners imported slaves or indentured servants. Often the plantation owners mixed laborers from different language backgrounds, and in these situations the adult laborers developed a simplified language, or **pidgin**, based on words from the language of their captors, but without much in the way of grammar. Their children were exposed to the pidgin, which served as the main language of the community. However, the children did not end up speaking the pidgin. Instead, they developed an entirely new language, called a **creole**, that had all the complexity found in long-established languages. According to Derek Bickerton, a linguist from the University of Hawaii, unrelated creoles from different parts of the world have many common features. Bickerton believes that such similarities arise because children create the creole in a single generation based on their innate knowledge of grammar. In long-established languages these similarities have been obliterated by thousands of years of linguistic change.

In recent years, a very similar process has been observed among deaf sign-language learners in Nicaragua. Analysis of different sign languages shows that they have all of the complexity of spoken language, and seem to depend on much of the same underlying mental machinery. Until 1979, when the Sandinista government took power, there were no schools for the deaf in Nicaragua. Deaf people were scattered and isolated, and there was no sign language except for the private languages used by the deaf within their own families. When the new government was installed, deaf children were brought together in special schools that did not teach sign language. Nonetheless, the children, mainly teenagers, invented a form of sign language based on pantomime and their private sign languages. Usually, utterances consisted of only a few signs, different people ordered the signs quite differently, and there was little grammar. This form of sign language was much like a pidgin. The next cohort of deaf children, starting at much younger ages, converted the pidgin to a full-fledged sign language in which complex, grammatical sentences were common.

Language Capacities Are Derived

Comparative evidence suggests that animals can associate concepts with arbitrary symbols but have only a limited mastery of syntax.

Language is based on the ability to attach meaning to arbitrary sounds, and allows us to communicate about things outside of our immediate experience. When one person smells smoke and yells fire in a crowded theater, the rest of the

Figure 15.2

Vervet monkeys associate particular calls with specific predators.

audience knows that it is time to head for the exits even if they can't smell the smoke themselves. Moreover, the audience knows that it is a fire, not a flood or an earthquake or the sighting of a celebrity.

Robert Seyfarth and Dorothy Cheney of the University of Pennsylvania have demonstrated that alarm calls in vervet monkeys work much the same way (Figure 15.2). It has long been known that when one monkey spots a predator and gives an alarm call, other members of the group take flight. Vervets give acoustically distinct calls when they encounter snakes, eagles, and leopards. They also respond differently to each of these predators. When they see a leopard, they rush up into a tree; when they see an eagle, they dive into dense bushes; and when they see a snake, they look down at the ground around them. These are clearly sensible responses, but it is not clear whether the alarm calls themselves convey any information to the monkeys. It is possible, for example, that monkeys look up whenever they hear a call, spot the predator themselves, and then take appropriate action. It is also possible that they may simply imitate the caller's actions.

To find out whether the alarm calls convey specific information to the vervets, Seyfarth and Cheney conducted a playback experiment. They recorded examples of alarm calls given to snakes, eagles, and leopards. Then they played one of the calls to the monkeys from a hidden speaker. When monkeys heard the alarm calls given to a leopard, they generally climbed into trees; when they heard an eagle alarm call, they most often ran into bushes or looked up; and, when they heard a snake alarm, they usually looked down (Table 15.1). These results demonstrate that vervets attach meaning to an arbitrary symbol. Since there was no predator actually present, only a tape recorder hidden in the bushes, the monkeys could not have responded because they saw the predator themselves. Moreover, since they heard only a tape-recorded call, they could not have imitated the response of an animal giving an alarm call. Finally, since different calls provoke different responses, the calls must have meaning for the monkeys.

Chimpanzees, gorillas, orangutans, and gray parrots also demonstrate these kinds of abilities in laboratory studies. For example, a gorilla named Koko was taught to

Table 15.1   Responses of vervet monkeys to playbacks of alarm calls. The data indicate that the monkeys recognize the meaning of different calls. (From Table 1 in R. M. Seyfarth, D. L. Cheney, and P. Marler, 1980, Monkey responses to three different alarm calls: evidence of predator classification and semantic communication, *Science* 210:801–803.)

Type of Alarm Call	Response to Playback			
	Climb into Tree	Run into Bushes	Look Up	Look Down
Leopard	8	2	4	1
Eagle	2	6	7	4
Snake	2	2	2	14

use American Sign Language to communicate (Figure 15.3); a bonobo named Kanzi points to abstract written symbols, or **lexigrams**, to tell his trainers when he wants a banana or a tickle (Figure 15.4); and a gray parrot named Alex has learned to count, label, and categorize objects as "the same" or "different" in spoken English. These individuals, and several others, have extensive vocabularies and seem to grasp the relationship between objects and arbitrary symbols.

None of the language-trained animals has mastered the full complexities of human syntax. They can string symbols together to make requests ("Sue tickle"), create novel terms ("water bird"), and express thoughts ("surprise hide"). However, nearly all of the strings are either short or repetitive. For example, they might sign "give orange give give orange give orange orange." Chimpanzees have been taught to use simple rules that specify the order between elements, such as action–agent (for example, "hug Roger"). Kanzi, the bonobo trained by Sue Savage-Rumbaugh (see Chapter 11), has gone one step further. Some of Kanzi's comments involve two lexigrams, or a lexigram and a gesture, and he has spontaneously generated simple rules to order these elements. For example, Kanzi is more likely to name an action ("keep away") first, and the agent or object second ("balloon"), than vice versa. He is also more likely to name something, and then point to the object, than he is to point to the object and then name it.

However, syntax is more than word order. Kanzi's combinations of elements and the most fluent signing apes both fail to meet several of the criteria for syntax

Figure 15.3

Koko has been trained by psychologist Francine Patterson to use American Sign Language.

Figure 15.4

Kanzi, a bonobo trained by Sue Savage-Rumbaugh and Duane Rumbaugh, uses abstract lexigrams to communicate.

that was defined earlier in the chapter. For example, the elements are not grouped into phrases, they provide no information about the timing of events, and much of the meaning must be derived from context. What do we make of the apes' achievements and limitations? Patricia Greenfield of the University of California, Los Angeles, emphasizes that ape abilities are comparable to those of very young children just beginning to acquire language, and she suggests that there is continuity between the linguistic abilities of apes and humans. Others, including Steven Pinker, emphasize differences between apes and young children in the ease with which they learn symbols, acquire grammatical rules, and naturally grasp the complexities of syntax. Those that emphasize this gap argue that linguistic abilities in humans are derived traits.

How Language Evolved

It is possible to imagine a sequence of intermediate languages linking the behavior of signing apes to modern human language.

Some people have argued that language, particularly grammar, is such a complex interdependent system of rules that it could not evolve bit by bit in the normal Darwinian fashion. Until the whole system of grammatical rules was in place, the argument runs, there would be no advantage to having any of the necessary structure.

This argument should be familiar. It is the same position that Darwin's critics sometimes take against the evolution of complex structures like the eye. It has no force as a general argument. As we saw in Chapter 1, there is ample evidence that the eye and many other complex structures did evolve bit by bit just as Darwinian theory requires. In any particular case, however, the argument could be true, so it is important to try to figure out plausible evolutionary pathways that connect the cognitive and communicative abilities of apes with those of modern humans.

Linguist Derek Bickerton suggested the following scenario for the evolution of language in his book *Language and Species*. Bickerton, remember, studies pidgins—primitive languages with little grammar. The typical utterance consists of a few words strung together in more or less random order. Pidgin speakers rely on context to sort among alternative meanings. If you have spent time in a country in which you did not speak the local language you are familiar with how this works. If someone who only knows a few words of German wants to buy a ticket on the 8:07 train to Hannover, she might say something like "Hannover seben oct," or, literally "Hannover seven eight." The man behind the counter infers that she wants a ticket from the fact that she is standing in the ticket line, and that it is the 8:07 and not the 7:08 because there is no 7:08 to Hannover. Remember, too, that language-trained apes can associate symbols with concepts and rearrange these symbols into utterances that resemble a simple pidgin.

Thus, Bickerton argues that language could evolve in several steps. First, humans could learn to associate particular vocalizations with particular concepts, much the way Kanzi associates abstract lexigrams with particular objects and concepts. This step could be based partly on a generalized learning ability and would not initially require any specialized learning mechanisms. The early hominid that attained this grade of language could derive substantial benefit without any syntax. For example, if your brother's dead body is lying on the ground, and a bystander says "kill Thag brother" you don't need grammatical rules to figure out that the action occurred in the past, that it was your brother that was killed, and that Thag was the perpetrator. You also have a much better idea about what happened than you would if you didn't have any language at all.

The next step would be to evolve the capacity for creating and understanding syntax. The advantage to having syntax is that it would reduce the dependence on context and increase accuracy. At first, syntactical rules could be acquired using general learning abilities. For example, people might use word order to distinguish the subject and object. Finally, once this pattern existed, selection could favor innate, special-purpose rules that made decoding sentences less arduous. Spoken language could have been elaborated using unspecialized mental abilities, and then special-purpose abilities could have evolved to automate the process.

The ability to communicate complex ideas must have been useful to early hominids.

Some people have argued that the richness and complexity of modern human language could not have arisen by natural selection. What possible advantage could an early hominid derive from the ability to understand the subtle and complex sentences of Proust or Faulkner? What use do foragers, who live in small groups and lead relatively simple lives, have for the great complexity of human language?

In fact, foragers use language the same way we do, and have just as much need for its powerful complexity. They face abstract, logical problems that are just as difficult as the problems that face Western scientists, and they use language to solve problems. For example, according to anthropologist Louis Liebenberg, tracking is very similar to scientific research. He studied the art under the tutelage of !Xõ hunter-gatherers who live in the vast and arid Kalahari Desert (Figure 15.5). The following anecdote illustrates his contention:

> While tracking down a solitary wildebeest spoor of the previous evening !Xõ trackers pointed out evidence of trampling which indicated that the animal had slept at that spot. They explained consequently that the spoor leaving the sleeping place had been made early that morning and was therefore relatively fresh. The spoor then followed a straight course, indicating that the animal was on its way to a specific destination. After a while, one tracker started to investigate several sets of footprints in a particular area. He pointed out that these footprints all belonged to the same animal, but were made during the previous days. He explained that the particular area was the feeding ground of that particular wildebeest. Since it was, by that time, about midday, it

Figure 15.5
!Kung men, like !Xõ men, are expert trackers and have an enormous amount of knowledge about their environment.

could be expected that the wildebeest may be resting in the shade in the near vicinity (From p. 80 in L. Liebenberg, 1990, *The Art of Tracking,* David Philip, Cape Town).

Acting on their hypothesis, the !Xõ followed the tracks and killed the wildebeest. Like scientists, they collected data, made careful observations of the spoor, and used existing knowledge to interpret their data. How do tracks change with time? How does this depend on soil type, recent wind conditions, and rainfall? And like modern scientists, they combined their data with other knowledge to generate a hypothesis. How do wildebeest behave? How does this depend on the temperature and on the time of year? !Xõ trackers stopped frequently to discuss their ideas. Different men offered competing hypotheses to explain the patterns that they saw, and there was much discussion of different views. In tracking, as in science, such discussion is extremely useful.

Even when they weren't tracking, the !Xõ spent a great deal of time talking about nature, sharing observations and hypotheses about the behavior of organisms they lived amongst and depended on. As a result, their knowledge of animal behavior frequently exceeded that of Western scientists. For example, !Xõ hunters told anthropologist H. Heinz in the 1970s that porcupines mated for life, a fact not known to zoologists until 1987. Think about how hard it would be to talk about tracks seen yesterday or how a particular feature of the tracks suggests why the kudu that was hunted during the last rains got away, without all of the grammatical machinery available to modern speakers.

As if these tasks aren't challenging enough, people living in small-scale societies face social dilemmas that are just as complex as the problems that plague any contemporary lawyer, priest, or psychotherapist. The welfare of people in small-scale societies depends crucially on the behavior of other group members. They cooperate in finding and sharing food, help each other when sick or injured, and mate with each other. Inevitably, such intimate interdependency gives rise to conflicts of interest. Rivalries, jealousies, feuds, and resentments arise. Among contemporary hunter-gatherers like the !Kung, language plays a crucial role in deter-

Figure 15.6

In a !Kung camp, people cluster around the fire and hold spirited conversations, tell stories, resolve disputes, and sing songs.

mining who did what, whether there were any mitigating circumstances, and what should be done (Figure 15.6). As anthropologist Melvin Konner of Emory University writes:

> Conflicts within the group are resolved by talking, sometimes half or all of the night, for nights and weeks on end. After two years with the San [another name for the !Kung], I came to think of the Pleistocene epoch of human history . . . as one interminable marathon encounter group. When we slept in a grass hut in one of their villages, there were many nights when its flimsy walls leaked charged exchanges of feeling and contention beginning when the dusk fires were lit and running on until the dawn (From p. 7 in M. Konner, 1982, *The Tangled Wing,* Harper & Row, New York).

When Language Evolved

Anthropologists are not certain when language evolved.

Language plays a crucial role in human ecology and social behavior. Suppose we had good evidence that early *Homo* was the first hominid to have language. This knowledge would change how we think about the lives of these creatures. It would then seem more likely that early *Homo* had complex social groups, food sharing, cooperative hunting, and a number of other features that characterize contemporary hunter-gatherers.

Unfortunately, we do not know when language evolved. The fossil and archaeological records provide some clues, but they are contradictory and inconclusive. Based on reconstructions of hominid brains, some anthropologists believe that the earliest hominids already had evolved significant language ability. Others, impressed with the apelike vocal tract of early humans and the startling innovations of the human revolution, believe that language evolved only in anatomically modern humans.

Did Language Arise Early?

Reconstructions of early hominid brain anatomy provide the best evidence that language arose early in human evolution—either with the australopithecines or early Homo.

The brains of modern hominids differ from those of apes. Certain derived features seen in human brains are associated with the production or comprehension of language. Some paleoanthropologists believe that detailed study of fossil hominid crania proves that these features first occurred in australopithecines or, at the latest, in early *Homo*.

To understand this line of argument, we need to discuss the relationship between speech and brain anatomy.

Language processing is located in the human brain.

Studies of brain injuries provided the first indication that brain functions are localized. Victims of strokes, brain tumors, and head injuries who suffer damage in a restricted part of the brain sometimes lose the ability to perform specific behaviors. By establishing an association between the location of damage and the nature of these behavioral deficits, neurobiologists have been able to determine the function of different parts of the brain.

These studies indicate that language processing is concentrated near the middle of the left hemisphere of the brain. This area is called the **perisylvian region** because of its proximity to the **sylvian fissure** (Figure 15.7). Within the perisylvian region, there are places where damage usually leads to specific linguistic deficits (Figure 15.8). For example, damage to Broca's area often leads to difficulty with syntax. Broca's aphasics, who have suffered damage to Broca's area, can remember individual words but have difficulty combining words into coherent, grammatically correct sentences. Thus, neurobiologists conclude that Broca's area has something to do with processing syntax. Damage in Wernicke's area usually causes people to have difficulty recognizing familiar words and to make mistakes

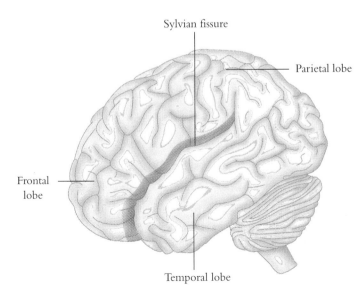

Sylvian fissure

Parietal lobe

Frontal lobe

Temporal lobe

Figure 15.7

The sylvian fissure is a deep involution on the surface of the brain that divides the frontal and parietal lobes from the temporal lobe. Language function is generally localized around the sylvian fissure.

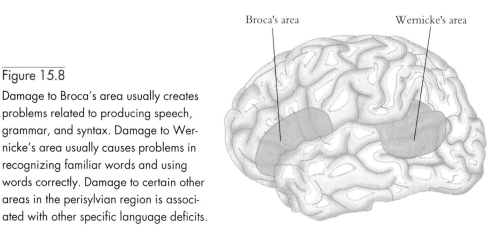

Figure 15.8

Damage to Broca's area usually creates problems related to producing speech, grammar, and syntax. Damage to Wernicke's area usually causes problems in recognizing familiar words and using words correctly. Damage to certain other areas in the perisylvian region is associated with other specific language deficits.

about their meaning. These people speak freely, but they inadvertently substitute incorrect sounds or words for correct ones, so they cannot be understood. People who suffer damage to both Broca's area and Wernicke's area lose the ability to produce or understand speech.

In the last decade, our knowledge of brain structure and function has been augmented by the use of sophisticated computer-aided imaging techniques. For example, positron emission tomography (PET) scanners allow neuroscientists to monitor the level of metabolic activity in particular regions of the brain. An experimental subject is put into the PET scanner and then asked to perform a specific linguistic task, such as listening to a list of words. The areas of the brain used to complete the task show increased metabolic activity. Different linguistic tasks produce metabolic activity in different parts of the brain (Figure 15.9).

It is tempting to think of specific sites in the brain as language "organs" that are analogous to other organs in the body, like the heart. However, there are problems with this analogy. The heart has only one function—to pump blood—and is in exactly the same place in every normal body. The language areas are not always located in exactly the same place in the brain, and a particular area in the brain may not always have the same function. For example, some people who suffer severe damage to Wernicke's area experience symptoms normally associated with damage to Broca's area, and vice versa. This does not mean that

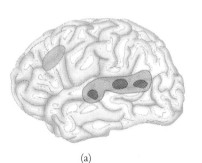

(a)

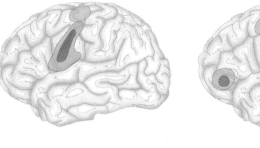

(b)

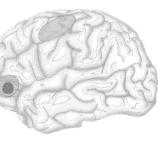

(c)

Figure 15.9

PET scans enable researchers to monitor brain activity while subjects perform specific linguistic tasks, such as (a) listening to a rapid sequence of words, (b) repeating out loud a series of rapidly presented words, and (c) associating a verb with a noun.

various language functions are not localized, only that we cannot map their locations precisely using the low-resolution techniques now available. As we shall see, this makes it harder to determine when language evolved.

Studies of the insides of fossil hominid crania convince some anatomists that language arose at least 2 million years ago.

We have seen that the brain size of hominids increased throughout the Pleistocene. Australopithecine brains were the same size as the brains of contemporary chimpanzees, while the brains of modern humans are much larger than australopithecine brains, even after accounting for the increased body size in humans.

There is good evidence that some regions of the brain expanded more than others. The relative sizes of different brain regions are similar among living apes. This suggests that the brains of the earliest hominids were like those of chimpanzees and gorillas. By comparing the relative size of various parts of human brains and chimpanzee brains, scientists can determine which parts of the human brain have expanded the most during the course of human evolution. The results of such comparisons indicate that the language areas in the perisylvian region became relatively larger, while other areas, like the visual cortex, became relatively smaller.

The human brain is a lumpy thing covered with folds separated by deep fissures called **sulci** (singular = sulcus). Sulci demarcate different brain regions. Thus, if we had a pickled australopithecine brain, we could use the sulci to determine whether the language areas had enlarged, and whether specific areas associated with language, like Broca's area, had attained the modern configuration. Of course, pickled australopithecine brains only exist in paleoanthropologists' dreams. However, we have a substitute. The fissures on the surface of the brain leave faint traces on the bone of the inside of the cranium. By making a plaster cast of the inside of a fossil cranium, called an **endocast**, many anatomists believe they can partly reconstruct the brain anatomy of extinct creatures.

The two principle researchers using endocasts to reconstruct hominid brains agree that language arose early in human evolution, but they disagree about exactly how early. Ralph Holloway of Columbia University believes that the language areas in the human brain had already increased substantially in the australopithecines. This belief is based on his reconstruction of the **lunate sulcus** in the Taung child. This prominent sulcus is located on the visual cortex of both apes and modern humans. However, because our visual cortex is relatively much smaller than that of apes, the lunate sulcus is closer to the back of our brains than it is in apes. Holloway argues that the Taung child's lunate sulcus is in the modern position. In contrast, Dean Falk of the State University of New York at Albany believes that the lunate sulcus in the Taung child and in a number of other australopithecine fossils was in the same position as in modern chimpanzees. She concludes that australopithecine brains had apelike proportions. Falk and Holloway agree that the first clear evidence for language-related brain structures comes with early *Homo*. In ER 1470, the large-brained specimen of early *Homo* from Lake Turkana, Falk detects the traces of two sulci that form the boundaries of Broca's area in human brains, but that are not found on ape brains.

Using fossil hominid endocasts as evidence for the origins of language has drawbacks.

Many anthropologists think the endocast evidence does not prove that language evolved early, for two reasons. First, different anatomists are unable to agree on the interpretation of the same endocast. For example, Falk and Holloway disagree about the position of the lunate sulcus on the Taung child endocast, and other anatomists who have examined the specimen think the position of the lunate sulcus is indeterminate. The reason for this controversy is that the traces of the sulci are very faint even in complete modern skulls. To complicate matters further, many fossil crania have been reconstructed from small pieces like three-dimensional jigsaw puzzles, a process that leaves many anomalous bumps on the endocast.

Second, language areas may have had different functions in early hominid brains than they have in modern humans. Remember that language function has not been fully mapped even in modern humans. Moreover, the area homologous to Broca's area exists in the brains of monkeys even though it does not have the same pattern of folds. The increase in brain size in the course of human evolution caused an increase in the number of folds all over the brain. Thus, the appearance of the sulci that presumably demarcate a language area could have been simply a side effect of brain enlargement. The increasingly upright stance of the bipedal body required that the brainstem shift its position also. This may have led to other changes in brain morphology, such as the movement of the lunate sulcus toward the back of the brain. Thus, the appearance of Broca's area in early *Homo* does not necessarily imply that these early hominids could talk.

Several other, less-than-convincing arguments have been put forward to support the hypothesis that language evolved early.

Scientists have cited a number of other kinds of evidence to show that language evolved early. None of the data is completely convincing.

- *Brain size* Early *Homo* had a large brain compared with those of apes and australopithecines. Those who believe that language is simply a consequence of greater intelligence argue that brain enlargement indicates that language capacities had evolved. However, the evidence that language ability is based on special-purpose brain machinery suggests this view is wrong. Moreover, there is no simple correlation between brain size and linguistic ability. Some contemporary individuals have brains about the same size as those of australopithecines, yet they are able to use language.
- *Brain asymmetry* In most modern humans, the left and right sides of the brain regulate different tasks: speech and fine motor control are localized on the left side of the brain, and facial recognition and other skills are found on the right side. Endocasts suggest that the brains of both australopithecines and early *Homo* were asymmetric. This is supported by the fact that the makers of Oldowan tools were mainly right-handed. Some authorities have argued that brain asymmetry implies language. However, recent evidence shows that different halves of ape and monkey brains also have different functions, though not separated by hemisphere to the same degree as in modern humans. Also, brain asymmetry is associated with many other brain functions beside language.
- *Complexity of language* Language requires a number of special-purpose adaptations. Both our brain and our vocal tract have been modified in complex, interdependent ways. Many authors feel it is unlikely that natural selection could have assembled such a complex adaptation in roughly 100,000 years, as required by the hypothesis that language arose late. The problem with this argument is that biologists do not have a very good idea of how long it takes for selection to create complex adaptations. There is a vague intuition that it probably takes a long time, but this intuition is not very well supported by theory or data.

Did Language Arise Late?

The morphology of the Homo erectus *boy, KNM-WT 15000, indicates that this species did not have the same control over its diaphragm and the muscles of it torso as modern humans do, which suggests that* Homo erectus *lacked language.*

The mammalian backbone contains many small bones, called **vertebrae**, stacked one on top of the other like dishes in a cupboard. Each vertebra has a hole in the middle like a doughnut. They line up to form a tube called the **vertebral canal**, which forms a conduit for the **spinal cord**, a massive bundle of nerves that connects the brain with the body. Overall, the vertebrae of WT 15000 are more similar to those of modern humans than they are to those of any other primate. In the thoracic region, however, WT 15000 is more apelike, and this bears on the question of language.

The ribs articulate (join) with **thoracic vertebrae**. The vertebral canal in the thoracic region is relatively much larger in modern humans than it is in apes, and it contains a proportionately thicker spinal cord. Detailed studies of the neuro-anatomy in the thoracic region suggest that all of the extra nerves that enlarge the spinal cord enervate the muscles of the rib cage and diaphragm. They do not carry messages to the arms and legs or other parts of the body. The thoracic vertebrae of WT 15000 are comparable to those of other primates, suggesting that this young male *Homo erectus* had less precise control over the muscles of his rib cage and diaphragm than modern humans do.

Anatomist Ann McLarnon of Whitelands College, England, argues that the enlargement of the thoracic vertebral canal is an adaptation for speech. When people talk, they subconsciously time their breathing to match the patterning of sentences. Sentences begin with a short, rapid intake of air, followed by a much slower outflow, and this allows production of the words of the sentence. Because sentences vary in length, the breathing pattern during speech is much more complex than normal rhythmic breathing is. McLarnon argued that the increased motor control of the diaphragm and thoracic muscles is an adaptation that allows people to manage this complex motor task. If McLarnon's hypothesis is correct, it follows that WT 15000 did not have the same speech abilities as modern humans. However, the evidence is not conclusive, and there may be alternate hypotheses to account for the enlarged thoracic vertebral canal in modern humans.

According to some authorities, reconstructions of fossil hominid vocal tracts indicate that language appeared only with modern humans, about 100,000 years ago.

Recall that the upper vocal tract of modern humans is quite different from the tracts of other primates. The larynx of modern humans is much lower than in other primates, an arrangement that allows us to produce the full range of vowels used in modern human languages. Based on reconstructions of the vocal tract of several fossil hominids, Brown University linguist Philip Lieberman argues that modern human vocal morphology first appeared in the fossil record when modern humans did, about 100,000 years ago. These reconstructions indicate that Neanderthals had vocal tracts much like those of other primates, and could not have produced the full range of sounds necessary for modern speech. To reconstruct the Neanderthal vocal tract, Lieberman first determined the shape of the modern human tongue by tracing X-ray images made while people were speaking. He then "fit" these drawings of modern human tongues to a drawing of a Neanderthal

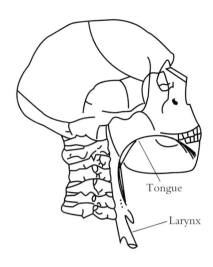

Figure 15.10

If a modern human tongue (forming the vowel "i," as in *hit*) is scaled to fit into a Neanderthal mouth, the larynx has to extend down to the chest. The lines represent the upper surface of a number of modern human tongues drawn by tracing X-ray images. These tongues were then enlarged to fill the Neanderthal mouth.

skull by enlarging the modern human tongue so that it was long enough to fill the very long Neanderthal mouth. When he did this, the part of the tongue that lies in the throat was so long that the larynx had to be placed in the chest (Figure 15.10). Since no primate has such an arrangement, Lieberman concluded that the Neanderthal vocal tract had proportions more like those of apes than of modern humans. Computer simulations of the reconstructed vocal tract suggest that Neanderthals would only have been able to produce a single vowel sound. The same procedure of fitting a modern human tongue into a fossil *H. sapiens* skeleton yields a modern vocal tract. Lieberman's conclusions are supported by studies of the hominid basicrania performed by anatomist Jeffrey Laitman of Mt. Sinai School of Medicine. (The basicranium is the base of the skull.) Laitman and his colleagues recorded the positions of several anatomical landmarks on the basicrania of humans and several other primates. In all primates the basicranium has a distinct hump, but the size of the hump differs (Figure 15.11). In infant humans, this basicranial hump is gentle, like those of other primates, but as humans mature, the bottom of the cranium develops a pronounced upward indentation. Laitman believes this relatively high hump helps to make room for the elongated human vocal tract. Australopithecine adults have apelike basicrania, while *H. erectus* is intermediate between apes and modern humans. Laitman argues that the initial shift in *H. erectus* may have facilitated breathing through the mouth during heavy aerobic exercise. The basicrania in Neanderthals show a less humped profile than those of *H. erectus*, suggesting that the Neanderthals' capacity for language was more restricted than the language capacities even of their own ancestors.

The discovery of a headless, but otherwise complete, Neanderthal skeleton at Kebara Cave by a team of French and Israeli anthropologists in 1983 has caused some researchers to question the conclusion that Neanderthals lacked the proper anatomy for language. This remarkable specimen included something never before found with a fossil hominid, a **hyoid bone**. The hyoid is a small U-shaped bone that sits in the jaw near the base of the tongue, and is part of the system that allows mammals to raise their larynx while swallowing. The Kebara hyoid is very similar to modern human hyoid bones, and is very different from the hyoid bones of apes. Careful study of the Kebara hyoid bone led University of Tel Aviv anatomist Baruch Arensburg to conclude that Lieberman was wrong: the Neanderthal vocal tract was indistinguishable from that of modern humans. However, Lieberman and Laitman strongly disagree, noting that the pig hyoid bone is also very similar to the human hyoid bone. As of this writing, the dispute remains unresolved.

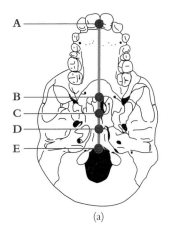

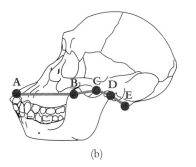

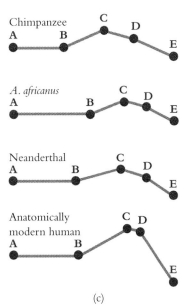

Figure 15.11

Jeffrey Laitman uses a number of anatomical landmarks to describe the shape of the basicranium. (a) The anatomical landmarks A–E are located on the basicranium here seen from below. (b) The path from one landmark to the next is traced in a side view. Notice that landmark C lies higher than landmarks B and D. (c) Using this pathway we can compare the shapes of the basicrania of humans, chimpanzees, and several early hominids. Notice that only the anatomically modern human has a sharply arched basicranium. Laitman believes this arching is necessary to make room for a humanlike vocal tract.

It is also possible that Lieberman and Laitman's reconstructions are correct, but that grammatically complex language still appeared long before modern humans evolved. It is possible for a creature with a high larynx to make all of the vowel sounds by using the nasal cavity. In modern humans, such nasalized vowels take longer to produce and are harder to understand. Thus, the lowering of the larynx might have served merely to decrease the error rate of understanding a language that was otherwise very modern. Second, as Steve Pinker points out, "e lengeege weth e smell nember ef vewels cen remeen quete expresseve" (*The Language Instinct*, p. 354). So, even if Lieberman is correct, early hominids might have had quite fancy languages.

> *Some authorities argue that dramatic changes in tools and art about 40,000 years ago suggest that modern language evolved about the same time.*

Remember that about 40,000 years ago there is an abrupt change in the archaeological record. Tools and other forms of technology became much more sophisticated, and there is widespread evidence of decoration, statuary, and other forms of symbolic activity. Many authors have argued that technology, art, and language are linked. The argument comes in several versions. Some researchers believe that the standardization of tool types and the existence of symbolic activity implies that modern humans were much better at dealing with abstract symbols than earlier hominids were. The ability to manipulate symbols more easily would facilitate language. Thus, these authors argue, the human revolution demarcates a major increase in linguistic ability. Another version of the argument focuses on the process of toolmaking. Standardized tools, it is argued, are manufactured in a series of steps using mental rules much like grammatical rules. Fancier, more standardized tools indicate that the toolmakers could master more complex rules, and this behavior could facilitate a more complex language.

These arguments assume that the same cognitive abilities that allow people to make tools and art are also responsible for language. However, as we have seen, there is good evidence that language is rooted in special-purpose mental abilities that have evolved to facilitate language. Language is based on a number of specialized cognitive skills, and there is no reason to believe that these skills are directly related to toolmaking or artistic ability.

Further Reading

Bickerton, D. 1990. *Language and Species*. University of Chicago Press, Chicago.
Lieberman, P. 1991. *Uniquely Human: The Evolution of Speech, Thought, and Selfless Behavior*. Harvard University Press, Cambridge, Mass.
Pinker, S. *The Language Instinct*. 1994. William Morrow, New York.
Pinker, S., and P. Bloom. 1990. Natural language and natural selection. *Behavioral and Brain Sciences* 13:707–784.

Study Questions

1. What evidence suggests that the universal cognitive capacities of human language are a product of genetic versus cultural evolution?
2. How does the anatomy of the throat and tongue of modern humans differ from that of apes? How do the derived features of the human vocal tract make speech possible?
3. Human speech is composed of many words that can be rearranged to make different meanings. What are the advantages of this system?
4. What is the evidence that language is based on special-purpose cognitive mechanisms?
5. Summarize the evidence that language evolved early in hominid evolution.

PART FOUR

Evolution and Modern Humans

C H A P T E R   1 6

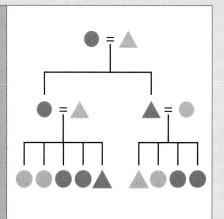

Human Genetic Diversity

Explaining Human Variation
Variation in Traits Influenced by Single Genes
    Causes of Genetic Variation within Groups
    Causes of Genetic Variation among Groups
Variation in Complex Phenotypic Traits
    Genetic Variation within Groups
    Genetic Variation among Groups
The Race Concept

Explaining Human Variation

Human beings vary in myriad ways. In any sizable group of people, there is variation in height, weight, hair color, eye color, food preferences, hobbies, musical tastes, skills, interests, and so on. Some people that you know are tall enough to dunk a basketball, some have to roll up the hems of all their pants; some have blue eyes and freckle in the sun, others have dark eyes and can get a terrific tan; some people have perfect pitch, others can't tell a flat from a sharp. Your friends may include heavy drinkers and teetotalers, great cooks and people who can't microwave popcorn, skilled gardeners and some who can't keep a geranium alive, some who play classical music and some who prefer heavy metal.

If we look around the world, we encounter an even wider range of variation. Some of the variation is easy to observe. Language, fashions, customs, religion, technology, architecture, and other aspects of behavior differ among societies. People in different parts of the world also look very different. For example, most of the people in northern Europe have blond hair and pale skin, while most of the

517

people in southern Asia have dark hair and dark skin. As we described in Chapter 13, Arctic peoples are generally shorter and stockier than people who live in the savannas of East Africa. Groups also differ in ways that cannot be detected as readily. For instance, the peoples of the world vary in blood type and the incidence of many genetically transmitted diseases. Figure 16.1 shows that the distributions of three debilitating diseases—PKU, cystic fibrosis, and Tay-Sachs—vary considerably among populations. Tay-Sachs, for example, is nearly 10 times more common among Ashkenazi Jews in New York than among other New Yorkers.

In this chapter we consider how much of this variation is due to genetic differences among people. We want to know how people vary genetically within and among societies, and to understand the processes that create and sustain this variation. We begin by describing the nature of variation in traits that are influenced by single genes with large effects. Next, we consider variation in traits that are influenced by many genes. As you will see, the methods that are used to assess variation in traits caused by single genes and multiple genes are quite different. In both cases, we will consider the processes that give rise to variation within and among populations. Finally, we will use our understanding of human genetic diversity to explore the significance and meaning of a concept that plays an important, albeit often negative, role in modern society—race. We will argue that a clear understanding of the nature and source of human genetic variation demonstrates that race is not a valid scientific construct.

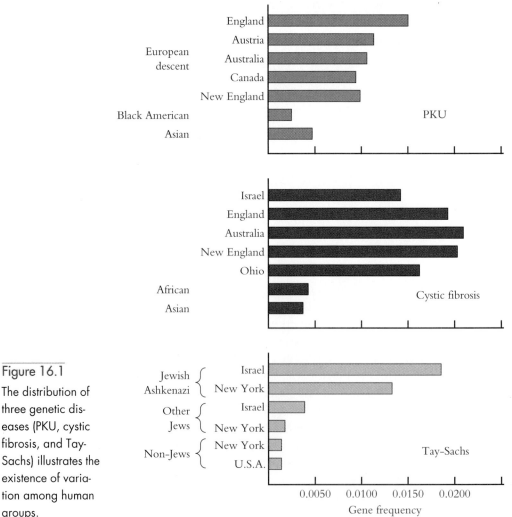

Figure 16.1

The distribution of three genetic diseases (PKU, cystic fibrosis, and Tay-Sachs) illustrates the existence of variation among human groups.

Scientists distinguish two sources of human variation: genetic and environmental.

Scientists conventionally divide the causes of human variation into two categories. **Genetic variation** refers to differences between individuals that are caused by the genes they inherited from their parents, while **environmental variation** refers to differences between individuals caused by environmental factors (such as climate, habitat, and competing species) on the organisms' phenotypes. For humans, culture is an important source of environmental variation.

A practical example—variation in body weight—will clarify this distinction. Many environmental factors affect body weight. Some factors, such as the availability of food, have an obvious and direct impact on body weight. The majority of people living under siege in Sarajevo in the mid-1990s were undoubtedly leaner than they were a decade earlier when Sarajevo was a rich, cosmopolitan city. Other environmental effects are more subtle. For example, culture can affect body weight because it influences our ideas about what constitutes an appropriate diet and shapes our standards of physical beauty. In the United States, many young women adopt strict diets and rigorous exercise regimes in order to maintain a slim figure because thinness is considered desirable. But in a number of West African societies, young women are secluded and force-fed large meals several times a day for the express purpose of gaining weight and becoming fat. In these societies, obesity is extremely desirable, and fat women are thought to be very beautiful. Body weight also appears to have an important genetic component. Recent research has shown that individuals with some genotypes are predisposed to be heavier than others, even when diet and levels of activity are controlled.

Genetic and environmental causes of variation may also interact in complicated ways. Consider, for example, two people who have acquired quite different ideas about physical beauty; one believes that good looks are extremely important and that it is desirable to be quite thin, while the other thinks that being thin is not particularly important and has a healthy appetite. Both individuals may be thin if they have to subsist on one cup of porridge a day, but only the one who is unconcerned about being thin will gain weight when Big Macs and fries are readily available.

It is difficult to determine the relative importance of genetic and environmental influences for particular phenotypic traits.

It is often difficult to separate the genetic and environmental causes of human variation in real situations. The problem is that both genetic transmission and shared environments cause parents and offspring to be similar. For example, suppose we were to measure the weights of parents and offspring in a series of families living in a range of environments. It is likely that the weight of parents and offspring (corrected for age) would be closely related. However, we would not know whether the association was an effect of genes or environment. Children might resemble their parents because they inherited genes that affect fat metabolism or because they learned eating habits and acquired food preferences from their parents.

Quite different processes create and maintain genetic and environmental variation among groups, and identifying the source of human differences will help us understand why people are the way they are. Genetic variation is governed by the processes of organic evolution—mutation, drift, recombination, and selection. Biologists and anthropologists know a great deal about how the various processes work to shape the living world and how evolutionary processes explain genetic differences among contemporary humans in particular cases.

It is also important to distinguish variation within human groups from variation among human groups.

Variation within groups refers to differences between individuals within a given group of people. In the National Basketball Association, for example, 1.60-m (5 ft 3 in.) Muggsy Bogues, now with the Golden State Warriors, competed against much larger players, like 2.11-m (6 ft 11 in.) Bill Lambeer, formerly of the Detroit Pistons (Figure 16.2). **Variation among groups** refers to differences between entire groups of people. For instance, the average height of NBA players is much greater than the average height of professional jockeys (Figure 16.3). It is important to distinguish these two levels of variation because, as we shall see, the source of the variation within groups can be very different from the source of variation among groups.

Variation in Traits Influenced by Single Genes

Scientists have shown that variation in some traits is genetic by establishing the connection between particular DNA sequences and specific traits.

Although it is sometimes difficult to establish the source of variation in human traits, there are some cases in which we are certain that variation arises from genetic differences between individuals. For example, recall from Chapter 2 that in West Africa many people suffer from **sickle-cell anemia**, a disease that causes their red blood cells to have a sickle shape instead of the more typical rounded shape. We know that people with this debilitating disease are homozygous for a gene that codes for one variant of hemoglobin, the protein that transports oxygen molecules in red blood cells. Hemoglobin is made up of two different protein subunits, labeled α (the Greek letter alpha) and β (the Greek letter beta). The DNA sequence of the most common hemoglobin gene, hemoglobin A, specifies the amino acid glutamic acid in the sixth position of the protein chain of the β subunit. But there is another hemoglobin gene, hemoglobin S, which specifies the amino acid valine at this position. People who suffer from sickle-cell anemia are homozygous for the hemoglobin S gene.

We can prove that traits that are controlled by genes at a single genetic locus by showing that their patterns of inheritance conform to Mendel's principles.

Although it is often difficult to distinguish between genetic and environmental sources of variation, we can sometimes do so when traits are affected by genes at a single genetic locus. In such cases, Mendel's laws make very detailed predictions about the patterns of inheritance (see Chapter 2). If scientists suspect that a trait is controlled by genes at a single genetic locus, they can test this idea by collecting data on the occurrence of the trait in families. If the pattern of inheritance shows a close fit to the pattern predicted by Mendel's principles (conventionally called "laws"), then we can be confident that the trait is affected by a single genetic locus.

Research by the McGill University linguist Myrna Gopnik on the genetic basis of a language disorder called **specific language impairment (SLI)** illustrates this approach. Children with SLI have difficulty learning to speak, and although most eventually learn, they continue to make frequent grammatical errors as adults. They have great difficulty, for example, forming the plurals of unfamiliar words. If most

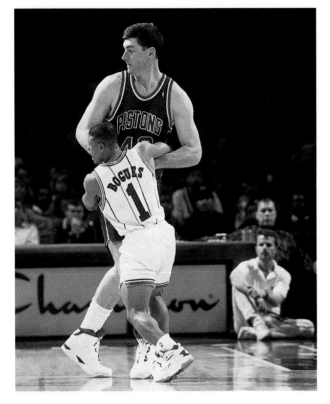

Figure 16.2

Muggsy Bogues, the 1.60-m (5 ft 3 in.) guard, now with the Golden State Warriors, lays a screen on the 2.11-m (6 ft 11 in.) former Detroit Pistons center, Bill Lambeer. The difference in their heights illustrates variation in this character within the group consisting of NBA players.

Figure 16.3

Former NBA star Wilt Chamberlain (left), at 2.21 m (7 ft 2 in.), and the 1.52-m (4 ft 11 in.) star jockey, Bill Shoemaker. The difference in height between these two men exemplifies the variation in height between NBA players and professional jockeys.

children are told a doll is called a *wug,* and asked what are two of the dolls called, they readily answer *wugs.* In contrast, children suffering from SLI find this problem very difficult and many cannot form the plural at all. The results of a battery of such tests have led Gopnik to conclude that sufferers of SLI have a specific cognitive deficit that prevents them from acquiring morphology, the rules that govern the way words are put together.

The patterns of inheritance of SLI in families suggest that at least some cases of SLI are caused by a single, dominant allele. Gopnik and her colleagues have collected data from a number of large, multigenerational families like the one diagrammed in Figure 16.4. The grandmother (shown as a black circle) had SLI, but her husband (a white triangle) did not. Four of her five children and 11 of her 24 grandchildren also had SLI. Suppose SLI is caused by a dominant gene. Then, since SLI is rare in the population as a whole, the Hardy-Weinberg equations tell us that almost all SLI sufferers will be heterozygotes. Of course, anyone not suffering SLI must be a homozygote for the normal allele at this locus. From Mendel's laws, on average half of the offspring of a mating between a person suffering SLI and a nonsufferer will suffer SLI and half will have normal linguistic skills. The families shown in this tree fit this prediction very well. Both of the children of the son without SLI are normal, and the rest of the matings produced approximately equal numbers of normal and language-impaired children.

While the data from families like the one in Figure 16.4 suggest that SLI is caused by a single dominant gene, it is possible that there is some environmental factor that causes SLI to run in families, and to mimic a pattern of genetic inheritance. Scientists search for two kinds of data to clinch the case. First, they collect data on more families. The larger the number of families that fit the pattern associated with the inheritance of a single-locus dominant gene, the more confident researchers can be that this pattern did not occur by chance. Second, researchers search for **genetic markers** (genes whose location in the genome is known) that show the same pattern of inheritance. Thus, if every individual who has SLI also has a specific marker on a particular chromosome, we can be confident that the gene that causes SLI lies close to that genetic marker. Gopnik and her colleagues have collected data on a number of large families from populations speaking several

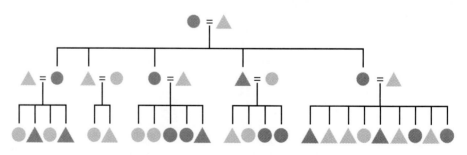

Figure 16.4

This family tree suggests that some cases of specific language impairment (SLI) are caused by a single dominant gene. Circles represent women, triangles represent men, and blue symbols represent people with SLI. If SLI is caused by a dominant gene, then, since SLI is rare in the population as a whole, we know from the Hardy-Weinberg equations that almost all people with SLI will be heterozygotes. Thus, Mendel's principles tell us that, on average, half of the offspring of a mating between a person with SLI and a person without it will have SLI, and half the offspring will have normal linguistic skills. Notice how well the families shown in this tree fit this prediction.

languages, and all show the single-locus dominant pattern. As of this writing, they are still searching for a marker that shows the same pattern of inheritance as SLI.

If it turns out that SLI is caused by a single gene, it does not necessarily mean that the gene is responsible for all of the psychological machinery in the human brain that implements morphological rules. It only means that there is a single gene that, if damaged, prevents that bit of cognitive machinery from working properly. If you cut the wire connecting the hard disk to the power supply in your computer, the hard disk will stop working, but that does not mean that the wire contains all of the machinery necessary for the operation of the hard disk.

Causes of Genetic Variation within Groups

Mutation can maintain deleterious genes in populations, but only at a low frequency.

There are many diseases that are caused by recessive genes. For example, only people who are homozygous for hemoglobin S are afflicted with sickle-cell anemia. Other diseases caused by recessive alleles include PKU (phenylketonuria, introduced in Chapter 3), Tay-Sachs disease, cystic fibrosis, and Huntington's chorea. All of these diseases are caused by mutant genes that code for proteins that do not serve their normal function, and all produce severe debilitation and sometimes death. Why haven't such deleterious genes been eliminated by natural selection?

One answer to this question is that natural selection steadily removes such genes, but they are constantly being reintroduced by mutation. Very low rates of mutation can maintain these deleterious genes because they are recessive traits, and so most individuals who carry the gene are heterozygous for the deleterious genes and do not suffer the disastrous consequences that homozygotes suffer. The observed frequency of many deleterious recessive genes is about 1 in 1000. Then, according to the Hardy-Weinberg equations, the frequency of newborns homozygous for the recessive allele will be $0.001 \times 0.001 = 0.000001$! Thus, only one in a million babies will carry the disease. This means that even if the disease is fatal, selection will remove only two copies of the deleterious gene for every 1 million people born. Since mutation rates for such deleterious genes are estimated to be a few mutations per million gametes produced, mutation will introduce enough new mutants to maintain a constant frequency of the gene. When this is true, we say that there is **selection-mutation balance**.

Selection can maintain variation within populations if heterozygotes have higher fitness than either of the two homozygotes.

Some deleterious lethal genes are too common to be the result of selection-mutation balance. For example, in West African populations the frequency of the hemoglobin S allele is typically about 1 in 10. How can we account for this? The answer in the case of hemoglobin S is that this allele increases the fitness of heterozygotes. It turns out that individuals who carry one copy of the sickling allele, *S*, and one copy of the normal allele, *A*, are partially protected against the most dangerous form of malaria, called **falciparum malaria** (Figure 16.5). As a consequence, where falciparum malaria is prevalent, heterozygous *AS* newborns are about 15% more likely to reach adulthood than *AA* infants.

When heterozygotes have a higher fitness than either homozygote, natural selection will maintain a **balanced polymorphism**, a steady state in which both

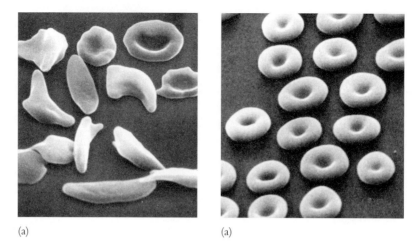

(a)   (a)

Figure 16.5

(a) Sufferers of sickle-cell anemia have abnormal red blood cells with a sickle shape. (b) Normal red blood cells are round. Sickle-cell anemia partially protects against falciparum malaria.

alleles persist in the population. To see why balanced polymorphisms exist, first consider what happens when the S allele is introduced into a population and is very rare. Let us say that its frequency is 0.001. The frequency of SS individuals will be 0.001×0.001, or about 1 in 1 million, while the frequency of AS individuals will be $2 \times 0.001 \times 0.999$, or about 2 in 1000. This means that for every individual who suffers the debilitating effects of sickle-cell anemia, there will be about 2000 heterozygotes who are partially immune to malaria. Thus, when the S allele is rare, most S alleles will occur in heterozygotes, and the S allele will increase in frequency. However, this process will not lead to the elimination of the A allele. To see why, let's consider what happens when the S allele is common and the A allele is rare. Now almost all of the A alleles will occur in AS heterozygotes, and be partially resistant to malaria, while almost all of the S alleles will occur in SS homozygotes and suffer debilitating anemia. The A allele has higher fitness than the S allele when the S allele is common (Box 16.1). The balance between these two processes depends on the fitness advantage of the heterozygotes and the disadvantage of the homozygotes. In this case, the equilibrium frequency for the hemoglobin S allele is about 0.1, approximately the frequency actually observed in West Africa.

Scientists suspect that the relatively high frequencies of genes that cause a number of other genetic diseases may also be the result of heterozygote advantage. For example, the gene that causes Tay-Sachs disease has a frequency as high as 0.05 in some Eastern European Jewish populations. Children who are homozygous for this gene seem normal for about the first six months of life. Then, over the next few years a gradual deterioration takes place, leading to blindness, convulsions, and finally death, usually by age four. There is some evidence that individuals who are heterozygous for the Tay-Sachs allele are partially resistant to tuberculosis. Jared Diamond of the University of California at Los Angeles points out that tuberculosis was much more prevalent in cities than in rural areas of Europe over the last 400 years. Jews, confined to the crowded urban ghettos of Eastern Europe, may have benefited more from increased resistance to tuberculosis than the other Europeans, most of whom lived in rural settings.

BOX 16.1

Calculating Gene Frequencies
for a Balanced Polymorphism

It is easy to calculate the frequency of hemoglobin S when selection has reached a stable, balanced polymorphism. Suppose the fitness of *AA* homozygotes is 1.0, the fitness of *AS* heterozygotes is 1.15, and the fitness of *SS* homozygotes is 0, and let p be the equilibrium frequency of the allele *S*. If individuals mate at random, a fraction p of the *S* alleles will unite with another *S* allele to form an *SS* heterozygote, and a fraction $1 - p$ will unite with an *A* allele to form an *AS* heterozygote. Thus, the average fitness of the *S* allele will be

$$0p + 1.15(1 - p)$$

By the same reasoning the average fitness of the *A* allele will be

$$1.15p + 1(1 - p)$$

The relationship between the average fitness of each allele and the frequency of hemoglobin S is shown in Figure 16.6, and confirms the reasoning given in the text. When *S* is common, so that p is close to 1, the average fitness of the *S* allele is close to zero, but when *S* is rare, its average fitness is almost 1.15. If one gene has a higher fitness than the other, natural selection will increase the frequency of that gene. Thus, a steady state will occur when the average fitnesses of the two alleles are equal; that is, when

$$1.15(1 - p) = 1.15p + (1 - p)$$

If you solve for p, you will find that

$$p = \frac{1.15 - 1.0}{1.15 + 1.15 - 1.0}$$
$$= \frac{0.15}{1.30} \approx 0.1$$

which is about the observed frequency of the sickle-cell allele in West Africa.

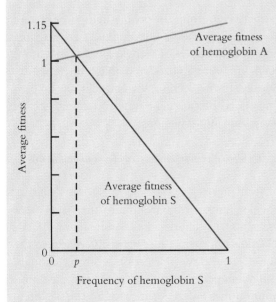

Figure 16.6

The average fitness of the *S* allele of hemoglobin S declines as the frequency of *S* increases because more and more *S* alleles are found in *SS* homozygotes. Similarly, the average fitness of the *A* allele of hemoglobin A increases as the frequency of *S* increases because more and more *A* alleles are found in *AS* heterozygotes. A balanced polymorphism occurs when the average fitness of the two alleles is equal.

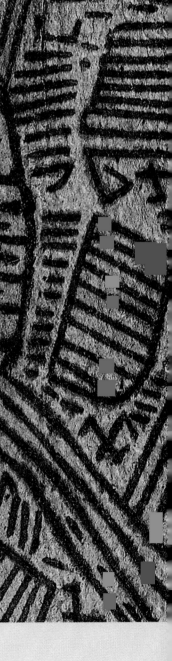

Variation may exist because environments have recently changed and genes that were previously beneficial have not yet been eliminated.

Some genetic diseases may be common because the symptoms they create have not always been deleterious. One form of diabetes, **noninsulin-dependent diabetes (NIDD)**, may be an example of such a disease. **Insulin** is a protein that controls the uptake of blood sugar by cells. In NIDD sufferers, blood-sugar levels rise above normal levels because the cells of the body do not respond properly to insulin in the blood. High blood-sugar levels cause a number of problems including heart disease, kidney damage, and impaired vision. NIDD is also known to have a genetic basis. (The other form of diabetes, insulin-dependent diabetes, occurs because the insulin-producing cells in the pancreas have been destroyed by the body's own immune system. It is unlikely that insulin-dependent diabetes was ever adaptive.)

In some contemporary populations the occurrence of NIDD is very high. For example, on the Micronesian island of Nauru more than 30% of people over 15 years of age now have the disease. Such high rates of NIDD are a recent phenomenon, although the genes that cause the disease are not new. Human geneticist J. V. Neel of the University of Michigan has suggested that the genes now leading to NIDD were beneficial in the past because they caused a rapid buildup of fat reserves during periods of plenty—fat reserves that would help people survive periods of famine in harsh environments. Traditionally, life on Nauru was very difficult. The inhabitants subsisted by fishing and farming, and famine was common. NIDD was virtually unknown during this period. However, Nauru was colonized by Britain, Australia, and New Zealand in recent times, and this brought many changes in the residents' lives. They obtained access to Western food, and prosperity derived from the island's phosphate deposits allowed them to adopt a sedentary lifestyle. NIDD became common. Genes that formerly conferred an advantage on the residents of Nauru now cause NIDD to be the leading cause of nonaccidental death.

Causes of Genetic Variation among Groups

There are many genetic differences between groups of people living in different parts of the world. The existence of genetic variation among groups is intriguing because we know that all living people are members of a single species, and, as we saw in Chapter 4, gene flow between different populations within a single species tends to make them uniform genetically. In this section, we consider several processes that oppose the homogenizing effects of such gene flow and thereby create and maintain genetic variation among human populations.

Selection that favors different genes in different environments creates and maintains variation among groups.

The human species inhabits a wider range of environments than any other mammal. We know that natural selection in different environments may favor different genes, and that natural selection can maintain genetic differences in the face of the homogenizing influence of gene flow if selection is strong enough. Variation in the distribution of hemoglobin genes provides a good example of this process. Hemoglobin S is most common in tropical Africa, around the Mediterranean Sea, and in southern India (Figure 16.7a). Elsewhere it is almost unknown. Generally, hemoglobin S is prevalent where falciparum malaria is common, and hemoglobin A is prevalent where this form of malaria is absent (Figure 16.7b).

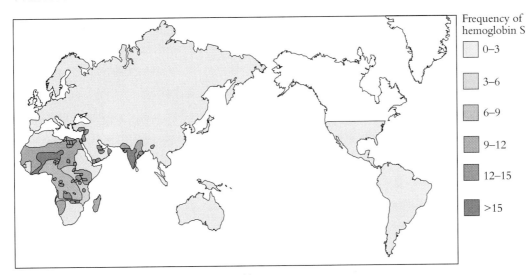

(a)

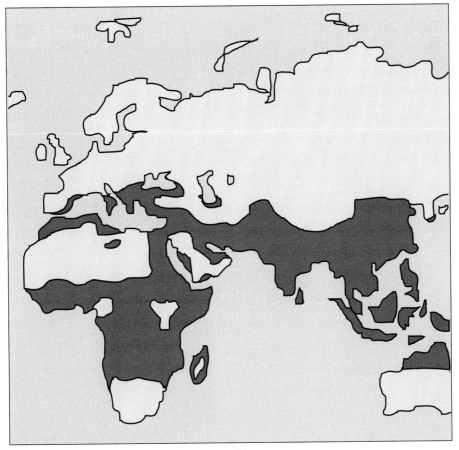

(b)

Figure 16.7

Hemoglobin S is only common in areas of the world in which falciparum malaria is prevalent. (a) The colors show the frequency of hemoglobin S throughout the world. (b) The regions of the Old World in which falciparum malaria is prevalent are in red.

Southeastern Asia represents an exception to this pattern, and it is possible that hemoglobin E, a hemoglobin gene that is common in that region, also provides resistance to malaria.

Table 16.1   The ability to digest lactose is controlled by a single genetic locus with two alleles, *LAC*P* and *LAC*R*.

GENOTYPE	PHENOTYPE
LAC*P/LAC*P	Digests lactose as adults (HDC)
LAC*P/LAC*R	Digests lactose as adults (HDC)
LAC*R/LAC*R	Can't digest lactose as adults (LDC)

The gene that controls the digestion of **lactose**, a sugar found in mammalian milk, provides another interesting example of genetic variation maintained by natural selection. Lactose is synthesized in the mammary glands, and occurs in large amounts only in mammalian milk. Most mammals have the ability to digest lactose as infants but gradually lose this ability after they are weaned. The vast majority of humans follow the mammalian pattern, and cannot digest lactose after the age of five. Such people are said to have **low digestive capacity (LDC)**. When people with LDC drink more than about a half a liter (just over 1 pint) of fresh milk at a sitting, they suffer gastric distress that ranges from mild discomfort to quite severe pain. However, most northern Europeans and members of a number of North African and Arabian populations retain the ability to digest lactose as adults, and are said to have **high digestive capacity (HDC)**. Evidence from family studies indicates that the ability to digest lactose as an adult is controlled by a single dominant gene, labeled **$LAC\star P$**. People who have one copy of $LAC\star P$ have HDC; people who are homozygous for the alternative allele, labeled **$LAC\star R$**, have LDC. The relationship between genotype and phenotype is summarized in Table 16.1.

The prevalence of $LAC\star P$ in the deserts of North Africa and Arabia is probably the result of natural selection for the ability to digest large amounts of fresh milk. Traditionally, nomadic pastoralism has been the main mode of subsistence in this part of the world, and fresh milk plays a crucial role in this way of life. Gebhard Flatz of the Medizinische Hochschule of Hannover, Germany, studied the Beja, a people who wander with their herds of camels and goats in the desert lands between the Nile and the Red Sea (Figure 16.8). During the nine-month dry season, the Beja rely almost entirely on the milk of their camels and goats. They drink about 3 liters of fresh milk a day, and they obtain virtually all of their energy, protein, and water from milk. Moreover, the milk must be consumed when it is fresh. High desert temperatures prevent the Beja from storing milk, and their nomadic lifestyle makes it difficult to produce dairy products, like cheese and yogurt, that lack lactose. The Beja probably could not survive without the ability to digest fresh milk. Thus, it comes as no surprise that $LAC\star P$ is common among them—84% of the Beja are able to digest lactose.

Scientists are less sure about why $LAC\star P$ is common in northern Europe. HDC frequencies are highest in Scandinavia and Great Britain, decline steadily to the south, and reach quite low values around the Mediterranean Sea. Traditionally, large amounts of fresh milk have not been an important part of the diet of the peoples in this region. Low temperatures make milk storage practical, and northern Europeans have long made use of cheese and other processed-milk products. Thus, it does not seem likely that the ability to digest fresh milk would provide a big enough advantage to account for the high frequency of $LAC\star P$.

Some scientists conjecture that modern northern Europeans are descended from

Figure 16.8

Pastoralists in northern Africa, like the Beja, herd camels, and during some parts of the year they obtain virtually all of their nourishment from fresh milk. A high proportion of the Beja are able to digest lactose as adults.

nomadic horse pastoralists who invaded Europe from the steppes of central Asia. Milk consumption may have played the same role in these ancient pastoral economies as it does today among pastoralists like the Beja. Others have argued that lactose digestion is favored in northern Europe because it enhances absorption of vitamin D. When sunlight penetrates the fatty tissue in human skin, vitamin D is synthesized. Consequently, in sunny environments humans rarely suffer vitamin D deficiency. In northern Europe, long, dark, cloudy winters create a potential for vitamin D deficiency. Laboratory studies suggest that the ability to digest lactose increases vitamin D absorption. Thus, the ability to digest lactose might be common in northern Europe because it helps prevent vitamin D deficiency.

Many people think of evolution by natural selection as a glacially slow process that acts only over millions of years. However, the ability to digest lactose has evolved since people began domesticating livestock in the Middle East and North Africa, about 7000 years ago. Moreover, there is no need to posit extraordinarily strong selection to account for this change. Figure 16.9 shows how fast HDC could increase in frequency, assuming that the ability to digest lactose leads to a 3% increase in fitness. As you can see, even this relatively small benefit from the ability to digest lactose can easily explain the high frequencies of HDC we now find in some parts of the world.

Genetic drift creates variation among isolated populations.

In Chapter 3 we saw that genetic drift causes random changes in gene frequencies. This means that if two populations become isolated from each other, both will change randomly, and over time the two populations will become genetically distinct. Because drift occurs more rapidly in small populations than in large ones, small populations will diverge from one another faster than large ones will. Genetic

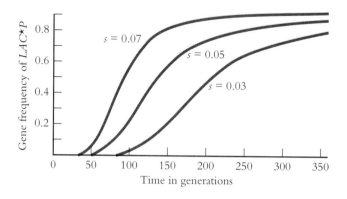

Figure 16.9

The *LAC*P* gene could have spread in the 7000 years (300–350 generations) since the origin of dairying if the ability to digest lactose as an adult leads to even as little as a 3% increase in fitness (s).

drift caused by the expansion of a small founding population is sometimes called the **founder effect**.

Genetic differences among members of three religious communities in North America demonstrate how this process can create variation among human groups. Two of these groups are Anabaptist sects—the Old Order Amish (Figure 16.10) and Hutterites—and the third group is the Utah Mormons. Each of these groups forms a well-defined population. About 2000 Mormons first arrived in the area of what is now Salt Lake City in 1847, but members of the church continued to arrive until 1890. Virtually all of the immigrants were of Northern European descent. At the turn of the century, there were about 250,000 people in this area, and about 70% of them belonged to the Mormon Church. In contrast, the two Anabaptist groups were much smaller. The founding population of the Old Order Amish was only 200 people, and gene flow from outside the group has been very

Figure 16.10

The Old Order Amish were founded by a group of 200 people. The Amish dress plainly and shun most forms of modern technology, including motor vehicles.

limited. Contemporary Hutterites are all descended from a population of only 443 people and, like the Amish, have been almost closed to immigration.

Researchers have studied the genetic composition of each of these populations. They have found that Mormons are genetically similar to other European populations. Thus, even though Mormon populations have been partly isolated from other European populations for over 150 years, genetic drift has led to very little change. This is just what we would expect given the size of the Mormon population. In contrast, the two Anabaptist populations are quite distinct from other European populations. Because their founding populations were small and the communities were genetically isolated, drift has created substantial genetic changes in the same period of time.

Genetic drift can also explain why certain genetic diseases are common in some populations but not in others. For example, Afrikaners in what is now the Republic of South Africa are the descendants of Dutch immigrants who arrived in the 17th century. By chance, this small group of early immigrants carried a number of rare genetic diseases, and these genes occurred at much higher frequency among members of the colonizing population than in the Dutch populations from which the immigrants were originally drawn. The Afrikaner population grew very rapidly and preserved these initially high frequencies, causing these genes to occur in higher frequencies among modern Afrikaners than in other populations. For example, sufferers of the genetic disease **porphyria variegata** develop a severe reaction to certain anesthetics. About 30,000 Afrikaners now carry the dominant gene that causes this disease, and every one of these people is descended from a single couple who arrived from Holland in the 1680s.

Current patterns of genetic variation reflect the history of migration and population growth in the human species.

Some of the genetic variation among human groups reflects the history of the peoples of the earth. In Chapter 14 we explained that the pattern of genetic variation in mitochondrial DNA indicates that the human species underwent a worldwide population expansion about 100,000 years ago. This was only one of several expansions of the world's population. The invention of agriculture led to expansions of farming peoples into Europe, eastern Asia, **Oceania** (the Pacific Islands groups of Polynesia, Melanesia, and Micronesia), and central Africa between 4000 and 1000 years ago; the domestication of the horse and associated military innovations led to several expansions of peoples living in the steppes of central Asia between 3000 and 500 years ago; and improvements in ships, navigation, and military organization led to the expansion of European populations during the last 500 years.

L. L. Cavalli-Sforza, a human geneticist at Stanford University, has argued that patterns of genetic variation preserve a record of these expansions. As human populations grow, they often expand into new geographic regions. As the distance grows between parts of the population, they become genetically isolated from one another and begin to accumulate genetic differences. If the population expansion continues, the initial "parent" population may eventually be split into several isolated and genetically differentiated "daughter" populations. If the daughter populations remained completely isolated from all other human populations, they would be just like new species, and we could use the techniques of phylogenetic reconstruction described in Chapter 4 to reconstruct the demographic history of the daughter populations. Such complete isolation is very rare—instead, daughter populations usually experience gene flow with each other and with other human

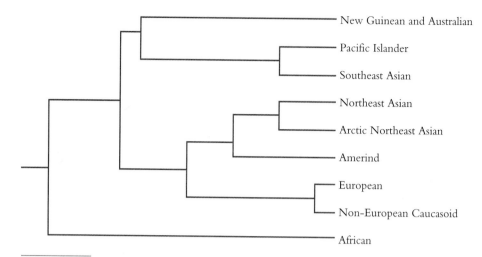

Figure 16.11

This tree, based on the frequencies of 120 genes in 42 populations from every part of the globe, is consistent with the hypothesis that humans originated in Africa and spread from there to the rest of the globe.

populations they encounter during their expansion, and such gene flow tends to obscure the history of the populations. Nonetheless, Cavalli-Sforza argues, if there is not too much gene flow, the present patterns of genetic variation can help us reconstruct the pattern of past migrations.

Cavalli-Sforza and his colleagues believe that the worldwide pattern of genetic variation results from the expansion of modern humans about 100,000 years ago. Recall from Chapter 14 that supporters of the replacement model believe this expansion began in Africa, and then spread north and east to the rest of Eurasia and Australasia, and finally reached the New World. Cavalli-Sforza and his colleagues have collected data on gene frequencies in populations across the world. Figure 16.11 shows a tree based on the frequencies of 120 genes in 42 populations from every part of the globe. This tree was constructed using the genetic distance methods described in Chapter 4, and assumes that the rate of genetic change along each branch of the tree has been constant. This tree is consistent with the view that modern humans originated in Africa and spread to the rest of the world. A simplified version of the tree (Figure 16.12) tells us that African populations first expanded into southern Asia. Then southern Asian populations expanded into northern Asia and Oceania. Finally, northern Asian populations spread into Europe. If at each split, daughter populations remained isolated, then genetic distances would have the general form shown in Figure 16.12b, and further differentiation within regions would produce the tree shown in Figure 16.11.

Other data suggest that the pattern of genetic variation in Asia and Europe has been influenced by subsequent demographic events. Figure 16.13 shows a tree constructed using data from a different set of over 100 genetic loci. Like the previous tree, this one suggests that there was an early split between African populations and Eurasian populations. However, because this tree was constructed using a method that does not assume a constant rate of genetic change, it also indicates that European populations evolved roughly 10 times more slowly than other human populations. Since the rate of genetic drift depends mainly on population size, it is possible that European populations changed slowly because they were much larger than other populations. However, archaeological data give little support to this idea. A more plausible explanation is that European populations

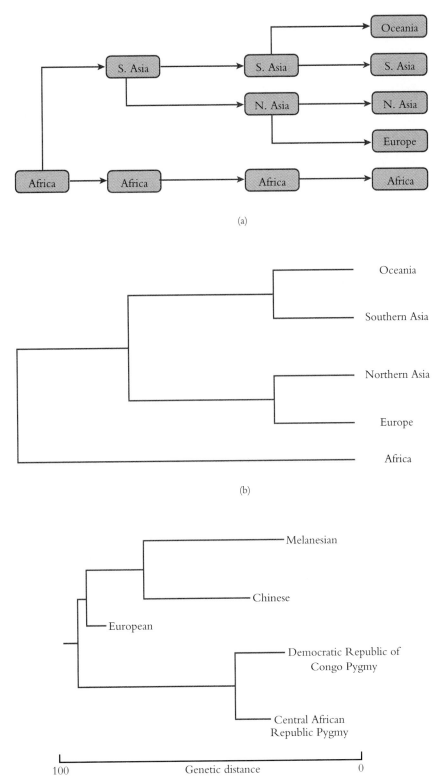

(a)

(b)

Figure 16.12

(a) This model of the expansion of early anatomically modern human populations generates the phylogenetic tree shown in (b) if geographically separate populations remain genetically isolated.

Figure 16.13

This tree, based on genes from over 100 loci, indicates that European populations have undergone less genetic change than Asian or African populations. The length of the path between any two contemporary populations represents the genetic distance between them. The fact that the genetic distance between African and Asian populations is much greater than the distances between African and European populations or between Asian and European populations indicates that European populations have experienced less genetic change than other populations.

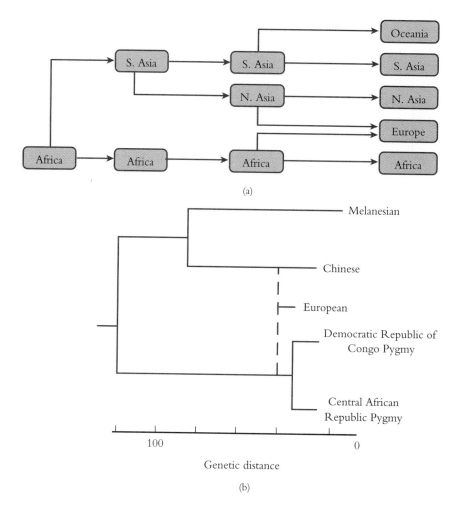

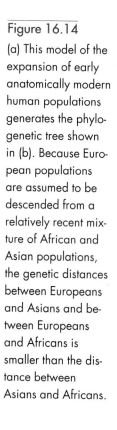

Figure 16.14

(a) This model of the expansion of early anatomically modern human populations generates the phylogenetic tree shown in (b). Because European populations are assumed to be descended from a relatively recent mixture of African and Asian populations, the genetic distances between Europeans and Asians and between Europeans and Africans is smaller than the distance between Asians and Africans.

have experienced substantial recent gene flow from Africa and Asia. A mixture of African and Asian populations would approximate the genetic composition of their parent populations, making it appear that European populations evolved very slowly (Figure 16.14). These genetic data are consistent with the archaeological and genetic data that suggest that farming peoples from North Africa and the Middle East expanded into Europe about 4000 years ago, and were followed by invasions of central Asian pastoralists.

Variation in Complex Phenotypic Traits

As we saw in Chapter 3, the vast majority of human traits are influenced by many genes, each alone having a relatively small effect. For such traits, it is usually impossible to detect the effects of single genes or to trace the patterns of Mendelian inheritance. Nonetheless, it is still possible to assess the influence of genes on such traits. Geneticists have developed statistical methods that enable them to estimate the relative importance of genetic and environmental components of variation within groups. Because the relative importance of genetic variation and environmental variation will affect the resemblance between parents and offspring, the measure that computes the proportion of variation due to the effects of genes is referred to as the **heritability** of phenotypic traits.

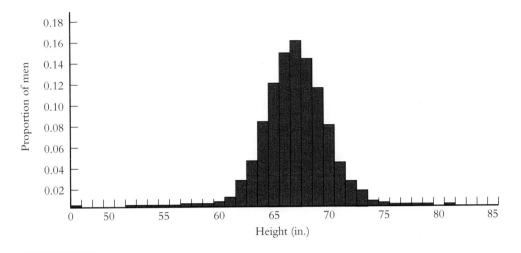

Figure 16.15

The height of men joining the British army in 1939 varied considerably, illustrating the range of variation in morphological characters within populations. The tallest men joining the army were more than 2 m (84 in., about 7 ft) tall, while others were less than 1.5 m (60 in., or 5 ft) tall.

This material will be easier to understand if we have a concrete example in mind. Height is an ideal trait: it is quite easy to measure, it is quite variable within and between populations, it is relatively stable once individuals reach adulthood, and there is a wealth of data on height of individuals in different populations. In any moderately large sample of people there will be a wide range of variation in height. For example, the data in Figure 16.15 are taken from men who joined the British army in 1939. Some of these young recruits were more than 2 m (around 7 ft) tall, while others were less than 1.5 m (5 ft) tall.

Genetic Variation within Groups

Under certain conditions, measuring the phenotypic similarities among relatives such as twins allows us to estimate the fraction of the variation within the population that is due to genes.

In Part I, we saw that the transmission of genes from parents to offspring causes children and parents to be phenotypically similar. If parents who are taller than average tend to have offspring who are taller than average, and parents who are shorter than average tend to have children who are shorter than average, then you might think that height is determined by genes. In contrast, if parents and offspring are no more similar to each other than to other individuals in the population, then you might think that genes have little effect on height. The problem with this reasoning is that nongenetic factors may also cause parents and offspring to be similar. It is known that many environmental factors, such as nutritional levels and the prevalence of infectious diseases, also affect height. Human parents directly affect their offsprings' environment in many ways: they provide food and shelter, arrange for their children to be inoculated against diseases, and shape their children's beliefs about nutrition. Similarity between the environments of parents and their offspring is called **environmental covariation** and is a serious complication in computing heritability.

Data from studies of twins can be used separate the effects of genetic transmission from environmental covariation. The technique involves comparing the similarity between monozygotic and dizygotic twins. **Monozygotic** (identical) **twins** begin life when the union of a sperm and an egg produces a single zygote. Then, very early in development, this zygote divides to form two separate, genetically identical individuals. **Dizygotic** (fraternal) **twins** begin life when two different eggs are fertilized by two different sperm to form two independent zygotes. Dizygotic twins share approximately one half of their genetic material, and are just like other pairs of full siblings except that they were conceived at the same time. Both monozygotic and dizygotic twins share a womb, and experience the same intrauterine environment. After they are born, most twins grow up together in the same family. Thus, if most of the variation in stature has a genetic origin, monozygotic twins are likely to be more similar to one another than dizygotic twins are because monozygotic twins are genetically identical. On the other hand, if most of the variation is due to the environment, and the similarity between parents and offspring is due to having a common family environment, then monozygotic twins will be no more similar to one another than dizygotic twins are. Population genetic theory provides a way to use comparisons of the similarity among dizygotic and monozygotic twins to adjust estimates of heritability for the effects of correlated environments.

Twin studies are useful in trying to estimate the relative magnitude of the effects of genetic variation and environmental variation on phenotypic characters, but the data may be biased in certain ways. For example, twin studies will overestimate heritability if the environments of monozygotic twins are more similar than the environments of dizygotic twins. There are several reasons that this may be the case. In the uterus, some monozygotic twins are more intimately associated than dizygotic twins are. Monozygotic twins are always the same sex. After they are born, monozygotic twins may be treated differently by their parents, family, teachers, and friends from dizygotic twins. It is not uncommon to see monozygotic twins dressed in identical outfits or given rhyming names, and it is inevitable that their physical similarities to one another will be pointed out to them over and over (Figure 16.16).

Figure 16.16

Identical, or monozygotic, twins are produced when a zygote splits and produces two genetically identical individuals.

Studies of monozygotic and dizygotic twins suggest that somewhat more than half of the variation in height in most human populations is due to genetic similarities between parents and their children.

Genetic Variation among Groups

There is variation in stature among human populations.

Just as people within groups vary in many ways, groups of people collectively vary in certain characteristics. There is, for example, a considerable amount of variation in average height among populations. People from northwestern Europe are tall, averaging about 1.75 m (5 ft 8 in.) in height. People in Italy and other parts of southern Europe are about 12 cm (5 in.) shorter on the average. African populations include very tall peoples like the Nuer and the Maasai, and very short peoples like the !Kung. Within the western hemisphere, Native Americans living on the Great Plains of North America and in Patagonia are relatively tall, while peoples in tropical regions of both continents are relatively short.

Some of the variation in body size among human groups appears to be adaptive.

Recall that in Chapter 13 we learned that larger body size is favored by natural selection in colder climates. Tall peoples, like the indigenous people of Patagonia and the Great Plains, usually live in relatively cold parts of the world, while shorter peoples typically live in warmer areas, like southern Europe or the New World tropics. This pattern suggests that variation in body size among groups may be adaptive. This conjecture is borne out by data on the relationship between body size and climate from a large number of human groups (Figure 16.17). Thus, at least some fraction of the variation in body size among groups is adaptive.

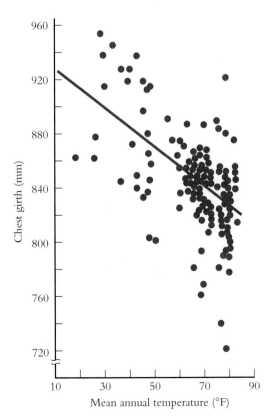

Figure 16.17

People living in colder climates have larger bodies. The vertical axis plots mean chest girth for numerous human groups and the horizontal axis plots the mean yearly temperature in the regions in which each group lives. Because chest girth is a measure of overall size, these data show that people living in colder climates have larger bodies.

The fact that variation within populations has a genetic component does not mean that differences between groups are caused solely by genetic differences.

We have seen that a significant proportion of the variation in height and body size within some populations reflects genetic variation. We also know that the variation in height and body shape among populations seems to be adaptive. From these two facts, it might seem logical to conclude that the variation in body size among populations is genetic, and that this variation represents a response to natural selection.

However, this logic is *fallacious!* All of the variation in height within populations could be genetic *and* all of the variation in height among populations could be adaptive, but this would not mean that there is variation among populations in the distribution of genes that influence height. In fact, all of the observed variation in height among groups could be solely due to differences in environmental conditions.

This point is frequently misunderstood—a misunderstanding that leads to serious misconceptions about the nature of genetic variation among human groups. A simple example may help you see why the existence of genetic variation within groups does not imply the existence of genetic variation among groups. Suppose Rob and his neighbor Pete both set out to plant a new lawn. They go to the garden store together, buy a big bag of seed, and divide the contents evenly. Pete, an avid gardener, goes home and plants the seed with great care. He fertilizes, balances the soil acidity, and provides just the right amount of water at just the right times. Rob, who lives next door, scatters the seed in his backyard, waters it infrequently and inadequately, and never even considers fertilizing it. After a few months, the two gardens are very different. Pete's lawn is thick, green, and vigorous while Rob's lawn hardly justifies the name (Figure 16.18). We know that the difference between the two lawns cannot be due to the genetic characteristics of the grass seed because Rob and Pete used seed from the same bag. Nonetheless, variation in the height and greenness of the grass within each lawn might be largely genetic because the seeds within each lawn experience very similar environmental conditions. All of the seeds in Pete's lawn get regular water and fertilizer, while

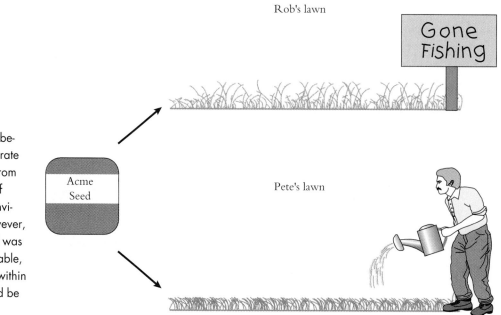

Figure 16.18

The differences between two separate lawns planted from the same bag of seed must be environmental. However, if the seed used was genetically variable, the differences within each lawn could be genetic.

all of the seeds in Rob's lawn are neglected to an equal extent. Thus, if the seed company has sold genetically variable seed, differences between individual plants within each lawn could be due mainly to genetic differences between the seeds themselves.

Exactly the same argument applies to variation in human stature. The fact that much of the variation in stature among Americans has a genetic component does not mean stature is determined entirely by genes. It means there is genetic variation that affects stature, and these effects are relatively large in comparison with the effects of environmental differences among Americans. It does not follow that the differences in stature between Americans and other peoples are the result of genetic differences between them. For example, although Americans are taller on average than citizens of Japan, these differences in height are not necessarily genetic. That would be true only if two quite different conditions held: First, there would have to be a difference between Americans and Japanese in the distribution of genes affecting stature. Second, this genetic difference would have to be large compared with the differences in culture and environment between the two groups. The fact that there is genetic variation among Americans does not tell us whether or not Americans are genetically different from Japanese. The relatively small effect of environmental and cultural variation on height among Americans tells us nothing about the average difference in environment between Americans and Japanese.

The increase in stature that coincided with modernization is evidence for the influence of environmental variation on stature.

For height, there is good reason to believe that variation among groups is at least partly environmental. There has been a striking effect of modernization on the average heights of many peoples. For example, Figure 16.19 plots the height of several groups of English boys between the ages of 5 and 21 in the 19th and 20th centuries. Nineteen-year-old factory workers in 1833 averaged about 160 cm

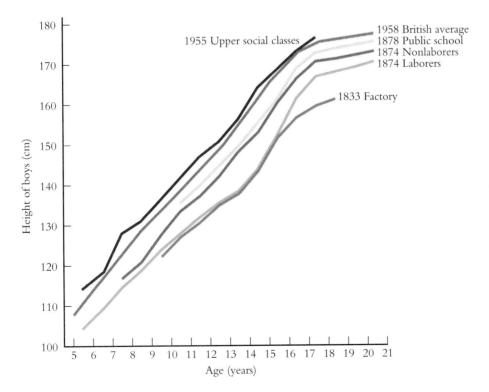

Figure 16.19

Height increases with time in English populations, but at any given time richer people are taller. Note that English ''public schools'' are the equivalent of American private schools.

Table 16.2 Japanese men who immigrated to Hawaii during the first part of the 20th century were shorter than their children who were born and raised in Hawaii. The immigrants were similar in height to the Japanese who remained in Japan, which indicates that the immigrants were a representative sample of the Japanese populations from which they came. The fact that the children of the immigrants were taller than their parents shows that environmental factors play an important role in creating variation in stature. (Data from Table 12.3 in G. A. Harrison, J. M. Tanner, D. R. Pilbeam, and P. T. Baker, 1988, *Human Biology: An Introduction to Human Evolution, Variation, Growth, and Adaptability*, 3rd ed., Oxford Science Publications, Oxford.)

	AVERAGE HEIGHT (CM)	SAMPLE SIZE
Japanese immigrants to Hawaii	158.7	171
People from same regions of Japan who remained in Japan	158.4	178
The immigrants' children born in Hawaii	162.8	188

(about 5 ft 2 in.). In 1874, laborers of the same age averaged about 167 cm (5 ft 5 in.). In 1958 the average British 19 year old stood 177 cm (5 ft 9 in.). Similar increases in height over the last hundred years can be seen among Swedish, German, Polish, and North American children. These changes have occurred very rapidly, probably too fast to be due solely to natural selection.

There also have been substantial changes in height among immigrants to the United States in the course of a few generations. During the first part of this century, when Japan had only begun to modernize, many Japanese came to Hawaii to work as laborers on sugar plantations. The immigrants were considerably shorter than their descendants who were born in Hawaii (Table 16.2). This change in height among immigrants and their children was so rapid that it cannot be the result of genetic change. Instead, it must be due to some environmental difference between Japan and Hawaii in the first part of this century. The underlying cause of this kind of environmental effect is not completely understood. In England, poverty was involved to some extent as relatively wealthy public school boys were taller than less affluent nonlaborers were, and nonlaborers were taller than poorer laborers in the 1870s. Observations like this have led some anthropologists to hypothesize that increases in the standard of living associated with modernization improve early childhood nutrition and increase children's growth rates. However, this cannot be the full explanation. Notice that even the richest people in England 120 years ago were shorter than the average person in England is today. Since it seems unlikely that wealthy Britons were malnourished in the 1870s, there must be other factors that contribute to the increase in stature during the last 120 years. Some authorities think the control of childhood diseases may play an important role in these changes.

The Race Concept

Race plays an important role in modern life, but the popular conception of race is flawed.

Race is part of everyday life. For better or worse, our race affects how we see the world and how the world sees us; it affects our social relationships, our choice of marriage partners, our educational opportunities, and our employment prospects. We may decry discrimination, but we cannot deny that race plays a major role in many aspects of our lives.

Like any widely used word, *race* means different things to different people. However, the understanding of race held by many North Americans is based on three fundamentally flawed propositions:

1. The human species can be naturally divided into a small number of distinct races. According to this view, almost every person is a member of exactly one race—the only exceptions are the offspring of the members of different races. For example, many people in the United States think that people belong to one of three races: descendants of people from Europe, North Africa and Western Asia; descendants of people from sub-Saharan Africa; and descendants of people from eastern Asia.

2. Members of different races are different in important ways, so that knowing a person's race gives you important information about what he or she is like. For example, biomedical researchers sometimes suggest that race predicts susceptibility to diseases like high blood pressure, heart disease, and infant mortality. Less benignly, some people believe that knowing a person's race tells you something about their intelligence or character.

3. The differences between races are due to biological heritage. Members of each race are genetically similar to each other, and genetically different from members of other races. Most Americans view African-Americans and European-Americans as members of different races because they are marked by genetically transmitted characters like skin color. In contrast, the Serbs and Croats of the former Yugoslavia are seen by most Americans as ethnic groups, rather than racial groups, because the differences between them are cultural rather than genetic.

While many people believe these propositions to be true, they are not consistent with scientific knowledge about human variation. There are genetic differences between groups of people living in different parts of the world. But, as we shall see, these differences do not mean that the human species can be meaningfully divided into a set of nonoverlapping categories called races. *The common view of race is bad biology.*

Because people vary, it is possible to create classification schemes in which similar peoples are grouped together.

People tend to be more genetically and phenotypically similar to people who live near them than to people who live farther away. As we have seen already in this chapter, this is true for many genes. For example, the sickle-cell gene is common in central Africa and in India but is quite rare elsewhere; the $LAC\star P$ gene, which allows adults to digest lactose, predominates in northern Europe and North Africa but is uncommon elsewhere; Ashkenazi Jews share a genetically inherited susceptibility to Tay-Sachs disease; and Afrikaners suffer a high incidence of Huntington's chorea. Overall genetic similarity is strongly associated with geographical proximity (Figure 16.20). Morphological similarities also link neighboring peoples. For example, most of the people who live near the equator have dark skin (Figure 16.21), and most of the people who live at high latitudes are stocky.

One consequence of these similarities between people is that it is possible to group people into geographically based categories based on their genetic or phenotypic similarities. For example, a classification scheme based on the ability to

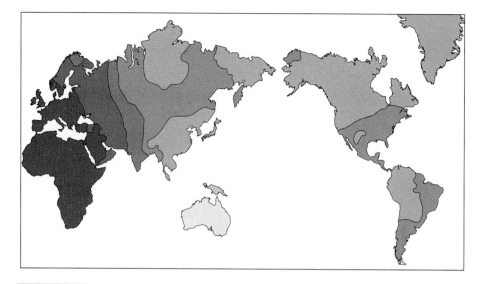

Figure 16.20

This contour map of overall genetic similarity is based on a sample of 120 genes from 42 populations assembled by Cavalli-Sforza and his colleagues. The fact that the contours of equal genetic similarity are roughly evenly spaced indicates that there is a smooth east-to-west gradient of overall genetic similarity—there are no sharp boundaries between groups. Sharp boundaries would produce a map in which many contour lines would be positioned closely together. This map is drawn from the same data used to construct the tree shown in Figure 16.11.

digest lactose would place northern Europeans and North Africans in one group and the rest of the world in a second group. On the other hand, the overall genetic similarity shown in Figure 16.11 could be used to create a classification system that puts Africans in one group; Europeans, northern Asians, and Native Americans in a second group; and southern Asians and people from Oceania in a third group.

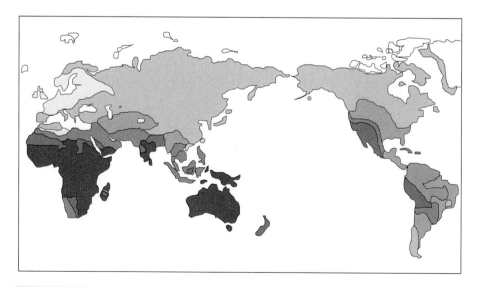

Figure 16.21

This map shows contours in skin color. Notice that there are smooth gradients away from the equator.

However, such classification schemes do not support two properties required by the common concept of race. First, classification schemes based on different characters lead to radically different groupings, and the placement of individuals within any single category is often arbitrary. Second, classification schemes are not very informative. The average difference between groups of people living in different parts of the world is much smaller than the differences among individuals within each group.

There is no single natural classification of the human species.

In Chapter 4, we argued that species are distinctive entities that can be unambiguously identified in nature. Racial classifications for humans are quite different, and there is no natural classification scheme for categorizing us. To see what is meant by a natural classification scheme, consider the following analogy. Suppose you are a clerk in a hardware store. Your boss assigns you to classify the contents of two large cabinets. In the first cabinet there are power drills made by a number of different manufacturers; there are many drills but only one model per manufacturer. The second cabinet holds a variety of different kind of screws. The power drills vary in many ways: they are different colors, shapes, weights, and power ratings. The screws also vary: they have different lengths, diameters, pitches, and heads. You will have little trouble sorting the drills into piles based on manufacturer because all the drills made by a single manufacturer are similar in all of their dimensions—they are the same color, shape, weight, and they have the same power rating. Moreover, each model is distinctly different; there are no intermediate types. This is a natural classification system. You will have a much harder time classifying the screws. Using length will produce one set of piles, diameter a second set of piles, and screw pitch a third set of piles. Moreover, even a classification based on a single characteristic like length will require arbitrary distinctions: should there be three or four piles? Should the 1.5-inch screws be put in the pile with the smallest screws, or with the next bigger sizes? There is no natural way to classify the screws.

Racial classification schemes based on different sets of characters don't result in the same groupings for all characters. For example, a classification scheme based on the ability to digest lactose would yield very different groupings from one based on resistance to malaria. A classification based on skin color would produce a different grouping from one based on height. This problem cannot be solved by using many different characters at the same time. As Figure 16.22 demonstrates, trees based on a large number of genes and trees based on a large number of morphological characters organize local groups in very different ways. The tree based on genetic traits groups aboriginal peoples from Australia and New Guinea together with people from Southeast Asia, while the one based on morphological features groups these aboriginal peoples with African pygmies and the !Kung. Also notice that the genetic tree says that people from northern China are more similar to Europeans than they are to Southeast Asians.

Racial classification schemes explain very little of the world's human genetic variation.

Geneticists have tried to account for the patterns of variation across the globe in biochemically detectable characters, such as blood group loci and other genetic polymorphisms. In these studies, the human species was categorized first into local groups of people belonging to the same ethnic group, linguistic group, or nationality. Then local groups were collected into larger geographically based categories

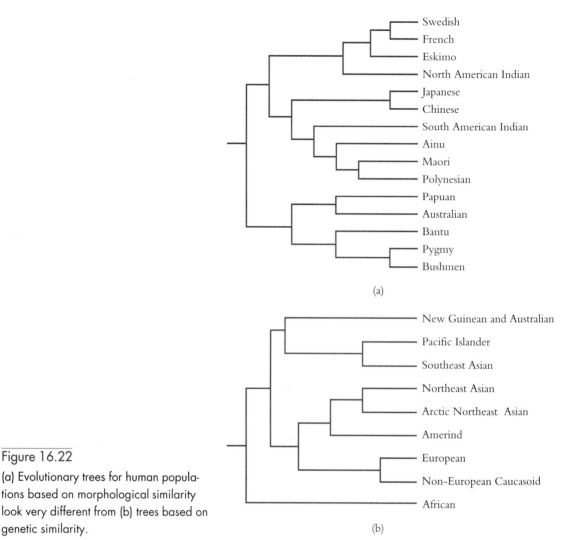

Swedish
French
Eskimo
North American Indian
Japanese
Chinese
South American Indian
Ainu
Maori
Polynesian
Papuan
Australian
Bantu
Pygmy
Bushmen

(a)

New Guinean and Australian
Pacific Islander
Southeast Asian
Northeast Asian
Arctic Northeast Asian
Amerind
European
Non-European Caucasoid
African

(b)

Figure 16.22

(a) Evolutionary trees for human populations based on morphological similarity look very different from (b) trees based on genetic similarity.

that roughly correspond to the usual races. Geneticists computed the amount of variation in these characters within each local group, among groups within each race, and among races. They found that there is much more genetic variation *within* local groups than there is *among* local groups or among races themselves. Differences within local groups account for about 85% of all the variation in the human species. To put this another way, suppose a malevolent extraterrestrial wiped out the entire human species except for one local group, which it preserved in an extraterrestrial zoo. The alien could pick any local group at random—the Efe, the Inuit, the citizens of Ames, Iowa, or the people of Patagonia—and then wipe out the rest of the humans on the planet. This group would still contain on average 85% of the genetic variation that exists in the entire human species. This means that standard racial classifications do not correspond to variation in genetic traits among the people of the world.

Racial classification schemes don't represent natural biological categories.

The bottom line is that you can classify people, and such classifications are not necessarily arbitrary, but they do not reflect any *natural* subdivision of the human species into biologically distinct groups. Nor does knowing a person's position in a classification tell you very much about what that person is like.

This conclusion is consistent with what we have learned about human evolution. It seems likely that anatomically modern humans are a very recently evolved species—both mitochrondrial DNA and fossil evidence suggest an age of less than 200,000 years. This means that it is less likely that natural selection and genetic drift have produced larger genetic differences within the human species than in other, older species. Recall that the genetic differences among different species of chimpanzees are much greater than among different groups of humans, probably because chimpanzees are a considerably older genus. Moreover, we know that gene flow tends to eliminate genetic differences between groups, and there has been extensive gene flow in human history.

Races may represent important cultural categories.

Many people find the conclusion that races aren't real completely implausible. They "know" from experience that race is real. Many view the claim that race is not a valid scientific construct as another example of "political correctness" that is now common in academia. This view has some support within anthropology. For example, in the November 1994 issue of *Discover* magazine, University of Colorado anthropologist Alice Brues stated:

> A popular political statement now is, "There is no such thing as race." I wonder what people think when they hear this. They would have to suppose that the speaker if he were dropped by parachute into downtown Nairobi, would be unable to tell, by looking around him, whether he was in Nairobi or Stockholm. This could only damage his credibility. The visible differences between different populations tell everyone that *there is something here* (p. 60).

Of course, our intuitions lead us astray in many ways. Our eyes tell us that the earth is flat, but it is really a sphere. Our intuition tells us that a bullet fired horizontally from a rifle will hit the ground long after a bullet dropped from the muzzle at the same instant, yet in reality they will hit the ground at the same time. Our mind tells us that it is impossible that an elephant should be descended from a shrewlike insectivore, even though that is exactly what happened. The intuition that race is real may be a similar illusion.

Professor Brues is undoubtedly correct that a parachute jumper who landed in Nairobi would be unlikely to think that he was in Stockholm, Tokyo, Bombay, or Honolulu. But would he be certain that he was in Nairobi, and not in Johannesburg, or Khartoum? Probably not. And what if we imagine an intrepid soul who attempts to bicycle from Nairobi to Stockholm? As she pedaled from Nairobi through Kinshasa, Khartoum, Cairo, Istanbul, Budapest and finally reached Stockholm, she would observe that people change in many ways, but the change would always be subtle—there would be no sharp boundaries with one kind of people on one side and a different kind of people on the other side. The cyclist would see that people vary, but this variation does not sort itself out into neat, discrete categories. Thus, our perception of race may depend on which of these two metaphors we think captures the essential features of human variation. However, the scientific evidence demonstrates quite conclusively that there is considerably more variation within human groups than among them.

In our opinion, race is a culturally constructed category, not a meaningful biological concept. The characters that North Americans conventionally use to sort people into racial categories tend to be biologically transmitted traits, like skin color and facial features. But we could just as well use culturally transmitted traits like religion, dialect, or class. Classifications based on a small set of biologically

Deconstructing Race

One November morning in 1984, Norm Sauer, a forensic anthropologist at Michigan State University, received a call from the state police. Somebody had found a body in the woods. The decomposed corpse displayed the typical mute profile of an unknown homicide victim: no clothing, no personal possessions at the scene, not even enough soft tissue left to readily identify its sex. The police only knew the body was human. They asked Sauer if he could recover the person's disintegrated identity—turn the "it" back into a he or a she.

Sauer got into his car and drove up to the hospital where the body was being kept. He examined the form and structure of the skeleton, concentrating on the skull and pelvis, then took a number of measurements with his calipers—the distance between the eye orbits, the length and width of the skull, for example—and plugged them into standard forensic equations. Within a few hours he was able to inform the police that the skeleton was that of a black female who had stood between 5 foot 2 and 5 foot 6 and was 18 to 23 years old at the time of her death. She had been dead somewhere between six weeks and six months. With that information in hand, the police were able to narrow their search through the files of missing persons to a handful of cases. Some unusual dental restorations completed the puzzle: the skeleton belonged to a woman who had lived two counties away and had been missing for three months. She'd been 5 foot 3, 19 years old, and black.

Age, sex, stature, and race are the cardinal points of a preliminary forensic report, the cornerstones that support the reconstruction of a specific human identity. Three out of four of these characteristics are firmly anchored in empirical fact. A person's sex, age, and height at any given moment are discrete quantities, not matters to be interpreted, revised, or dissected into their constituent parts. Whether I am 6 foot 1 or 5 foot 3 does not depend on who is holding the ruler. If I am male in Milwaukee, I remain male in Mobile. My age, like it or not, is 43; no amount of investigation into my personal history will reveal that I am mostly 43, with some 64 mixed in, and just a trace of 19 from my mother's side.

But the fourth cornerstone—race—is mired in a biological, cultural, and semantical swamp. In the United States, most people who are considered to be black trace their ancestry to West Africa; biologically speaking, though, some 20 to 30 percent of the average African American's genetic material was contributed by ancestors who were either European or American Indian. Different jurisdictions, government bureaucracies, and social institutions tend to classify race in different ways—as do different individuals. Most Americans get to decide which race box to check on a form, and their decision may depend on whether they are filling out a financial-aid application or a country club membership form. A recent study found that in the early 1970s, 34 percent of the people participating in a census survey in two consecutive years changed racial groups from one year to the next.

The classifications themselves are eminently changeable: the Office of Management and Budget, which is responsible for overseeing the collection of statistics for the federal government, recently held public hearings and is currently reading written commentary on the categories used by the Census Bureau. In addition to the racial categories now in place—white, black, American Indian, Eskimo, Aleut, Asian or Pacific Islander, and "other"—the OMB is considering adding slots for native Hawaiians, Middle Easterners, and people who consider themselves multiracial. If such categories are added, they should be in place for the census in the year 2000.

"Race is supposed to be a strictly biological category, equivalent to an animal subspecies," says anthropologist Jonathan Marks of Yale. "The problem is that humans also use it as a cultural category, and it is difficult, if not impossible, to separate those two things from each other."

So what is race? How important is it? Is it a notion rooted in our culture, or a reality living in our genes? Should the word be abandoned by scientists, or would banishing it simply cripple any attempt to help the public understand the true nature of human diversity, forcing us to seek our definitions on the street, in the jaundiced folklore of prejudice?

Everyone agrees that all human beings are members of a single biological species, *Homo sapiens*. Since we are all one species, by definition we are all capable of interbreeding with all other humans of the opposite sex to produce fertile offspring. In practice, however, people do not mate randomly; they normally choose their partners from within the social group or population immediately at hand and have been doing so for hundreds of generations. As a result, the physical expressions of the genes inherited from an expanding chain of parents and grandparents—most of whom lived in the same region as one another—also tend to cluster, so that there can be a great deal of variation from one geographic region to another in skin color, hair form, facial morphology, body proportions, and a host of less immediately obvious traits. Roughly speaking, then, race is the part of one person's variation on the theme of humanity created by the interplay of geography and inheritance.

The problem with this definition rests in the way patterns of human variation have traditionally been packaged and perceived. In the past, most anthropologists unquestioningly accepted the concept of races as fixed entities or types, each of which was pure and distinct. These types were seen as gigantic genetic bushel baskets into which people could be sorted. Admittedly, the rims of the bushel baskets might not be stiff enough to keep some of their contents from spilling out and mingling with geographically adjacent baskets. In the sixteenth century, European colonialism began flicking genes from one basket into other parts of the world; soon afterward the forced importation of large numbers of Africans into the Americas had a similar effect. But until recent decades, anthropologists believed that no amount of interracial mixing could ever dilute the purity of the racial ideals themselves.

In the bushel-basket scheme, races are defined by sets of physical characteristics that cluster together with some degree of predictability in particular geographic regions. Asians, for instance, are typically supposed to have "yellow" skin, wide, flat cheekbones, epicanthic folds (those little webs of skin over the corners of the eyes), straight black hair, sparse body hair, and "shovel-shaped" incisor teeth, to name just a few such distinctive traits. And sure enough, if you were to walk down a street in Beijing, stopping every once in a while to peer into people's mouths, you would find a high frequency of these features.

But try the same test in Manila, Tehran, or Irkutsk—all cities in Asia—and your Asian bushel basket begins to fall apart. When we think of an "Asian race," we in fact have in mind people from only one limited part of that vast continent. You could, of course, replace that worn-out, overloaded bushel basket with a selection of smaller baskets, each representing a more localized region and its population. A quick scan through some supposedly Asian traits, though, shows why any number of subcontinental baskets would be hopelessly inadequate for the job. Most inhabitants of the Far East have epicanthic folds on their eyes, for instance—but so do the Khoisan (the "Bushmen") of southern Africa. Shovel-shaped incisors—the term refers to the slightly scooped-out shape of the back side of the front teeth—do indeed show up in Asian and American Indian mouths more often than in other people's, but they also pop up a lot in Sweden, where very few people have coarse, straight black hair, epicanthic folds, or short body stature.

The straightforward biological fact of human variation is that there are no traits that are inherently, inevitably associated with one another. Morphological features *do* vary from region to region, but they do so independently, not in packaged sets. "I tell my students that I could divide the whole world into two groups: the fat-nose people and the skinny-nose people," says Norm Sauer. "But then I start adding in other traits to consider, like skin color, eye color, stature, blood type, fingerprints, whatever. It doesn't take long before somebody in the class gets the point and says, 'Wait a minute! Pretty soon you're going to have a race with only one person in it.'"

Indeed, despite the obvious physical differences between people from different areas, the vast majority of human genetic variation occurs *within* populations, not *between* them, with only some 6 percent accounted for by race, according to a classic study done in 1972 by geneticist Richard Lewontin of Harvard. Put another way, most of what separates me genetically from a typical African or Eskimo *also* separates me from another average American of European ancestry.

But if the bushel-basket view of race is insupportable, does that mean that the concept of race possesses no biological reality? "If I took a hundred people from sub-Saharan Africa, a hundred from Europe, and a hundred from Southeast Asia, took away their clothing and other cultural markers, and asked somebody at random to go sort them out, I don't think they'd have any trouble at all," says Vincent Sarich of the University of California at Berkeley, a controversial figure in biological anthropology since the late 1960s, most recently because of his views on the race issue. "It's fashionable to say there are no races. But it's silly."

It's certainly true that indigenous Nigerians, for instance, look different from native Norwegians, who look different from Armenians and Australian aborigines. But would these differences be just as obvious if you could see the whole spectrum of humanity? Since people tend to mate with others in their immediate geographic area, there should be only a gradual change from one region to the next in the frequency of various genes and the morphological features they code for. In this scenario, human variation is the result of a seamless continuum of genetic change across space. The race concept, on the other hand, lumps people into clearly delineated groups. This, says anthropologist Loring Brace of the University of Michigan, is a purely historical phenomenon.

"The concept of race didn't exist until the invention of oceangoing transport in the Renaissance," Brace explains. Even the most peripatetic world travelers—people like Marco Polo or the fourteenth-century Arabian explorer Ibn Battutah—never thought in racial terms, because traveling by foot and camelback rarely allowed them to traverse more than 25 miles in a day. "It never occurred to them to categorize people, because they had seen everything in between," says Brace. "That changed when you could get into a boat, sail for months, and wind up on a different continent entirely. When you got off, boy, did everybody look different! Our traditional racial groupings aren't definitive types of people. They are simply the end points of the old mercantile trade networks."

Sarich, however, is not so willing to dismiss race as an accident of history. "I don't know whether Marco Polo referred to race or not," he says. "But I'll bet that if you were able to ask him where this person or that one came from just by looking at their physical features, he would be able to tell you."

If populations were uniform in density all over the world, adds Sarich, then the whole panoply of human variation would indeed be distributed smoothly, and race would not exist. But populations are not so evenly dispersed. Between large areas of relatively high density there are geographic barriers—mountain ranges, deserts, oceans—where population densities are necessarily low. These low-population zones have acted as filters, impeding the flow of genes and allowing distinct, discernible patterns of inheritance—races—to develop on either side. The Sahara, for instance, represents a formidable obstacle to gene flow between areas to the north and south. Such geographic filters have not completely blocked gene flow, notes Sarich—if they had, separate human species would have developed—but their influence on the pattern of human variation is obvious.

Given the layered confusion surrounding the term *race*—and its political volatility—it's no wonder that scientists struggle over its definition and question its usefulness. Surveys of physical anthropologists have found that almost half no longer believe that biological races exist. . . .

Medical researchers, unlike the anthropologists, seem to have little question about the reality of racial categories. Race, it seems, is quite useful for organizing data; each year dozens of reports in health journals use it to show purported clear differences between the races in susceptibility to disease, infant mortality rates, life expectancy, and other markers of public health. Black men are supposedly 40 percent more likely to suffer from lung cancer than are white men, and a number of recent studies on breast cancer seem to show that black women tend to develop tumors that are more malignant than those found in white women. In the United States, black infants are almost two and a half times more likely to die within the first 11 months of life than white infants. And it's been shown that American Indians are far more likely than blacks or whites to carry an enzyme that makes it harder for them to metabolize alcohol; this would leave them genetically more vulnerable to alcoholism. Other studies claim to demonstrate racial differences in rates of cardiovascular disease, diabetes, kidney ailments, venereal disease, and a host of other pathologies.

Are these studies pointing at genetic differences between the races, or are they using race as a convenient scapegoat for health deficiencies whose causes should be sought in a person's socioeconomic status and environment? The lung cancer statistic, for instance, should really be considered along with the numbers that show that black men are much more likely to smoke than are white men.

A recent study of hypertension in black Americans, conducted by Randall Tackett and his colleagues at the University of Georgia, exemplifies the difficulties found in trying to tease out a single answer to such a question. It's been known for some 30 years that blacks in the United States are almost twice as likely as whites to suffer from hypertension, or high blood pressure—a condition that carries with it an increased risk of heart failure, stroke, hardening of the arteries, and other cardiovascular diseases. Black men are reported to have a 27 percent higher death rate from cardiovascular disease than white men, and black women a 55 percent higher rate than white women. What causes this discrepancy is still unknown: some investigators have attributed the higher incidence of hypertension in blacks to socioeconomic factors such as psychosocial stress, poor diet, and limited access to health care, while others have suggested a genetic predisposition to the disorder, which is often taken to mean a racial predisposition. Trying to track down a genetic cause, however, has proved even more confounding than it might otherwise have been, since high blood pressure can be the result of a number of factors, ranging from higher dietary sodium levels to increased exposure to psychological insult.

Yet last June, Tackett and his associates reported on a possible physiological mechanism underlying the higher incidence of hypertension in blacks. They exposed veins obtained during heart-bypass operations to chemicals that stressed the tissues and caused them to constrict, and found that veins from blacks were slower to return to normal size than those taken from whites. Veins that stay constricted longer in response to stress allow less blood to flow through and require the heart to work harder—the essence of hypertension. "This is the first direct demonstration that there are racial differences at the level of the vasculature," says Tackett.

The hope is that these findings will lead the medical community to treat hypertension in blacks even more aggressively, and that they will thus save lives. But whether the findings really say anything about the role of race in disease is another matter altogether. Tackett's sample of African Americans was limited to 22 individuals from southern Georgia; would blacks from Los Angeles or New York, living in different circumstances and with different genetic histories, show the same blood vessel impairment? What about native Africans, who unlike their American counterparts generally have remarkably *low* rates of hypertension? And what about the Finns and the Russians, who have high rates? What do the findings say about their race? And even if American blacks have a greater susceptibility to hypertension primarily because of their blood vessels and not the inequities in their socioeconomic status, who's to say that those inequities—environmental stresses that American whites never have to face—aren't the trigger for the prolonged, potentially lethal constriction? Isn't it possible that the chain of causation leading from blood vessels to blood pressure to heart disease is anchored not in race, but in racism?

After all, Lewontin's study, done more than two decades ago, showed that the concept of race doesn't really have much of a genetic punch. "I'm not denying that the difference Tackett sees is there," says [Emory anthropologist George] Armelagos [an outspoken critic of the biological concept of race]. "But race only explains 6 percent of human biological variation. How can he be sure that the 6 percent accounts for the pathology?"

The techniques used in genetic analysis have improved vastly since Lewontin's 1972 study; though race is responsible for just a small amount of genetic difference, it is now somewhat easier to distinguish one population from another and place an individual by looking at a sample of DNA. Of course, there are still limits. "If you ask me to look at a sample and say whether it came from Wales or Scotland, that would be tough," says Peter Smouse, a population geneticist at Rutgers. "But ask me if somebody is from Norway or Taiwan, sure, I could do that. Humans are tremendously variable genetically across the planet, almost surely representing how long we've been out there and spreading around. Now, whether the piles are nice and neat is not so clear; they're probably not as neat as would be convenient for someone who wanted to make piles."

In the end, says Smouse, no one would deny that there are genetic differences between groups of people. But in comparison with the differences between, say, chimps and humans, those dissimilarities shrink to "totally nothing." It's all a matter of perspective.

"What you make of race depends on what the question is," says Smouse. "And who wants to know."

Source: From pp. 57–63 in J. Shreve, 1994, Terms of estrangement, *Discover*, November. James Shreeve/Copyright 1994 The Walt Disney Company. Reprinted with permission of *Discover* Magazine.

transmitted traits have no more scientific validity than classifications based on religion, language, or political affiliation. While skin color and facial features are salient characteristics in the United States, there are societies where other types of characters matter much more. Religion is the basis of bitter animosity in Northern Ireland, although the genetic differences between Protestant and Catholic Irish are even more microscopic than are the genetic differences between African-Americans and European-Americans. In the ethnic conflicts that have seared the globe in the last decade, it is literally a matter of life and death whether you are Azeri or Armenian in Ngorno-Karabach; Hutu or Tutsi in Rwanda; or Serb or Muslim in Bosnia. Yet, in each case, people are divided by culture, not genes.

Further Reading

Bodmer, W., and L. L. Cavalli-Sforza. 1976. *Genetics, Evolution, and Man*. W. H. Freeman, New York.

Cavalli-Sforza, L. L., P. Menozzi, and A. Piazza. 1994. *The History and Geography of Human Genes*. Princeton University Press, Princeton, N.J.

Falconer, D. R. 1981. *Introduction to Quantitative Genetics*. Longman, London.

Flatz, G. 1987. The genetics of lactose digestion in humans. *Advances in Human Genetics* 16:1–77.

Harrison, G. A., J. M. Tanner, D. R. Pilbeam, and P. T. Baker. 1988. *Human Biology: An Introduction to Human Evolution, Variation, Growth, and Adaptability*. 3rd ed. Oxford Science Publications, Oxford.

Study Questions

1. What sources of human variation are described in this chapter? Why is it important to distinguish among them?
2. Consider the phenotype of human finger number. Most people have exactly five fingers on each hand, but some people have fewer. What is the source of variation in finger number?
3. Why is it hard to determine the source of variation in human phenotypes? Why might it be easier to determine the source of variation for other animals?
4. Explain how studies with human twins allow researchers to estimate the effects of genetic and environmental differences on phenotypic traits.
5. What kind of body size and shape is best in a humid climate? Why?
6. Suppose you are told that differences in IQ scores among white Americans have a genetic basis. What would that tell you about the differences in average scores between white Americans and Americans belonging to other ethnic groups? Why?
7. Explain why races are not biologically meaningful categories of classification.

C H A P T E R   1 7

Evolution and the Human Life Cycle

Maternal-Fetal Conflict during Pregnancy
    Why There Is Parent-Offspring Conflict
    Spontaneous Abortion
    Blood Sugar
The Evolution of Senescence
    Two Evolutionary Theories of Senescence
    Evidence for the Theories
The Evolution of Menopause

Every human life begins the same way: an egg and sperm unite and pool their genetic contents (Figure 17.1). Human lives are marked by the passage through birth, infancy, childhood, adolescence, adulthood, and finally old age. These stages define the human life cycle, and have been shaped by evolutionary forces over the course of evolutionary history. For example, we know from careful analysis of hominid teeth that the period of juvenile development has been extended in humans and that this change arose with *Homo erectus* less than 2 mya. We also know that life expectancy has increased. Neanderthals rarely lived past 45 years, while Upper Paleolithic peoples sometimes survived until the age of 60. In this chapter, we consider some of the ways that natural selection has shaped several different elements of the human life cycle. We have chosen four specific examples because they illustrate in some detail the power of the evolutionary approach for illuminating critical components of our lives.

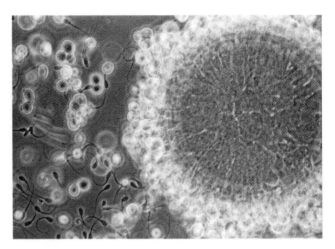

Figure 17.1

Every human life begins when an egg and sperm unite to form a zygote.

Maternal–Fetal Conflict during Pregnancy

Pregnancy is potentially a period of conflict between mothers and their fetuses.

Everyone knows that parents and children are often in conflict. The parent wants the toddler to eat the pureed spinach, and the child (reasonably) tosses the spinach on the floor. The child wants to stay awake until midnight, and the exhausted parent wants him to go to sleep. The teenager wants to get a driver's license on her 16th birthday, and the parents wince as their insurance payments skyrocket. Such conflicts seem like a common and natural part of raising children.

In reality, conflict begins at conception, even before parents and infants begin to negotiate about sleeping through the night, taking formula from a bottle, and napping regularly. This may seem surprising because the relationship between a mother and her unborn infant is obviously one of great intimacy. For nine months, the human fetus grows and develops in the security of its mother's uterus, relying entirely on the mother for its own survival. Conscientious mothers modify their diets, and avoid cigarettes, alcohol, and other drugs to protect the fetus during the critical stages of development. However, the intimacy of the maternal-fetal bond does not mean that the interests of the mother and fetus are entirely in synch. In fact, evidence recently summarized by Harvard biologist David Haig suggests that pregnancy is a period of finely orchestrated conflict between mothers and their unborn infants over both the quantity and the quality of maternal investment.

Why There Is Parent-Offspring Conflict

Conflict between mothers and their infants arises because they share only some genes.

In sexually reproducing species, like humans, a mother passes half of her genes on to each of her offspring. From a genetic point of view, all of her children are equivalent. Her children, on the other hand, are likely to view the matter very differently. If a pair of children share the same mother and the same father, there is only a 50% chance that a gene carried in one child will also be carried in the other. When children are sired by different fathers, they have only a 25% chance of obtaining the same gene. Thus, a child is more closely related to itself than to its siblings. While a mother may be indifferent about which of her offspring she invests in, her offspring have an interest in securing investment for themselves rather than for their siblings. This asymmetry leads to the evolution of conflict between mothers and their children. Natural selection shapes the physiology and the behavior of children so that they demand more investment from their mothers than their mothers are willing to give. Selection will shape the physiology and behavior of mothers so that they resist demands that conflict with their own interests. This phenomena has been called **parent–offspring conflict** by Rutgers University biologist Robert Trivers, who was the first to recognize the evolutionary rationale underlying the conflict between parents and their offspring.

The amount of investment mothers contribute to a fetus is determined by the same kind of trade-off between quality and quantity that we described in Chapter 7. Increasing the investment in the current fetus will make it bigger and healthier, and this will make the child more likely to survive and to reproduce successfully when it grows up. However, this investment imposes costs on the mother because she has only a finite amount of time and energy to invest in her children. Any increase in investment in one child inevitably reduces her ability to invest in other children.

To understand how evolution shapes parent-offspring conflict, imagine a mutation that increases the amount of maternal investment in the current fetus a small amount, and therefore reduces investment in future children by some amount. According to Hamilton's rule for the evolution of traits by kin selection (Chapter 8), such a gene expressed in mothers will be favored by selection if

$$0.5 \times \left(\begin{array}{c} \text{increase in fitness} \\ \text{of current fetus} \end{array} \right) > 0.5 \times \left(\begin{array}{c} \text{decrease in fitness} \\ \text{of future offspring} \end{array} \right)$$

Since the mother shares half of her genes with each of her offspring, 0.5 appears on both sides of the inequality. The inequality tells us that selection will increase investment in the fetus until the benefits to the fetus are equal to the costs to future offspring.

The result is quite different if the genes expressed in the fetus control the amount of maternal investment. This time, consider a gene that is expressed in the fetus that increases investment that the fetus receives by a small amount. Once again, we use Hamilton's rule, this time from the perspective of the fetus:

$$1.0 \times \left(\begin{array}{c} \text{increase in fitness} \\ \text{of current fetus} \end{array} \right) > 0.5 \times \left(\begin{array}{c} \text{decrease in fitness} \\ \text{of future offspring} \end{array} \right)$$

In this case, the fetus is related to itself by 1.0 and to its full sibling by 0.5. Now selection will increase the amount of maternal investment until the incremental benefit of another unit of investment in the current fetus is twice the cost to future brothers and sisters of the fetus (and four times for half siblings) (Figure 17.2). Thus, genetic asymmetries lead to a conflict of interest between mothers and their children. Selection will favor mothers that provide less investment than their infants desire, and selection will favor children who demand more investment than their mothers are willing to give. As we shall shortly see, this conflict begins in the womb.

At the outset, the notion of conflict between a mother and her fetus seems preposterous. How can the fetus, an immature creature that cannot survive outside the womb, exert effective control over its mother? How can the interests of a few barely differentiated cells compete with the interests of a complete and mature individual? As with David and Goliath, we will see that it is a mistake to under-

Figure 17.2

In many species, siblings actively compete for food and other forms of parental investment.

estimate the power of the underdog. The fetus plays an active and effective role in determining whether the pregnancy is sustained, and in regulating the quantity and quality of nutrients that it receives from its mother.

Spontaneous Abortion

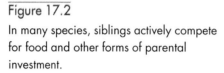

Selection sometimes favors the termination of pregnancy when either the mother or the fetus is in poor condition.

Most people think of miscarriages as pathologies—maladaptive and unpredictable aberrations that cause the fetus to die and the pregnancy to be terminated. For those who suffer miscarriages, the loss of a fetus is often deeply disconcerting. In such cases, parents may never know why their fetus died or whether their next one is likely to survive. They may always wonder what kind of child it might have been if it had survived. As upsetting as miscarriages are for parents, there is good reason to believe that some of the mechanisms that produce miscarriages may be adaptive responses that have been favored by natural selection over the course of human history.

In some cases, miscarriages appear to provide a maternal mechanism for maintaining quality control over pregnancy. It has been estimated that only 22% of all conceptions are carried to term. The rest are miscarried, usually before the 12th week of pregnancy (Figure 17.3). In the majority of cases, miscarriages are associated with overt chromosomal abnormalities in the fetus. There are good reasons to expect that natural selection acting on mothers would favor miscarriages under these circumstances.

For women, every pregnancy is a costly investment. Pregnancy lasts nine months, and during their pregnancies, women need more food and work somewhat less efficiently. Childbirth is also a major source of mortality among women. The magnitude of these costs, and their impact on the mother's future reproductive

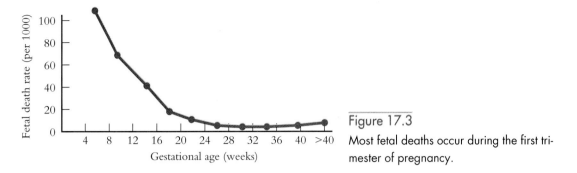

Figure 17.3

Most fetal deaths occur during the first trimester of pregnancy.

success, will depend partly on the mother's condition at conception. Women who are in poor health or badly nourished will be less able to meet the demands of pregnancy and will take longer to recover from their pregnancies than will healthy, well-nourished women.

The costs of pregnancy must be weighed against the potential benefits. In evolutionary terms, the benefit is a child who carries copies of maternal genes. The benefit the mother will gain depends largely on the quality of the fetus that she carries. If the fetus is sickly or suffers from a genetic defect, it will be unlikely to survive. If the mother is in poor condition, she may be unable to provide the resources that her fetus or infant needs to survive. In both such cases, the costs of the pregnancy will outweigh the benefits.

Thus, we would expect natural selection to favor physiological mechanisms that terminate a pregnancy when there is something wrong with the fetus or the mother is in very poor condition. When pregnancies are terminated early, time and energy that can be invested in future offspring are conserved. Thus, it makes sense from an adaptive point of view that many miscarriages are associated with chromosomal anomalies and that most miscarriages occur during the first trimester.

There is conflict between mother and fetus over whether a given pregnancy should be sustained.

To this point we have considered miscarriages from the point of view of selection acting on the mother. The picture looks different from the perspective of the fetus. The fetus has a greater interest in its own survival (and a weaker interest in any future sibling's survival) than does its mother. Thus, there will be a certain scope for conflict between the mother and her fetus over whether a given pregnancy should continue. Natural selection will favor genes expressed in the fetus that prevent those spontaneous abortions that are induced by genes expressed in the mother. The fetus plays an active role in promoting its own survival.

In humans, pregnancy is maintained by the production of **leutenizing hormone (LH)** by the anterior pituitary gland. Leutenizing hormone, in turn, stimulates the production of another hormone, **progesterone**, by the **corpus luteum**, a mass of endocrine cells formed in the mother's ovary after ovulation. Progesterone inhibits uterine contractions that would terminate the pregnancy.

From very early in pregnancy, the fetus influences the levels of progesterone in its mother's bloodstream. Initially, the fetus releases **human chorionic gonadatropin (hCG)**, which stimulates the production of progesterone by the mother's corpus luteum, directly into the mother's bloodstream. However, by the eighth week of pregnancy, the fetus produces enough progesterone to sustain the pregnancy, even if the corpus luteum is removed from the mother. Thus, the

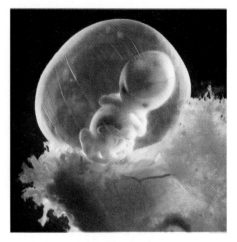

Figure 17.4

Eight weeks after conception, the human fetus is able to produce enough progesterone to sustain pregnancy.

eight-week-old fetus is able to manipulate the maternal endocrine system to enhance its own viability (Figure 17.4). It seems quite plausible that the ability of the fetus to manipulate the level of progesterone in its mother's bloodstream has been favored by natural selection because it allows the fetus to protect itself from premature termination of the pregnancy.

Blood Sugar

There is also conflict between the mother and fetus over levels of sugar in the mother's blood.

The fetus receives all of its nourishment from its mother and, in a sense, it shares each of her meals. Mothers literally "eat for two." There is evidence of maternal-fetal conflict over distribution of the nutrients in these meals. The fetus tries to increase the amount of sugar present in the mother's blood, thereby getting more food for itself, while the mother's body tries to thwart the fetus's efforts to commandeer her resources (Figure 17.5).

To understand this conflict, you need to know something about the normal relationship between blood sugar and insulin production. Glucose, a simple form of sugar, is the major source of energy for cells. It is extracted from food and

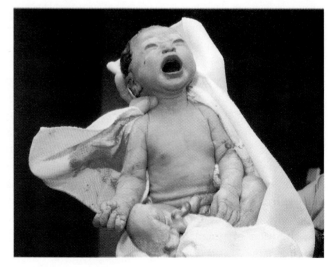

Figure 17.5

Although human infants seem totally dependent on their parents for survival, they are veterans of a persistent campaign to increase the quantity and quality of the resources they obtain from their mother.

Got Milk?

All mammals feed their young milk . . . and all milk is a mixture of water, fats, proteins, sugars, and sprinklings of various minerals, vitamins, and hormones. The proportions of those ingredients vary greatly from species to species. Sheep milk, for example, is about 80 percent water, 9 percent fat, and 5 percent protein; the milk of the great blue whale is 50 percent water, 30 percent fat, and 12 percent protein. Infants, as might be expected, do best on the milk of their species. This much biologists have known for a very long time, but they did not even suspect how milk composition related to parenting styles until Devorah Ben Shaul, an Israeli biologist employed by the Jerusalem Biblical Zoo in the 1950s and '60s, began pondering some unusual findings.

Ben Shaul became interested in milk and its composition because she often had to hand-rear wild animals and wanted to be able to concoct solutions that would simulate an infant's normal diet. In order to do this, she needed to know what that diet was. So she set out to collect and analyze samples of milk from hundreds of wild animals—humpback whales, Arabian camels, hippopotamuses, water shrews, water buffalo, one dead spiny anteater, and almost everything in between.

"I expected, of course, that the major correlations could be made on the basis of species relationship," Ben Shaul wrote in her groundbreaking paper in the *International Zoo Yearbook* in 1962, "but [I] soon found that this was not so. I found myself confronted with irrational results such as the fact that a grizzly bear and a kangaroo had virtually the same milk composition or, as another example, the reindeer and the lion."

Some researchers might have been tempted to put these findings away in a drawer, to chalk them up to contaminated glassware or an incompetent technician, but Ben Shaul persevered. As the number of her samples grew, she began to see a pattern—a connection between the composition of an animal's milk and its nursing behaviour. Perhaps, after all, it made sense that the milk of a grizzly bear and a kangaroo should be so similar, since in both species mothers are always with their young and the young can nurse at any time. The milk of both is very dilute compared with that of many other animals. It consists of about 88.9 percent water, 3.0 percent fat, and 3.8 percent protein.

Dogs, cats, and rodents, by contrast, all have milk that is more concentrated and higher in fat. But they all leave their young in nests for hours at a time, so perhaps that also makes sense. Fat, after all, has the highest satiety value of any food. Goats, sheep, and primates, whose young follow or are carried by their mothers at all times, have, like grizzlies and kangaroos, milk that is relatively dilute and low in fat. . . .

In the decades since Ben Shaul's paper was published, her data on milk composition have been augmented and refined with samples from many

more animals and from animals at different points in their lactation periods, but her basic insight has stood. It has been used to explain even more bizarre nursing phenomena than those of which she was aware. Kangaroos, for example, have since been found to express very different types of milk from their two teats. One type, a low-fat variety, is for the newly born kangaroo, that embryoniclike thing that will be a permanent resident of the pouch for many months. The second type, a higher-fat brand, is for the young-at-foot, a joey that is old enough to leave the pouch but will continue to suckle for five to eight months, putting its head inside the pouch to do so. What is most astonishing is that a kangaroo mother can express these two types of milk at the same time. She can nourish both a young-at-foot and a pouch baby. The joey is always sure to drink from the correct tap because its embryonic sibling is actually attached to its teat (by its mouth) and will be for some time.

In her original paper on the rationale of milk composition, Ben Shaul also recognized that milk can reflect factors in addition to child care, factors relating to the ecology of a species, to the environment in which it lives and how it lives in that environment. The high-fat milk of whales, seals, and dolphins, she pointed out, is a result of the fact that these animals spend a good portion of their time in cold water. Since then, bats, mammals in whom flight has placed a premium on weight reduction, have been found to have milk that is high in fat and dry matter but very low in water. Primates have been found to have milk high in carbohydrates because carbohydrates are necessary for the rapid brain growth that primate young experience after birth. Animals whose brains are almost completely developed at birth—seals, for instance—have milk that is very low in carbohydrates.

Seals also provide a striking example of just how intimate the connection is between milk composition and the environment. Small ringed seals, for instance, give birth on the fast, permanent ice of the Arctic seas, and mothers nurse their infants for two months on milk that is about 45 percent fat. By contrast, hooded seals (also known as creasted or bladdernose seals because of black sacs on top of the males' heads) give birth on the impermanent sea ice of the North Atlantic regions. Because their cradle is always in danger of being broken up by storms, currents, or warm weather, these animals have the shortest lactation period of any mammal, just four days, but their milk is the richest of that of any mammal, more than 60 percent fat. During the four days that hooded seal infants spend with their mothers, they nurse frequently, about every half hour, and put on about fifteen pounds per day.

Milk is some kind of miracle brew all right, carefully concocted to meet the physiological needs of the young in the environment in which they are raised; concocted, too, to reflect how a mother takes care of her young, whether she is a nester, cacher, carrier, or follower, as biologists have dubbed some of Ben Shaul's groups. The nutritional needs of infants and the feeding styles of mothers evolved together to produce, for each species, a finely tuned solution of proteins, vitamins, fats, minerals, and carbohydrates. An infant best grows and thrives on its particular solution, and improper growth—or consequences even more severe—results when infants are fed on the milk of almost any other species.

Not long after Ben Shaul published her findings, researchers began to wonder what, if anything, the composition of human milk could tell us about humans, about the circumstances in which humans evolved and the way human infants first suckled.

Nicholas Blurton Jones, an English ethologist, was one of the first to look at this question, and he began by simplifying Ben Shaul's categories. He divided terrestrial mothers into just two basic types: continuous feeders, mothers who carry or are followed by their young and are, therefore, in constant contact with them, and spaced feeders, mothers who cache their young or keep them in nests. These two groups, as Blurton Jones pointed out, differ in certain predictable ways. Spaced feeders, as their name implies, feed their young at widely spaced intervals, and their milk has a high protein and fat content; also, their infants suckle at a rapid rate. Continuous feeders, on the other hand, carriers like primates, marsupials, and certain bats, as well as followers like sheep and goats, feed their young more or less continuously, and their milk is low in fat and protein; their infants suckle slowly.

So where do humans fit in this scheme? With a fat content of 4.2 percent and a protein content of 0.9 percent, human milk clearly puts us in the category of continuous feeders. This fits in well with what we know of infant care in the few remaining hunter-gatherer societies, the !Kung of the Kalahari Desert or the Papua New Guineans, whose mothers carry their infants (on their hips or in a sling) and nurse them very frequently (during the day as often as every fifteen minutes, and at night at least once until they are about three years old). It also makes sense of some of the idiosyncrasies of the modern human infant's behavior, such as the fact that a crying baby is quieted by rocking movements in the range of sixty cycles per minute, just the speed of a human female as she walks slowly, looking for food, perhaps, and carrying an infant on her hip. Or the fact that today's infant is noisy, in sharp contrast to the silence of most primate babies. The wailing that we've come to expect from our infants may not always have been part of their behavior pattern. Infants who are in constant skin-to-skin contact with their mothers rarely get so hungry that they cry for food. Their mothers are able to read their early hunger signals—moving, gurgling, fretting—and help their infants to the breast long before they get to the point of crying.

Colic, too, may be the result of treating our continuously feeding infants as if they were spaced feeders. When hand-reared rhesus monkeys are fed on a two-hour schedule, they vomit and burp frequently, something that rhesus monkeys fed by their mothers never do. "Perhaps the frequent vomiting and 'posseting' of human babies is a result of our insistence on the very early development of a four-hourly schedule rather than the quarter-hour and two-hour interval suggested by comparative data on milk composition," Blurton Jones has observed.

Blurton Jones is probably right, yet it is so easy to see how this change might have come about, how women, even women who loved their infants dearly, must have leapt at the opportunity to put their babies down for a while in some safe spot—a house, a bed, the care of an older sibling—and go about their business unimpeded. It is easy to see how they might have stretched the nursing interval, not to the point where the infant's health and growth obviously suffered and milk production was curtailed but to some point where the infant continued to thrive *and* the mother had a little time to herself. Baboon mothers would certainly understand this. It is the nature of every lactating female as she juggles her needs and the needs of her infant. It is the same balancing act that has led to crèches and aunting behavior, baby-sitters and communal nursing.

Source: From pp. 155–160 in S. Allport, 1998, *A Natural History of Parenting,* Three Rivers Press, New York. Copyright © 1997 by Susan Allport. Reprinted by permission of Harmony Books, a division of Crown Publishers, Inc.

transported through the bloodstream. However, glucose is poisonous to the body in high concentrations. Glucose levels are controlled by a protein called insulin, which is produced by cells in the pancreas. When the body is unable to produce enough insulin, diabetes occurs. In modern societies, diabetes is effectively controlled by diet or artificial sources of insulin.

During pregnancy there is a major change in the events that follow meals. In nonpregnant women, blood glucose levels rise after eating, then insulin levels rise, and finally blood glucose levels fall. But late in pregnancy, levels of both glucose and insulin rise higher after meals and remain elevated longer. Thus, pregnant women secrete more insulin, but it is less effective than it is in nonpregnant women. The ineffectiveness of the maternal insulin response may be related to the production of high levels of **human placental lactogen (hPL)** by the fetus. This hormone is closely related to **human growth hormone (hGH)**, which is believed to block the effects of insulin. Thus, it seems plausible that the fetus attempts to increase blood sugar levels by secreting hPL into the mother's bloodstream, and the mother counters by increasing the amount of insulin she secretes.

These skirmishes appear to have real consequences for both parties. Mothers with higher glucose levels produce heavier and healthier infants. At the same time, however, mothers who have high glucose levels during pregnancy are more likely to develop diabetes later in life.

The Evolution of Senescence

Aging and death are probably adaptive.

Like most other animals, humans experience **senescence**—that is, we age. As we get older, every part of our bodies begins to deteriorate—we can't run as fast, see as well, heal as quickly, or remember as much as we could when we were younger. This deterioration progresses until eventually some important physiological system fails, and we die.

At first glance, aging and death seem just to be the inevitable effects of wear and tear on bodies (Figure 17.6). Organisms are complicated machines, and as with

Figure 17.6

Virtually all organisms experience senescence. When this photograph was taken this old male chimpanzee, named Hugo, was missing a lot of hair on his shoulders and back, he had lost a considerable amount of weight, and his teeth had worn to the gums. Hugo died a few weeks later.

all complicated machines, there are many components that must function together in order for the organism to survive. It seems logical that, like the components in any other machine, the components in our bodies simply wear out. Transmissions, clutches, and valves break down, and eventually a car no longer runs. So, it seems, do people. Teeth wear down, cartilage tears, cataracts grow, and arteries accumulate fatty deposits; eventually, something crucial breaks and the human body stops functioning.

On a little reflection, you will see that this explanation of aging is suspect because the analogy between organisms and machines is not really apt. In every cell, organisms contain all of the genetic information necessary to build a complete, new body, and this genetic information can be used to repair damage. In humans, cuts heal and bones mend, and in some organisms like frogs, entire limbs can be regenerated. We know that in some circumstances repair mechanisms allow organisms to live indefinitely. The most obvious examples are the **germ cells** (the cells that give rise to gametes) in your own body: each one is the descendant in a line of cells millions of years old. It also turns out that some organisms that reproduce asexually by budding or fission do not undergo senescence.

The fact that some organisms can evade aging raises an intriguing question. All other things being equal, longer life would increase reproductive success, and we have repeatedly seen that natural selection favors adaptations that increase reproductive success. If death is not inevitable, why hasn't natural selection favored mechanisms that allow humans to live and reproduce indefinitely? The answer, oddly enough, seems to be that genes causing aging and death have been favored by natural selection because these genes increase fitness.

Two Evolutionary Theories of Senescence

Natural selection becomes weaker with age.

The key to understanding how aging and death can be favored by selection is to realize that selective pressures are much weaker on traits that affect only the old. Carriers of a mutant allele that kills 10-year-old children have a fitness of zero because none of the individuals who carry the allele survive long enough to transmit the gene to their descendants. Therefore, there will be strong selection against alleles with deleterious effects on the young. In contrast, selection will have much less impact on a mutant allele that kills 70-year-olds. By the time they have reached 70, people have already had as many children as they are going to have, and most of their children are probably grown. The gene that kills 70-year-old people has almost no effect on reproductive performance, and thus there will be little or no selection against this mutation.

To get an idea of how important the effect of age on selection has been during human evolution, we need information about the survival and reproduction of early humans. Of course, such information is not available, but we can use data from living foragers as a model. There are many reasons why foragers provide a useful point of reference. For the most part, they do not have access to Western medicine, they do not wean their children onto cow's milk or supplement breast milk with cereals at an early age, and they do not live at high enough densities to sustain infectious disease pathogens that plague more settled peoples.

Figure 17.7 provides what demographers call a **life table** for one group of foragers, the Dobe !Kung of the Kalahari Desert. Using information about births and deaths in this group, demographer Nancy Howell calculated the probability

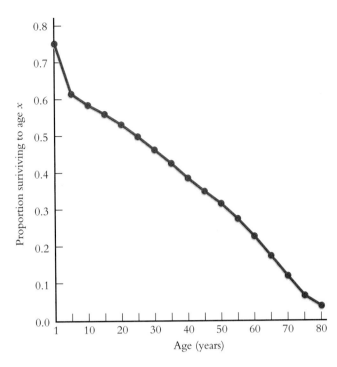

Figure 17.7

Demographic information is often presented in the form of life tables, like this one for the !Kung. The vertical axis gives the probability of surviving to a given age. Among the !Kung, mortality is highest for young children. Only 75% of all children born survive their first year of life, and 60% are still alive at age 5. After age 5, there is a steady, but more gradual, decline in the probability of surviving. About 25% of all individuals survive to the age of 60.

of surviving to various ages. As you can see, the rate of mortality is highest in the early years of life and declines at a steady rate through adulthood. About 25% of the !Kung survive to the age of 60 (Figure 17.8). Howell also computed age-specific fertility rates (the probability of reproducing at particular ages) for men and women (Figure 17.9). Because the !Kung population in this study was quite small, there are large uncertainties in these data. However, other hunter-gatherer populations have roughly similar demographic statistics, and so it is reasonable to assume that for most of human evolutionary history relatively few people have reached old age.

W. D. Hamilton has shown how to use the data in life tables to calculate the

Figure 17.8

An old !Kung man holds his grandchild. About 25% of the !Kung survive to the age of 60. (Photograph by Nicholas Blurton Jones.)

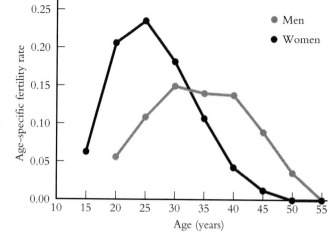

Figure 17.9

Age-specific fertility rates give the probability of producing a child at particular ages. !Kung women have their first child between the ages of 15 and 19 and the highest fertility in their 20s. !Kung men do not begin to reproduce until their early 20s, and their fertility rates are fairly stable in their 30s and 40s, dropping to low levels in their 50s.

strength of selection against a mutation that causes a small decrease in survival at particular ages. We can use data from the !Kung life table along with Hamilton's formula to assess the strength of selection against mutations that decrease survival at different ages. It is clear from Figure 17.10 that as age increases, the strength of selection against mutations that decrease survival declines substantially. Selection against a mutation leading to a 1% decrease in survival from the age of 20 to 21 years is 30 times stronger than selection against a mutation that decreases survival 1% between the ages of 50 and 51 years. Mutations that reduce survival at later ages face even weaker selective pressures. The strength of selection declines with age because genes expressed late in life have a smaller effect on lifetime fitness than do genes expressed early in life.

This weakening of selection does not depend on the existence of senescence. If they do not fall ill or suffer accidents, humans regularly reach their seventies. However, among the !Kung, few people actually survive to old age; many are killed in accidents and suffer from fatal diseases, and some women die in childbirth. Thus, even if the !Kung were potentially immortal, selection against genes that decrease the survival of the old still would be much weaker than selection against genes that decrease the survival of the young.

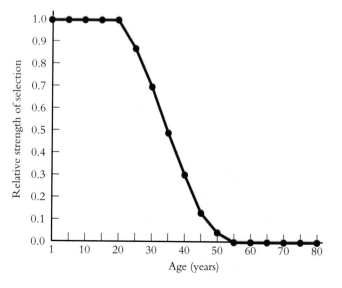

Figure 17.10

Demographic data for the !Kung (Figure 17.7) have been plugged into Hamilton's formula to estimate the strength of selection against mutations that influence mortality at different ages. There is much stronger selection against mutations causing small increases in mortality at young ages than older ages.

Aging may result from selection for characteristics that increase survivorship among the young but decrease survivorship of the elderly.

There is good reason to believe there is a tradeoff between survivorship and reproduction at different ages. We all know there is a tradeoff between cost and quality in machines. A Subaru costs much less to build than a Lexus, but also is of lower quality. As a consequence, a Lexus is expected to break down less often than a Subaru is. The same tradeoff applies to organisms. If more was invested in the construction of organisms, they would last longer. However, building higher-quality organisms consumes time and resources, and this will reduce the organism's growth rate and early fertility. Thus, longer-lived individuals will, on average, have lower fitness when they are younger.

Our teeth provide a good example of this tradeoff. One of the effects of age is that teeth get worn down by use and degraded by decay. As a result, young people usually have better teeth than older people do. The rate at which teeth wear down and decay can potentially be reduced in a number of ways. As we have seen, organisms that feed mainly on coarse material have large teeth and thick enamel. If humans had larger teeth and thicker enamel, their teeth might last longer. Second, organisms can be equipped with multiple sets of teeth so that one set can be replaced with another as the organism ages. Humans have two sets of teeth, and some of the permanent molars do not erupt until late adolescence or early adulthood. Reptiles, on the other hand, have an endless supply of teeth. Why don't we have thicker enamel and more sets of teeth? The answer is that each of these mechanisms has costs: larger teeth with thicker enamel take longer to grow and require more nutrients, particularly calcium. Replacement teeth also require resources, necessitate more complicated developmental mechanisms, and make precise occlusion (the orientation of opposing upper and lower teeth) more difficult to achieve. It seems likely that people with smaller teeth and fewer replacement teeth would grow faster and have more children when they were young, but would suffer more problems with their teeth as they age.

The fact that selection weakens with age means this tradeoff is greatly biased against characteristics that prolong life at the expense of early survival or reproduction. For example, suppose people with a gene that caused thicker enamel had a 1% greater chance of dying during childhood because they were more prone to calcium deficiency, but they were more likely to live past 50 years of age. Among people with a life table like the !Kung, this gene would be favored by selection only if having thicker enamel caused a 30% increase in survivorship after age 50!

This explanation of the evolution of aging is called the **antagonistic pleiotropy hypothesis** because it posits the existence of genes with pleiotropic effects that are positive at early ages and negative at older ages. According to this hypothesis, aging is favored by selection because traits that affect fertility at young ages are favored at the expense of traits that increase longevity.

Aging also may result from the accumulation of mutations that affect only the old.

As we saw in Chapter 16, there is always a balance between mutation and selection. Mutation introduces new, usually deleterious, variants, and natural selection removes these variants. The frequency of deleterious mutants present in the population at any given time depends on the strength of selection. If selection is strong, most variants will be eliminated immediately, so few variants will be present in the population at any given time. If selection is weak, mutations will build up to higher frequencies.

Because selection becomes weaker with age, mutations that affect only older people will occur at higher frequencies than mutations affecting younger people. Calculations by population geneticists suggest that this effect is quite substantial. Thus, since mutations are almost always deleterious to the functioning of the organism, aging may be caused by the accumulation of mutations that affect only the old. Such mutations may be very deleterious, causing conditions such as cancer, osteoporosis, and heart disease. These mutations have not been eliminated by natural selection because humans did not often reach old age, so until recently, these traits have rarely been expressed.

This mechanism for the evolution of aging is called the **mutation accumulation hypothesis** because it posits that aging results from the accumulation of mutations that affect only older individuals. According to this hypothesis, aging is not an adaptation but rather a side effect of the fact that selection on traits that affect the old is relatively weak.

Evidence for the Theories

Laboratory experiments suggest that both antagonistic pleiotropy and mutation accumulation are responsible for aging.

The antagonistic pleiotropy hypothesis assumes that genes causing longer life will also reduce early reproduction, and that genes increasing early reproduction will also lead to a reduced life span. In other words, the hypothesis assumes there is a negative genetic correlation between early reproduction and longevity. Thus, we can test for the existence of antagonistic pleiotropy in the laboratory by selecting for one character and seeing whether we get a correlated response to selection in the other character. For example, if we select against earlier reproductive capabilities, then we should see an increase in longevity; if we select for prolonged life, then we should see a decrease in early fertility. Experiments with several different organisms show the predicted results. The work of University of California, Irvine, biologist Michael Rose provides a good example. Rose began with laboratory populations of fruit flies *(Drosophila melanogaster)* that had stable patterns of reproduction and aging. He then began selecting for longevity. As predicted by the antagonistic pleiotropy hypothesis, the line that was selected for longer life span produced fewer eggs at younger ages and more eggs at later ages (Figure 17.11).

Other experiments conducted by Rose and his colleagues suggest that aging in fruit flies is due to both antagonistic pleiotropy and the accumulation of mutations. In fruit flies, three characters are associated with increased longevity: starvation resistance, resistance to drying out (desiccation), and resistance to low levels of ethanol (an alcohol that damages cells). Rose and his colleagues selected for longevity in one population of fruit flies. They found that the flies lived longer because they became more resistant to disease, desiccation, and ethanol toxicity. However, the flies also showed a decrease in early reproductive rates, which suggests that some of the aging was the result of antagonistic pleiotropy.

To see whether some aging was also due to the accumulation of mutations, Rose and his colleagues stopped selecting for longevity and returned the flies to their original rearing conditions. If antagonistic pleiotropy were responsible for all aging, then there should be strong selective pressure to alter the traits associated with longevity (resistance to desiccation, starvation, and ethanol toxicity) because these traits reduce early fertility. On the other hand, if the accumulation of mu-

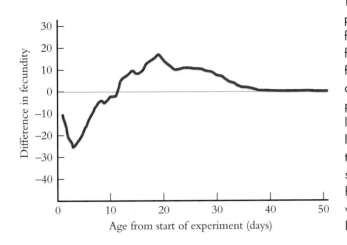

Figure 17.11

This graph compares fecundity (the ability to reproduce) in two lines of *Drosophila,* one selected for increased longevity and the other not selected for increased longevity. On the vertical axis, differences in fecundity between the long-lived line and the short-lived line at particular ages are plotted. Values greater than 0 indicate the long-lived line has greater fecundity than the short-lived one, while values less than 0 indicate that the long-lived line has lower fecundity than the short-lived one. At early ages, the short-lived line has higher fecundity than the long-lived line, while the pattern is reversed at later ages. Thus, longevity depresses early fecundity.

tations was responsible for aging, then these characters should change gradually because mutations accumulate very slowly. In fact, Rose and his coworkers found that starvation resistance rapidly returned to the original state, while resistance to desiccation and resistance to ethanol showed no change after 50 generations. These results suggest that aging caused by starvation resistance results from antagonistic pleiotropy, while aging due to resistance to desiccation and ethanol results from the accumulation of mutations.

> *If aging is due to antagonistic pleiotropy, it will be hard to find a single "cure" for senescence, but if aging is due to mutation accumulation it may be easier to find a cure.*

People have always dreamed of finding the "fountain of youth," a magic potion that would cure senescence and allow them to live forever. In recent times, **gerontologists** (researchers who study aging) have actively searched for a medical cure for senescence. The hope has been to discover that just a few simple biochemical mechanisms cause aging. Then, we could radically extend human life by finding medical interventions that block them.

If aging is due to antagonistic pleiotropy, there will be many different synchronized causes of aging. Organisms are complex systems with many different, partially independent subsystems, each potentially subject to aging. The kinds of failures leading to aging of the teeth are likely to be quite different from the kinds of failures leading to aging of the heart, eye, or brain. To see why these processes should be synchronized, suppose that one cause of aging, such as heart disease, acts at much earlier ages than all of the other causes of aging. Then selection would either favor the postponement of the expression of genes that cause heart disease so that heart disease becomes synchronized with other forms of aging, or it would favor earlier action of all of the other forms of aging so that they become synchronized with heart disease. In either case selection would cause all forms of aging to occur simultaneously. Thus, if aging is due to antagonistic pleiotropy, it is unlikely that curing one, or only a few processes, would lead to an indefinitely long life.

If aging is due to mutation accumulation, there will also be many causes of aging, but there is no reason to believe they should be synchronized. Mutation is a random process, and it could be by chance that much human aging is caused by the accumulation of only a few mutations whose effects occur relatively early compared with most mutations. If this is the case, then curing the effects of these mutations might substantially lengthen human lives.

Although the theory of senescence provides a powerful explanation for the decline in the performance of many bodily systems, it does not account for the complete termination in reproductive function in women, which we consider in the next section.

The Evolution of Menopause

Emma Wedgwood Darwin, the wife of Charles Darwin, gave birth to the last of her 10 children, Charles Waring, when she was 48 years old (Figure 17.12). Emma had produced her first nine children at fairly regular intervals, but there was a five-year gap between the birth of Emma's ninth child and her tenth child (Figure 17.13). Sadly, baby Charles did not thrive and died of scarlet fever before his second birthday. Emma outlived her husband and several of her children; she died in 1896 at the age of 88.

Unlike other animals, human females live for a long time after they are no longer able to reproduce.

Like Emma Darwin, most women's reproductive careers end well before the end of their life span. As women reach the age of 40, their menstrual cycles usually become less regular in length and the number of nonovulatory cycles increases. Fertility declines to very low levels for women over the age of 40. We know this is not just a recent effect due to contraception and family planning because the pattern is evident in groups that value large families and don't practice contraception, such as the Hutterites, an Anabaptist sect in North America (Figure 17.14). It is quite unusual for women to give birth after the age of 45, as Emma did. Moreover, older women who do conceive suffer more spontaneous abortions than younger mothers, and the fetuses they carry are more likely to have chromosomal

Figure 17.12
This photograph of Emma Wedgwood Darwin was taken when she was about 50 years old. During her marriage, Emma gave birth to 10 children. Her last child was born in 1856 when she was 48 years old.

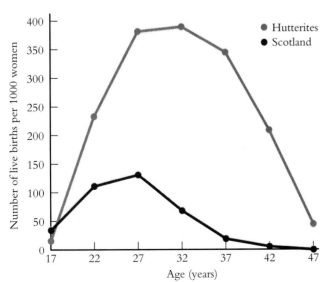

Figure 17.13

Charles and Emma Wedgwood Darwin had 10 children. Their last child was born when Emma was nearly 50 years old.

anomalies. Most women in developed countries experience **menopause** (the end of menstrual cycles) at about the age of 50, while life expectancy for women who survive to reproductive age is about 70 years.

Thus, we see that women's reproductive careers end well before the end of their life spans. This raises the puzzling question: why do women stop reproducing so early in their lives?

All of the egg cells a woman will ever have are present in her ovaries at birth, and then decline in number to zero by about age 50.

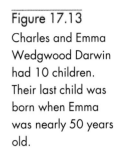

Before we consider the evolutionary forces that influence menopause, we need to know more about what happens to a woman's body during menopause. Surprisingly, the events that lead to menopause begin before birth. In females, primordial germ cells emerge as distinct cell lines in the developing embryo only 24 days after conception. Within the next few weeks, the germ cells, which then number a few thousand, migrate to the site of the embryonic tissues that will eventually become the ovaries. Then the germ cells begin to multiply rapidly through mitosis, so that by the end of the sixth month after conception several million germ cells have been created. During the final months before birth, the germ cells begin to divide by meiosis, reducing the genetic material in each germ cell by half. Meiosis is suspended in an early phase of its cycle, and the germ cells remain in this incomplete state until puberty. After puberty, a small number of germ cells resume meiosis during each menstrual cycle, but normally only one germ cell completes the process in each cycle (see Box 17.1).

Figure 17.14

Female fertility declines as women age. The blue circles represent Hutterite women, members of an Anabaptist sect that values marriage and large families and does not condone contraception. The red circles are based on data from the General Register Office in Scotland in 1980.

BOX 17.1

The Evolution of Menstruation

Menarche and menopause mark the beginning and end points of women's reproductive careers. During the years between menarche and menopause, women undergo cyclic changes in their ability to conceive and sustain a pregnancy. In humans, these monthly cycles are marked by conspicuous vaginal bleeding, or menstruation, and are therefore known as menstrual cycles. For many women, menstruation is an innocuous reminder of their reproductive status. For others, menstruation is more problematic. Menstruation is sometimes associated with unpleasant symptoms, ranging from bloating to severe cramps and anemia. Some women complain of irritability, moodiness, fatigue, headaches, and other problems just before menstruation begins, a collection of symptoms that has become known as **premenstrual syndrome**, or **PMS**. Recently, behavioral ecologists have begun to consider why these cycles occur and why women menstruate. To understand their hypotheses, we need to know something about the events that characterize the menstrual cycle.

All mammalian females experience cyclic changes in their ability to conceive and to become pregnant. These cycles are generally known as ovarian cycles, rather than menstrual cycles, because external bleeding is the exception rather than the rule among mammals. The central event in each ovarian cycle is ovulation, the development and release of a mature egg cell from the ovaries. Although the full complement of egg cells is present at birth, the development of egg cells in the ovaries is halted at an intermediate stage. During each ovarian cycle, a single egg cell is somehow selected for development. But before the lucky egg is ready to be released from the ovary, certain preparations must be completed: the egg cell must resume its arrested activity, and the uterus must be prepared to sustain a pregnancy. Hormones secreted by the pituitary gland and the ovaries regulate these processes.

The pituitary hormones, **follicle-stimulating hormone (FSH)** and **luteinizing hormone (LH)**, stimulate the development of the ovarian steroid **estradiol** (a form of estrogen) by the follicle, a packet of cells that surrounds and nurtures each egg cell. As levels of estradiol rise, the follicle increases greatly in size and complexity. Just before ovulation occurs, levels of FSH, LH, and estradiol peak (Figure 17.15).

Estradiol and another ovarian steroid, **progesterone**, play an important role in preparing the uterus to sustain a fetus in the event of successful conception. Rising levels of estradiol stimulate the growth and development of the uterine lining, or endometrium, and the thickening of the **myometrium**, the layer of smooth muscle that lies underneath the endometrium. Estrogen also stimulates the endometrium to produce receptors for progesterone. Progesterone, in turn, promotes the development of glands in the endometrium that secrete glycogen and vital enzymes into the uterus. At the same time, the arteries that supply the endometrium join together, forming pools of blood in the endometrial tissue (Figure 17.16). If pregnancy does not occur, the tissue, secretions, and blood that have been accumulated in anticipation of pregnancy are shed. In most animals, these materials are reabsorbed into the body. In a few species, including humans, they are discharged from the body through the vagina. This is the

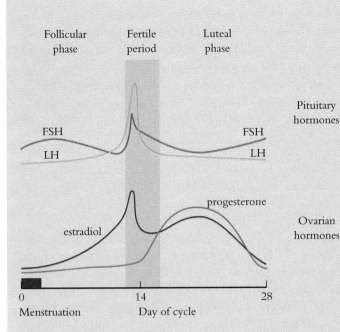

Follicular phase | Fertile period | Luteal phase

FSH
LH

FSH
LH

Pituitary hormones

estradiol

progesterone

Ovarian hormones

0
14
28

Menstruation

Day of cycle

Figure 17.15

The human menstrual cycle is divided into two parts. The follicular phase precedes ovulation and the luteal phase follows ovulation. Women are fertile for a short period around the time of ovulation, which typically occurs at the midpoint of the cycle (Day 14). Levels of the pituitary hormones, FSH and LH, and the ovarian hormone estradiol peak immediately before ovulation occurs. Levels of progesterone rise during the luteal phase and remain elevated if pregnancy occurs.

phenomena that we know as menstruation.

At first, it may seem inefficient to prepare the uterus repeatedly for pregnancy and then to abandon these preparations if pregnancy does not occur. After all, it is cheaper to leave a house standing that is intermittently occupied than to destroy and rebuild it every month. However, University of Michigan anthropologist Beverly Strassman argues that this is a poor metaphor because metabolic tissue is not made of inert substances, like brick or wood. Living tissue must be nourished, and this nourishment requires energy.

Strassman suggests that it would actually be more energetically expensive to maintain the uterus in a state of continual readiness for pregnancy than it is to cycle regularly. Metabolic rates are 7% lower during the follicular phase, which precedes ovulation, than during the luteal phase, which follows ovulation. Women consume 11% to 35% more calories per day during the luteal phase than during the follicular phase. Similar patterns are seen in other mammalian species. Thus, if the uterus was always maintained in the luteal phase condition, females would need to increase their food intake substantially. Strass–

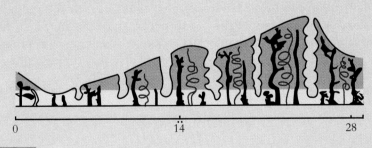

0
14
28

Figure 17.16

During the human menstrual cycle, a number of changes occur in the endometrium, the cells lining the uterus. The endometrium becomes progressively thicker, and glands that secrete glycogen and enzymes into the uterus become larger. Arteries that supply the endometrium join together, forming pools of blood in the endometrial tissue. If pregnancy does not occur, accumulated tissue and blood is shed through menstruation.

man estimates that women save six days' worth of food over four menstrual cycles. During the 12 months of lactational amenorrhea, women save nearly half a months' food. Since females' ability to reproduce successfully is typically limited by the availability of food (see Chapter 7), cyclical changes in readiness for pregnancy offer females a way to reduce their energetic requirements.

Although Strassman's argument provides a plausible explanation for ovarian cycling, it does not necessarily explain why women menstruate. After all, all mammalian females experience ovarian cycles, but conspicuous menstrual bleeding is relatively uncommon. Among primates, overt menstrual bleeding occurs only in Old World monkeys and apes. Two hypotheses have been proposed to explain why menstruation occurs. Marjorie Profet has suggested that menstruation evolved to rid the uterus of pathogens carried by sperm. Strassman has criticized Profet's hypothesis and contends instead that menstrual bleeding is simply a side effect of ovarian cyclicity and has no adaptive function.

Profet's hypothesis is that menstruation evolved to protect the uterus and other reproductive organs from being colonized by pathogens carried by sperm. Strassman has examined several predictions derived from this hypothesis. First, if menstruation defends the uterus against pathogens, then uterine infections should be more common before menstruation than after menstruation. In fact, much the opposite is true. Menstruation seems to increase the risk of infection, perhaps because blood is an excellent medium for bacteria. Second, Profet argues that menstruation tracks women's exposure to sperm-borne pathogens, occurring during the portion of women's lives when they are sexually active. However, in traditional societies women do not menstruate for long periods when they are pregnant and lactating, even though they are sexually active during these periods. And women continue to engage in sexual activity after they have reached menopause. Thus, menstruation provides little protection to women during much of their sexually active lives. Third, Profet predicts that menstruation will be most copious among primate species in which females have multiple mating partners. However, comparative analyses across primates provide no support for this prediction. Thus, there is little evidence to support Profet's hypothesis that menstruation reduces the risk of infection by sperm.

Strassman contends that menstruation is not maintained by natural selection, but is instead a byproduct of ovarian cycling. She suggests that variation in the degree of menstrual bleeding among primates may be related to anatomical differences in 1) the structure of blood vessels that supply the endometrial tissue with nutrients, 2) the thickness of the endometrium, or 3) the volume of the uterus. She points out that in Old World monkeys and apes conspicuous menstrual bleeding tends to occur in species in which females produce relatively large infants. In such species, there may simply be too much blood to reabsorb.

A few months before birth, a female fetus carries nearly 7 million germ cells in her ovaries (Figure 17.17). By birth, this number has dropped to about 1 million as the remaining germ cells degenerate and die in a process called **atresia**. The number of egg cells continues to decline, dropping to approximately 250,000 at puberty. Roughly one egg cell is lost every 4 minutes over the course of women's lives. The rate of egg loss accelerates as women near 40, and the supply of egg cells is nearly depleted over the next decade, until at 50 less than 1000 egg cells

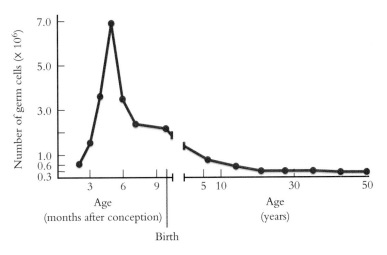

Figure 17.17

The number of primordial egg cells reaches a peak three months before birth and then declines. The rate of decline may increase after the age of 40, and the supply of eggs is exhausted at menopause.

remain. It is at this point that the ovaries stop producing the hormones that regulate ovulation, the menstrual cycles stop, and menopause occurs.

Menopause is universal, but its timing varies around the world.

Women in every society experience menopause, but there is considerable variation within and between societies in the timing of menopause. In the United States, some women experience menopause in their mid-30s while others continue to menstruate until their mid-50s (Figure 17.18). On average, women in developed nations like the United States undergo menopause between the ages of 49 and 51. In some other regions, like Bangladesh and Central Africa, the average age at menopause is 43 or 44 years of age. We do not fully understand the reason that the age of menopause varies within or between regions, although some evidence suggests that good nutrition delays menopause.

Menopause occurs earlier and more abruptly than the senescence of other bodily systems.

Menopause differs from other forms of aging in two important ways. First, menopause precedes senescence in other bodily systems. In Figure 17.19, you can see that women's fertility begins to decline in their late 30s and early 40s, well before declines in the function of the heart, kidneys, and lungs are seen. Women's

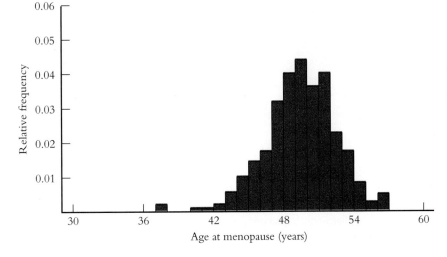

Figure 17.18

The age of menopause ranges from the late 30s to the early 50s among 324 women in the United States who were born between 1910 and 1925. The average age of menopause among these women was 49.5 years.

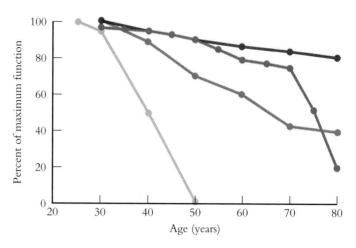

Figure 17.19

Female reproductive capacity declines more rapidly than other physiological functions. Red = basal metabolic rate, blue = maximum breathing capacity, green = male reproductive capacity, yellow = female reproductive capacity.

fertility also begins to decline much sooner than men's reproductive capacity declines. Second, unlike other bodily systems that show gradual losses in efficiency, the cessation in women's reproductive capacity is much more abrupt.

Menopause might be a nonadaptive artifact of recent extensions of the human life cycle.

Menopause is also interesting because it is not part of the life history of other primates. Among some nonhuman primates, interbirth intervals increase in length and fertility declines as females age, but a substantial fraction of females in all other primate species continue to reproduce throughout their lives (Figure 17.20). Even among chimpanzees and gorillas, whose life spans are most similar to our own, some females continue reproducing until advanced ages (Figure 17.21). The pilot whale is the only other species we know of in which all females stop reproducing well before they die.

What could account for this recent development in hominid evolution? Some researchers suggest that the human reproductive system was simply designed for a much shorter life span than modern humans now enjoy. Remember that few of the Neanderthals survived past the age of 45, and women rarely reached the age of 40 in the Upper Paleolithic. Careful analysis of the skeletons of a population of

(a)   (b)

Figure 17.20

In other primate species, females do not experience menopause. This female chimpanzee, Fifi, was (a) about 14 when her first son, Freud, was born, and (b) about 34 years old when her fifth infant, Ferdinand, was born. (b, Photograph courtesy of Craig Stanford.)

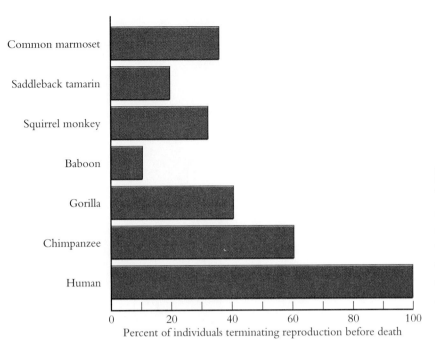

Figure 17.21
The majority of primate females continue to reproduce regularly until death, although fertility generally declines among older females. However, the complete and universal termination of reproduction occurs only in humans and pilot whales.

prehistoric people living about a thousand years ago in what is now Ohio indicate that the average life expectancy for those who survived to the age of 15 was only 34 years.

But this hypothesis suffers from two important problems. First, Figure 17.19 indicates that women's reproductive function declines at a different rate than other bodily functions. If menopause simply reflects the length of the evolved human life span, we must ask why women's reproductive systems are designed for a 50-year life span if their hearts and kidneys are designed for a 70-year life span. And we must ask why men's reproductive systems are designed for much longer life spans than are women's reproductive systems.

Second, in traditional food-foraging societies, men and women routinely live well beyond the age of menopause. People living in these societies have limited access to Western technology, medicine, and education, but they enjoy much greater longevity than people living in prehistoric communities apparently did. Among the !Kung and other foraging peoples, a substantial fraction of individuals live beyond the age of menopause (see Figure 17.7).

It is also possible that menopause simply reflects a constraint imposed by the biology of egg development. Figure 17.17 indicates that egg cells degenerate and die at a fixed rate, so we might conclude that the length of a woman's reproductive career is determined directly by the number of egg cells present initially and the rate of deterioration. If this is the case, then we must ask why evolution has not provided women with a larger initial supply of eggs or reduced the rate of egg loss. We must also ask why reproduction continues throughout the life span in other long-lived species like elephants.

Menopause may have been favored by natural selection because postmenopausal women were better able to care for their existing children.

Although it seems paradoxical that women would benefit from terminating their reproductive careers before death, a number of researchers have explored the idea that natural selection may have favored menopause because women of a certain age benefited more by investing in their existing children or their grandchildren

Figure 17.22

An old !Kung woman looks after her grandchild. (Photograph by Nicholas Blurton Jones.)

than by producing additional children themselves. This idea rests on the fact that human children require extensive investment from their parents for many years. In most human societies, children do not become completely independent of their parents until they reach their teens. Women who produce children as they reach the end of their life spans would not live long enough to assure their survival. These women might be better off, in an evolutionary sense, if they devoted their efforts to caring for their existing children or for their children's children rather than producing additional children of their own (Figure 17.22).

This general idea is supported by evidence that postmenopausal women work very hard at subsistence activities in some societies even though they have no dependent children of their own. Among the Hadza, a food-foraging group living near Lake Eyasi in northern Tanzania, women spend on average four to six hours each day gathering wild plant foods. They dig up roots and tubers, pick berries, collect honey, and gather baobab fruits. Kristin Hawkes and Jim O'Connell of the University of Utah and Nicholas Blurton Jones of the University of California, Los Angeles, describe women's efforts to excavate tubers:

> Excavations were sometimes more than a metre deep, and often involved moving large rocks through complex sequences of levering. Particularly intricate engineering problems drew the attention of several spectators. However, about three-quarters of the time spent in //*ekwa* (tuber) patches was devoted to active digging. Women who had carried nursing infants sometimes enlisted older children to sit with them and hold the infant, interrupting their digging periodically to nurse [p. 344 in K. Hawkes, J. F. O'Connell, and N. G. Blurton Jones, 1989, Hardworking Hadza grandmothers, pp. 341–365 in *Comparative Socioecology: The Behavioural Ecology of Humans and Other Mammals,* ed. by V. Standen and R. A. Foley, Blackwell Scientific, Oxford].

Hadza women work hard, but the oldest women work the hardest. Figure 17.23 compares the amount of food brought back to camp by young, unmarried women who have no children of their own; women of child-bearing age; and women who have stopped reproducing. In general, the oldest women spent the most time

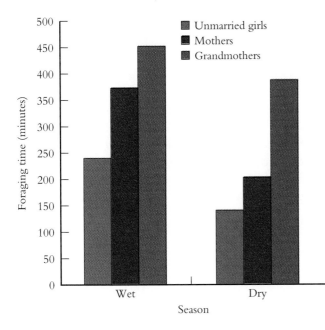

Figure 17.23

Hadza grandmothers who have no dependent offspring of their own spend more time per day foraging for wild plant foods than do mothers of young children or unmarried girls. The same pattern is seen in the wet season as in the dry season.

foraging and brought the most food back to camp, even though none of these women had any children under the age of 15 to feed.

There have been a number of attempts to develop formal mathematical models of several variants of the grandmother hypothesis. These models specify the range of conditions under which natural selection will favor the evolution and maintenance of menopause. However, these conditions do not match realistic values of demographic variables, such as maternal mortality and infant survivorship. The current models do not support the idea that menopause was favored because of the selective benefits that females conferred on their existing descendants. Thus, the intriguing question of why women reach the end of their reproductive careers long before they reach the end of their life span remains unanswered.

Further Reading

Allport, S. 1998. *A Natural History of Parenting*. Three Rivers Press, New York.

Austad, S. N. 1994. Menopause: an evolutionary perspective. *Experimental Gerontology* 29:255–263.

Haig, D. 1994. Genetic conflicts in human pregnancy. *Quarterly Review of Biology* 68:495–532.

Rose, M. 1991. *The Evolutionary Biology of Aging*. Oxford University Press, Oxford.

Strassmann, B. I. 1996. Energy economy in the evolution of menstruation. *Evolutionary Anthropology* 5:157–164.

Study Questions

1. Why does natural selection lead to conflict between parents and offspring over the amount of time and energy that parents devote to children?

2. Explain why natural selection acting on genes expressed in the mother may favor physiological mechanisms that cause miscarriages. Could selection ever favor genes expressed in the fetus that cause it to be spontaneously aborted?

3. There is evidence that chromosomes are chemically imprinted in such a way that the expression of particular genes can depend on whether the chromosome carrying the gene came from the mother or the father. It is as if genes "know" whether they have been inherited from the mother or from the father. Suppose such a gene was expressed in the fetus and affected maternal investment. Describe the difference in the effect of such a gene depending on whether it was inherited from the mother or from the father.

4. Why do the forces of natural selection weaken as an organism ages?

5. Explain why natural selection favors genes that cause aging and death. In what kinds of environments should selection favor especially long life?

6. Why is the theory of senescence not sufficient to account for the evolution of menopause?

C H A P T E R   1 8

Evolution and Human Behavior

Why Evolution Is Relevant to Human Behavior
Evolutionary Psychology
    The Logic of Evolutionary Psychology
    Reasoning about Reciprocity
Evolutionary Psychology and Human Universals
    Color Terms
    Inbreeding Avoidance
Evolution and Human Culture
    Culture Is a Derived Trait in Humans
    Culture Is an Adaptation
Human Behavioral Ecology

Why Evolution Is Relevant to Human Behavior

The application of evolutionary principles to understanding human behavior is controversial.

The theory of evolution is at the core of our understanding of the natural world. There is no doubt that every aspect of all living creatures is the product of evolution. By studying how natural selection, recombination, mutation, drift, and other evolutionary processes interact to produce evolutionary change, we come to understand why organisms are the way they are. Of course, our understanding of evolution is far from perfect, and there are other disciplines, most notably chemistry and physics, that contribute greatly to our understanding of life. Nonetheless, evolutionary theory is an essential part of biology and anthropology.

So far, the way we have applied evolutionary theory in this book is not controversial. We are principally interested in the evolutionary history of our own species, *Homo sapiens,* but we began by using evolutionary theory to understand the behavior of our closest relatives, the nonhuman primates. Twenty years ago, when evolutionary theory was new to primatology, this approach generated some controversy, but now most primatologists are committed to evolutionary explanations of behavior. In the section on hominid evolution, we used evolutionary theory to develop models of the patterns of behavior that might have characterized early hominids. While some researchers might debate the fine points of this analysis, the use of evolutionary theory in this context creates little controversy. Perhaps this is because the early hominids were simply "bipedal apes," with brains the size of modern chimpanzees. Not many people object to evolutionary analyses of human physiology like those presented in the last chapter. While scientists argue the merits of particular evolutionary explanations of senescence, few doubt that human physiology has been shaped by natural selection, and that we may gain important insights about how our bodies work if we understand these processes more fully.

It is mainly when we enter the domain of contemporary human behavior that evolutionary analyses provoke intense controversy. Most social scientists acknowledge that evolution has shaped our bodies, our minds, and our behavior to some extent. However, many social scientists are critical of the application of evolutionary theory to contemporary human behavior because they think this implies that contemporary human behavior is determined genetically. Genetic determinism of behavior seems inconsistent with the fact that so much of human behavior is acquired through learning, and that so much of our behavior and beliefs is strongly influenced by our culture and environment.

All phenotypic traits, including behavioral ones, result from interactions between genes and the environment.

The controversy arises, in part, because many people view learning and genetic transmission as mutually exclusive alternatives: behaviors are either genetic and immutable, or learned and controlled entirely by environmental contingencies. This assumption lies at the heart of the "nature-nurture question," a debate that has raged in the social sciences for many years. Despite its persistence, the debate is based on a fundamental misunderstanding of how the natural world works.

The nature-nurture debate assumes there is a clear distinction between the effects of genes (nature) and the effects of the environment (nurture). People often think that genes are like engineering drawings for a finished machine, and people vary simply because their genes carry different specifications. For example, they would think that Bill Lambeer is tall because his genes specified an adult height of 2.11 m (6 ft 11 in.), while Muggsy Bogues is short because his genes specified an adult height of 1.60 m (5 ft 3 in.).

However, this way of thinking about genes is wrong. Genes are not like blueprints that precisely specify the details of the adult phenotype. Every trait results from the *interaction* of some genetic program with the environment. Thus, genes are more like recipes in the hands of a creative cook, sets of instructions for the construction of an organism using materials available in the environment. Genes control the production of a large number of chemicals, mainly enzymes, that interact with each other and the environment to generate the adult phenotype. At each step, this very complex process depends on the nature of environmental conditions. For example, the rate of chemical reactions depends on temperature, the availability of certain chemical substrates, the presence of pathogens, and myriad other environmental factors. As a result, the expression of any genotype always depends on the environment. Thus, we might discover that people are short because of poor nourishment during childhood or because they contracted a disease that limited their growth.

The expression of behavioral traits is usually more sensitive to environmental conditions than the expressions of morphological and physiological traits are. Therefore, we expect language to vary widely among human societies, while we expect finger number and the size of canine teeth to vary little among societies. As we saw in Chapter 3, traits that develop uniformly in a wide range of environments, like finger number, are said to be canalized. Traits that vary in response to environmental cues are said to be plastic. The point to remember is that no trait is purely genetic or purely environmental. Every trait, plastic or canalized, results from the unfolding of a developmental program in a particular environment. Even highly canalized characters can be modified by environmental factors, such as fetal exposure to mutagenic agents.

Natural selection can shape developmental processes so that organisms develop different adaptive behaviors in different environments.

Some people accept that all traits are influenced by a combination of genes and environment, but they reject evolutionary explanations of human behavior because they have fallen prey to a second, more subtle misunderstanding—the belief that evolutionary explanations imply that behavioral differences between individuals must be caused by genetic differences because natural selection cannot create adaptations unless such differences between individuals exist. According to this view, we cannot argue that variation in foraging strategies in different human groups is adaptive without implying that particular foraging strategies have a genetic basis. Since there is considerable evidence that human differences are the product of learning and culture, adaptive explanations must be invalid.

Of course, it *is* true that evolution of adaptations requires genetic variation. However, it does not follow that observed adaptive differences between individuals are the result of genetic differences. In Chapter 3 we saw that male mate-guarding behavior varies adaptively among populations of the soapberry bug in Oklahoma. Where there are many more males than females, males guard females for extended periods, but males do not guard where females are more common. Recall that most of this variation is not genetic; instead, individual males vary their behavior adaptively in response to the local sex ratio. In order for this behavior to evolve, there had to be some genetic variation affecting male mate-guarding behavior—small genetic differences in the male propensity to guard a mated female, and small genetic differences in how this propensity changes with the local sex ratio. If such variation exists, then natural selection can mold the response of males so that it is locally adaptive. However, in any given population, most of the observed behavioral variation is due to the fact that individual males respond adaptively to environmental cues.

Behavior in the soapberry bug is relatively simple. Human learning and decision making are immensely more complex and flexible. People living in varied environments and cultures have profoundly different subsistence strategies, domestic arrangements, child-rearing practices, religious beliefs, political systems, and languages. We know much less about the mechanisms that produce such flexibility in humans than we do about the mechanisms that produce flexibility in mate guarding among soapberry bugs. Nonetheless, such mechanisms must exist, and it is reasonable to assume that they have been shaped by natural selection. This assumption may be wrong, because evolution does not produce adaptation in every case, as we discussed in Chapter 3. However, there is no reason to assume that differences in behavior among humans are the product of genetic differences.

In the remainder of this chapter we survey several ways that evolutionary theory has been used to understand the behavior of modern humans. We begin by discussing attempts to use evolutionary ideas to understand the psychological mechanisms that give rise to human behavior. Next, we consider the relationship between evolutionary theory and human culture, and finally we apply evolutionary theory to understanding the structure of contemporary societies.

Evolutionary Psychology

The Logic of Evolutionary Psychology

In recent years, a new field called evolutionary psychology has developed. Its practitioners are committed to using evolutionary theory to understand human psychology and behavior. This research program is based on the following precepts:

- Minds are built up out of a large number of special-purpose mechanisms that solve particular kinds of problems.
- Since brains are expensive to build and to maintain, organisms can't be good at all cognitive tasks. For example, organic computing power that is devoted to reasoning about social problems can't also be used to solve abstract, logical puzzles.
- Natural selection determines the kinds of problems that the brains of particular species are good at solving. To understand the psychology of any species we must know what kinds of problems its members need to solve in nature.
- Complex adaptations evolve relatively slowly.
- People have lived in societies with agriculture, high population density, and stratified social organization for only a few thousand years. They lived in small-scale foraging societies for a much longer stretch of human history.
- To understand human psychology we must determine the kinds of problems humans needed to solve when they lived in small groups as Pleistocene foragers (Figure 18.1).

Even the most flexible strategies are based on special-purpose psychological mechanisms.

Psychologists once thought that people and other animals had a few general-purpose learning mechanisms that allowed them to modify any aspect of their phenotypes adaptively. However, a considerable body of empirical evidence indicates that animals are predisposed to learn some things and not others. Rats learn very quickly to avoid novel foods that make them ill. Moreover, rats' food aversions are based solely on the taste of foods that have made them sick, not the food's size, shape, color, or other attributes. This learning rule seems to be sensible because rats live in a very wide range of environments, and usually forage at night. Since their environments vary, they frequently encounter new foods. In order to determine whether a new food is edible, they taste a small amount first, and then wait several hours. If it is poisonous, they soon become ill, and they do not eat it again. Rats may pay attention to the taste of foods, instead of other attributes,

Figure 18.1

Evolutionary psychologists believe the human mind has evolved to solve the adaptive challenges that confront food foragers because this is the subsistence strategy that humans have practiced for most of our evolutionary history.

because it is often too dark to see what they are eating (Figure 18.2). However, there are limits to the flexibility of this learning mechanism. There are certain items that rats will never sample, and in this way their diet is rigidly controlled by genes. Moreover, the learning process is not equally affected by all environmental contingencies. Thus, rats are affected more by the association of novel tastes with gastric distress than they are with other possible associations.

It is hard to determine the kind of environment that has shaped the evolution of human reasoning.

Evolutionary psychologists use the term **environment of evolutionary adaptedness (EEA)** to refer to the social, technological, and ecological conditions under which human mental abilities evolved. Many evolutionary psychologists envision the EEA as being much like the world of contemporary hunter-gatherers.

There are two problems with this notion. First, there is great uncertainty about the rate at which complex adaptations, like psychological mechanisms, can arise.

Figure 18.2

Rats initially sample small amounts of unfamiliar foods, and if they become ill soon after eating something, they will not eat it again.

In Chapter 1, we saw that natural selection can sometimes lead to rapid change, but we also saw that vast spans of time can pass without any change at all. We cannot exclude the possibility that new mental mechanisms might have evolved since the origin of agriculture just 10,000 years ago, nor can we reject the possibility that some mental mechanisms might have been shaped by selection long before the origin of humans.

The second problem with the idea that the EEA resembled the environment of modern human foragers is that there is great uncertainty about the ecology and behavior of extinct hominids. As we have seen, some authorities believe that *H. erectus* and perhaps even earlier *Homo* were much like contemporary human foragers: They lived in small bands and subsisted by hunting and gathering. They controlled fire, had home bases, and shared food. They could talk to one another and had a body of inherited culture. There are other authorities who think *H. erectus* and even the Neanderthals lived lives completely unlike modern hunter-gatherers. These researchers think that Neanderthal speech was primitive, and that Neanderthals didn't hunt large game, share food, or have home bases. If early hominids lived lives that resembled the lives of contemporary foragers, then it is reasonable to think that the human brain has evolved to solve the kinds of problems that also confront modern foragers. For example, since food sharing is important for foragers, it would make sense for humans to have evolved psychological mechanisms to detect and punish freeloaders. On the other hand, if lifeways that characterize contemporary foragers did not emerge until 40,000 years ago, then we might doubt that there was enough time for selection to assemble specialized psychological mechanisms to manage food sharing.

Reasoning about Reciprocity

Despite the problems in determining the nature of the EEA, evolutionary psychology is a rapidly growing research area. It has provided novel and productive insights about the nature of human psychology, insights that promise to solve long standing problems in the social sciences. The following example shows how evolutionary reasoning helps us to understand why people can easily solve social problems but find it much more difficult to solve nearly identical problems in other domains.

Human reasoning abilities are content-dependent.

At one time it was thought that human reasoning was governed by the principles of formal logic. However, research by cognitive psychologists has shown that this assumption is not correct. People reason logically only in limited domains, and human reasoning is **content-dependent**, meaning that the subject matter that people are asked to reason about seems to regulate how they reason.

The following problems, drawn from the work of John Tooby and Leda Cosmides of the University of California, Santa Barbara, will convince you that human reasoning is content-dependent. First, consider the clerical problem in Box 18.1. Study this question for a few minutes, and write down the answer. Next, consider the bartender's problem in Box 18.2, and write down your answer on a piece of paper. If you are like most people, you selected only the "D" card, or perhaps the "D" and "3" cards, for the clerical problem. If you are like most people, you selected the cards labeled "Beer" and "17" for the bartender problem. And, finally, if you are like us and most other people, you got the first problem wrong and the

BOX 18.1

Clerical Problem

Part of your new clerical job at the local high school is to make sure that student documents have been processed correctly. Your job is to make sure that the documents conform to the following rule:

If a person has a 'D' rating, then his documents must be marked code '3'.

You suspect that the secretary whom you replaced did not categorize the stu-dents' documents correctly. The cards below have information about the documents of four people who are enrolled at this high school. Each card represents one person. One side of a card tells a person's letter rating and the other side of the card tells that person's number code.

Indicate only those card(s) you definitely need to turn over to see if the documents of any of these people violate this rule.

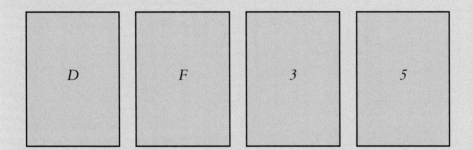

| D | F | 3 | 5 |

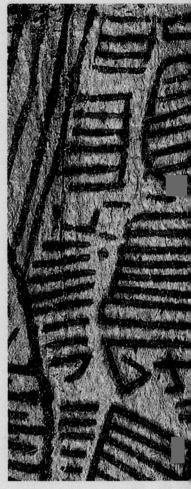

From Figure 3.3 in L. Cosmides and J. Tooby, 1992, Cognitive adaptations for social exchange, in *The Adapted Mind, Evolutionary Psychology and the Generation of Culture*, J. Barkow, L. Cosmides, and J. Tooby, eds. Oxford University Press, New York.

second problem right. The correct answer for the clerical problem is "D" and "5." The correct answer for the bartender's problem I is "Beer" and "17."

You will probably be surprised to find out that these two problems are logically identical. In each case you are given the statement

If P, then Q

and in both problems the four cards state

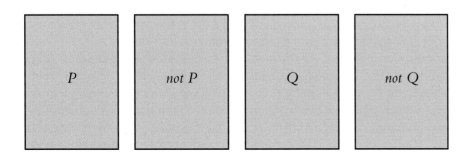

| P | not P | Q | not Q |

BOX 18.2

Bartender's Problem I

In their crackdown against drunk drivers, state law enforcement officials are revoking liquor licenses left and right. You are a bouncer in a local bar, and you'll lose your job unless you enforce the following law:

If a person is drinking beer, then he must be over 21 years old.

The cards below have information about four people sitting at a table in your bar. Each card represents one person. One side of a card tells what a person is drinking and the other side of the card tells that person's age.

Indicate only those card(s) you definitely need to turn over to see if any of these people are breaking this law.

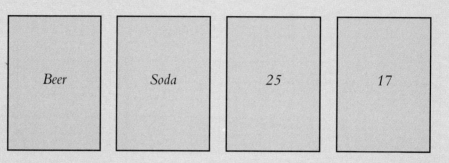

| Beer | Soda | 25 | 17 |

Adapted from Figure 3.3 in L. Cosmides and J. Tooby, 1992, cognitive adaptations for social exchange, in *The Adapted Mind, Evolutionary Psychology and the Generation of Culture,* J. Barkow, L. Cosmides, and J. Tooby, eds. Oxford University Press, New York.

Then you are asked which cards you need to turn over to determine whether the statement is true for that card. The laws of logic say that the statement *If P, then Q* can only be falsified by observing *P* and *not Q*. Thus the only two cards that need to be turned over are *P*, to see if the other side is *not Q*, and *not Q*, to see if the other side is *P*. Most people can easily see this reasoning is correct in the case of the bartender problem, but not in the case of the clerical problem.

Some recent data suggest that humans are particularly able to solve problems involving social exchange.

The psychologists who first discovered this effect believed that people were better at solving familiar problems than unfamiliar problems. Since most people have been carded in bars, and few people have done clerical work, the bartender problem is easier to solve. However, additional experiments show that familiarity does not predict which problems people will find easy to solve.

Recently, Cosmides and Tooby have provided a different explanation. They contend that people are good at solving problems involving reciprocal altruism and other forms of social exchange, and this skill was crucial for success in the small groups that have characterized human societies for most of our history. Food sharing, which is an essential part of hunter-gatherer ecology, is a form of reciprocal altruism. The big problem with reciprocal altruism is that it is costly to interact

BOX 18.3

Bartender's Problem II

In their crackdown against drunk drivers, state law enforcement officials are revoking liquor licenses left and right. You are a bouncer in a local bar, and you'll lose your job unless you enforce the following law:

If a person is over 21 years old, then he may drink beer.

The cards below have information about four people sitting at a table in your bar. Each card represents one person. One side of a card tells what a person is drinking and the other side of the card tells that person's age.

Indicate only those card(s) you definitely need to turn over to see if any of these people are breaking this law.

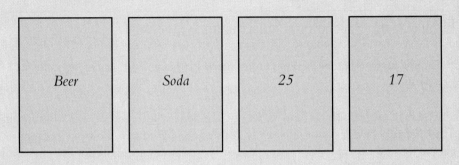

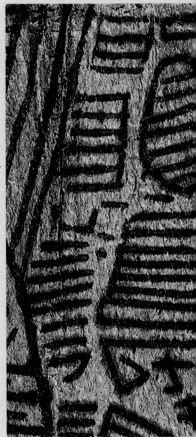

Adapted from Figure 3.3 in L. Cosmides and J. Tooby, 1992, Cognitive adaptations for social exchange, in *The Adapted Mind, Evolutionary Psychology and the Generation of Culture*, J. Barkow, L. Cosmides, and J. Tooby, eds. Oxford University Press, New York.

with individuals who do not reciprocate. Thus, human cognition should be tuned to look for cheaters in social exchanges. The bartender problem that you were asked to solve above has the form "If you take the benefit, then you pay the cost." In this case, beer is the benefit, and being over 21 is the cost. Tooby and Cosmides argue that people are tuned to attend to situations in which people take the benefit without paying the cost. Thus, they look for people who are drinking beer even though they are under 21.

According to Cosmides and Tooby, most people get the logically correct answer on the bartender problem because social reasoning and logic just happen to coincide, not because most people are familiar with being carded in bars. If the logical answer and the social reasoning solution are opposed, what solution would people choose? To answer this question, Cosmides and Tooby also asked students to solve what they called the "switched social contract," which takes the form "If you pay the cost, then you get the benefit." For example, "If a person is over 21, then he may drink beer." This problem still has the logical form *If P, then Q,* but now *P* is "21" and *Q* is "drink beer." Box 18.3 portrays this problem. Again, try to solve this problem yourself.

According to the laws of formal logic, to falsify this statement, you must observe *P* and *not Q.* Thus, in this version of the problem, the logical solution requires you to turn over the "25" card to see if the other side is "not drinking beer," and the "Soda" card to see if the other side is "over 21." However, if you are like most people, you turned over the cards "17" and "Beer." According to Tooby and Cosmides, people get this problem and other problems like this wrong because they have a strong tendency to look for cheaters in social exchange, people who take the benefit (drink beer) without paying the cost (being over 21). People choose the cards that the bartender would need to turn over to detect underage drinkers, not the cards that meet rules of formal logic. (If this is hard to grasp, don't despair. If Cosmides and Tooby are right, then this is simply not a distinction that the human mind has evolved to understand.)

Although Cosmides and Tooby's results are intriguing, some psychologists think that there may be other explanations for them. However, all of the alternate explanations that have been offered presume that people are predisposed to reason about different problems in different ways.

Evolutionary Psychology and Human Universals

Evolved psychological mechanisms cause human societies to share many universal characteristics.

Much of anthropology (and other social sciences) is based on the assumption that human behavior is not effectively constrained by human biology. To be sure, people have to obtain food, shelter, and other resources necessary for their survival and reproduction. But beyond that, human behavior is not constrained by biology.

This assumption is not very plausible from an evolutionary perspective. Evolutionary reasoning suggests that humans should be better at learning some kinds of tasks, making some kinds of decisions, and solving some kinds of problems than others. From the perspective of evolutionary psychology, it is likely that evolved mechanisms in the human brain channel the evolution of human societies and human culture, making some outcomes much more likely than others. It is not that natural selection has led to genetic differences among different societies. Instead, the mental mechanisms that underly human behavior are held in common among people all over the world; they are universal features of the human species. Tooby and Cosmides liken these universal cognitive features to organs like the liver or the kidneys. Everyone has a liver and two kidneys, and these organs perform the same function from Berkeley to Bora Bora. Similarly, everyone's brain is composed of the same set of **mental organs**, and in every society these mental organs constrain and shape peoples' thoughts, perceptions, and behaviors.

In the remainder of this section, we give two examples of such **human universals**.

Color Terms

Basic color terms in all languages show a regular pattern of variation.

People in every society have linguistic terms that refer to color. In English, we use words like *red, blue,* and *green;* compound color terms like *blue-green;* and metaphorical descriptions like *Carolina blue.* In addition, people who need to make finer distinctions about colors (such as painters, interior decorators, and fashion designers) have a more extensive vocabulary of color terms. For example, they might know the difference between taupe and beige, or puce and mauve.

In 1969 two anthropologists, Brent Berlin and Paul Kay, published a landmark study of cross-cultural variation in what they called **basic color terms**. These are words that refer to color and meet the following conditions:

- They consist of only one word, such as *green,* not two words, such as *forest green* or *kelly green.*
- They refer to colors that are not contained within another color. *Scarlet* is not a basic color term because it is contained within *red.*
- They must not be used only for certain objects. For example, *blond* is used to describe hair and wood in English, but not to describe houses or cars.
- They must be widely known. Most people don't know the difference between aquamarine and turquoise, for example.

The number of basic color terms varies from language to language. All languages have at least two, and some have as many as 11. However, Berlin and Kay were able to show that there were several interesting regularities in how people classify color. After showing people living in a variety of different cultures a chart with 320 different color chips (Figure 18.3), Berlin and Kay asked them to explain what basic color term described each chip and to select the color chip that was the best example of each basic color term.

There was little pattern to the range of colors included in basic color terms. For example, there are many chips that English speakers from the United States labeled as *blue* that Spanish speakers from Mexico did not include in *azul* (Figure 18.4). However, in all cultures with 11 color terms, people picked very similar colors as the best examples of the basic colors. More precisely, the colors chosen as the best exemplar of each color term, called a **focal color**, varied more among individuals within cultures than among cultures (Figure 18.5). In English these focal colors are black, white, red, yellow, green, blue, brown, purple, pink, orange, and gray. In languages with fewer terms, there are color terms that include two or more such focal colors. For example, some languages have a word that includes the best example of blue and green. Speakers of such languages pick either the best example of blue or the best example of green. They don't pick an intermediate color like aqua.

There is also a regular relationship between how many color terms a language has and which color terms these are. For example, some languages have only two color terms—these are always black and white. In languages that have three color terms, they are always black, white, and red. When a language has four color terms, they are black, white, red, and yellow, blue, or green. The general pattern is enumerated in Table 18.1.

There is a simple relationship between the basic color terms and the physiology of color vision.

Macaques and humans have very similar visual systems, and studies of macaque visual systems have helped us understand how humans perceive color. There are six types of nerve cells that link the eye to the brain. Two of these cells determine the perception of brightness, while the other four, called **opponent response cells**, determine the perception of hue. Two of these produce the perception of blue and yellow. Both fire at a baseline rate if there is no external stimulus. A blue stimulus causes the $+B - Y$ cells to fire above their baseline rate, and a yellow stimulus causes them to fire below their baseline rate. The $+Y - B$ cells react in the opposite manner, increasing their firing rate with a yellow stimulus and de-

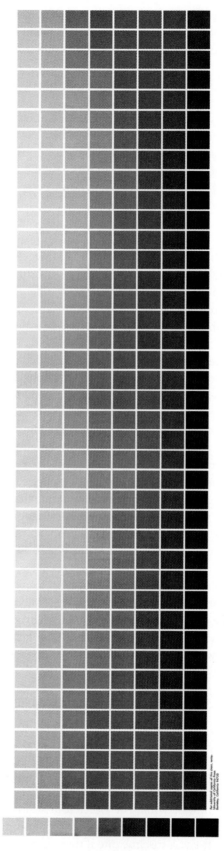

Figure 18.3
This is the Munsell color chart used by Berlin and Kay in their experiments.

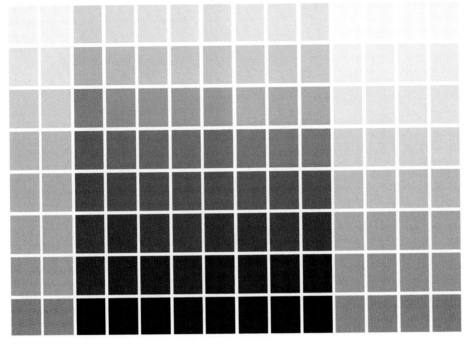

(a)

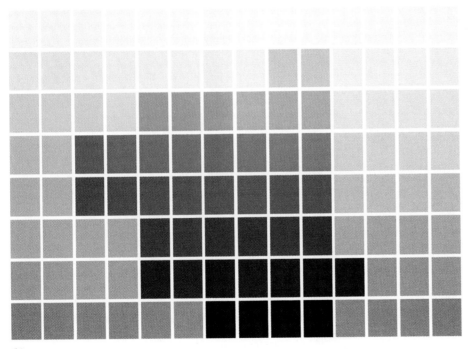

(b)

Figure 18.4

(a) The range of colors labeled *blue* by English speakers from the United States is somewhat different from (b) the range of colors labeled *azul* by Spanish speakers from Mexico.

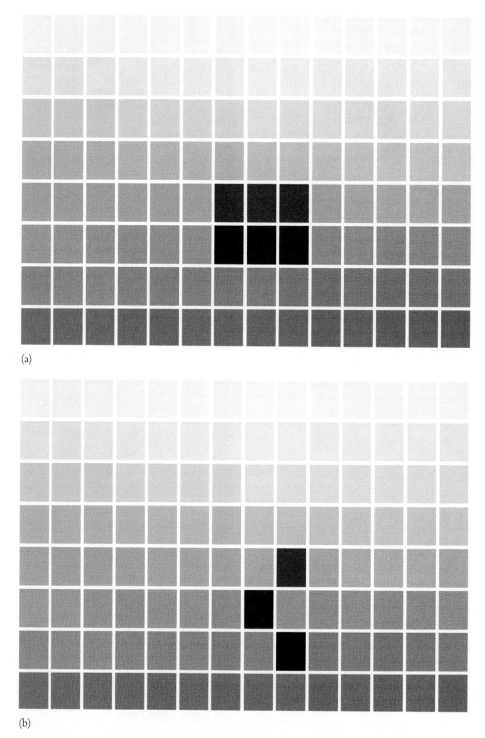

(a)

(b)

Figure 18.5

(a) The colors chosen as best examples of *blue* by English speakers from the United States are largely the same as (b) the colors chosen as the best example of *azul* by Spanish speakers from Mexico.

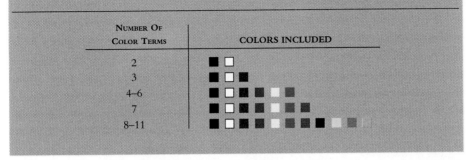

Table 18.1 The relationship between the number of basic color terms in a language and their identity. This relationship holds true across all cultures.

NUMBER OF COLOR TERMS	COLORS INCLUDED
2	▪ ▫
3	▪ ▫ ▪
4–6	▪ ▫ ▪ ▪ ▪
7	▪ ▫ ▪ ▪ ▪ ▪
8–11	▪ ▫ ▪ ▪ ▪ ▪ ▪ ▪ ▪ ▪

creasing with a blue stimulus. A second pair of opponent response cells, labeled $+R - G$ and $+G - R$, behave in an analogous way for red and green stimuli.

There is a simple relationship between the physiology of color perception and the colors chosen as foci for basic color terms around the world. A focal blue stimulus causes the blue-yellow cells to give the maximum blue response but has no effect on the red-green opponent cells. Similarly, a focal yellow causes blue-yellow cells to give a yellow response, but red-green cells continue to fire at their base rates. Focal red causes red-green opponent cells to give a red response, but does not excite blue-yellow cells. Black stimuli cause the darkness-sensitive cells to fire at their maximum rate, lightness cells to fire at their minimum rate, and all four opponent cells to fire at their base rate.

The nonprimary focal colors, purple, orange, pink, gray, and brown, occur when two of the pairs are firing at their maximum rate. For example, focal purple stimuli cause blue-yellow cells to give the maximum blue response and red-green cells to give the maximum red response.

The cross-cultural patterns of color terms may result from the fact that focal colors are easier to remember and to manipulate cognitively.

The Dani are a group of people living in Irian Jaya (Indonesia), which constitutes the western half of the island of New Guinea (Figure 18.6). The Dani language has only two basic color terms, *mili* (dark-cool, including the focal colors black, green, and blue) and *mola* (light-warm, including white, red, and yellow). In the 1960s, University of California, Berkeley, psychologist Eleanor Rosch did a famous series of experiments to find out how the Dani think about color. Her results suggest that the Dani are able to use and to remember names for focal colors with much greater ease than names for nonfocal colors. For example, she taught one group of Dani eight made-up names for the focal colors. She taught a second group eight made-up names for nonfocal colors. The Dani were much better at learning and remembering the new words for focal colors than the new words for nonfocal colors. Even very young children showed greater facility with names for the focal colors.

Rosch argued that color categories are represented by a **prototype**, or model, and membership in the category is related to the similarity to the prototype. People do not store lists of colors in their mind, and then categorize them. In the case of color terms, the prototypes seem to arise from the physiology of color vision. All people have the same visual systems, and they have the same set of prototypes for color, even though not all cultures attach separate words to each of them. When

Figure 18.6

The Dani live in Irian Jaya. Their language contains only two basic color terms.

languages add more basic color terms, they attach words to preexisting color categories.

These studies suggest that the biology of color perception and the machinery of cognition *channel* cultural evolution, precluding a wide range of possibilities. This does not mean, however, that culture is *determined* by biology. The number of basic color terms varies from two to 11 among societies, and there does not seem to be any simple biological explanation of why the Maasai have five color terms and the Tiv only three. The range of colors grouped with particular color terms also varies widely from culture to culture.

Inbreeding Avoidance

Incest and its avoidance play a crucial role in many influential theories of human society. Thinkers as diverse as Sigmund Freud, the founder of psychoanalysis, and Claude Levi-Strauss, the father of structuralist anthropology, have asserted that people harbor a deep desire to have sex with members of their immediate family. According to this view, only the existence of culturally imposed rules against incest save society from these destructive passions.

This view is not very plausible from an evolutionary perspective. There are compelling theoretical reasons to expect that natural selection will erect psychological barriers to incest, and good evidence that it has done so in humans and other primates. Both theory and observation suggest that the family is not the focus of desire; it's a tiny island of sexual indifference.

The offspring of genetically related parents have lower fitness than the offspring of unrelated parents do.

Geneticists refer to matings between relatives as **inbred matings**, and contrast them with **outbred matings** between unrelated individuals. The offspring of inbred matings are much more likely to be homozygous for deleterious recessive alleles than are offspring of outbred matings. As a consequence, they are less robust and have higher mortality than the offspring of outbred matings. In Chapter 16 we discussed a number of genetic diseases like PKU, Tay-Sachs, and cystic fibrosis that are caused by a recessive gene. People who are heterozygous for such deleterious recessive alleles are completely normal, while people who are homozygous suffer severe, often fatal consequences. Recall that such alleles occur at low frequencies in most human populations. However, there are many loci in the human genome. Thus, even if the frequency of deleterious recessives at each locus is very small, there is a good chance that everybody has at least one lethal recessive gene somewhere in their genome. Geneticists have estimated that each person carries the equivalent of two to five lethal recessives. Mating with close relatives is deleterious because it greatly increases the chance that both partners will carry a deleterious recessive at the same locus. Theory and experimental data indicate that inbreeding leads to substantial decreases in fitness (Box 18.4). This, in turn, suggests that natural selection should favor behavioral adaptations that reduce the chance that inbreeding occurs.

Matings between close relatives are very rare among nonhuman primates.

Remember from Chapter 7 that in all species of nonhuman primates, members of one or both sexes leave their natal groups near the time of puberty. It is very likely that dispersal is an adaptation to prevent inbreeding. In principle, primates could remain in natal groups and simply avoid mating with close kin. In practice, this is problematic because males and females are likely to know their maternal kin but unlikely to know their paternal kin. To avoid the deleterious consequences of inbreeding, members of one sex must leave the group and seek mating opportunities elsewhere.

Natural selection has provided at least some primates with another form of protection against inbreeding—a strong inhibition against mating with close kin. In a large captive colony of Barbary macaques, German primatologists Jutta Kuester, Andreas Paul, and their colleagues have demonstrated that maternal kin ties have a strong influence on sexual attraction (Figure 18.7). Maternal kin who had lived together in the group since birth very rarely mated, and very few infants were conceived in such unions. However, there was no aversion to mating among maternal relatives who had been separated at some point in their lives, among paternal kin, or among members of different matrilineages. Interestingly, males did not mate with females they had cared for as infants. At Gombe Stream National

Figure 18.7

Barbary macaques avoid mating with close kin, and this aversion is apparently based on familiarity during early life.

BOX 18.4

Why Inbred Matings Are Bad News

The following simple example illustrates why inbreeding increases the chance that offspring will be homozygous for a deleterious recessive gene. Suppose Cleo has exactly one lethal recessive somewhere in her genome. If she mates with a nonrelative, Mark, the chance that he carries the same deleterious recessive is simply equal to the frequency of heterozygotes carrying that gene in the population. (We don't need to consider the frequency of individuals who are homozygous for the recessive allele, because it is lethal in homozygotes.) Let p be the frequency of deleterious recessives in the population. Using the Hardy-Weinberg law, we know that the frequency of heterozygotes is $2p(1 - p)$. The frequency of lethal recessives is typically about 0.001. Thus, there is only about one chance in 500 that Mark and Cleo carry the same deleterious recessive. If Cleo mates with her brother, Ptolemy, the story is quite different. Remember from Chapter 8 that r, the coefficient of relationship,

gives the probability that two individuals will inherit the same gene through common descent. For full siblings $r = 0.5$. Thus, if Cleo mates with her brother, there is one chance in two that he will carry the same lethal recessive that she does, a value that is about 100 times greater than for an outbred mating. Since Cleo and Ptolemy are both heterozygotes for the lethal recessive, Mendel's laws say that 25% of their offspring will be homozygotes and therefore will die. Thus on average, Cleo will experience a 12.5% decrease in fitness by mating with her brother Ptolemy compared with an outbred mating with Mark.

This simple example actually understates the cost of inbreeding because people often carry more than one deleterious recessive. Oxford University biologist Robert May derived the estimates in Figure 18.8 using a more realistic model. As you can see, inbreeding is quite bad. Matings between siblings or between parents and their

Figure 18.8

Population genetics theory predicts that close inbreeding can lead to serious reductions in fitness. The vertical axis plots the percent reduction in fitness due to inbreeding, and the horizontal axis plots the coefficient of relatedness among spouses. The curve is based on the conservative assumption that people carry the equivalent of 2.2 lethal recessives.

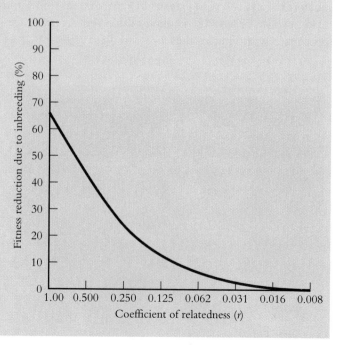

offspring ($r = 0.5$) produce about 40% fewer offspring than do outbred matings. The effects for other inbred matings are smaller but still serious.

Empirical studies strongly support these predictions. Data from 38 species of captive mammals indicate that the offspring of matings among siblings and among parents and their offspring are 33% less likely to survive to adulthood than the offspring of outbred matings. Studies in wild populations, though less conclusive, tell the same story. Several studies suggest that humans are affected by inbreeding in the same way. For example, the 161 children of father-daughter or brother-sister matings studied by geneticist Eva Seemanová were twice as likely to die during their first year as their maternal half-siblings, and 10 times as likely to suffer serious congenital defects. Other evidence comes from studies of Moroccan Jews living in Israel who do not consider marriages between men and their nieces to be incestuous. A study of 131 children from such marriages indicates that they suffer a 20% reduction in fitness compared with a control group from the same population.

Park, Tanzania, where chimpanzees have been studied for more than three decades by Jane Goodall and her coworkers, many adult females have contact with their sons or brothers, and some young females in the study group may have been sired by older male residents. Thus, when these females come into estrus, they have opportunities to mate with close kin. In fact, they rarely do. Although mothers have relaxed and affectionate interactions with their sons, matings among them are very rare. There are more brother-sister matings than mother-son matings, but these too are uncommon and generally unwelcome to females. Female chimpanzees seem to have a general aversion to mating with males much older than they are, and males seem to be generally uninterested in females much younger than themselves. These mechanisms may protect females from mating with their fathers, and vice versa.

Humans rarely mate with close relatives.

During the first half of the 20th century, cultural anthropologists fanned out across the world to study the lives of exotic peoples. Their hard and sometimes dangerous work has given us an enormous trove of information about the spectacular variety of human lifeways. They found that domestic arrangements vary greatly across cultures: some groups are polygynous, some monogamous, and a few polyandrous. Some people reckon descent through the female line and are subject to the authority of their mother's brother. In some societies, married couples live with the husband's kin, in others they live with the wife's kin, and in some they set up their own households. Some people must marry their mother's brothers' children, while others are not allowed to do so.

In all of this almost endless variety of domestic arrangements, there is not a single ethnographically documented case of a society in which brothers and sisters regularly marry, or one in which parents regularly mate with their own children. The only known case of regular brother-sister mating comes from census data collected by Roman governors of Egypt from 20 A.D. to 258 A.D. From the 172 census returns that have survived, it is possible to reconstruct the composition of 113 marriages: 12 were between full siblings and eight between half-siblings. These marriages seem to have been both legal and socially approved, as both prenuptial agreements and wedding invitations survive.

The pattern for more distant kin is much more variable. Some societies permit both sex and marriage with nieces and nephews or between first cousins, while other societies prohibit sex and marriage among even distant relatives. Moreover, the pattern of incest prohibitions in many societies does not conform to genetic categories. For example, even distant kin on the father's side may be taboo in a given society, while maternal cousins may be the most desirable marriage partners in the same society. Sometimes the rules about who can have sex are different from the rules governing who can marry.

Adults are not sexually attracted to the people with whom they grew up.

Until recently, most cultural anthropologists thought that cultural rules against incest were the only thing preventing people from doing what they might do otherwise. However, this seems unlikely. The fact that inbreeding avoidance is universal among primates indicates that our human ancestors also had psychological mechanisms preventing them from mating with close kin. These psychological mechanisms could disappear during human evolution only if they were selected against. However, mating with close relatives is highly deleterious in humans and in other primates. Thus, both theory and data predict that modern humans should have psychological mechanisms that reduce the chance of close inbreeding, at least in the small-scale societies in which human psychology was shaped.

There is evidence that such psychological mechanisms exist. In the late 19th century the Finnish sociologist Edward Westermark speculated that childhood propinquity stifles desire. By this he meant that people who live in intimate association as small children do not find each other sexually attractive as adults. Three natural experiments indicate that this mechanism causes inbreeding avoidance in humans.

- *Taiwanese minor marriage.* Until recently, an unusual form of marriage was widespread in China. In **minor marriages**, children were betrothed, and the prospective bride was adopted into the family of her future husband during infancy. There, the betrothed couple grew up together like brother and sister. According to Taiwanese informants interviewed by Stanford University anthropologist Arthur Wolf, the partners in minor marriages found each other sexually unexciting. Sexual disinterest was so great that fathers-in-law sometimes had to beat the newlyweds to convince them to consummate their marriage. Wolf's data indicate that minor marriages produced about 30% fewer children than did other arranged marriages (Figure 18.9a), and were much more likely to end in separation or divorce (Figure 18.9b). Infidelity was also more common in minor marriages. When modernization reduced parental authority, many young men and women who were betrothed in minor marriages broke their engagements and married others.
- *Lebanese parallel-cousin marriage.* Anthropologist Justine McCabe studied life in a small town in southern Lebanon during the 1970s. The people in this village did not consider marriage between cousins to be incestuous. In fact, they thought that the ideal marriage was between the offspring of brothers, called **patrilateral parallel cousins** by anthropologists (Figure 18.10). Social life in the traditional Arab village studied by McCabe was organized around groups of relatives related through the male line. Brothers often lived together, and even if they did not live in a communal household, they were often best friends and confidants, and

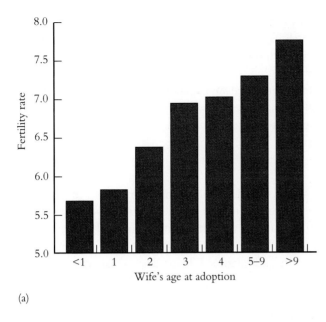

(a)

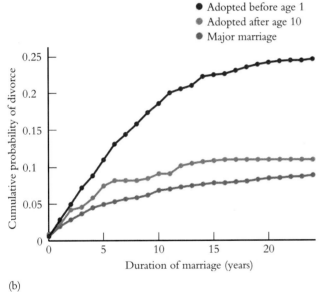

(b)

Figure 18.9

In minor marriages, the age of the wife when she arrives into her future husband's household (age of adoption) affects both fertility and the likelihood of divorce.
(a) The fertility of women adopted at young ages is depressed. (b) The younger women are when they arrive in their husband's household, the less likely it is that their marriages will survive.

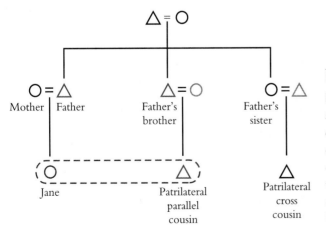

Figure 18.10

Patrilateral parallel cousins are the offspring of siblings of the same sex. In this hypothetical lineage, there are three children, two males (triangles) and a female (circle). The brothers marry and produce children, and their children are patrilateral parallel cousins. Their sister also marries and has children, but her children are not patrilateral parallel cousins of her brothers' children.

their families maintained daily contact with each other. Thus, patrilateral parallel cousins usually grew up much as brothers and sisters do. McCabe compared a large sample of marriages between different kinds of cousins (offspring of sisters, offspring of a brother and a sister, and offspring of brothers). Like Chinese minor marriages, the marriages between patrilateral parallel cousins produced about 20% fewer offspring than did marriages between other kinds of cousins, and were also more likely to end in divorce than were other kinds of marriages. McCabe discovered that people in the town thought that patrilateral parallel cousins who married were less sexually attracted to each other than were other married couples.

- *Kibbutz age-mates.* Before World War II, many Jewish immigrants to Israel organized themselves into utopian communities called **kibbutzim** (plural of **kibbutz**). In these communities, children were raised in communal nurseries, and they lived intimately with a small group of unrelated age-mates from infancy to adulthood. The ideology of the kibbutzim did not discourage sexual experimentation or marriage by children in such peer groups, but neither behavior occurred. Israeli sociologist Joseph Sepher, himself a kibbutznik, collected data on 2769 marriages in 211 kibbutzim. Only 14 of them were between members of the same peer group, and in all of these cases, one partner joined the peer group after the age of six. From data collected in his own kibbutz, Sepher found no instance of premarital sex among members of the same peer group.

Taken together, all of these data suggest that close inbreeding is almost unknown in human societies because people are not sexually attracted to close relatives. These data suggest that people have an innate psychological mechanism causing them to find their childhood companions sexually unattractive. Since brothers and sisters are raised together almost everywhere, people look beyond their parents' household for mates.

The relationship between an aversion to inbreeding and culturally transmitted rules against incest is unclear.

As we discussed earlier, virtually every society has rules that specify acceptable sexual partners for every person. Having sex within the family is incestuous in almost every society, but the rules for more distant kin are quite variable. Taboos sometimes prohibit sex with some very distant relatives, while allowing it with other relatives who are much more closely related.

The relationship between the propinquity effect and the cultural evolution of incest rules is controversial. Westermark thought that this mechanism made incest rare and unattractive. Incest generated feelings of disgust, which in turn caused people to adopt cultural rules against it. The obvious problem with this argument is that the same mechanism ought to lead to an aversion to minor marriage in China and to patrilateral-parallel-cousin marriage in the Middle East. Yet both these practices have persisted for a long time. Other authorities think that people recognize the deleterious effects of inbreeding and consciously adopt rules against it. This view is supported by the observation that people in many societies believe that incest leads to sickness and deformity. However, this explanation does not explain why in some societies marriages between virtually unrelated people are often prohibited while marriages among first cousins are allowed. Finally, many anthropologists think that incest rules are really about marriage and alliances between families, and not about sex at all.

Evolution and Human Culture

For many anthropologists, culture is what makes us human. Each of us is immersed in a cultural milieu that influences the way we see the world, that shapes our beliefs about right and wrong, and that endows us with the knowledge and technical skills to get along in our own environment. Despite the central importance of culture in anthropology, there is little consensus about how or why culture arose in the evolution of the human lineage. In what follows, we present a view of the evolution of human culture developed by one of us (R. B.). Although we believe strongly in this approach to understanding the evolution of culture, there is not a broad consensus among anthropologists that this, or any other particular view of the origins of culture, is correct.

Before we begin, we must provide some definitions of terms. We define **culture** as information acquired by individuals through some form of social learning. Cultural variation is the product of differences among individuals that exist because they have acquired different behaviors as a result of some form of social learning. Culture sometimes has quite different properties from other forms of environmental variation. If people acquire behavior from others through teaching or imitation, then culturally transmitted adaptations can gradually accumulate over many generations (Figures 18.11 and 18.12). For example, one man may learn to fletch his arrows from his father, and his son may learn to dip the arrow point in poison from a neighbor; his grandsons may both fletch their arrows and dip them in poison. To understand culture, we need to take its cumulative nature into account.

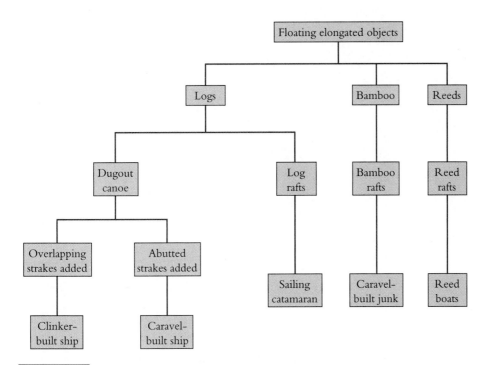

Figure 18.11

We can trace the development of certain technological innovations through time. In China, the first boats were elongated structures that floated on the water. Some were made of logs, others of bamboo or reeds. Floating logs were transformed into canoes with the addition of a keel. Other types of ships, including Chinese junks, have a square hull and no keel.

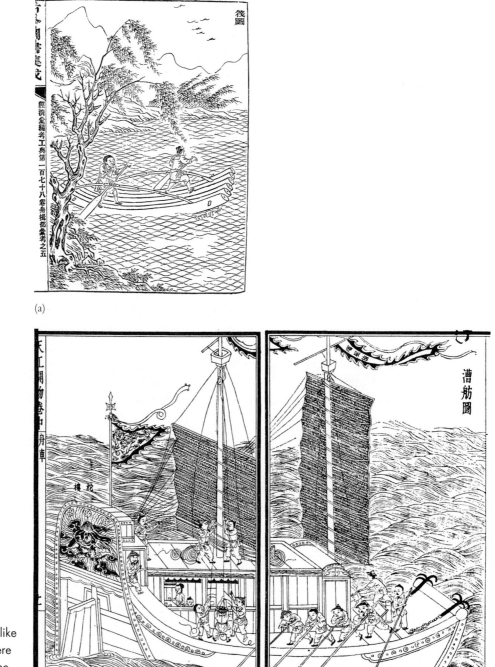

(a)

Figure 18.12
(a) Bamboo rafts like
the one shown here
may have been the
precursors of (b) the
great Chinese junks.   (b)

Culture Is a Derived Trait in Humans

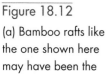

There has been much debate about whether other animals have culture.

Anthropologists and psychologists have long debated whether other animals have culture. Some authors deny culture to other animals on the grounds that traditions observed in other animals lack essential features of human culture, such as symbolic coding. Others argue that those who deny culture to nonhuman animals are applying a double standard—if the kind of behavioral variation observed

Cultural Diversity and Human Universals

Many anthropologists, probably most of them, are skeptical of statements that generalize about what all peoples do. But are there not generalizations of that sort that really do hold for the wide array of human populations? There are—and not enough has been said about them. This skepticism and neglect of human universals is the entrenched legacy of an "era of particularism" in which the observation that something *doesn't* occur among the Bongo Bongo counted as a major contribution to anthropology. . . . The truth of the matter is, however, that anthropologists probably always take for granted an indefinite collection of traits that add up to a very complex view of human nature. Let me give some examples.

In a course that I teach on the peoples and cultures of Southeast Asia I have often illustrated the cultural elaboration of rank that is found in many Southeast Asian societies—and certainly among the Brunei Malays with whom I did my doctoral research—with the following anecdote. In the course of my research I was once seated with two young men on a wooden bench at the front of the house that my wife and I rented in a ward of the Brunei capital. A third young man was seated just a few feet away on the rung of a ladder but at the same height as the rest of us. There was no one else around. Tiring of sitting on the bench, I slipped down from it to sit on the walkway. I was followed almost instantly by all three of the young men. Just as quickly I realized that they had done it not because they too were uncomfortable on the bench (I had been there longer than they) but because in the Brunei scheme of things it is not polite to sit higher than another person, unless you considerably outrank that other person. So I protested, urging them to please remain seated on the bench. They said it wouldn't look nice. I said there was no one but us around to notice. One of them closed the matter by noting that people across the river—to which he gestured (it was about a quarter mile away)—just *might* see what was going on. The clear implication was that he and his fellows weren't about to let anyone see them apparently breaking one of the important rules in the etiquette of rank, even though they knew they wouldn't be offending me.

I always told this story to illustrate difference, to show the extremity to which Bruneis concerned themselves with rank, and it always seemed to be a very effective message. As a teacher of anthropology I know very well that cultural differences elicit some sort of inherent interest. Ruth Benedict's *Patterns of Culture* (1934) is an all-time anthropological best-seller, and its essential message is the astonishing variability of human customs. No one teaching anthropology can ignore the way students react to revelations about the amazing ways other peoples act and think. And no one teaching anthropology can fail to sense the wheels turning in students' minds when they use these revelations to rethink the ways people act and think in their own society. Teachers of anthropology not only see this in students, they cultivate it. But are the differences all that should be of concern to anthropology? Does an emphasis on differences present a true image of humanity?

I now realize that the story I have told my students is pervaded with evidence of similarities: above all, the young men were concerned with what other people would think about them; they were also concerned with politeness in particular, rules in general; even their concern with rank was only a matter of difference in degree. I could go on, mentioning their use of language and gestures; the smooth conversational turn taking; the concepts of question, answer, explanation; the use of highness/lowness to symbolize rank; and much more.

At a more subtle level, I believe, some amazing things were happening that I took no note of. Without my explaining things in detail, in my broken Malay, the young men had instantly grasped my point: the setting was informal and I wanted them to treat me as they would treat each other (they would not have moved down or up in unison for each other in those circumstances); furthermore, it was "not my custom" to be offended by people sitting higher than me. I think that my companions sized up these aspects of the immediate situation just as I had.

But they also saw a wider context in which their behavior could be misinterpreted by others, and with what seemed like a few words and a gesture, they explained their position to me and closed the matter. There were more than a few words and gestures: there were tone of voice, facial expressions, body language, and an enormously complex context of past, present, and future. And there were four human minds, each observing, computing, and reacting to the "implicature" . . . of the bare words so silently and automatically as to occasion no notice. All this—from the conscious concern with what others would think to the unconscious assessments of implications—formed a plainly human background, from which I in my lectures had pulled out a quantitative difference as the focus of attention.

I use the word "quantitative" because, although it may not be *my* custom to think that the height of one's seat should match one's rank, the idea is not foreign to western culture. There are some wonderful examples of the equation between seating height and rank, or dominance, in Charlie Chaplin's film "The Great Dictator." What distinguishes the Bruneis from us is the greater frequency of day-to-day contexts in which the equation is observed among Bruneis.

Now it might be objected that the Brunei Malays are so westernized that of course they are similar to us in many ways; one needs a pristine, uncontacted people to see the real exceptions. This is an assumption that I would have taken quite seriously at one time, and that was acted upon by my university schoolmate Lyle Steadman. . . . Like me, he received his anthropological training in the 1960s and was steeped in cultural determinism. In order to fully explore the consequences of having a nonwestern worldview, he did his work among a New Guinean people, the Hewa, who had had no more than the most fleeting and widely spaced contacts with European patrols. At the time Steadman studied the Hewa they lived in one of the last "restricted" areas of New Guinea. This meant that the area was "uncontrolled," and Europeans, including missionaries, were forbidden to enter it. One of the reasons the Hewa were essentially uncontacted was that they lived so sparsely on the land that from one family's household to another was typically a grueling 2-hour walk over a rugged terrain covered by dense rain forest.

Steadman had to learn the Hewa language in the field, but long before he was conversant in it he discovered—somewhat to his surprise, because it didn't jibe with his assumptions about the influence of differing world views—that he and the Hewa "could understand each other well enough to live together." . . . As time went by, and he learned more about the ways in which the world is put together differently in Hewa than in English, he was led to observe that the differences were largely superficial: "This fact of experiencing the world in a similar way," in spite of its being carved up differently in different languages, "became increasingly obvious as I acquired greater proficiency in the language." At the deeper level of why language might be used in the first place, at the level of motives, the similarities were just as evident: "Living, travelling, working and hunting with the Hewa, made it clear to me that their basic concerns, the concerns motivating their behaviour, were similar to my own." . . .

Source: From pp. 1–3 in D. Brown, 1991, *Human Universals*, McGraw-Hill, New York. Reproduced by permission of The McGraw-Hill Companies.

among some primate populations were observed among human populations, they argue, anthropologists would regard it as cultural.

Such debates make little sense from an evolutionary perspective. The psychological capacities that give rise to human culture are likely to have homologies in the brains of other primates, and the function of cultural transmission in humans could well be related to its function in other species. The study of the evolution of human culture must be based on categories that allow human cultural behavior to be compared to potentially homologous, functionally related behavior of other organisms. At the same time, such categories should be able to distinguish between human behavior and the behavior of other organisms because it is quite plausible that human culture is different in important ways from related behavior in other species.

Culture is common among other animals, but cumulative cultural change is rare.

There are many examples of cultural variation in nature, particularly among primates (Box 18.5). There is little evidence, however, of cumulative cultural evolution in other species. In most cases, social learning only leads to the spread of behaviors that individuals could have learned on their own.

Scientists who study cultural behavior in other organisms distinguish two different classes of mechanisms that can account for cultural differences between populations:

1. **Social facilitation** occurs when the activity of older animals indirectly increases the chance that younger animals will learn the behavior on their own. Young individuals do not acquire the behavior by observing older individuals. Social facilitation would account for the persistence of tool use in the following scenario. In populations where chimpanzees use tools to crack nuts, young chimpanzees spend a lot of time in proximity to both nuts and hammer stones. Nuts are a greatly desired food, and young chimpanzees find eating nut meats highly reinforcing. They experiment with stone hammers and anvils until they master the skill of opening the nuts. In populations in which chimpanzees do not use stones to open nuts, young chimpanzees never spend enough time in proximity to both nuts and hammer stones to acquire the skill.
2. **Observational learning** occurs when younger animals observe the behavior of older animals and specifically learn how to perform an action by watching others. In this case, the tool tradition is preserved because young chimpanzees actually imitate the behavior of the older ones.

Social facilitation and observational learning are similar in that they both can lead to persistent behavioral differences between populations. However, there is an important distinction between the two processes. Social facilitation can only preserve variation in behavior that organisms can learn on their own, albeit in favorable circumstances. Observational learning allows cumulative cultural change. To see the difference, consider the cultural transmission of stone-tool use. Suppose that on her own, in especially favorable circumstances, an early hominid learned to strike rocks together to make useful flakes. Her companions, who spend time near her, would be exposed to the same kinds of conditions, and some of them might learn to make flakes too, entirely on their own. This behavior could be preserved by social facilitation because groups in which tools were used would spend more time in proximity to the appropriate stones. However, that would be as far as it would go. If an especially talented individual found a way to improve

BOX 18.5

Examples of Culture in Other Animals

These are just a few of the hundreds of well-documented examples of cultural variation in nonhuman species.

POD DIPPING Anthropologist Marc Hauser, of Harvard University, observed an old female vervet monkey dip an *Acacia* pod into a pool of liquid that had collected in a cavity in a tree trunk. She soaked it for several minutes and then ate the pod. This behavior had never been seen before, though this group of monkeys had been observed regularly for many years. Within nine days, three other members of the old female's family had dipped pods themselves, and so had one unrelated adult female. Ultimately, seven of the 10 members of the group learned to dip pods.

GROOMING POSTURES Chimpanzees in the Mahale Mountains of Tanzania often adopt a unique grooming posture (Figure 18.13). During grooming sessions, both partners may simultaneously extend one arm over their heads. The two partners clasp hands, and then groom one another's exposed armpits. These grooming handclasps occur often, and are performed by all members of the group. Chimpanzees at Gombe, who live less than 100 km away in a similar type of habitat, groom often but never perform this behavior.

TERMITING TOOLS Chimpanzees at several sites use probes made from plant materials to fish for termites (Figure 18.14). At Gombe in Tanzania and Mt. Assirik in Senegal, chimpanzees sometimes use woody vegetation to make termiting tools. At Mt. Assirik, woody twigs and vines are peeled to remove the bark, the bark is discarded, and the peeled twig is used to fish for termites. At Gombe, twigs and vines are sometimes peeled, but when they are peeled, the bark is used to fish for termites. In this case, both populations have access to similar kinds of vegetation for termiting tools, and chimpanzees in both populations know how to peel the twigs. But in one population, the bark is used as the tool, and in the other population the bark is discarded.

Figure 18.13

Chimpanzees in the Mahale Mountains often hold their hands above their head and clasp their partner's hand as they groom. This grooming posture has never been seen at Gombe, just 100 km away. (Photograph courtesy of William C. McGrew.)

Figure 18.14

A female chimpanzee plucks termites from her termiting tool. Some workers think that regional variation in tool use is the product of cultural transmission. (Photograph courtesy of William C. McGrew.)

NUTCRACKING Chimpanzees in the Taï Forest crack open hard-shelled nuts with stone hammers that they pound against other stones and exposed tree roots. This technique is difficult for young chimpanzees to master, and they do not become proficient at cracking nuts for many years. Infants often watch as their mothers crack nuts; mothers often share their hammers and their nut meats with their offspring. One mother watched her five-year-old daughter struggle unsuccessfully to crack open nuts with a hammer stone that she held awkwardly. After several minutes, the mother approached and her daughter gave her the hammer stone. Then the mother very slowly and deliberately ro-

Figure 18.15

A young Japanese macaque learned to wash her sweet potatoes, and many members of her group subsequently began to wash their potatoes as well. For many years, this was thought to be a good example of social learning. However, we now think each monkey learned the behavior by trial and error.

tated the hammer into the correct position. She cracked open several nuts and shared the contents with her daughter. Then the mother left and the daughter picked up the hammer stone, oriented it in the same position as her mother had, and began cracking nuts. This kind of teaching has not been observed in other wild chimpanzee populations, and is quite rare even at Taï.

POTATO WASHING This behavior was invented when researchers studying a group of Japanese macaques, whose range included a sandy beach, provisioned them with sweet potatoes. A young female macaque accidentally dropped her sweet potato into the sea as she was trying to rub sand off it. She must have liked the result, as she began to carry all of her potatoes to the sea to wash them (Figure 18.15). Other monkeys followed suit. However, it took other members of the group quite some time to acquire the behavior, and many monkeys never washed their potatoes.

the flakes, this innovation would not spread to other members of the group because each individual had to learn the behavior by himself. With observational learning, on the other hand, innovations can persist as long as younger individuals are able to acquire the modified behavior by observing the actions of others. As a result, observational learning can lead to the cumulative evolution of behaviors that no single individual could invent on its own.

Several lines of evidence suggest that social facilitation, not observational learning, is responsible for most cultural traditions in other primates. First, many of the behaviors described in Box 18.5, like potato washing and pod dipping, are relatively simple and could be learned independently by individuals in each generation. Second, new behaviors like potato washing often take a long time to spread through the group, a pace more consistent with the idea that each individual had to learn the behavior on its own. Finally, experiments with capuchin monkeys, known for their ability to use tools and to manipulate objects, suggest that even very clever monkeys like capuchins cannot learn by observation. Primatologist Elisabetta Visalberghi, of the Consiglio Nazionale delle Ricerche in Rome, gave capuchins the opportunity to learn to use a stick to push a peanut out of a horizontal, clear plastic tube. She then allowed a second set of monkeys to watch the skilled monkeys get the highly desired peanut out of the tube. If observational learning were important to this species, we would expect these monkeys to learn the behavior much more rapidly than monkeys who did not have skilled models to imitate. However, observing the skilled monkeys didn't speed up the learning process in the new participants. As Visalberghi and her colleague, Dorothy Fragazy, of Washington State University, put it, "Monkeys don't ape."

We can make a slightly better case for observational learning among apes. It is possible that young chimpanzees, for example, learn to fish for termites and to crack nuts by watching their mothers and imitating their behavior, though this is far from certain. These behaviors seem to be more complicated and harder to learn than the cultural behaviors of monkeys. Psychologist Anne Russon of York University has compiled several anecdotes about the acquisition of human behaviors by orangutans at a rehabilitation center operated by primatologist Biruté Galdikas. Orangutans have extensive contact with humans at the rehabilitation center and observe them closely. At various times, orangutans have stolen rowboats and paddled away down the river or attempted to siphon gas from discarded gas cans. One orangutan followed along behind a worker who was trimming plants growing over the path. The orangutan pulled up plants alongside the path and made his own neat piles in the middle of the path. What is interesting about these anecdotes is that these are behaviors that are not part of the orangutan's natural repertoire and are unlikely to have been intrinsically rewarding to the orangutans.

Nonetheless, it seems clear that no other primate relies on observational learning to the same extent that humans do, and that their behavior is much less variable from group to group or from region to region than the behavior of humans is.

Culture Is an Adaptation

Observational learning is not a byproduct of intelligence and social life.

Chimpanzees and capuchins are among the world's cleverest creatures. In nature, they use tools and perform many complex behaviors; in captivity, they can be taught extremely demanding tasks. Chimpanzees and capuchins live in social groups and have ample opportunity to observe the behavior of other individuals, and yet the best evidence suggests that neither chimpanzees nor capuchins make much use of observational learning.

This indicates that observational learning is not simply a byproduct of intelligence and having opportunities for observation. Instead, observational learning seems to require special psychological mechanisms. This conclusion suggests, in turn, that the psychological mechanisms that enable humans to learn by observation are adaptations that have been shaped by natural selection because culture is beneficial (Figure 18.16). Of course, this need not be the case. Observational learning could be a byproduct of some other adaptation that is unique to humans, such as language. However, given the great importance of culture in human affairs, it is reasonable to think about the possible adaptive advantages of culture.

Psychological mechanisms that allow cumulative cultural change may have been favored by selection because they allowed Pleistocene humans to exploit a much wider range of habitats than any other animal species.

The archaeological record suggests that during the Pleistocene, humans occupied virtually all of Africa, Eurasia, and Australia. Modern hunter-gatherers developed an astounding variety of subsistence practices and social systems. Consider

Figure 18.16

Infants are prone to spontaneously imitating behaviors they observe. Here, a 13-month-old infant flosses her two teeth.

just a few examples. The Copper Eskimos lived in the high Arctic, spending their summers hunting near the mouth of the MacKenzie River and the long dark months of the winter living on the sea ice, hunting seals. Groups were small and heavily dependent on hunting. The !Xõ lived in the central Kalahari, collecting seeds, tubers, and melons, hunting impala and gemsbok, enduring fierce heat, and living without surface water for months at a time. Their small, nomadic bands were linked together in large band clusters organized along male kinship lines. The Chumash lived on the southern California coast, gathering shellfish and seeds and fishing the Pacific from great plank boats. They lived in large permanent villages with pronounced division of labor and extensive social stratification.

The fact that the !Xõ could acquire the knowledge, tools, and skills necessary to survive the rigors of the Kalahari is not so surprising—many other species can do the same. What is amazing is that the same brain that allowed the !Xõ to survive in the Kalahari, also permitted the Copper Eskimo to acquire the very different knowledge, tools, and skills necessary to live on the tundra and ice north of the Arctic Circle, and the Chumash to acquire the skills necessary to cope with life in crowded, hierarchical Chumash settlements. There is no other animal that occupies a comparable range of habitats or utilizes a comparable range of subsistence techniques and social structures. For example, savanna baboons, the most widespread primate species, are limited to Africa and Arabia, and the diet, group size, and social systems of these far-flung baboon populations vary to a much smaller degree than the diet, group size, and social systems of human hunter-gatherers.

Humans can adapt to a wider range of environments than baboons because they acquire knowledge about the environment culturally, not genetically. Baboons cannot acquire behavior by observational learning because their behavior is limited to things that individuals can learn on their own. As we have seen, general-purpose learning mechanisms are costly and inefficient compared with mechanisms that work in a relatively narrow range of environments or solve limited kinds of problems. Thus, it is plausible that the innate psychological machinery of baboons allows them to adapt only to a relatively narrow range of environments. In contrast, because humans *can* acquire behavior by observational learning, cultural variation can accumulate gradually, and humans are often able to make use of skills and knowledge that cannot be acquired by any single individual. In this way, less specialized psychological mechanisms allow humans to adapt to a wider range of environments than baboons can.

Culture can lead to evolutionary outcomes not predicted by ordinary evolutionary theory.

Once culture becomes established, it is like a second system of inheritance. We acquire genes from our parents, and we acquire culturally transmitted beliefs and values from parents, relatives, friends, and, perhaps, even from professors. To emphasize this similarity, biologist Richard Dawkins refers to culturally transmitted beliefs and values as **memes**.

The logic of natural selection applies to memes in the same way that it does to genes. Memes compete for our memory and for our attention, and not all memes can survive. Some memes are more likely to survive and be transmitted than others; memes are heritable, often passing from one individual to another without major change. As a result, some memes spread and others are lost.

However, the fact that memes are subject to natural selection does not mean that genetically adaptive memes will spread. Because the rules of cultural transmission are different from the rules of genetic transmission, the outcome of selection on memes can be different from the outcome of selection on genes. The rules of genetic transmission are simple. With some exceptions, every gene that an individual carries in her body is equally likely to be incorporated into her gametes, and the only way those genes will be transmitted is if she produces children who carry those gametes. Thus, only genes that increase reproductive success will spread. Cultural transmission, by contrast, is Byzantine in its complexity. Memes are acquired and transmitted piecemeal throughout an individual's entire life, not just from parents to offspring, but from grandparents, siblings, friends, coworkers, teachers, and even from completely impersonal sources like books, television, and the Internet. Unlike genes, which are transmitted with little modification from parents to offspring, memes may spread along many different pathways. Ideas about dangerous hobbies like rock climbing or heroin use can spread from friend to friend, even though these behaviors are likely to reduce the chance of reproducing successfully (Figure 18.17). Beliefs about heaven and hell can spread from priest to parishioner, even if the priest is celibate. To add to the complexity, much heritable cultural variation accumulates among groups of people who form clans, fraternities, business firms, religious sects, or political parties. Memes that improve, say, quality assurance, can spread through a business community because they cause firms to earn profits, grow, and spawn new firms with the same methods of quality assurance.

The fact that culture can lead to outcomes not predicted by conventional evolutionary theory does not mean that human behavior has somehow transcended biology. The idea that culture is separate from biology is a popular misconception that cannot withstand scrutiny. Culture cannot transcend biology because it is as much a part of human biology as bipedal locomotion. Culture is generated from organic structures in the brain that were produced by the processes of organic evolution. However, cultural transmission leads to novel evolutionary processes. Thus, to understand the whole of human behavior, evolutionary theory must be

Figure 18.17

Cultural evolution may permit the spread of ideas and behaviors that do not contribute to reproductive success. Dangerous sports, like rock climbing, may be examples of such behaviors.

modified to account for the complexities introduced by these, as yet poorly understood, processes.

The fact that culture can lead to outcomes that would not be predicted by conventional evolutionary theory does not mean that ordinary evolutionary reasoning is useless. The fact that there are processes that lead to the spread of risky behaviors like rock climbing does not mean that these are the only processes that influence cultural behavior. Earlier in this chapter we saw that it is likely that many aspects of human psychology have been shaped by natural selection so that people learn to behave adaptively. We love our children, prefer not to mate with relatives, and think adeptly about social contracts. There is every reason to suspect that these predispositions play an important role in determining which memes spread and which don't. To the extent that this observation is true, ordinary evolutionary reasoning will be useful for understanding human behavior.

Human Behavioral Ecology

Human behavioral ecologists view humans as rational actors who seek to maximize their fitness.

The workings of the human mind are not a central problem for most social scientists. Instead, they want to understand human behavior and social organization. They ask, why do some people subsist via hunting and gathering while others rely on agriculture? Why are some societies egalitarian while other societies are hierarchical? Why are some societies polygynous and others monogamous? Although in principle we might answer such questions by identifying special-purpose modules in the brain that evolved to solve specific adaptive problems faced by Pleistocene foragers, this is not a practical alternative at present. Evolutionary psychology is a young discipline, and we know very little about how natural selection has shaped specific aspects of human psychology.

Some researchers have, instead, adopted the same kinds of tactics that evolutionary biologists use when they want to understand particular aspects of the morphology and behavior of other organisms. As we have seen throughout this book, the assumption that phenotypes have been shaped by natural selection provides a powerful tool for understanding organisms. It can help to explain why different primates have very different kinds of teeth and why male monkeys regularly commit infanticide in one-male, multifemale groups but not in other types of groups. However, we know from Chapter 3 that there are several reasons why evolution may not produce adaptations in every case: selection acting on one character may affect other characters, genetic drift may lead to maladaptation, or constraints may prevent adaptation from occurring. To be sure that adaptive reasoning is justified in any particular case, a biologist would need to know enough about the genetics, development, and history of the organism of interest to discount maladaptive processes. However, biologists almost never know much about these things. Many biologists simply ignore these complications and assume that observed phenotypes are adaptive, a research tactic that Oxford biologist Alan Grafen calls the **phenotypic gambit**. Biologists use the phenotypic gambit, knowing that it may not be correct in every case, because experience in many areas of biology suggests that adaptive reasoning is often very useful.

A group of researchers called human behavioral ecologists have applied the phenotypic gambit to the human species. Their theory assumes that contemporary humans act as if they are attempting to maximize their genetic fitness. The assumption that humans are rational actors who maximize their own selfish interests is widely used in the social sciences. Economists have used this idea to build an elaborate and successful theory that explains how people behave in market settings. Economists and other social scientists have extended rational-actor models to account for political behavior, marriage, decisions about family size, and many other kinds of behavior in nonmarket settings. While these applications are controversial in sociology and political science, there is little doubt that the rational-actor approach is important in all of the social sciences. Human behavioral ecology is simply rational-actor theory that goes one step further. It predicts that peoples' interests are defined in evolutionary terms—humans want to maximize their fitness. As we will see, this step provides human behavioral ecologists with a rich source of hypotheses about human behavior, particularly for traditional societies in which markets play a relatively minor role.

Because animals must acquire enough food to meet their energy needs, natural selection tends to favor adaptations that increase foraging efficiency.

No matter what kind of food an animal eats, or how it obtains its food, natural selection usually favors foraging efficiency. Sometimes, morphological adaptations enhance foraging abilities. For example, some folivorous monkeys have complex stomachs that increase their ability to extract nutrients from leaves. Foraging efficiency, measured in terms of the amount of energy obtained per unit time, may also be enhanced by reducing the amount of time it takes to locate food, by selecting foods that have high nutritional value and low processing time, or by abandoning food patches when they are depleted.

In behavioral ecology, the evolution of these types of adaptations is the domain of **optimal foraging theory**. This theory is based on the assumption that natural selection shapes foraging behavior so as to maximize the amount of energy gained per unit time. Of course this assumption is not exactly true because getting food is not the only problem facing animals. For example, primates might be more efficient if they foraged alone, but they forage in groups to reduce the risk of predation. However, foraging efficiency is so important that it is often reasonable to ignore other factors. Optimal foraging theory generates quantitative predictions about foraging behavior that can be tested in real-life situations. This body of theory has been applied to the Aché, a hunter-gatherer people living in the forests of Paraguay.

Optimal foraging theory has been used to understand the economy of the Aché.

Until the mid-1970s, the Aché were nomadic foragers who subsisted entirely on resources that they obtained in the forests of Paraguay. Like many other indigenous peoples, the Aché's lives have been radically altered by contact with Western culture. They have been encouraged to settle near missions, to grow crops, and to raise livestock. Nonetheless, small bands of Aché continue to make periodic foraging trips into the forest, much as they did before European contact. The Aché are proficient hunters (Figure 18.18) who take many animal species, including armadillos, monkeys, and peccaries (small piglike mammals). Men sometimes head off into the forest together, fanning out in different directions to search for game, but remain close enough to call out when they spot prey, and to coordinate the chase and capture. Men also collaborate in locating and extracting honey, which is greatly prized. Women make camp, tend children, collect wood for fires, help

Figure 18.18

Aché men are skilled hunters, but they do not always forage optimally. If they did, they would spend less time hunting and more time processing palm starch. (Photograph courtesy of Kim Hill.)

in hunts, and gather plant foods (Figure 18.19). In the forest, the Aché obtain about 3600 calories per person each day, and about 70% of the calories come from meat.

A group of anthropologists from the University of Utah (that included Kristin Hawkes, Kim Hill, Magdalena Hurtado, and Hillard Kaplan) carefully monitored the time that hunters spent in search and pursuit of prey and the amount of time that they needed to process their kills. They also kept track of the amount of time gatherers spent searching for and processing plant foods. They tried to weigh all of the food that each member of the foraging group obtained. Using this information, along with information about the caloric values of various foods, they were able to estimate the number of calories obtained per hour of effort, as well as the rate of return for particular foods and for individual foragers.

Optimal foraging theory predicts that foragers will exploit only those resources that provide a greater rate of return after encounter than the overall return rate for all resources.

On an average day, a forager can expect to obtain a given number of calories per hour. This is based on all the food she obtains and all the time that it takes to

Figure 18.19

Aché women spend much of their time gathering plant foods and processing palm starch. (Photograph courtesy of Kim Hill.)

find, pursue, and process these food items. Now, as the forager moves through the forest she encounters a potential food item, say a palm tree. Should she continue walking, or should she begin to process palm to extract the starch? If more productive resources are nearby, then she should keep walking. But how does she know whether there are more productive resources in the vicinity? Her best estimate of the return for alternative foods is simply the average rate of return for foraging. Thus, her decision will be based on the difference between the rate of return from the resource she has just found and the overall mean rate of return. If the difference between these rates is positive, then she will on average do best by beginning to process palm starch. If the difference is negative, then she will be better off to continue her search for food.

With two important exceptions, Aché behavior conforms to the predictions of optimal foraging theory.

The Aché concentrated their foraging efforts on resources that generated a higher rate of return than the mean (Figure 18.20). They tended to take the most productive resources most often. However, the University of Utah researchers discovered two major discrepancies between the predictions of optimal foraging theory and the Aché's foraging behavior. First, the men routinely ignore plant foods, even the ones that generate relatively high rates of return (Figure 18.20a). Men could greatly increase the number of calories they obtain per hour if they devoted themselves to extracting and processing fiber from the trunks of palm trees. Instead, men hunt.

Second, women routinely ignore resources, like honey and meat, that bring high rates of return and concentrate their efforts on plant foods with lower return rates (Figure 18.20b). These observations clearly violate the predictions of the optimal foraging theory model.

Evolutionary theory suggests reasons why Aché men hunt rather than gather and why women gather rather than hunt.

The results of the University of Utah group's analysis clearly show that if men were trying to maximize the rate of acquiring calories, they would give up their bows and shotguns and become full-time gatherers. And the women would occasionally pass up palms to hunt game. Evolutionary theory provides some clues to understand why men and women don't forage optimally.

One possibility is that the optimal foraging models ignore things that matter to Aché foragers. For example, the model assumes all calories are of equal value. However, calories obtained from meat may be more valuable than calories obtained from plant foods. Meat provides both protein and fat, and may be of greater nutritional value than equivalent amounts of carbohydrates. The problem with this explanation is that the Aché eat much more meat than most people do. They obtain nearly 70% of their calories from meat, and their consumption of protein and fat greatly exceeds the body's daily requirements for these nutrients. Thus, Aché men could hunt less and gather more, without jeopardizing their nutritional status.

Hunting may also increase men's reproductive success because better hunters get social payoffs that increase their reproductive success. There are substantial differences among men in foraging efficiency. In general, young and old men forage less efficiently than men in their prime do. Among middle-aged men, some

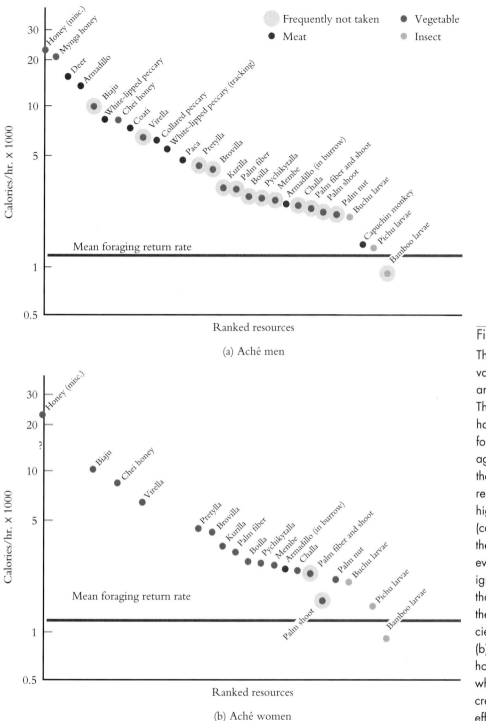

Figure 18.20

The Aché gather a variety of plant foods and hunt wild game. Their foraging behavior generally conforms to optimal foraging models, as they concentrate on resources that bring higher rates of return (calories/hour) than the mean. (a) However, men routinely ignore plant foods that would increase their foraging efficiency; and (b) women ignore honey and meat, which would increase their foraging efficiency.

are better hunters than others, bringing back substantially more meat per unit time spent hunting. These differences are apparently the result of skill, not luck, as the differences among men persist over several years. Aché men's reproductive success is correlated with their foraging efficiency, as the men with the highest return rate had the largest number of surviving children by their wives and the greatest number of illegitimate children. Interestingly, this correlation does not arise because these men are able to provide more food for their own children than other men do.

The Aché share their kills with all members of their foraging party, and the hunter's family does not get a disproportionate share of his kills. Instead, the University of Utah group speculated that successful hunters may gain reproductive advantages in other ways. As an inducement for successful hunters to remain in the group, other members of the band may treat a good hunter's children better, being more accommodating when they are ill, grooming them, or protecting them from the many hazards in the forest. Members of the group benefit from the presence of a successful hunter who shares his kills; the hunter benefits by increasing the chance that his children will survive.

This example illustrates why evolutionary theory is a rich source of hypotheses about human behavior. The University of Utah group started with a simple picture of the Aché economy that generated several detailed predictions about foraging patterns. They collected the requisite data, and it turned out that much of Aché foraging behavior conformed to the model, suggesting that the Aché forage efficiently. However, the model failed to account for why men don't process palms and why women don't hunt. Because evolutionary theory specifies what people should value, it allows us to determine what was left out of the model, such as other nutrients. By modifying the model to include these factors, we can generate new predictions and make new observations to test these predictions. This cycle of theory and observation is a potent process for generating a richly textured picture of the Aché subsistence economy.

Further Reading

Barkow, J., L. Cosmides, and J. Tooby, eds. 1992. *The Adapted Mind, Evolutionary Psychology and the Generation of Culture.* Oxford University Press, New York.

Berlin, B., and P. Kay. 1991 (1969). *Basic Color Terms.* University of California Press, Berkeley. (The 1991 paperback edition has a useful supplementary review of developments since the original publication in 1969.)

Boyd, R., and P. Richerson. 1985. *Culture and the Evolutionary Process.* University of Chicago Press, Chicago.

Brown, D. 1991. *Human Universals.* McGraw-Hill, New York.

Smith, E. A., and B. Winterhalder. 1992. *Evolutionary Ecology and Human Behavior.* Aldine de Gruyter, Hawthorne, N.Y.

Study Questions

1. Much of the behavior of all primates is learned. Nonetheless, we have suggested many times that primate behavior has been shaped by natural selection. How can natural selection shape behaviors that are learned?

2. Many of the things that we do are consistent with general predictions derived from evolutionary theory. We love our children, help our relatives, and avoid sex with close kin. But there are also many aspects of the behavior of members of our own society that seem unlikely to increase individual fitness. What are some of these behaviors?

3. In some species of primates there seems to be an aversion to mating with close kin. The aversion seems to be stronger for females than for males. Why do you think this might be the case? Under what conditions would you expect this gender difference to disappear?

4. The verb *to ape* means "to copy or imitate." Is its meaning consistent with what we now know about the learning processes of other primates?

5. Evolutionary psychologists and human behavioral ecologists take a very different view of the adaptive value of contemporary human behavior. Explain how researchers working in each of these areas view contemporary behavior. What aspects of modern behavior seem to be consistent with each point of view?

C H A P T E R   1 9

Human Mate Choice and Parenting

The Psychology of Human Mate Preferences
Some Social Consequences of Mate Preferences
    Kipsigis Bridewealth
    Nyinba Polyandry
Raising Children
    Child Abuse and Infanticide
        Cross-Cultural Patterns of Infanticide
        Child Abuse in the United States and Canada
    Adoption
        Adoption in Oceania
        Adoption in Industrialized Societies
    Family Size
Is Human Evolution Over?

You should be convinced by now that every aspect of the human phenotype is the product of evolutionary processes. We humans are large, nearly hairless, bipedal primates, having grasping hands and large brains because natural selection, mutation, and genetic drift made us that way over the last 10 million years. The same evolutionary processes have molded the psychological mechanisms that influence human behavior, causing us to behave differently than other primates in some contexts. There is simply no other explanation.

However, the fact that modern humans are the product of evolutionary processes does not necessarily mean that evolutionary theory will help us understand contemporary human behavior. Constraints on adaptation, which we discussed in Chapter 3, may have predominated in the evolution of human behavior, limiting the usefulness of adaptive reasoning. However, in Chapter 18 we saw that evo-lutionary theory has provided useful insights into certain aspects of human psychology, social organization, and culture. In this chapter, we present several more examples that show how evolutionary thinking can help us understand the day-to-day behavior of humans in modern societies.

In choosing these examples, we have focused on reproductive behavior because mating and parenting strongly affect fitness. The unfortunate person who chooses a mate who is lazy, infertile, or unfaithful is likely to have many fewer children than the person whose spouse is healthy, hardworking, and faithful. The inattentive or abusive parent will have fewer children than the parent who carefully nurtures his or her offspring. Because reproductive decisions are likely to have a marked effect on fitness, there is good reason to expect that our psychology has been shaped by natural selection to improve the chances of making good choices. We begin by presenting research on the psychology of mate preferences, and then consider two examples that illustrate how such preferences work themselves out in real social settings. We then turn to studies that illustrate how evolutionary theory can help us to understand the range of variation in human parenting strategies. We do not have enough space to provide a comprehensive analysis of these topics. Instead, we have focused on certain examples in which evolutionary theory provides novel and fundamental insights into human behavior.

The Psychology of Human Mate Preferences

Marry

Children—(if it Please God)—Constant companion, (& friend in old age) who will feel interested in one, — object *to be* beloved and played with. better than a dog anyhow.— Home, & someone to take care of house— Charms of music & female chit-chat. — These things good for one's health. — *but terrible loss of time.*—

My God, it is intolerable to think of spending one's whole life, like a neuter bee, working, working, & nothing after all.— No, no won't do.—Imagine living all one's day solitary in smoky dirty London house.— Only picture to yourself nice soft wife on a sofa with good fire, & books, & music perhaps—Compare this vision with the dingy reality of Grt. Marlbro St.

Marry—Mary—Marry Q.E.D.

Not Marry

Freedom to go where one liked— choice of Society & *little of it.*—Conversation of clever men at clubs—Not forced to visit relatives, & to bend in every trifle.—to have the expense & anxiety of children—perhaps quarelling— **Loss of time**.—cannot read in the Evenings—fatness & idleness—Anxiety & responsibility—less money for books & c—if many children forced to gain one's bread.—(But then it is very bad for ones health to work too much)

Perhaps my wife wont like London; then the sentence is banishment & degradation into indolent, idle fool.

[p. 444 in F. Burkhardt and S. Smith, eds., 1986, *The Correspondence of Charles Darwin,* Vol. 2, 1837–1843, Cambridge University Press, Cambridge, U.K.]

These are the thoughts of 29-year-old Charles Darwin, recently returned from his five-year voyage on the *Beagle*. Eventually, Darwin married his cousin, Emma, the daughter of Josiah Wedgwood, the progressive and immensely wealthy manufacturer of Wedgwood china (Figure 19.1). They were, by all accounts, a devoted couple. Emma bore Charles 10 children and nursed him through countless bouts of illness. Charles toiled over his work and astutely managed his investments,

Figure 19.1

Charles Darwin courted and married his cousin, Emma Wedgwood. This portrait was painted when Emma was 32, just after the birth of her first child.

parlaying his modest inheritance and his wife's more substantial one into a considerable fortune.

Darwin's frank reflections on marriage were very much those of a conventional, upper-class, Victorian gentleman. But people of every culture, class, and gender have faced the problem of choosing mates. Sometimes people choose their own mates, and at other times parents arrange their children's marriages. But everywhere, people care about the kind of person they will marry.

Evolutionary theory generates a number of testable predictions about the psychology of human mate preferences.

David Buss, a psychologist at the University of Texas, has conducted a series of studies designed to assess whether predictions derived from evolutionary theory fit patterns of human mate preferences in contemporary human societies. Buss reasons that the psychology underlying contemporary human mate preferences was shaped during the Pleistocene when humans were hunters and gatherers. Using this approach, he has generated the following predictions about evolved mate preferences:

- *Women prefer males who are able and willing to provide resources.* In the environment of evolutionary adaptedness, it may have been difficult for anyone to accumulate resources, but some men were probably better providers than others. Women may have preferred men who could provide the most resources for themselves and for their offspring. As foraging gave way to horticulture, and then to agriculture, the capacity for accumulating resources increased dramatically. In many contemporary societies, there are great disparities in wealth between individuals. Females' preference for good providers may now be expressed as a preference for wealthy men.
- *Males prefer females who will reproduce successfully.* A man's reproductive success will depend partly on the reproductive success of his partner. This, in turn, will depend on his partner's reproductive value and fertility. A woman's **reproductive value** (her expected contribution to future generations) peaks before she reaches 20, steadily declines until she reaches menopause, and then drops to zero.

Female **fertility**, or birth rate, is highest in the early 20s. In general, men are expected to prefer younger women to older women. Some researchers have suggested that cross-cultural standards of beauty reflect an evolved preference for physical traits that are generally associated with youth, such as smooth skin, good muscle tone, and shiny hair. Therefore, men are expected to value beauty in prospective mates.

• *Both men and women value fidelity, but men value it more highly than women.* When both parents provide care for offspring, both males and females have reason to value fidelity. Men should not want to invest resources in another man's children, and women should not want their mates to divert resources to another female's offspring. However, the value of fidelity is likely to be greater for men than for women because men can be cuckolded. A man can only be sure that he is the father of all his wife's children if he is certain she has been completely faithful during their marriage, while a woman knows that she is the mother of all her children, no matter what her husband does. Thus, men are likely to value fidelity more than women do.

Although Buss focused mainly on traits that men and women are likely to value differently, there are also traits that we might expect both sexes to consider important. Since parental investment lasts for many years, we would expect both men and women to value traits in their partners that help them sustain their relationships. Both are likely to value personal qualities such as compatibility, kindness, helpfulness, and tolerance.

Buss tested the importance of these factors to men and women across the world. He enlisted colleagues in 33 countries to administer standardized questionnaires about the qualities of desirable mates to more than 10,000 men and women. Most of the data were collected in Western, industrialized nations, and most samples represent urban populations within those countries (Figure 19.2). In the questionnaires, people were asked to rate 18 traits of potential mates, such as good looks, financial prospects, compatibility, and so on, according to their desirability. Respondents were also asked about their preferred age at marriage, the preferred age difference between themselves and their spouse, and the number of children that they wanted to have. The results provide many interesting insights about human mate choice, but we will limit our account to a few key points.

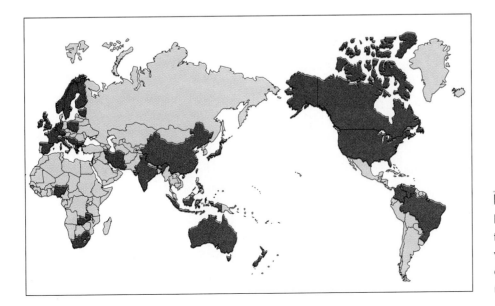

Figure 19.2

People in 33 countries (red) were surveyed about the qualities of an ideal mate.

People generally care most about the personal qualities of their mates.

People around the world rate mutual attraction or love above all other traits (Table 19.1). The next most highly desired traits are personal attributes, such as dependability, emotional stability and maturity, and a pleasing disposition. Good health is the fifth most highly rated trait for men and the seventh for women. Good financial prospect is the 13th most highly rated trait by men and the 12th by women. Good looks are rated 10th by men and 13th by women. It is interesting, and somewhat surprising, that neither sex seems to value chastity highly. Perhaps this is because people were asked to evaluate the desirability of sexual experience before marriage (that is, virginity), not fidelity during their marriage.

Men and women show the differences in mate preferences predicted by evolutionary theory.

You can see that there is a high degree of concordance in the rankings of the scores that men and women assign to the traits in Table 19.1. For example, men and women both rate love more highly than any other trait, and give the lowest ratings to political and religious affinities. Even though the ranking of the scores

Table 19.1   People from more than 30 countries around the world were asked to rate the desirability of a variety of traits in prospective mates. The rankings of the values assigned to each trait, on average, by men and women in countries around the world are given here. Subjects were asked to rate each trait from 0 (irrelevant or unimportant) to 3 (indispensable). Thus, high ranks (low numbers) represent traits that were generally thought to be important. (From Table 4 in D. M. Buss et al., 1990, International preferences in selecting mates: A study of 37 cultures, *Journal of Cross-Cultural Psychology* 21:5–47.)

	RANKING OF TRAIT BY:	
TRAIT	MALES	FEMALES
Mutual attraction-love	1	1
Dependable character	2	2
Emotional stability and maturity	3	3
Pleasing disposition	4	4
Good health	5	7
Education and intelligence	6	5
Sociability	7	6
Desire for home and children	8	8
Refinement, neatness	9	10
Good looks	10	13
Ambition and industrious	11	9
Good cook and housekeeper	12	15
Good financial prospect	13	12
Similar education	14	11
Favorable social status or rating	15	14
Chastity★	16	18
Similar religious background	17	16
Similar political background	18	17

★Chastity was defined in this study as having no sexual experience before marriage.

assigned to these traits is similar for men and women, Buss emphasized that there are still important and consistent differences between men and women in how desirable each gender thinks these traits are. Buss found that a person's gender had the greatest effect on their ratings of the following traits: "good financial prospect," "good looks," "good cook and housekeeper," "ambition and industrious." As the evolutionary model predicts, women value good financial prospects and ambition more than men do, while men value good looks more than women do. Gender has a smaller and somewhat less uniform effect on ratings of chastity. In 23 populations, men value chastity significantly more than women do, while in the remaining populations men and women value chastity equally. There are no populations in which women value chastity significantly more than men do. Men and women also differ about the preferred age differences between themselves and their spouses. Men invariably want to marry women who are younger than themselves, while women want to marry men who are older than they are.

Buss's analysis shows that culture exerts a stronger influence on people's mate preferences than gender does.

The country of residence has a greater effect than gender does on variation in all of the 18 traits in Table 19.1, except for "good financial prospects." This means that knowing where a person lives tells you more about what he or she values in a mate than knowing the person's gender. Of the 18 traits, chastity (defined in Buss's study as no sexual experience before marriage) shows the greatest variability among populations. In Sweden, men rate chastity at 0.25, and women rate it at 0.28 on a scale of 0 (irrelevant) to 3 (indispensable). In contrast, Chinese men rate chastity 2.54, and Chinese women rate it 2.61 (Figure 19.3). Buss and his colleagues

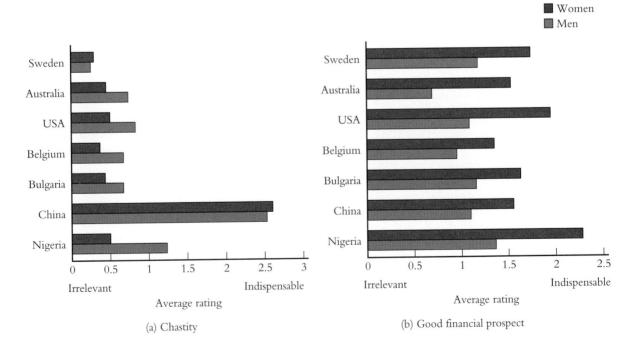

Figure 19.3

Culture accounts for substantial variation in mate preferences. The average ratings given by men and women in several countries surveyed are shown for (a) the trait with the highest interpopulation variability ("chastity") and (b) the trait with the lowest interpopulation variability ("good financial prospect").

interpreted these results to mean that there is more similarity between men and women from the same population than there is among members of each sex from different populations.

This result illustrates an important point: evolutionary explanations that invoke an evolved psychology and cultural explanations that are based on the social and cultural milieu are not mutually exclusive. Buss's results suggest there are cross-cultural uniformities in people's mate preferences that are the result of evolved psychological mechanisms. People want to marry kind, caring, trustworthy people. Men want to marry younger women and women want to marry older men. But this is not the whole story. Buss's data also suggest that human mate preferences are strongly influenced by the cultural and economic environment in which we live. Ultimately, culture also arises out of our evolved psychology, and the cultural variation in mate preferences that Buss observed must therefore also be explicable in evolutionary terms. However, the way our evolved psychology shapes the cultures in which we live is complicated and poorly understood, and many interesting questions remain unresolved. Evolutionary theory does not yet explain, for example, why chastity is essential in China but undesirable in Sweden.

Some Social Consequences of Mate Preferences

You might wonder how people's preferences for certain kinds of traits in prospective mates influence their actual decisions and choices about marriage partners. In this section, we describe the findings from one ethnographic study that suggest these kinds of preferences actually influence people's behavior in social situations, and consequently shape the societies in which they live.

Kipsigis Bridewealth

Evolutionary theory explains marriage patterns among the Kipsigis, a group of East African pastoralists.

Among the Kipsigis, a group of Kalenjin-speaking people who live in the Rift Valley Province of Kenya, women usually marry in their late teens, men usually marry for the first time when they are in their early 20s, and it is common for men to have several wives, a practice called **polygyny**. As in many societies, the groom's father makes a **bridewealth** payment to the father of the bride at the time of marriage. The payment, tendered in livestock and cash, compensates the bride's family for the loss of her labor and gives the groom rights to her labor and the children she bears during her marriage. The amount of the payment is settled through protracted negotiations between the father of the groom and the father of the bride. The average bridewealth consists of six cows, six goats or sheep, and 800 Kenyan shillings. This is about one-third of the average man's cattle holdings, one-half of his goat and sheep herd, and two months' wages for men who hold salaried positions. Since men marry polygynously, there is a competition over eligible women. Often, the bride's father entertains several competing marriage offers before he chooses a groom for his daughter. The prospective bride and groom have little voice in the decisions their fathers make.

Anthropologist Monique Borgerhoff Mulder, at the University of California, Davis, reasoned that Kipsigis bridewealth payments provide a concrete index of the qualities that each party values in prospective spouses. The groom's father is likely to prefer a bride who will bear his son many healthy children. His bride-wealth offer is expected to reflect the potential reproductive value of the prospective bride. The groom's father is also expected to prefer that his son marry a woman who will devote her labor to his household. Kipsigis women who remain near their own family's households are likely to be called on to help their mothers with the harvest and to assist their mothers in childbirth. Thus, the groom's father may prefer a woman whose natal family is distant from his son's household. The bride's father is likely to have a different perspective on the negotiations. Since wealthy men can provide their wives with larger plots of land to farm and more resources, the bride's father is expected to prefer that his daughter marry a relatively wealthy man. At the same time, since the bride's family will be deprived of her labor and assistance if she moves far away from their land, the bride's father is likely to prefer a groom who lives nearby. The fathers of the bride and groom are expected to weigh the costs and benefits of prospective unions in their negotiations over bride-wealth payments. For example, while the bride's father may prefer a high bride-wealth payment, he may settle for a lower payment if the groom is particularly desirable. In order to determine whether these preferences affected bridewealth payments, Borgerhoff Mulder recorded the number of cows, sheep, goats, and the amount of money that each groom's family paid to the bride's family.

Plump women whose menarche occurred at an early age fetched the highest bride payments.

Borgerhoff Mulder found that bridewealth increased as the bride's age of **menarche** (her first menstruation) decreased (Figure 19.4). That is, the highest bride-

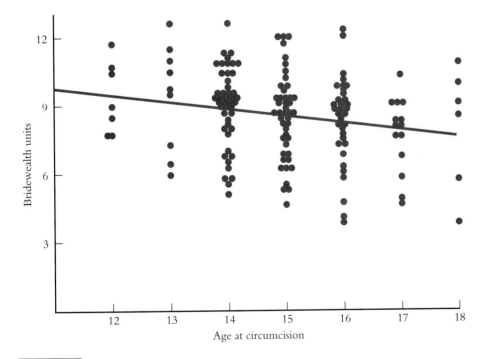

Figure 19.4

Kipsigis girls who mature early fetch larger bridewealth payments than older girls do. Among the Kipsigis, girls undergo circumcision (removal of the clitoris) within a year of menarche. The largest bridewealths were paid for the girls that underwent menarche and circumcision at the youngest ages. Bridewealth is transformed into standardized units to account for the fact that the value of livestock varies over time.

Figure 19.5

These Kipsigis women are eligible for marriage. Their fathers will negotiate with the fathers of their prospective husbands over bridewealth. (Photograph courtesy of Monique Borgerhoff Mulder.)

wealths were paid for the women who were youngest when they first menstruated (Figure 19.5). Among the Kipsigis, age at menarche is a reliable index of women's reproductive potential. Kipsigis women who reach menarche early have longer reproductive life spans, higher annual fertility, and higher survivorship among their offspring than do women who mature at later ages.

Borgerhoff Mulder wondered how a man assessed his prospective bride's reproductive potential, since men often do not know their bride's exact age, nor her age of reaching menarche. One way to be sure of a woman's ability to produce children would be to select one who had already demonstrated her fertility by becoming pregnant or producing a child. However, bridewealths for such women were typically lower than bridewealths for women who had never conceived. Instead, bridewealth payments were associated with the physical attributes of women. The fathers of brides who were considered by the Kipsigis to be plump received significantly higher bridewealth payments than did the fathers of brides considered to be skinny. The plumpness of prospective brides may be a reliable correlate of the age at menarche, since menarcheal age is determined partly by body weight. Plumpness may also be valued because a woman's ability to conceive is partly determined by her nutritional status.

Bridewealth payments are also related to the distance between the bride's home and the groom's home; the further she moves, the less likely she is to provide help to her mother, and the higher the bridewealth payment. However, there is no relationship between the wealth of the father of the groom and the bridewealth payment. The bride's father does not lower the bridewealth payment to secure a wealthy husband for his daughter. Although this finding was unexpected, Borgerhoff Mulder suggested that it may be related to the fact that differences in wealth among the Kipsigis are unstable over time. A wealthy man who has large livestock herds may become relatively poor if his herds are raided, decimated by disease, or diverted to pay for another wife. While land is not subject to these vicissitudes, the Kipsigis traditionally have not held legal title to their lands.

Nyinba Polyandry

Polyandry is rare in humans and other mammals.

Since mammalian females are generally the limiting resource for males, males usually compete for access to females. As a result, polygyny is much more common among mammalian species than polyandry. As you may recall from Chapter 7, polyandry is a mating system in which one female is paired with two or more males. Polyandry is rare because, all other things being equal, males who share access to a single female produce fewer offspring than males who maintain exclusive access to one or more females. The reproductive costs of polyandry may be somewhat reduced if brothers share access to a female, an arrangement called **fraternal polyandry**. Thus in accordance with Hamilton's rule, it is plausible to assume that natural selection has shaped human psychology so that a man is more willing to care for his brother's offspring than for children who are unrelated to him.

Polyandrous marriage systems are as rare among human societies as they are in other mammalian species. In a sample of 862 societies compiled by anthropologist George Peter Murdock, polygyny is the ideal form of marriage in 83% of societies, 16% are exclusively monogamous, and only 0.5% are polyandrous.

Polyandrous marriage occurs among several societies in the Himalayas.

Although polyandry is generally rare among human societies, there are several societies in the highlands of the Himalayas where fraternal polyandry is the preferred form of marriage. In these societies, several brothers marry one woman and establish a communal household. Since polyandry seems to limit a man's reproductive opportunities, we might ask why men tolerate polyandrous marriages. Nancy Levine of the University of California, Los Angeles, studied one of these polyandrous societies, the Nyinba of northwestern Nepal (Figure 19.6). Her data shed some interesting light on this question.

The Nyinba live in four prosperous villages, nestled between 9500 and 11,000 feet on gently sloping hillsides. Nyinba families support themselves through a combination of agriculture, herding, and long-distance trade. The Nyinba believe that marriages of three brothers are most desirable, allowing one husband to farm, another to herd livestock, and the third to engage in trade. In practice, however, all brothers marry jointly, no matter how many there are. The Nyinba are quite concerned with the paternity of their children. Women identify the fathers of each

Figure 19.6
Polyandry is the preferred form of marriage among the Nyinba. In this family, three brothers are married to one woman. The Nyinba consider this to be the ideal number of co-husbands. (Photograph courtesy of Nancy Levine.)

of their children, a task made easier by the fact that one or more of their husbands is often away from home for lengthy periods tending to family business. Although it is impossible to be certain how accurate these paternity assessments are, it is clear that the Nyinba place great stock in them. Men develop particularly close relationships with the children they have fathered, and fathers bequeath their share in the family's landholdings to their own sons.

Evolutionary theory helps to explain why some polyandrous marriages are successful and others fail.

Polyandry is the ideal form of marriage among the Nyinba, but reality does not always conform to this ideal. While all men marry jointly, some marriages break down when one or more of the brothers brings another wife into the household or leaves to set up an independent household. When households break up, each man receives a share of the estate, and takes with him all the children he has fathered during the marriage. Levine reasoned that a comparison of marriages that remain intact and marriages that break up would provide clues about the kinds of problems that arise in polyandrous marriages. Levine together with one of us (J. B. S.) has used the data to determine whether male decisions conform to predictions derived from evolutionary theory.

All other things being equal, the more men who are married to a single woman, the lower each man's reproductive success is likely to be. Thus, we might expect marriages with many co-husbands to be less stable than marriages with fewer co-husbands. This turns out to be the case; few of the largest communal Nyinba marriages remained intact. Only 10% of the marriages that began with two co-husbands partitioned, while 58% of the marriages that began with four co-husbands did (Figure 19.7).

Some Nyinba marriages are arranged by parents, while others are set up by the oldest brother. In either case, the wife is usually a few years younger than the oldest co-husband. In some marriages, the youngest co-husbands are considerably younger than their wives. Remember that Buss's cross-cultural data suggest that men prefer to marry younger women. Thus, we might expect that men would be dissatisfied with marriages to women much older than themselves. In fact, the

Figure 19.7

Among the Nyinba, marriages that contain more than three brothers are more likely to break up than marriages that include only two or three co-husbands. Polyandrous households are those that maintain intact polyandrous marriages. Conjoint households are those that added a second wife, which often effectively created two separate marriages within a single household. Partitioned households are those in which one or more men left the fraternal marriage and formed a new household.

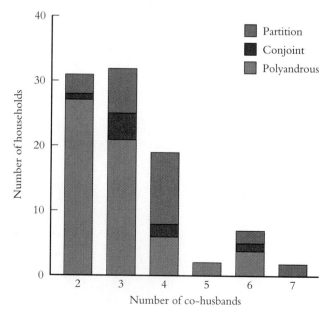

Nyinba men who were most junior to their wives were the ones most likely to initiate partitions. When these men remarried, they invariably chose women who were younger than themselves and who were younger than their first wives.

If the basic problem with polyandry is that it limits male reproductive opportunities, then we should expect males' satisfaction with their marriages to be linked to their reproductive performance. In most marriages, the oldest brother fathers the wife's first child. The second-oldest brother often fathers the second child, and so on. However, disparities in reproductive success among co-husbands often persist. A man's place in the birth order is directly related to his reproductive success, and the oldest brothers have more children than the youngest brothers. Men's decisions to terminate their marriages are associated with their reproductive performance; men who remained in stable polyandrous marriages fathered 0.1 children per year, while men who terminated their marriages had produced only 0.04 children per year during the relationship.

Finally, kin selection theory would lead us to predict that close kinship among co-husbands would help to stabilize polyandrous marriages because men should be more willing to invest in their brothers' children. We can test this prediction because there is a considerable amount of variation in the degrees of relatedness among co-husbands within households. Some households, known as sibships, are composed entirely of full siblings. In others, the co-husbands may have multiple fathers, who are related themselves, and multiple mothers, who may also be related. This produces a complicated web of relationships among co-husbands. Contrary to predictions based on kin selection, there is no difference in the degree of relatedness among co-husbands in households that remained intact and households that dissolved. Moreover, when men remarried, they showed no inclination to align themselves with the co-husbands to whom they were more closely related.

Evolutionary theory explains certain aspects of polyandry but not others.

Polyandry is rare among human societies, as we would expect from an evolutionary perspective. When marriages dissolve, they seem to do so in ways that are consistent with predictions derived from evolutionary theory. Men leave marriages in which the women are much older, and they leave marriages when they have not fathered many children. However, men seem to ignore kinship to their co-husbands in their decisions about whether to maintain their marriages, a surprising result from an evolutionary perspective.

Raising Children

Like other mammalian females, women are predisposed to invest heavily in their offspring. In humans, pregnancy lasts nine months; in traditional societies, lactation lasts at least one year and often longer. Parental investment extends well beyond weaning, as children do not become fully independent until they reach their teens. In most human societies, both mothers and fathers help to raise their own children. All this seems so normal and so obvious that it is easy to overlook the possibility that societies could be organized in many other ways. For example, babies could be bought and sold like pets. Such a system would have lots of practical advantages. It would be easy to adjust the size of your family, the ratio or order of boys and girls, and to regulate the spacing between children. Women who preferred not to undergo pregnancy and delivery, or who were unable to bear children, could have a family just as easily as anyone else. Some people might even opt to skip the dubious pleasures of frequent diaper changes and midnight feedings, and acquire children who were two or three years of age. But there are no societies in which this happens. Instead, in virtually every society, most people raise their own children. It seems likely that people do this because their evolved psychology causes them to value their own children very differently from other people's children.

There is, however, much more cross-cultural variation in human parenting behavior than in the parenting behavior of other primate species. In some societies, most men are devoted fathers, while in others men take little interest in children and rarely interact with their own offspring. In some societies, parents supervise their children's every waking moment, while in other societies, children are mainly reared by older siblings and cousins. Moreover, some aspects of parenting behavior in some societies seem hard to reconcile with the idea that human psychology has been shaped by natural selection. There are cultures where parents sometimes kill or physically abuse their own infant children, and cultures where people regularly raise children who are not their own. It is also common for people to intentionally limit their own fertility, and to produce fewer children than they are capable of raising. Because such behaviors seem inconsistent with the idea that human behavior is controlled by evolved predispositions, their existence has been cited as evidence that evolutionary reasoning has little relevance to understanding modern human behavior.

In the remainder of this chapter we consider data suggesting that observed patterns of infanticide, child abuse, adoption, and family planning are consistent with the idea that human behavior is influenced by evolved predispositions. Moreover, evolutionary reasoning can give fresh and useful insights about when and why such behaviors occur. However, we will also see that there are aspects of contemporary parenting behavior that evolutionary reasoning has not yet explained.

Child Abuse and Infanticide

Evolutionary theory predicts that parents should terminate investment in offspring if their prospects are poor.

Children have been deliberately killed, fatally neglected, or abandoned by their parents through recorded history and across the world's cultures. It is tempting to dismiss infanticide by parents as a pathological side effect of modern life. After all, a child is a parent's main vehicle for perpetuating his or her own genetic material. However, evolutionary analyses suggest that parents who kill their own children may, under some circumstances, have higher fitness than parents who do not do so. Natural selection may have molded our psychology so that we are sometimes prone to harm or neglect children under very special circumstances. If this assertion seems preposterous, remember from Chapter 17 that natural selection favors physiological mechanisms that terminate pregnancy when fetal prospects for survival are poor. The same adaptive logic applies to infants after they are born. Evolutionary reasoning suggests that parents should be predisposed to care only for offspring that are likely to survive and to reproduce successfully. Although this may seem cold, calculating, and cruel, it must be viewed in the context of human evolutionary history. In every plausible environment of evolutionary adaptedness, parental ability to invest in their offspring would have been limited, and natural selection would have favored psychological mechanisms that cause parents to terminate investment in offspring that are not likely to survive, and to channel investment toward the children that are most likely to live to maturity and to reproduce successfully. We expect these psychological mechanisms to be sensitive to the child's condition, the parents' economic circumstances, and the parents' alternative opportunities to pass on their genetic material. This phenomenon has been labeled **discriminative parental solicitude** by psychologists Martin Daly and Margo Wilson of McMaster University in Ontario, Canada.

CROSS-CULTURAL PATTERNS OF INFANTICIDE

Cross-cultural analyses indicate that infanticide occurs when a child is unlikely to survive, parents cannot care for the child, or the child is not sired by the mother's husband.

Daly and Wilson used the Human Relations Area Files, a vast compendium of ethnographic information from societies around the world, to compile a list of the reasons that parents commit infanticide. They based their study on a standard, randomly selected sample of 60 societies that represent traditional societies around the world. In 39 of these societies, infanticide is mentioned in ethnographic accounts, and in 35 of the societies Daly and Wilson found information about the circumstances under which infanticide occurs. There are three main classes of reasons that parents commit infanticide: (1) the child is seriously ill or deformed, (2) the parents' circumstances do not allow them to raise the child, or (3) the child is not sired by the mother's husband (Table 19.2).

In 21 societies, children are killed if they are born with major deformities or if they are very ill. In traditional societies, children with major deformities or serious illnesses require substantial amounts of care and impose a considerable burden on their families. They are unlikely to survive, and if they do, they are unlikely to be

Table 19.2 In a representative cross-cultural sample of societies, the rationales for infanticide are consistent with predictions derived from evolutionary theory. Parents mainly commit infanticide when the child is unlikely to thrive, when their own circumstances make it difficult to raise the child, or when the child was not fathered by the mother's spouse. (Data from Table 3.1 in M. Daly and M. Wilson, 1988, *Homicide*, Aldine de Gruyter, New York.)

RATIONALE	NUMBER OF SOCIETIES
Poor quality of child	
Deformed or ill	21
Twins born	14
Children too closely spaced or numerous	11
No male support for child	6
Mother died	6
Mother unmarried	14
Economic hardship	3
Born in wrong season	1
Quarrel with husband	1
Paternity not assigned to wife's husband	
Adulterous conception	15
Nontribal sire	3
Sired by mother's previous husband	2
Other reasons	15

able to support themselves, marry, or produce children of their own. Thus, it is plausible that selection has shaped human psychology so that we are predisposed to terminate investment in such children and to reserve resources for healthy children.

In some societies, children are killed when the parents' present circumstances make it too difficult or dangerous to raise them. This includes cases in which births are too closely spaced, twins are born, the mother has died, there is no male present to support the children, or the mother is unmarried. It is easy to see why these factors might make it very difficult to raise children. Twins, for example, strain a mother's abilities to nurse her children and to provide adequate care for them. Moreover, twins often are born prematurely and are of below-average birth weight, making them less likely to survive than single infants. In traditional societies, it is not always possible to obtain substitutes for breast milk. This dooms some children whose mothers die when they are very young. If a child's father dies before it is born or when it is an infant, its mother may have great difficulty supporting herself and her older children. The child's presence might also reduce her chances of remarrying and producing additional children in the future. When circumstances are not favorable, parents who attempt to rear children may squander precious resources that might better be reserved for older children or for children born at more favorable times in the future.

In 20 societies, infanticide occurs when a child is sired during an extramarital relationship, by a previous husband, or by a member of a different tribe who is not married to the mother. Recall from Chapter 8 that the theory of kin selection predicts that altruism will be restricted to kin. Males should not be predisposed to support other men's children.

Before you condemn those who kill their children, keep in mind that the options for parents in traditional societies are very different from the options available to people in our own society who have access to modern health-care systems, social services, and so on. Decisions by parents to kill their children are often reluctant responses to the painful realities of life. Whether these considerations are sufficient to absolve the parents in your eyes depends, of course, on your moral beliefs. However, there is little doubt that parents usually deeply regret the need to kill a newborn infant, and may grieve over the loss for many years.

CHILD ABUSE IN THE UNITED STATES AND CANADA

Evolutionary reasoning predicts that child abuse in North America will occur under the same circumstances that infanticide occurs cross-culturally.

Richard Gelles, of the University of Rhode Island and a national expert on child abuse, estimates that at least 1 million children are physically, emotionally, or sexually abused in the United States each year. Since many instances of abuse are not reported to authorities, the actual incidence of abuse may be much higher. It seems unlikely that child abuse in our own society is actually adaptive in the sense that it increases the abusive parent's reproductive success. Very few North Americans face the harsh necessity of killing a child in order to allow others to survive. However, it is plausible that the same psychological mechanisms that lead to infanticide in simpler, more resource-constrained societies may lead to child abuse in our own society. Evolutionary theory predicts that people have an evolved predisposition to cherish, protect, and provide for their own children. However, it also predicts that these bonds will weaken when families are stressed, when the child is sickly or handicapped, or when the child's paternity is uncertain. Evolutionary reasoning predicts that child abuse is most likely to occur in these circumstances.

Martin Daly and Margo Wilson applied this reasoning to an analysis of the factors that lead to child abuse in industrialized societies like the United States. Detailed, quantitative information about child abuse was difficult to come by when they began their work. They pored through descriptions of child homicides in police files, and combed statistical reports on child abuse maintained by governmental agencies and private organizations. They looked for records that would provide information relevant to evolutionary hypotheses about the conditions that lead to mistreatment and abuse of children, such as current parental circumstances and the abused child's relationship to members of its household. In the end, they were able to draw several important conclusions from the data.

Children living in households with a biological parent and an unrelated adult are at much higher risk of child abuse than are children living with both biological parents.

Cross-cultural research indicates that infanticide often occurs when a child is sired by someone other than the mother's current husband. We have seen that this makes sense from an evolutionary point of view, since men are not expected to invest in other men's children. Similar factors influence the risk of child abuse in Western industrialized societies. Due to divorce and other factors, a substantial number of children in the United States and other industrialized nations live in households with an unrelated **substitute parent**, such as a stepparent. Adoptive parents are not included in this category. Daly and Wilson, and others, have found that the risk of abuse is much lower in households that include two biological parents than in households that contain one natural parent and a stepparent. The data in Figure 19.8 come from Canada in 1983, but data from the United States, a much more violent country than Canada, show essentially the same pattern.

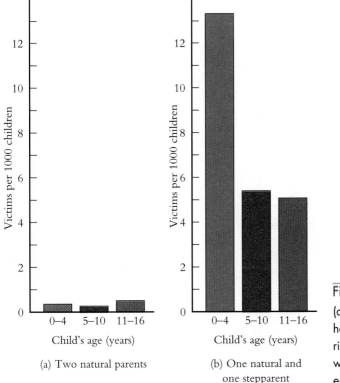

Figure 19.8

(a) In Canada, children living in households with two natural parents are at lower risk for child abuse than (b) children living with one natural parent and one stepparent at all ages.

(a) Two natural parents

(b) One natural and one stepparent

Evolutionary reasoning predicts that the substitute parent is usually responsible for the child abuse.

The statistics are silent about the identities and motives of the individuals responsible for abuse. However, evolutionary reasoning implicates the substitute parents. If, as Daly and Wilson suggest, evolution has predisposed us to form bonds with the children that we know intimately as infants, then it may be difficult for substitute parents to establish loving relationships with their partner's children. Unlike most biological parents and virtually all adoptive parents, substitute parents may be unprepared or reluctant to take on the responsibilities and expenses associated with parenthood. They may find their partner's children noisy, demanding, and irritating, and resent the time and money that their partners devote to them. Even the best-intentioned substitute parents may find it difficult to establish loving bonds with their partner's children, particularly if the substitute comes on the scene late in the child's life.

Adoption

In many societies children are adopted and raised by adults other than their biological parents.

Adoption is the flip side of infanticide. In some regions of the world, particularly Oceania and the Arctic, a substantial fraction of all children are raised in adoptive households. Adoption can be thought of as a form of altruism, because it takes considerable time, energy, and resources to raise children. In other animal species, including other primates, voluntary extended care of others' offspring is uncommon. There are, however, examples of *in*voluntary extended care: some species of parasitic birds, like cuckoos, routinely lay their eggs in other birds' nests, tricking the unwitting hosts into rearing cuckoo young as their own (Figure 19.9). The

Figure 19.9

Cuckoos lay their eggs in the nests of other species of birds. The hosts unwittingly rear the alien chicks as their own.

fact that we label such behavior **nest parasitism** reflects the fact that this is a form of exploitation, enhancing the fitness of the parasitic cuckoo at the expense of its host. In human societies, there is no need to trick adoptive parents into caring for children, as they are typically eager to assume responsibility for other people's children.

ADOPTION IN OCEANIA

Adoption is very common in the societies of the Pacific Islands. The pattern of adoption in those societies is consistent with the predictions of evolutionary theory.

At first glance, adoption seems inconsistent with evolutionary theory, which predicts that people should be reluctant to invest time and energy in unrelated children. This apparent inconsistency prompted Marshall Sahlins, a cultural anthropologist at the University of Chicago, to cite adoption as an example of a human behavior that contradicts the logic of evolutionary theory. In *The Use and Abuse of Biology,* Sahlins pointed out that adoption is very common in the societies of the Pacific Islands. In fact, it is so common that the majority of households in many of these island societies include at least one adopted child. According to Sahlins, such widespread altruism toward nonkin shows that evolutionary reasoning does not apply to contemporary humans, and demonstrates that human societies are free to invent almost any social arrangements.

Sahlins's challenge prompted one of us (J. B. S.) to take a closer look at the pattern of adoption in a number of the societies of Oceania. The data tell a very interesting story. Adoption seems to provide an adaptive means for birth parents to regulate the sizes of their families and to enhance the quality of care they give to each of their offspring. Moreover, many of the features of adoption transactions in Oceania are consistent with predictions derived from kin selection theory:

• Adoptive parents are usually close kin, such as grandparents, aunts, or uncles (Figure 19.10). Close kin participate in adoptive transactions much more often than would be expected if children were adopted at random, without respect to kinship.

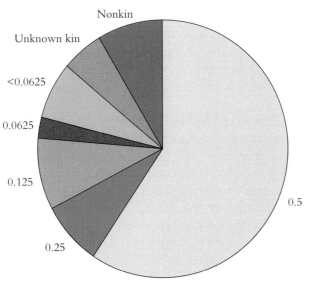

Figure 19.10

People in Oceania mainly adopt kin. Information about the numbers of adoptive parents and their relationships with their adoptive children in a number of communities in Oceania have been collected by ethnographers. Most adoptions involve close kin, such as aunts, uncles, and grandparents.

- Natural parents give up children when they cannot afford to raise them, and they rarely give up firstborn children for adoption.
- Adoptive parents generally have no dependent children; they are usually childless or the parents of grown children, but sometimes they are simply wealthy enough to be able to raise an additional child.
- Natural parents are often reluctant to give up their children for adoption, and regret the need to do so.
- Natural parents maintain contact with their children after adoption, and terminate adoptions if children are neglected, mistreated, or unhappy.
- Natural parents prefer to have their children adopted by well-to-do individuals who can provide adequately for their daily needs or bestow property on them.
- Sometimes asymmetries in investment exist between natural and adoptive children, as adopted children often inherit less property from their adoptive parents than their biological offspring do.

The same pattern characterizes adoptions in some traditional societies elsewhere.

Many of the same features characterize adoption in the North American Arctic, where many children are adopted. Moreover, adoption transactions in Oceania and the North American Arctic bear a striking similarity to more temporary fosterage arrangements common in West Africa. In each of these cases, birth parents seem to act in ways that increase the health, security, and welfare of their children, and care for children is preferentially delegated to close kin.

Adoption transactions seem to fit predictions derived from the theory of kin selection, but this does not mean that we have fully explained human adoption transactions. Not all adoption transactions in Oceania fit the evolutionary model. We have not accounted for the fact that adoption is more common in some societies than in others, or that there are some societies, including our own, in which adoption is not restricted to relatives. It is to this problem we now turn.

ADOPTION IN INDUSTRIALIZED SOCIETIES

In contemporary industrial societies, adoption often involves strangers.

As you read the description of adoption in the traditional societies of Oceania, you may have been struck by how different these transactions are from adoptions in the United States. In the United States and in other industrialized nations, adoption is often a formal, legal process involving strangers. However, as you compare the features of adoption transactions in Oceania and the United States in Table 19.3, you will see that there are similarities as well as differences between them.

In Oceania and in the United States, children are given up for adoption mainly when their parents are unable to raise them, and they are frequently adopted by people who have no dependent children of their own, or by people who can afford to raise additional children. In both Oceania and the United States, biological parents often give up their children with considerable reluctance, and hope to place their children in homes where their prospects will be improved.

The main difference between adoption in Oceania and in the United States is that in Oceania, adoption is a relatively informal, open transaction between close kin, while in the United States adoption is often a formal, legal transaction between strangers. This has a number of ramifications. When the identity of the biological

Table 19.3 The patterns of adoption in the United States and Oceania show both similarities and differences.

FEATURE	OCEANIA	UNITED STATES
Economic difficulties figure in the adoption decision.	Yes	Yes
Adoptive parents are usually childless or wealthy enough to raise another child.	Yes	Yes
Children are adopted by close kin.	Usually	Sometimes★
Biological parents often regret the need for adoption.	Yes	Yes
Legal authority for the child is transferred exclusively to the adoptive parents.	No	Yes
The identities of the natural parents are known to the adoptee.	Yes	No
Contact is maintained between biological parent and child after adoption.	Yes	No
Biological parents want children to be adopted by well-to-do people.	Yes	Yes
Asymmetries exist in the care of adoptive children and biological children.	Yes	?

★In the United States from 1952 to 1971, approximately one-quarter of adoptions involved kin, another quarter involved stepparents, and one-half involved unrelated individuals.

parents is not disclosed to the adoptive parents or the adopted child, contact between biological parents and their children is broken. In Oceania, adopted children's interests are protected by their biological parents, who maintain contact with them, and reserve the right to terminate the adoption if the children are mistreated or unhappy. In the United States, governmental agencies are responsible for protecting the interests of adopted children. In recent years, some of these differences have been eroded as open adoptions have become more common, and the courts have given adoptees the right to find out the identities of their birth parents.

The different patterns of adoption in Oceania and the United States may result from the same basic evolved psychological motivations.

Overall, it seems that the similarities in adoption are related to people's motivations about children; the differences are related to how the transactions are organized in each culture. In Oceania and the United States, people seem to share concerns about their children. We are deeply concerned about the welfare of our own children, and many people have deep desires to raise children. These feelings are likely to be the product of evolved psychological predispositions that motivate us to cherish and protect children. Although people may have very similar feelings about children in Oceania and the United States, the adoption process is clearly different. Nepotism (favoring relatives over nonrelatives) is a central element in adoption transactions in Oceania but not in the United States. People in Oceania rely on their families to help them find homes for their children and to obtain children, while people in the United States turn to adoption agencies, private attorneys, and strangers.

We are not sure why kinship plays such a fundamental role in adoption transactions in Oceania, but not in industrialized societies like the United States and Canada. We can speculate that it is due to differences in the availability of children who are eligible for adoption, or to the tendency of family members to be dispersed geographically. However, nepotism generally seems to play a more important role in the societies of Oceania than it does in industrialized societies.

Family Size

!Kung women deliberately space their children's births at long intervals and thus seem to have fewer children than they could during their reproductive years.

As we have repeatedly emphasized, natural selection favors adaptations that increase an individual's reproductive success. Parents are expected to produce as many offspring as they can. In many foraging societies, however, children are spaced carefully and born at relatively long intervals. The Dobe !Kung, one of the best-studied foraging groups, provide a good example. (Dobe is a place on the Botswana-Namibia border, and !Kung is a language group.) !Kung women give birth to their first child in their late teens and their last child in their mid-40s. The interval between births of surviving children is usually four years (Figure 19.11). Thus, during their 20-year reproductive careers, women give birth to about five children. If interbirth intervals were shortened, women would be able to have more children. From an evolutionary perspective, we must ask why the Dobe !Kung have children at such long intervals.

!Kung women space their births to avoid having to carry more than one young child.

The lives of Dobe !Kung women are described in rich detail by Richard Lee of the University of Toronto. The Dobe !Kung live in the vast reaches of the Kalahari Desert, and subsist on a variety of plant foods and wild game. In this

Figure 19.11

Among the !Kung, children are spaced about four years apart. This woman is pregnant and is flanked by her husband and two children.

Figure 19.12

The !Kung now live in the Kalahari Desert, a harsh and inhospitable habitat. (Photograph courtesy of Nicholas Blurton Jones.)

inhospitable terrain (Figure 19.12), women work hard to provide food for themselves and their families. The bulk of the calories in the Dobe !Kung diet come from plant foods gathered mainly by the women. During the driest part of the year, the Dobe !Kung camp near a few widely spaced water holes. The women travel long distances in search of food, foraging in mongongo nut groves that may be up to 10 km (6 miles) from their camps and water holes. The Dobe !Kung do not have donkeys or horses, so women forage on foot and carry loads on their backs. When women go out to forage, they take their children with them. They carry their babies for the entire journey, and children up to the age of four need to be carried at least part of the way (Figure 19.13). Children obtain very little of their own food until they reach their teens. Some of the children's calories come from plant foods gathered by their mothers, and some from food brought back and shared by hunters. Overall, women provide about 60% of their children's calories until the offspring reach the age of 15 years.

Nicholas Blurton Jones of the University of California, Los Angeles, has tried to determine why the !Kung have so few children. Following up on Lee's observation that carrying children is very hard work, Blurton Jones hypothesized that for Dobe !Kung women, interbirth intervals may be limited by the weight of the

(a)   (b)

Figure 19.13

Women carry heavy loads of nuts and other plant foods back to camp to share with their families. (a) A woman fills her sack with nuts, and then (b) her toddler climbs on top. (Photographs courtesy of Nicholas Blurton Jones.)

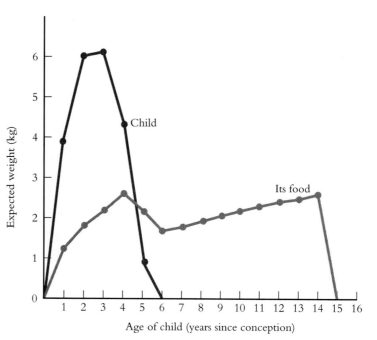

Figure 19.14

Loads carried by !Kung women decrease with the child's age. There are two components to the mother's load: The load due to the weight of the child is plotted in red, and the weight of the child's food is plotted in blue.

loads they can carry on their backs. Since children cannot walk long distances until they are four years or older, women who had children at short intervals would have to carry more than one child. One way for the women to restrict their loads to manageable limits would be to extend their interbirth intervals.

The dynamics of the problem are portrayed in Figure 19.14. The graph shows that the load a woman must carry has two components: the weight of the child and the weight of the food the child eats. From the time of a child's birth until the child is four years of age, the weight of the mother's load increases, because the child gets heavier as it gets older. As the child matures, it begins walking longer distances on its own, thereby lightening the mother's load. Of course, as children grow, they need more food. Hence, the weight of food carried back for children increases as they get older. However, children weigh considerably more than the food required to sustain them each day. By combining the weight of the child with the weight of the food carried back to feed the child, Blurton Jones calculated the total weight of the load that a woman must carry to support her child. A woman with a two-year-old child carries about 7 kg (15 lb), while a woman with a mobile six-year-old child carries less than 2.25 kg (about 5 lb).

These data can then be used to compare the weight of women's loads if they have children at different intervals (Figure 19.15). When women give birth at short intervals (such as every two years), their loads are very heavy (more than 22.5 kg, or 50 lb) because they must carry two children, along with the food needed to feed them both. Women who give birth at longer intervals have lighter loads because their older children don't need to be carried as much of the time. It is important to notice that while women's loads become steadily lighter as inter-birth intervals are extended to four years, the loads do not become appreciably lighter after this point. Thus, women may space their children at four-year intervals because this is the spacing that minimizes women's loads and maximizes their reproductive rate. The average interbirth interval is four years, just as the model predicts. Dobe !Kung women are actually having children as often as they can, given the immense effort associated with child care.

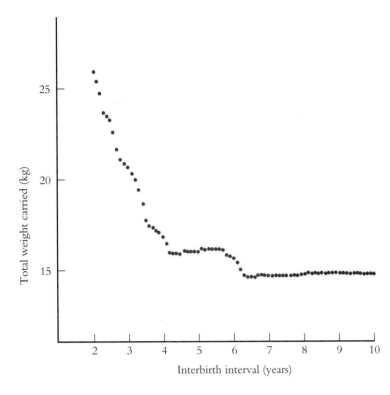

Figure 19.15

The total load carried by !Kung women is minimized when births are spaced at approximately four-year intervals. When interbirth intervals are shorter, women are forced to carry heavy loads. When interbirth intervals are longer than four years, women do not appreciably lighten their load.

Blurton Jones's model is supported by the fact that !Kung women who do not need to forage have shorter interbirth intervals.

One prediction of the "back-load" model is that when a Dobe !Kung woman's workload is reduced, her interbirth intervals will decline. There is evidence that this is the case. Figures 19.14 to 19.16 are based on women who live in the bush and rely on foraging for subsistence. However, some Dobe !Kung families have settled at cattle posts (as cattle-raising operations are called in southern Africa) where they work for their Herero and Tswana neighbors. Women who live at the cattle posts do not forage in the bush as often and do not have to carry their children over long distances. As the model would predict, Dobe !Kung women living at cattle posts have much shorter interbirth intervals than foraging women. Another implication of the back-load model is that women who live in the bush and have children at shorter intervals will reproduce less successfully than women who have children at longer intervals. In the bush, infants born after relatively short interbirth intervals are less likely to survive than those born after longer intervals.

This evolutionary explanation of birth spacing leaves unexplained many factors that affect birth spacing.

The analysis of !Kung birth spacing takes as given many features of !Kung society described by Lee and others. For example, it assumes that children do not acquire very much of their own food, that men don't contribute very much food to the family pot, and that women don't cooperate more in caring for each other's children. If any one of these features was changed, !Kung women might be able to have more children. Moreover, it is known that some of these features vary cross-culturally. For example, Blurton Jones has studied a second group of foragers,

Figure 19.16

Hadza children gather much of their own food from an early age, perhaps because they can find food close to their camps. Here several children draw water out of cracks in the rocks. (Photograph courtesy of Nicholas Blurton Jones.)

the Hadza of Tanzania. Hadza children acquire a much larger amount of their own food than do !Kung children (Figure 19.16). It may be that each of these features has an adaptive explanation. For example, Blurton Jones has suggested that Hadza children forage more than !Kung children do because there is much more food in the immediate vicinity of Hadza camps. On the other hand, these features may be maintained as the result of culturally transmitted norms, and people with a different cultural history might behave quite differently under the same conditions.

Is Human Evolution Over?

This question is often raised by students in our courses, and it seems to be a sensible question to consider as we come to the end of the story of human evolution. As we have seen, modern humans represent the product of millions of years of evolutionary change. Of course, so do cockroaches, peacocks, and orchids. All of the organisms that we see around us, including people, are the products of evolution, but they are not finished products. They are simply works-in-progress.

There is a sense, however, in which human evolution is over. Because cultural change is much faster than genetic change, most of the changes in human societies since at least the origin of agriculture, almost 10,000 years ago, have been the result of cultural, not genetic evolution. Most of the evolution of human behavior and human societies is not driven by natural selection and the other processes of organic evolution. It is driven by learning and other psychological mechanisms that shape cultural evolution. However, this fact does not mean that evolutionary theory or human evolutionary history is irrelevant to understanding contemporary human behavior. Natural selection has shaped the physiological mechanisms and psychological machinery that govern learning and other mechanisms of cultural change, and an understanding of human evolution can provide important insights into human nature and the behavior of modern peoples.

The Relationship between Science and Morality

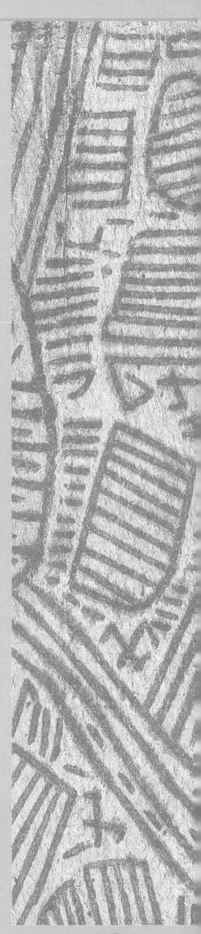

Recently, after giving a broadcast on Charles Darwin, I received through the post a pamphlet, 'Why are there "Gays" at all? Why hasn't evolution eliminated "Gayness" millions of years ago?', by Don Smith. The pamphlet points to a genuine problem; the prevalence of sexually ambiguous behaviour in our species is not understood, and is certainly not something that would be predicted from Darwinian theory. Smith's motive for writing the pamphlet was as follows. He believes that the persecution of gays has been strengthened and justified by the existence of a theory of evolution which asserts that gays are of low fitness because they do not reproduce their kind. He also believes that gays can only be protected from future persecution if it can be shown that they have played an essential and creative role in evolution.

I do not find the evolutionary theory he offers in the place of Darwinism particularly persuasive, although it is neither dull nor silly. However, that is not the point I want to make here. I think he would have been better advised to say:

> If people have despised gays because gayness does not contribute to biological fitness, they have been wrong to do so. It would be as sensible to persecute mathematicians because an ability to solve differential equations does not contribute to fitness. A scientific theory—Darwinism or any other—has nothing to say about the value of a human being.

The point I am making is that Smith is demanding of evolutionary biology that it be a myth; that is, that it be a story with a moral message. He is not alone in this. Elaine Morgan recently wrote an account of the origin of *H. sapiens* intended to dignify the role of women and of the mother–child bond (a relationship about which Don Smith is silent). Earlier, Shaw wrote *Back to Methuselah* avowedly as an evolutionary myth, because he found in Darwinism a justification of selfishness and brutality, and in Lamarckism a theory which justified free will and individual endeavour.

We should not be surprised at Don Smith, Elaine Morgan and Bernard Shaw. In all societies men have constructed myths about the origins of the universe and of man. The function of these myths is to define man's place in nature, and thus to give him a sense of purpose and value. Darwinism is, among other things, an account of man's origins. Is it to be wondered at that it is expected to carry a moral message?. . .

. . . We do not find it easy to distinguish science and myth. One reaction to this difficulty is to assert that there *is* no difference. Evolution theory has no more claim to objective truth than Genesis. Many scientists would be enraged by such an assertion, but rage is no substitute for argument. In the last century, it was widely held that the scientific method, conceived of as establishing theories by induction from observation, led to certain knowledge. If that were so, then there would indeed be a way to distinguish science from myth, because the truth of a myth certainly cannot be established by induction. However, Darwin and Einstein have robbed us of that confidence in induction—or have liberated us from that prison. By establishing the mutability of species, Darwin showed that there is not a fixed and finite number of kinds of things in the universe, each with a knowable essence; there is no 'Platonic idea' for each species. But induction can only lead to certainty if there is a finite and knowable set of objects, so that one can check that one's theory is true of every kind. If Darwin demonstrated the impossibility of acquiring certainty through induction, Einstein showed that what scientists had been most certain of—classical mechanics—was at worst false and at best a special case of a more general theory. After this twin blow, sure and certain knowledge is something we can expect only at our funerals.

But it is one thing to say that scientific knowledge cannot be certain, and quite another that there is no difference between science and myth. . . .

. . . Indeed, they have much in common. Both are constructs of the human mind, and both are intended to have a significance wider than the direct assertions they contain. . . . It is the function of a scientific theory to account for experience—often, it is true, the rather esoteric experience emerging from deliberate experiment. It is the function of a myth to provide a source and justification for values. What should be the relation between them?

Three views are tenable. The first, sometimes expressed as a demand for 'normative science', is that the same mental constructs should serve both as myths and as scientific theories. It is widely held. If I am right, it underlies the criticisms of Darwinism from gays, from the women's movement, from socialists, and so on. It explains the preference expressed by some churchmen for 'big bang' as opposed to 'steady state' theories in cosmology. Although well-intentioned, it seems to me pernicious in its effects. If we insist that scientific theories convey moral messages, the result will be bad morality or bad science, and most probably both. The danger is most apparent in evolutionary biology. Darwinism *is* an account of human origins. In all previous cultures, accounts of origins have been myths, so that it is to be expected that people will treat Darwinism as a myth. If one accepts the Darwinian account, then it is easy to equate 'natural' with 'successful in the struggle for existence'; if one treats the account as a myth, it is equally easy to equate 'natural' with 'right'. The consequence is that people either embrace Darwinism and draw from it the conclusion that gays are unnatural, social services impolitic, and charity wicked, or they are so disgusted by these conclusions that they embrace Lamarckism whether or not the evidence supports it. The first choice is bad morality and the second is bad science. There is no escape from this dilemma, so long as we insist on treating scientific theories as if they were myths. However difficult it may be to convince ourselves and others that 'natural' is *not* equivalent to 'right', the attempts must be made.

The second view is that we should do without myths and confine ourselves to scientific theories. This is the view I held at the age of twenty, but it really won't do. If, as I believe, scientific theories say nothing about what is right, but only about what is possible, we need some other source of values, and that source has to be myth in the broadest sense of the term.

The third view, and I think the only sensible one, is that we need both myths and scientific theories, but that we must be as clear as we can which is which. . . . To do science, one must first be committed to some values— not least to the value of seeking the truth. Since this value cannot be derived from science, it must be seen as a prior moral commitment, needed before science is possible.

Source: From pp. 39–40, 42–43, and 49–50 in J. M. Smith, 1989, *Did Darwin Get It Right? Essays on Games, Sex and Evolution*, Chapman & Hall, New York.

Further Reading

Blurton Jones, N. 1989. The costs of children and the adaptive scheduling of births: towards a sociobiological perspective on demography. Pp. 265–282 in *The Sociobiology of Sexual and Reproductive Strategies,* ed. by A. E. Rasa, C. Vogel, and E. Voland. Chapman & Hall, London.

Borgerhoff Mulder, M. 1988. Kipsigis bridewealth payments. Pp. 65–82 in *Human Reproductive Behavior,* ed. by L. Betzig, M. Borgerhoff Mulder, and P. Turke. Cambridge University Press, Cambridge.

Buss, D. 1994. *The Evolution of Desire.* Basic Books, New York.

Daly, M., and M. Wilson. 1988. *Homicide.* Aldine de Gruyter, Hawthorne, N.Y.

Levine, N. E. 1988. *The Dynamics of Polyandry.* University of Chicago Press, Chicago.

Silk, J. B. 1990. Human adoption in evolutionary perspective. *Human Nature* 1:25–52.

Study Questions

1. In Chapter 7 we said that the reproductive success of most male primates depends on the number of females with which they mate. Here Buss argues that a man's reproductive success will depend mainly on the health and fertility of his mate. Why are humans different from most primates? Among what other primate species should we expect males to attend to the physical characteristics of females when choosing mates?

2. Why should men value fidelity in prospective mates more than women do?

3. In Buss's cross-cultural survey, what was the most important attribute in a mate for both men and women? Does this result falsify his evolutionary reasoning?

4. Are Borgerhoff Mulder's observations about the Kipsigis consistent with Buss's cross-cultural results? Explain why or why not.

5. Explain why polyandry is so rare among mammalian species.

6. Explain why individual organisms may sometimes increase their fitness by killing their own offspring. What does your reasoning predict about the contexts in which infanticide will occur?

7. Newborn babies attract an incredible amount of attention. Complete strangers stop to chuck them under the chin, coo at them, and comment on how cute they are. If natural selection has favored discriminative parental solicitude, as the data on infanticide and child abuse seem to suggest, how can you explain the unselective attraction to newborn infants?

8. Why is adoption altruistic from an evolutionary perspective? How can adopting other people's babies be adaptive?

9. Explain why !Kung women have so few children.

There Is Grandeur in This View of Life . . .

Here we end our account of how humans evolved. As we promised in the Prologue, the story has not been a simple one. We began, in Part One, by explaining how evolution works—how evolutionary processes create the exquisite complexity of organic design, and how these processes give rise to the stunning diversity of life. Next, we used these ideas in Part Two to understand the ecology and behavior of nonhuman primates—why they live in groups, why the behavior of males and females differs, why animals compete and cooperate, and why primates are so smart compared to other kinds of animals. Then in Part Three we combined our understanding of how evolution works and our knowledge of the behavior of other primates with information gleaned from the fossil record to reconstruct the history of the human lineage. We traced each step in the transformation from a shrewlike insectivore living at the time of dinosaurs; to a monkeylike creature inhabiting in the Oligocene swamps of northern Africa; to an apelike creature living in the canopy of the Miocene forests; to the small-brained, bipedal hominids who ranged over Pliocene woodlands and savannas; to the large-brained and technically more skilled *Homo erectus,* who migrated to most of the Old World; and finally, to creatures much like ourselves who created spectacular art, constructed simple structures, and hunted large and dangerous game just 100,000 years ago. In the final part of the book, we turned to look at ourselves—to assess the magnitude and significance of genetic variation in the human species, to ask why we grow old and die, and to consider how we choose our mates and how we raise our children.

Evolutionary analyses of human behavior are not always well received. In Darwin's day, many were deeply troubled by the implications of his theory. One Victorian matron, informed that Darwin believed humans to be descended from apes, is reported to have said, "Let us hope that it is not true, and if it is true, that it does not become widely known." Darwin's theory profoundly changed the way we see ourselves. Before Darwin, most people believed that humans were fundamentally different from other animals. Human uniqueness and human superiority were unquestioned. But we now know that all aspects of the human phenotype are products of organic evolution, exactly the same processes that create the diversity of life around us. Nonetheless, many people still feel that we diminish ourselves by explaining human behavior in the same terms we use to explain the behavior of chimpanzees or soapberry bugs or finches.

651

Figure 1

Charles Darwin died in 1882 and was buried in Westminster Abbey beneath the monument to Isaac Newton.

In contrast, we think the story of human evolution is breathtaking in its grandeur. With a few simple processes, we can explain how we arose, why we are the way we are, and how we relate to the rest of the universe. It is an amazing story. But perhaps Darwin (Figure 1) himself put it best in the final passage of *On the Origin of Species:*

> It is interesting to contemplate an entangled bank, clothed with many plants of many kinds, with birds singing on the bushes, with various insects flitting about, and with worms crawling through the damp earth, and to reflect that these elaborately constructed forms, so different from each other, and dependent on each other in so complex a manner, have all been produced by laws acting around us. These laws, taken in the largest sense, being Growth with Reproduction; Inheritance which is almost implied by reproduction; Variability from the indirect and direct action of the external conditions of life, and from use and disuse; a Ratio of Increase so high as to lead to a Struggle for Life, and as a consequence, Natural Selection, entailing Divergence of Character and the Extinction of less-improved forms. Thus, from the war of nature, from famine and death, the most exalted object which we are capable of conceiving, namely, the production of the higher animals, directly follows. There is grandeur in this view of life, with it several powers having been originally breathed into a few forms or only one; and that, whilst this planet has gone cycling on according to the fixed law of gravity, from so simple a beginning endless forms most beautiful and most wonderful have been, and are being evolved (p. 490 in C. Darwin, 1964 (1859), *On the Origin of Species,* 1st ed., Harvard University Press, Cambridge, Mass.).

Appendix
The Skeletal Anatomy of Primates

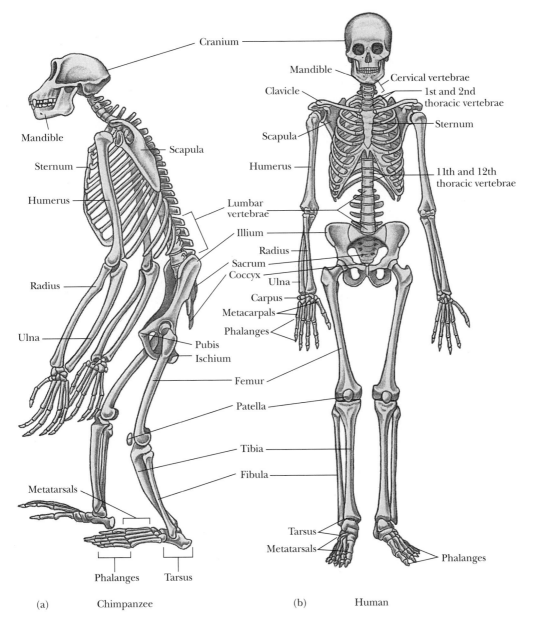

Cranium

Mandible

Clavicle

Scapula

Humerus

Lumbar vertebrae

Illium

Radius

Sacrum

Coccyx

Carpus

Metacarpals

Phalanges

Femur

Patella

Tibia

Fibula

Tarsus

Metatarsals

Phalanges

Mandible

Cervical vertebrae

1st and 2nd thoracic vertebrae

Sternum

11th and 12th thoracic vertebrae

Ulna

Pubis

Ischium

Mandible

Sternum

Humerus

Radius

Ulna

Scapula

Metatarsals

Phalanges   Tarsus

(a)   Chimpanzee

(b)   Human

Glossary

abductors   muscles whose contraction moves a limb away from the midline of the body. The abductors that connect the pelvis to the femur act to keep the body upright during bipedal walking.

Acheulean   a tool industry found at sites dated at 0.3 to 1.5 mya and associated with *Homo erectus* and some archaic *Homo sapiens*. Named after the French village of St. Acheul, where it was first discovered, the Acheulean industry is dominated by teardrop-shaped hand axes and blunt cleavers.

achondroplasia   a genetic disease caused by a dominant gene that leads to the development of short stature and disproportionately short arms and legs.

actor   in models of social behavior, the individual performing the behavior. *See also* recipient.

adaptation   a feature of an organism created by the process of natural selection.

adaptive radiation   the process in which a single lineage diversifies into a number of species, each characterized by distinctive adaptations. The diversification of the mammals at the beginning of the Cenozoic era is an example of an adaptive radiation.

adenine   one of the four bases of the DNA molecule. The complementary base of adenine is thymine.

affiliative   friendly.

alkaloids   secondary compounds produced and kept in plant tissues to make the plant distasteful or even poisonous to herbivores.

allele   one of two or more alternative forms of a gene. For example, the *A* and *S* alleles are two forms of the gene controlling the amino acid sequence of one of the subunits of hemoglobin.

allele frequency   *see* gene frequency.

alliances   interactions in which two or more animals jointly initiate against, or respond to aggression from, one or more other animals. Also called **coalitions**.

allopatric speciation   speciation that occurs when two or more populations of a single species are geographically isolated from each other and then diverge to form two or more new species.

altruism (altruistic, adj.**)**   behavior that reduces the fitness of the individual performing the behavior (the actor) but increases the fitness of the individual affected by the behavior (the recipient).

amino acids   molecules that are linked in a chain to form proteins. There are 20 different amino acids that share the same molecular backbone, but have a different side chain attached.

aminoacyl-tRNA synthetases   enzymes that link amino acids to the appropriate varieties of tRNA as part of protein synthesis.

analogy (analogous, adj.**)**   traits that are similar because of convergent evolution, not common descent. For example, the fact that humans and kangaroos are both bipedal is an analogy, not a homology. *See also* homology.

anatomically modern *Homo sapiens*   Human beings that first appear in the fossil record 100,000 years ago. Their bodies share some important derived features with present-day humans but not always the cultural traditions, symbolic behavior, and complex technologies of later peoples. *See also* archaic *Homo sapiens,* Neanderthals.

ancestral trait   a trait that appears earlier in the evolution of a lineage or clade. Ancestral traits are contrasted with derived traits, which appear later in the evolution of a lineage or clade. For example, the presence of a tail is ancestral in the primate lineage, while the absence of a tail is derived. Systematists must avoid using ancestral similarities when constructing phylogenies.

angiosperms   the flowering plants. The radiation of the angiosperms during the Cretaceous period may have played an important role in the evolution of the primates.

antagonistic pleiotropy   the condition in which a gene has a positive effect on one component of fitness and a negative effect on some other component of fitness. For example, a gene might increase fecundity early in life but reduce the probability of survival later in life. Antagonistic pleiotropy is the basis of one explanation for the evolution of senescence.

anthropoid   a member of the primate suborder that includes the monkeys and apes. The only other primate suborder (the prosimians) includes lemurs, lorises, and tarsiers.

anticodon   the sequences of bases on a transfer RNA molecule that binds complementarily to a particular codon. For example, for the codon ATC the corresponding anticodon is TAG because A binds to T and G to C.

apatite crystals   a crystalline material found in tooth enamel.

arboreal   predominantly active in trees.

archaeologist   a scientist who studies the material remains of past cultures and peoples.

archaic *homo sapiens*   hominids with larger brains and more modern crania that appear in the fossil record about 500 kya in Africa and Europe, and somewhat later in eastern Asia.

argon-argon dating   a sophisticated variant of the potassium-argon dating method that allows very small samples to be accurately dated.

atlatl   a spear-thrower—a device that lengthens the arm and allows the spear to be thrown with much greater force.

atresia   the decline in the number of viable germ cells that occurs as a woman gets older.

attribution   the ability to form a mental model of the mind of another individual. For example, adult humans understand that other people may have different knowledge about the world than they do. Small children do not have this ability.

australopithecines   a subfamily (Australopithecinae) consisting of the single genus *(Australopithecus)* of extinct hominids that lived from 4.2 mya to about 1 mya. Australopithecines were characterized by bipedal locomotion, robust teeth and jaws, and ape-sized brains. Most taxonomists recognize six species: *A. anamensis, A. afarensis, A. africanus, A. aethiopicus, A. robustus,* and *A. boisei.*

bachelor males   males who have not been able to establish residence in bisexual groups. Bachelor males may live alone or reside in all-male groups.

balanced polymorphism   a steady state in which two or more alleles coexist in a population. This state occurs when heterozygotes have a higher fitness than any homozygote.

basal metabolic rate   the rate of energy use required to maintain life when an animal is at rest.

bases   one of four molecules—adenine, guanine, cytosine, and thymine—that are bound to the DNA backbone. Different sequences of bases encode the information necessary for protein synthesis.

basic color term   a single word denoting color that is known by most speakers of a given language (for example, *blue* but not *cerulean*), can be used for all objects (*yellow* but not *blond*), and is not contained within another color term (*red* but not *crimson*).

basicranium   the base or underside of the cranium.

biface   a flat stone tool made by working both sides of a core until there is an edge along the entire circumference. *See also* hand ax.

bilaterally symmetric   describes an animal when the morphology on one side of its midline is a mirror image of the morphology on the other side.

binocular vision   when both eyes can focus together on a distant object to produce three-dimensional images. *See also* stereoscopic.

biochemical pathways   the chains of chemical reactions by which organisms regulate their structure and chemistry.

biological species concept   the concept that species are defined as a group of organisms that cannot interbreed in nature. Adherents of the biological species concept believe that the resulting lack of gene flow is necessary to maintain differences between closely related species. *See also* ecological species concept.

bipedal   walking upright on two (hind) legs.

blades   stone tools made from flakes that are at least twice as long as they are wide. Blades dominate the tool traditions of the Upper Paleolithic.

blending inheritance   a model of inheritance, widely held during the 19th century, in which the hereditary material of the mother and father was thought to blend irreversibly together in the offspring.

bottleneck   a severe but temporary reduction in population size that reduces the amount of genetic variation present in the population.

brachiation   a form of movement in which the body is propelled by the arms alone with a phase of free flight between handholds. True brachiation is found only in gibbons and siamangs.

bridewealth   valuable items transferred from the groom's family to the bride's family at the time of marriage.

Broca's area   a region of the human brain located in the left hemisphere. People who suffer damage to Broca's area often have difficulty with syntax.

camera-type eye   an eye in which light passes through a transparent opening and is then focused by a lens on photosensitive tissue. Camera-type eyes are found in vertebrates, mollusks, and some arthropods.

canalized   describes a trait that is very insensitive to environmental conditions during development, resulting in similar phenotypes in a wide range of environments. *See also* plastic.

canine   sharp, pointed tooth that lies between the incisors and the premolars in primates.

carbon-14 dating   a dating method based on an unstable isotope of carbon with atomic weight of 14. Carbon-14 is produced in the atmosphere by cosmic radiation and is taken up by living organisms. After organisms die the carbon-14 present in their bodies decays to the stable isotope nitrogen-14 at a constant rate. By measuring the ratio of carbon-14 to the stable isotope of carbon (carbon-12) in organic remains, scientists can estimate the length of time that has passed since the organism died. Also called **radiocarbon dating**.

carbohydrates   certain organic molecules with the formula $C_nH_{2n}O_n$ including common sugars and starches.

catalysis   the process by which a catalyst increases the rate at which a chemical reaction occurs.

catalyst   a substance that increases the rate at which a particular chemical reaction occurs at a given temperature. In organisms, catalysts are proteins called enzymes.

cathemeral   active both during the day and at night.

character   a trait or attribute of the phenotype of an organism.

character displacement   the result of competition between two species that causes the members of different species to become morphologically or behaviorally more different from each other.

Châtelperronian   an Upper Paleolithic tool industry found in France and Spain that dates from 36,000 to 32,000 years ago and is associated with Neanderthal fossil remains.

chopper   a simple tool made by removing a few flakes from a stone to produce an edge. Choppers dominate the Oldowan tool industry.

chromosomes   linear bodies in the cell nucleus that carry genes and that appear during cell division. Staining cells with dyes reveals that different chromosomes are marked by different banding patterns.

cladistic systematics   a system for classifying organisms in which patterns of descent are the only criteria used in classification. *See also* evolutionary systematics.

cleaver   a biface stone tool with a broad, flat edge. Cleavers are common at Acheulean sites.

coalitions   *see* alliances.

codon   a sequence of three DNA bases on a DNA molecule that comprises one "word" in the message used to create a specific protein. There are 64 different codons. *See also* anticodon.

coefficient of relatedness   an index measuring the degree of genetic closeness between two individuals. The index ranges from 0 (for no relation) to 1 (which occurs only between an individual and itself, or between identical twins). For example, the coefficient of relatedness between an individual and its parents or its siblings is 0.5.

cognitive map   a mental representation of the location of objects in space and time that allows for efficient navigation.

comparative method   a method for establishing the function of a phenotypic trait by comparing different species.

compound eye   an eye in which the image is formed by a large number of discrete photoreceptors. Compound eyes are found in insects and other arthropods.

computed tomography   an X-ray technique that generates three-dimensional images.

conspecifics   members of the same species.

content-dependent reasoning   modes of reasoning that depend on what is being reasoned about. The laws of formal logic are supposed to apply to any subject. However, humans seem to reason differently when given problems that are logically identical but deal with different subjects.

continental drift   the movement over the surface of the globe of the immense plates of relatively light material that make up the continents.

continuous variation   phenotypic variation in which there is a continuum of types. Height in humans is an example of continuous variation.

convergence   the evolution of similar adaptations in unrelated species. The evolution of camera-type eyes in both vertebrates and mollusks is an example of convergence. *See also* analogy.

core   a piece of stone from which smaller flakes are removed. Cores and/or flakes may be useful tools.

cornea   the clear covering of the front of the eye that serves to admit light.

corpus luteum   a mass of cells that forms in a woman's ovary after ovulation that in turn produces the hormone progesterone.

correlated characters   traits that are statistically associated in a population. For example, arm length and leg length are correlated if people with long arms also tend to have long legs and vice versa. Correlated characters arise when particular genes affect multiple characters. *See also* pleiotropic effects.

correlated response   an evolutionary change in one character caused by selection on a second, correlated character. For example, selection favoring only long legs will also increase arm length if arm length and leg length are positively correlated.

cortex   the original, unmodified surface of a stone used to make stone tools.

cranium   the braincase and bones of the face. Together the cranium and the lower jaw (mandible) make up the skull.

creole   a gramatically complete new language that has arisen on plantations and in other situations in which people speaking many different languages live closely together. *See also* pidgin.

cross   in genetics, a mating between chosen parents.

crossing over   the exchange of genetic material between homologous chromosomes during meiosis. Crossing over causes recombination of genes carried on the same chromosome.

crural index   the ratio of the length of the shin bone (tibia) to the length of the thigh bone (femur).

culture   information stored in human brains that is acquired by imitation, teaching, or some other form of social learning and that is capable of affecting behavior or some other aspect of the individual's phenotype.

cusps   projections on the biting surface of molars and premolars.

cytoplasm   the material inside the cell, but outside of the nucleus.

cytosine   one of the four bases of the DNA molecule. The complementary base of cytosine is guanine.

deciduous teeth   teeth that are shed as part of normal development and replaced with other teeth. Commonly called "baby teeth" or "milk teeth."

dental arcade   the row of teeth on either the upper or lower jaw.

dental formula   the number of incisors, canines, premolars, and molars in the upper and lower jaws.

denticulate   a flake tool on which at least one edge has been chipped to produce a serrated edge—that is, an edge with a series of adjacent indentations.

deoxyribonucleic acid   *see* DNA.

derived trait   a trait that appears later in the evolution of a lineage or clade. Derived traits are contrasted with ancestral traits, which appear earlier in the evolution of a lineage or clade. For example, the absence of a tail is derived in the hominid lineage, while the presence of a tail is ancestral. Systematists seek to use derived similarities when constructing phylogenies.

development   all of the processes by which the single-celled zygote is transformed into a multicellular adult.

diastema   a gap between adjacent teeth.

diploid   refers to cells containing pairs of homologous chromosomes, in which one chromosome of each pair is inherited from each parent. Also refers to organisms whose somatic (body) cells are diploid; all primates are diploid.

discontinuous variation   phenotypic variation in which there are a discrete number of phenotypes with no intermediate types. Pea color in Mendel's experiments is an example of discontinuous variation.

discriminative parental solicitude   a tendency among parents to adjust their investment in offspring according to cues that predict (or did predict in past environments) the likelihood that the offspring will survive and reproduce.

diurnal   describes an animal that is active only during the day. *See also* nocturnal.

dizygotic twins   twins that result from the fertilization of two separate ova by two separate sperm. Dizygotic twins are no more closely related than other full siblings. *See also* monozygotic twins.

DNA (deoxyribonucleic acid)   the molecule that carries hereditary information in almost all living organisms. It consists of two very long phosphate-sugar backbones to which the bases adenine, cytosine, guanine, and thymine are bound. Hydrogen bonds between the bases bind the two strands together.

DNA hybridization   a laboratory technique in which the similarity in sequence between DNA taken from different sources is assessed by measuring the temperature at which the two types of DNA dissociate.

dominance the ability of one individual to intimidate or defeat another individual in a pairwise (dyadic) encounter. In some cases, dominance is assessed from the outcome of aggressive encounters, and in other cases dominance is assessed from the outcome of competitive encounters.

dominance hierarchy a ranking of individuals in a group that reflects their relative dominance.

dominance matrix a square table constructed to keep track of dominance interactions among a group of individuals. Usually, winners are listed down the left side and losers are listed across the top, and the number of times each individual defeats another is entered in the cells of the matrix. Individuals are ordered in the matrix so as to minimize the number of entries below the diagonal. This ordering is then used to construct the dominance hierarchy.

dominant describes an allele that results in the same phenotype whether in the homozygous or the heterozygous state.

dyadic describes an interaction that involves two individuals. Also called **pairwise**.

ecological species concept the concept that natural selection plays an important role in maintaining the differences between species, and that the absence of interbreeding between two populations is not a necessary condition for defining them as separate species. *See also* biological species concept.

EEA (environment of evolutionary adaptedness) the past environment(s) in which currently observed adaptations were shaped. For example, the psychological mechanisms that cause contemporary humans to overeat were likely shaped in an environment of evolutionary adaptedness in which overeating was rarely a problem.

electron-spin-resonance dating a technique used to date fossil teeth by measuring the density of electrons trapped in apatite crystals in teeth. This method is important for sites that are too young to be dated using potassium-argon dating (less than 500,000 years ago) and too old to be dated with carbon-14 dating (more than 40,000 years ago).

endangered species a species that is in danger of becoming extinct due to its small population size or destruction of its habitat

endocast a cast of the inside of the braincase.

endocranial volume the volume inside the braincase.

endometrium the lining of the uterus.

environmental covariation the effect on phenotypes that occurs when the environments of parents and offspring are similar: environmental covariation causes the phenotypes of parents and offspring to be similar, and so can falsely increase estimates of heritability.

environmental variation phenotypic differences between individuals that exist because individuals developed in different environments.

environment of evolutionary adaptedness *see* EEA.

enzyme a protein that serves as a catalyst. Enzymes can control the chemical composition of cells by causing some chemical reactions to occur much faster than others.

equilibrium a steady state in which either gene or genotypic frequencies do not change.

estradiol a form of estrogen.

estrus a period during the reproductive cycle of most mammals (and most primates) when the female is receptive to mating and is capable of conceiving.

eukaryotes organisms whose cells have cellular organelles, cell nuclei, and chromosomes. All plants and animals are eukaryotes. *See also* prokaryotes.

evolutionary systematics a system for classifying organisms in which both patterns of descent and patterns of overall similarity are used in classification. *See also* cladistic systematics.

exons segments of the DNA in eukaryotes that are translated into protein. Exons contrast with intervening introns, which are not translated into protein.

extant still living—a term applied to taxons, especially species.

extractive foods foods that are embedded in a matrix, encased in hard shells, or otherwise difficult to extract. Such foods require complicated, carefully coordinated techniques to process.

F_0, F_1, and F_2 generations a system for keeping track of generations in breeding experiments. The initial generation is called the F_0 generation, the offspring of the F_0 generation constitute the F_1 generation, and the offspring of the F_1 generation comprise the F_2 generation.

falciparum malaria a severe form of malaria. The sickle-cell allele for hemoglobin is common in West Africa because it confers resistance to falciparum malaria in the heterozygous state.

family a taxonomic level above genus but below order. Thus a family may contain several genera, and an order may contain several families. Humans belong to the family Hominidae, and the other great apes belong to the family Pongidae.

fecundity   the biological capacity to reproduce. In humans, fecundity may be greater than fertility (the actual number of children produced) when people limit family size.

female-bonded kin groups   cooperative groups of related primate females, usually resulting from female philopatry and male dispersal.

femur   the thigh bone.

fertility   the number of children produced by a person in a given period of time. Fertility may be less than fecundity if people limit family size.

fitness   a measure of an individual's genetic contribution to subsequent generations.

fixation   a state that occurs when all of the individuals in a population are homozygous for the same allele at a particular locus.

flake   a smaller chip of stone knocked from a larger stone core.

focal color   the shade chosen as the best exemplar of a basic color term. There is a direct relationship between focal colors and the physiology of color vision.

follicle-stimulating hormone (FSH)   a pituitary hormone that stimulates the development of estradiol by the follicle cells.

folivore (folivorous, adj.**)**   an animal whose diet emphasizes leaves.

food sharing   the practice of sharing food within a social group. Food sharing has been widely observed among contemporary human foragers but is relatively uncommon in other primate species.

foragers   people who subsist by gathering, fishing, and hunting. Until about 10,000 years ago all humans were foragers. Foragers are also called **hunter-gatherers.**

foramen magnum   the large hole in the bottom of the cranium through which the spinal cord passes.

fossil   traces of life more than 10,000 years old preserved in rock. Fossils can be mineralized bones, plant parts, impressions of soft body parts, or tracks.

founder effect   a form of genetic drift that occurs when a small population colonizes a new habitat and subsequently greatly increases in number. Random genetic changes due to the small size of the initial population are amplified by subsequent population growth.

fraternal polyandry   a marriage system in which two or more brothers marry a single wife. Fraternal polyandry is practiced among the Nyinba, a Tibetan-speaking group living in western Nepal.

frugivore (frugivorous, adj.**)**   an animal whose diet emphasizes fruit.

gametes   in animals, eggs and sperm.

gene   a segment of the chromosome that produces a recognizable effect on phenotype and segregates as a unit during gamete formation.

gene flow   the movement of genes from one population to another, or one part of a population to another, as the result of interbreeding.

gene frequency   the fraction of the genes at a genetic locus that are a particular allele (and thus the same as allele frequency). For example, in a population in which there are 250 AA individuals, 200 AS individuals and 50 SS individuals, there are 700 copies of the A allele and 300 copies of the S allele, and therefore the frequency of the S allele is 0.3.

genetic distance   a measure of the overall genetic similarity of individuals or species. The best estimates of genetic distance utilize large numbers of genes.

genetic drift   random change in gene frequencies due to sampling variation that occur in any finite population. Genetic drift is more rapid in small populations than in large populations.

genetic markers   genes whose position on a chromosome is known and can be used to locate other nearby, linked genes.

genetic variation   phenotypic differences between individuals that result from the fact that they have inherited different genes from their parents.

gene tree   a phylogenetic tree tracing the pattern of descent for a particular gene.

genotype   the combination of alleles that characterizes an individual at some set of genetic loci. For example, in populations with only the A and S alleles at the hemoglobin locus, there are three possible genotypes at that locus: $AA, AS,$ and $SS.$ (SA is the same as $AS.$)

genotypic frequency   the fraction of individuals in a population who have a particular genotype.

genome   all of the genetic information carried by the organism.

genus   a taxonomic category below the family and above the species. There may be several species in a genus, and several genera (the plural of genus) in a family.

germ cells   cells in reproductive organs that give rise to gametes.

gerontologist   a scientist who studies aging.

gestational hypertension   a form of hypertension (high blood pressure) that occurs during pregnancy. Gestational hypertension is associated with high birthweight, and may be caused by manipulation of the mother's body by the fetus. *See also* preeclampsia.

glucose   a simple sugar that is used as a medium of energy transport in many animals.

Gondwanaland   the more southerly of the two supercontinents that existed about 180 mya. Gondwanaland included the continental plates that now make up Africa, South America, Antarctica, Australia, New Guinea, Madagascar, and the Indian subcontinent. *See also* Laurasia.

grooming   the process of picking through hair to remove dirt, dead skin, ectoparasites, and other material; a common form of affiliative behavior among primates.

guanine   one of the four bases of the DNA molecule. The complementary base of guanine is cytosine.

grammar   all of the rules of meaning our brains use to interpret information provided by language.

gum   a sticky carbohydrate produced by some trees in response to physical damage. Gum is an important food for many primates.

gummivore (gummivorous, adj.**)**   an animal that eats predominately gum.

gymnosperms   a group of plants that reproduce without flowering. Modern gymnosperms include pines, redwoods, and firs.

haft   to attach a spear point, ax head, or similar implement to a handle. Hafting greatly increases the force that can be applied to the tool.

Hamilton's rule   a rule predicting that altruistic behavior among relatives will be favored by natural selection if $rb > c$, where r is the **coefficient of relatedness** between actor and recipient, b is the sum of the benefits of performing the behavior on the fitness of the recipient(s), and c is the cost, in decreased fitness of the donor, of performing the behavior.

hand ax   The most common type of biface stone tool found in Acheulean sites. It is flat and teardrop-shaped, with a sharp point at the narrow end.

haploid   a cell with only one copy of each chromosome. Gametes are haploid, as are the cells of some asexual organisms.

haplorhines   members of the group containing tarsiers and anthropoid primates. The system that classifies primates into haplorhines and strepsirhines is a cladistic alternative to the system used in this text, in which primates are divided into prosimians and anthropoids, and tarsiers are grouped with prosimians. *See also* strepsirhines.

Hardy–Weinberg equilibrium   the unchanging frequency of genotypes that results from sexual reproduction and in the absence of other evolutionary forces such as natural selection, mutation, or genetic drift.

HDC (high digestive capacity)   the ability to digest lactose in adulthood.

hemoglobin   a protein in blood that carries oxygen, including two α (alpha) and two β (beta) subunits.

heritability   the fraction of the phenotypic variation in the population that is the result of genetic variation.

heterozygous   refers to a diploid organism whose cells carry two different alleles for a particular genetic locus. Organisms that are heterozygous are called "heterozygotes."

hindlimb dominated   describes an animal that habitually carries most of its weight on its hindlimbs.

homeotherms   organisms that maintain constant body temperature via internal physiological mechanisms.

home range   the area in which an individual or a group of animals travels, feeds, rests, and socializes. Territorial species actively defend the borders of their home ranges.

hominids   any member of the family Hominidae, including all species of *Australopithecus* and *Homo*.

hominoids   members of the superfamily Hominoidea, which includes humans, all the living apes, and numerous extinct ape and humanlike species from the Miocene, Pliocene, and Pleistocene epochs.

homologous chromosomes   a pair of chromosomes in a diploid cell in which one member of the pair is derived from the father and one is derived from the mother.

homology (homologous, adj.**)**   traits that are similar because of common ancestry, not convergence. For example, the fact that gorillas and baboons are both quadrupedal is due to the fact that they are both descended from a quadrupedal ancestor. *See also* analogy.

homozygous   refers to a diploid organism whose cells carry two copies of the same allele at a single genetic locus. Organisms that are homozygous are called "homozygotes."

hormone   a chemical substance produced by the endocrine system that is transported through the bloodstream to another part of the body where it produces a particular physiological response.

human chorionic gonadatropin (hCG)   a hormone, secreted by the embryo during the early part of pregnancy, that stimulates the mother's body to produce progesterone.

human growth hormone (hGH)   a hormone with a number of effects on growth and that blocks the effect of insulin.

human placental lactogen (hPL)   a hormone secreted by the fetus that is similar to hGH and is suspected of blocking the effects of insulin in the mother's body.

human universals   features that characterize humans all over the world.

humerus   the bone in the upper part of the forelimb (arm).

hunter-gatherers   people who subsist by gathering, fishing, and hunting. Until about 10,000 years ago all humans were hunter-gatherers. Also called **foragers**.

hybrid zone   a geographic region where two or more populations of the same species or two different species overlap and interbreed. Hybrid zones usually occur at the habitat margins of the respective populations.

hyoid bone   a small bone in the neck, located between the larynx and the tongue, which provides attachment for tongue and neck muscles.

ilium   one of the three bones in the pelvis.

inbreeding   mating between closely related individuals. Also called "inbred mating." *See also* outbred mating.

incisors   the front teeth in mammals. In anthropoid primates, incisors are used for cutting, and there are two on each side of the upper and lower jaw.

independent assortment   refers to Mendel's discovery that each of the genes at a single locus on a pair of homologous chromosomes is equally likely to be transmitted when gametes (eggs and sperm) are formed. This is because during meiosis the probability that a particular chromosome enters a gamete is 0.5 and is independent of whether other nonhomologous chromosomes enter the same gamete. Thus, knowing that an individual received a particular chromosome from its mother (and thus a particular allele) tells you nothing about the probability that it received other, nonhomologous chromosomes from its mother.

infraorder   the taxonomic level between order and superfamily.

insectivore (insectivorous, adj.**)**   an animal whose diet consists mostly of insects.

insulin   a substance that is created by the pancreas and is involved in the regulation of blood sugar.

intersexual selection   a form of sexual selection in which females choose who they mate with. The result is that traits making males more attractive to females are selected for.

intrasexual selection   a form of sexual selection in which males compete with other males for access to females. The result is that traits making males more successful in such competition, like large body size or large canines, are selected for.

introns   segments of the DNA in eukaryotes that are not translated into protein. Introns contrast with intervening exons, which are translated into protein.

ischium   a bone in the pelvis.

isotope   a chemical element with the same atomic number and properties as another element but having a different atomic weight. Unstable isotopes spontaneously change into more stable isotopes. *See also* radioactive decay.

kibbutz (pl. **kibbutzim)**   agricultural settlements in Israel, usually organized under collectivist principles.

kin selection   a theory stating that altruistic acts will be favored by selection if the product of the benefit to the recipient and the degree of relatedness, *r*, between the actor and recipient exceeds the cost to the actor. *See also* Hamilton's rule.

knapper   a stone toolmaker.

knuckle walking   a form of quadrupedal locomotion in which, in the forelimbs, weight is supported by the knuckles, rather than the palm or outstretched fingers. Chimpanzees and gorillas are knuckle walkers.

LAC*P   the dominant allele in humans that leads to high digestive capacity for lactose in either the homozygous or heterozygous state.

LAC*R   the recessive allele in humans that leads to low digestive capacity for lactose in the homozygous state.

lactation (lactate, v.**)**   production of milk by the mammary glands in females; a period in which milk is produced for nursing offspring. A characteristic feature of mammals.

lactose   a sugar present in mammalian milk. Most mammals—including most humans—lose the ability to digest lactose as adults. *See also LAC*P* and *LAC*R*.

larynx   a structure in the front of the neck that is composed of pieces of cartilage bound together with ligaments and membranes that protect the opening of the windpipe (trachea). The larynx houses the vocal cords, prevents food and other items from entering the lungs, and controls the flow of air into and out of the lungs.

Later Stone Age (LSA)   the tool industries found in Africa that correspond in age and type to the Upper Paleolithic industries in Europe.

Laurasia   the more northerly of the two supercontinents that existed about 180 mya. Laurasia included what is now North America, Greenland, Europe, and parts of Asia. *See also* Gondwanaland.

LDC (low digestive capacity)   the inability to digest lactose in adulthood.

lek   an arena in which males defend small territories in which they mount displays to attract females for mating. Males compete for particular, often central, positions on the lek, and males that hold such positions attract the most mates.

leutenizing hormone (LH)   a hormone secreted by the pituitary gland. In females, LH stimulates ovulation, formation and maintenance of the corpus luteum, and production of the hormone progesterone. In males, LH stimulates production of the hormone androgen.

Levallois technique   a three-step toolmaking method used by Neanderthals. First the knapper makes a core having a precisely shaped convex surface. Then the knapper makes a striking platform at one end of the core. Finally, the knapper knocks a flake off the striking platform.

lexigrams   abstract symbols the bonobo Kanzi was taught to use for communicating with his human trainers.

life table   a mortality schedule for individuals of different ages in a population.

linkage   the tendency for alleles at certain loci to be transmitted to the same gamete as a unit because they are located on the same chromosome. The closer together two loci are, the more likely they are to be linked.

locomotion   form of movement, such as vertical clinging and leaping or brachiation.

locus (loci, pl.)   the position on a chromosome that is occupied by a particular gene.

loess   fine dust produced by glaciers. During the last glaciation, parts of Europe were covered with dunes of loess.

Lower Pleistocene   the period from 1.64 mya to 900 kya. The Pleistocene is divided into three parts: the Lower, Middle, and Upper Pleistocenes. The beginning of the Lower Pleistocene coincides with a sharp cooling of the earth's climate, and the end is marked by the growth of continental glaciers in Europe.

lunate sulcus   a prominent sulcus that is located on the visual cortex of both apes and modern humans. Some researchers believe the position of the lunate sulcus in australopithecines is evidence that their brains had the same proportions as the brains of modern humans.

macroevolution   evolution of new species, families and higher taxa. *See also* microevolution.

maladaptive   detrimental to fitness.

mandible   the lower jaw. *See also* maxilla.

marsupial   mammals that give birth to live young that continue their development in a pouch equipped with mammary glands. Marsupials include kangaroos and opposums.

mate guarding   a form of mating in which the male defends his mate after copulation to prevent other males from mating with her.

mating systems   forms of courtship, mating, and parenting behavior that characterize particular species or populations. An example is polygyny.

matrilineage   individuals related through the maternal line.

maxilla   the upper jaw. *See also* mandible.

meiosis   the process of cell division in which haploid gametes (eggs and sperm) are created.

memes   a term coined by Richard Dawkins to refer to units of cultural information (beliefs and values) transmitted by imitation and teaching.

menarche   first menstruation.

Mendel's principles   the principles derived from Gregor Mendel's plant-breeding experiments. The first principle states that the observed characteristics of organisms are jointly determined by two particles (genes), one inherited from the mother and one inherited from the father. The second principle states that each of these genes is equally likely to be transmitted when gametes are formed.

menopause   the time in a woman's reproductive career when menstrual cycles stop, usually around the age of 50 in the United States.

mental organs   a term coined by evolutionary psychologists to refer to special-purpose modules that solve particular, circumscribed problems such as keeping accounts in social exchange.

messenger RNA (mRNA)   a form of RNA that carries specifications for protein synthesis from DNA to the ribosomes.

microevolution   evolution of populations within a species. *See also* macroevolution.

Middle Pleistocene   the period from 900 kya to 137 kya. The Pleistocene is divided into three parts: the Lower, Middle, and Upper Pleistocenes. The beginning of the Middle Pleistocene coincides with the initial growth of continental glaciers in Europe and ends with the temination of the next-to-last glacial period.

Middle Stone Age (MSA)   the stone-tool industries of sub-Saharan Africa and southern and eastern Asia that existed 150,000 to 30,000 years ago. The MSA is the counterpart of the Middle Paleolithic (Mousterian) in Europe. The MSA industries varied, but flake tools were manufactured in all of them.

mineralization   the process by which organic material in the bones of dead animals is replaced by minerals from the surrounding rock, creating fossils.

minor marriage   a form of marriage, formerly widespread in China, in which children were betrothed in infancy and then raised together in the household of the prospective groom.

mismatch distribution   the distribution of pairwise genetic differences within a population, used by Rogers and Harpending to make inferences about human demographic history.

mitochondria (mitochondrion, sing.**)**   cellular organelles involved in basic energy processing. *See also* mitochondrial DNA.

mitochondrial DNA (mtDNA)   DNA in the mitochondria that is particularly useful for evolutionary analyses, for two reasons: (1) mitochondria are inherited only from the mother, and thus there is no recombination, and (2) mtDNA accumulates mutations at relatively high rates, thus serving as a more accurate molecular clock for recent changes (last few million years).

mitosis   the process of division of somatic (normal body) cells through which new diploid cells are created.

modern synthesis   an explanation for the evolution of continuously varying traits that combines the theory and empirical evidence of both Mendelian genetics and Darwinism.

molars   the broad, square back teeth that are generally adapted for crushing and grinding in primates. Anthropoid primates have three molars on each side of the upper and lower jaw.

molecular clock   the hypothesis that genetic change occurs at a constant rate and thus can be used to measure the time elapsed since two species shared a common ancestor. It is based on observed regularities in the rate of genetic change along different phylogenetic lines.

monogamy   a mating system in which one female forms a stable pair-bond with a single male. Monogamy is common among birds, but relatively rare among primates.

monozygotic twins   twins that result from the fertilization of one ova by a single sperm. Early in development the fertilized egg splits to create two zygotes. *See also* dizygotic twins.

morphology   (1) the form and structure of an organism; also, a field of study that focuses on the form and structure of organisms. (2) A part of grammar that governs the way that words are put together.

Mousterian   a stone-tool industry characterized by points, side scrapers, and denticulates but an absence of hand axes. The Mousterian is generally associated with Neanderthals in Europe.

Mousterian variants   variations in the Mousterian tool industry that were once believed to be the product of ethnically distinct groups. Now they are generally thought to represent changes in tool manufacture over time, local variation in the type of tasks performed, or regional differences in the pattern of wear and extent of retouching.

mRNA   *See* messenger RNA.

mtDNA   *See* mitochondrial DNA.

multiregional model   a model of the evolution of anatomically modern humans that holds that populations of archaic *Homo sapiens* throughout the world were linked by gene flow. This allowed humans to evolve from *Homo erectus* into anatomically modern *Homo sapiens* as a single species all over the Old World. However, there was considerable regional variation in morphology among populations.

mutation   a spontaneous change in the chemical structure of DNA.

mutation accumulation   a nonadaptive theory of aging that considers aging the result of the accumulation of mutations that affect only older individuals.

mutualism (mutualistic, adj.**)**   behavior that increases the fitness of both actor and recipient.

myometrium   the layer of smooth muscle that lies underneath the endometrium, the lining of the uterus.

natal group   the group into which an individual is born. In many primate species the females remain in their natal groups throughout their lives, while the males emigrate and join new groups.

natural selection the process that produces adaptation. Natural selection is based on three postulates: (1) the availability of resources is limited; (2) organisms vary in the ability to survive and reproduce; and (3) traits that influence survival and reproduction are transmitted from parents to offspring. When these three postulates hold, natural selection produces adaptation.

Neanderthals a form of archaic *Homo sapiens* found in western Eurasia from about 300 kya to about 30 kya. Neanderthals had large brains and elongated skulls with very large faces. They also were characterized by very robust bodies.

negatively correlated description of a statistical relationship between two variables, in which positive values of one variable tend to co-occur with negative values of the other variable. For example, the size and number of seeds produced by an individual plant are negatively correlated in some plant populations.

neocortex part of the cerebral cortex and generally thought to be most closely associated with problem solving and behavioral flexibility. In mammals, the neocortex covers virtually the entire surface of the forebrain.

neocortex ratio the size of the neocortex in relation to the rest of the brain.

nest parasitism a behavior seen in some bird species in which females lay eggs in the nests of other birds who then unwittingly raise the alien chicks as their own.

neutral theory the theory that describes genetic change that is caused only by mutation and drift.

niche the way of life, or "trade," of a particular species—what foods it eats and how the food is acquired.

NIDD *see* noninsulin-dependent diabetes.

nocturnal describes an animal active only during the nighttime. *See also* diurnal.

noninsulin-dependent diabetes (NIDD) a form of diabetes in which cells of the body do not respond properly to levels of insulin in the blood. NIDD is known to have a genetic basis.

nucleus the distinct part of the cell that contains the chromosomes. Eukaryotes (fungi, protozoans, plants, and animals) all have nucleated cells, while prokaryotes (bacteria) do not.

observational learning a form of learning in which animals observe the behavior of other individuals and thereby learn to perform a new behavior.

occipital torus a horizontal ridge at the back of the skull in *H. erectus* and archaic *H. sapiens*.

Oceania a region of the world that includes Polynesia, Melanesia, and Micronesia.

olfaction (adj. **olfactory)** the sense of smell.

opponent response cells two pairs of nerve cells that link the eye to the brain and determine the perception of a color's hue. One pair responds oppositely to blue and yellow stimuli, while the other pair responds oppositely to red and green stimuli.

opposable a property of the thumb or big toe that enables some primates (Old World monkeys and apes) to touch the tip of their thumb and forefinger together.

optimal foraging theory a theory that predicts how animals will forage for food based on the assumption that they maximize rate of energy intake or some other measure that is closely related to fitness.

orbits the bony sockets of the eye.

organelle a portion of the cell that is enclosed in a membrane and has a specific function; examples are mitochondria and the nucleus.

outbred mating a mating between unrelated individuals. *See also* inbreeding.

Out of Africa hypothesis holds that anatomically modern humans evolved in Africa sometime during the last 100,000 years and spread out from that continent to replace other existing hominids in Europe and Asia.

outgroups taxonomic groups that are related to a group of interest and can be used to determine which traits are ancestral and which are derived.

pairwise describes an interaction that involves two individuals. Also called **dyadic.**

paleontologist a scientist who studies fossilized remains of plant and animal species.

Pangaea the massive single continent that contained all of the earth's dry land until about 180 mya.

parapatric speciation a two-step process of speciation that occurs when (1) selection causes the differentiation of geographically separate, partially isolated populations of a species, and (2) then subsequently the populations become reproductively isolated as a result of reinforcement.

parent-offspring conflict conflicts that arise between parents and offspring over how much the parents will invest in their offspring. These conflicts stem from the opposing genetic interests of parents and offspring.

patrilateral parallel cousins   an anthropological term for the offspring of brothers.

perisylvian region   the region of the brain close to the sylvian fissure. Language production is strongly affected by damage to the perisylvian region of the left hemisphere in humans.

phenotype   the observable characteristics of organisms. Individuals with the same phenotype may have different genotypes.

phenotypic gambit   the research tactic used by evolutionary biologists to generate hypotheses about behavior by assuming that behavior is adaptive. Purposely neglected in these hypotheses are constraints, correlated characters, genetic drift, and other factors that lead to maladaptive outcomes.

philopatry   the tendency in some animals to remain in their natal (birth) groups throughout their lives. In many Old World monkey species, females are philopatric.

phonemes   the basic unit of speech perception. Phonemes are the smallest bits of sound that we recognize as meaningful elements of language. People who speak different languages recognize slightly different sets of phonemes.

phylogeny   the evolutionary relationships among a group of species, usually diagrammed as a "family tree."

pick   a triangular-shaped biface stone tool found in Acheulean sites.

pidgin   a simple language with little syntax that is created when people who speak different languages come into contact. *See also* creole.

placental mammal   a mammal that gives birth to live young that developed for a period of time in the uterus and were nourished by blood delivered to a placenta.

plastic   describes traits that are very sensitive to environmental conditions during development, so that different phenotypes are produced in different environments. *See also* canalized.

pleiotropic effects   phenotypic effects created by genes that influence multiple characters. *See also* correlated characters.

plesiadapiforms   a group of primatelike mammals that lived during the Paleocene (65–55 mya). Although the pleisadapiforms are not considered to have been primates by most paleontologists, they probably were similar to the earliest primates, who lived about the same time.

pneumatized   containing cavities filled with air.

poikilotherms   animals that do not regulate their body temperature internally.

polyandry   a mating system in which a single female forms a stable pair-bond with two different males at the same time. Polyandry is generally rare among mammals, but is thought to occur in some species of marmosets and tamarins.

polygyny   a mating system in which a single male mates with many females. This is the most common mating system among primate species.

polymerase chain reaction (PCR)   a technique used in molecular genetics to make copies of a rare DNA molecule. PCR is important for paleoanthropology because it allows analysis of very small amounts of DNA found in fossils.

population genetics   the branch of biology dealing with the processes that change the genetic composition of populations through time.

porphyria variegata   a genetic disease caused by a dominant gene in which carriers of the gene develop a severe reaction to certain anesthetics.

positively correlated   description of a statistical relationship between two variables, in which positive values of one variable tend to co-occur with positive values of the other variable. For example, the height and weight of individuals are positively correlated in human populations.

postcranial   the skeleton excluding the skull.

potassium-argon dating   a method used for dating volcanic rocks.

preeclampsia   a severe form of hypertension (high blood pressure) that sometimes occurs during pregnancy and can produce convulsions, cerebral hemorrhage, kidney failure, and other complications in mothers. *See also* gestational hypertension.

prehensile   describes the ability of hands, feet, or tails to grasp objects, such as food items or branches.

premenstrual syndrome (PMS)   a collection of symptoms that occur in some women just before menstruation begins, including irritability, moodiness, fatigue, and headaches.

premolars   the teeth that lie between the canines and molars.

primary structure   the sequence of amino acids that make up a protein.

Proconsulidae   a group of early Miocene hominoids that includes the genus *Proconsul*.

progesterone   an ovarian steroid that plays an important role in preparing the fetus to sustain a pregnancy. Progesterone promotes the development of glands in the endometrium that secrete glycogen and important enzymes into the uterus.

prognathic   *see* subnasal prognathism.

prokaryotes   organisms without a cell nucleus or separate chromosomes. Bacteria are prokaryotes.

prosimian   a member of the primate suborder that contains the lemurs, lorises, and tarsiers. *See also* anthropoid.

proteins   large molecules that consist of a long chain of amino acids. Many proteins are enzyme catalysts, while other proteins perform structural functions.

prototype   the best example of a category. According to the theory of categorization developed by Eleanor Rosch, judgments about the membership in a category are based on similarity to the prototype.

Punnett square   a diagram that uses gene (or allele) frequencies to calculate the genotypic frequencies for the next generation.

quadrupedal   a form of locomotion in which the animal moves on all four limbs.

radioactive decay   spontaneous change from one isotope of an element to another isotope of the same element or to an entirely different element. Radioactive decay occurs at a constant rate that can be measured precisely in the laboratory.

radiocarbon dating   *see* carbon-14 dating.

radiometric methods   dating methods taking advantage of the fact that isotopes of certain elements change spontaneously from one isotope to another at a constant rate.

rain shadow   an area of reduced rainfall found on the lee (downwind) side of large mountains and mountain ranges.

recessive   describes an allele that is expressed in the phenotype only when it is in the homozygous state.

recipient   an individual to whom a particular behavior is directed. *See also* actor.

reciprocal altruism   a theory that altruism can evolve if pairs of individuals take turns giving and receiving altruism over the course of many encounters.

recombination   the creation of novel genotypes as a result of the random segregation of chromosomes and of crossing over.

redirected aggression   a behavior in which the recipient of aggression threatens or attacks a previously uninvolved party. Thus, if A attacks B and B then attacks C, B's attacks are an example of redirected aggression.

reinforcement   the process in which selection acts against the likelihood of hybrids occurring between members of two phenotypically distinctive populations leading to the evolution of mechanisms that prevent interbreeding.

replacement model   a model of the evolution of anatomically modern humans that holds that archaic *Homo sapiens* populations were genetically isolated from each other and evolved independently. Anatomically modern humans arose in Africa 100,000 to 200,000 years ago, and then spread outward replacing other populations of archaic *H. sapiens*.

reproductive value   a measure of the average contribution to subsequent generations of organisms of different ages. In most human populations, reproductive value increases through early childhood and then declines throughout the rest of life.

reproductive isolation   the relationship between two populations when there is no gene flow between them.

ribonucleic acid (RNA)   a long molecule that plays several different important roles in protein synthesis. RNA differs from DNA in having a slightly different chemical backbone and the base uracil is substituted for thymine.

ribosomes   small organelles composed of protein and nucleic acid that temporarily hold the messenger RNA and transfer RNAs together during protein synthesis.

robust   sturdy, heavy-boned. Description applied especially to certain australopithecine species to distinguish them from lighter (gracile) species.

rock shelters   sites sheltered by an overhang of rock.

sagittal crest   a sharp fin of bone that runs along the midline of the skull and serves to increase the area available for the attachment of chewing muscles.

sagittal keel   a feature running along the midline of the skull shaped like a shallow, upside-down V. The sagittal keel is a derived characteristic of *Homo erectus*.

sampling variation   the variation in the composition of small samples drawn from a large population.

scapulae (scapula, sing.**)**   the shoulder blades.

secondary compounds   toxic (poisonous) chemical compounds produced by plants and concentrated in plant tissues to prevent animals from eating the plant.

segregate independently   *see* independent assortment.

selection-mutation balance   an equilibrium that occurs when the rate at which selection removes a deleterious gene is balanced by the rate at which mutation introduces that gene. The frequency of genes at selection-mutation balance is typically quite low.

selfish behavior　behavior that increases the fitness of the donor and decreases the fitness of the recipient.

senescence　the physiological processes of aging.

sex ratio　the number of individuals of one sex in relation to the number of the opposite sex. By convention, sex ratios are generally expressed as the number of males to the number of females. "Primary sex ratio" refers to the sex ratio at conception, while "secondary sex ratio" refers to the sex ratio at birth.

sexual dimorphism　differences between sexually mature males and females in body size or morphology.

sexual selection　a form of natural selection that results from differential mating success in one gender. In mammals, sexual selection usually occurs in males and may be due to male-male competition (*see also* intrasexual selection) or female choice (*see also* intersexual selection).

sickle-cell anemia　a severe form of anemia that afflicts people who are homozygous for the sickle-cell gene.

side scraper　a flake tool on which at least one edge has been retouched to create a continuous smooth edge.

silverbacks　mature male gorillas. The term derives from the fact that the hair on the backs and shoulders turns silvery gray in mature males.

SLI　*see* specific language impairment.

social facilitation　the situation that occurs when the performance of a behavior by older individuals increases the probability that younger individuals will acquire that behavior on their own. Social facilitation does not mean that young individuals copy the behavior of older individuals. For example, the feeding behavior of older individuals may bring younger individuals in contact with the foods that adults are eating and, therefore, increase the chance that they acquire a preference for those foods.

social intelligence hypothesis　the hypothesis that the relatively sophisticated cognitive abilities of higher primates are the outcome of selective pressures that favored intelligence as a means to gain advantages in social groups.

special relationships　exclusive, stable, affiliative, and sexual relationships among particular pairs of unrelated male and female baboons.

species (sing. and pl.)　a group of organisms classified together at the lowest level of the taxonomic hierarchy. Biologists disagree about how to define a species. *See also* biological species concept; ecological species concept.

specific language impairment (SLI)　a language disorder in which the affected person experiences difficulty with morphology, the rules that govern the formation of words. For example, English speakers with SLI have difficulty forming regular plurals. There is evidence that at least some cases of SLI are hereditary.

spherical gradient lens　a spherical lens in which the index of refraction increases smoothly from the surface of the lens to its center. This type of lens is found in many aquatic organisms.

spinal cord　the massive bundle of nerves that connects the brain to most of the rest of the body. The spinal cord passes down the vertebral canal.

spiral arteries　the arteries that carry blood from the mother's body to the placenta.

spite (spiteful, adj.**)**　behavior that is costly both to the actor and to the recipient.

stabilizing selection　selection pressures that favor average phenotypes. Stabilizing selection reduces the amount of variation in the population but does not alter the mean value of the trait.

stereoscopic　describes vision in which three-dimensional images are produced because each eye sends a signal of the visual image to both hemispheres in the brain. Stereoscopic vision requires binocular vision.

strategy　a complex of behaviors deployed in a specific functional context, such as mating, parenting, or foraging.

stratum　a geological layer.

strepsirhines　members of the group containing lemurs and lorises. The system classifying primates into haplorhines and strepsirhines is a cladistic alternative to the system used in this text, in which primates are divided into prosimians and anthropoids and tarsiers are grouped with prosimians. *See also* haplorhines.

subnasal prognathism　the condition in which the part of the face below the nose is pushed out.

substitute parent　an adult who lives in the same household as a child but is not the child's biological mother or biological father, for example, a stepparent.

sulci (sulcus, sing.**)**　deep fissures or infoldings that demarcate different brain regions.

superfamily　the taxonomic level that lies between infraorder and family. Thus a superfamily can contain several families. For example, humans are a member of the superfamily Hominoidea, which contains the families Hominidae and Pongidae.

sutures   wavy joints between bones that mesh together and are separated by fibrous tissue.

sylvian fissure   a fissure roughly in the middle of the brain that marks the boundary between the frontal lobe and the parietal and temporal lobes.

sympatric speciation   a hypothesis that speciation can result from selective pressures favoring different phenotypes within a population, without positing geographical isolation as a factor.

syntax   the grammatical rules that allow people to assign meaning to strings of words.

systematics   a branch of biology that is concerned with the procedures for constructing phylogenies. *See also* taxonomy.

talus   a large bone in the back part of the foot.

taphonomy   the study of the processes that affect the state of the remains of organisms from the time that the organism dies until it is fossilized.

taurodont root   a single broad tooth root in molars, resulting from the fusion of three roots. Taurodont roots were characteristic of Neanderthals.

taxonomy   a branch of biology that is concerned with the use of phylogenies for naming and classifying organisms. *See also* systematics.

temporalis muscles   large muscles involved in chewing. They attach to the side of the cranium and to the mandible.

terrestrial   predominantly active on the ground.

territories   fixed areas occupied by animals who defend the boundaries against intrusion by other individuals or groups of the same species.

tertiary structure   the three-dimensional folded shape of proteins.

testes   the male organs responsible for sperm production.

theory of mind   the capacity to be aware of the thoughts, knowledge, or perceptions of other individuals. A theory of mind may be a prerequisite for deception, imitation, teaching, and empathy. It is generally thought that humans, and possibly chimpanzees, are the only primates to possess a theory of mind.

thermoluminescence dating   a technique used to date fossil teeth by measuring the density of trapped electrons in apatite crystals in teeth. This method is important for sites that are too young to be dated using potassium-argon dating (less than 500,000 years ago) and too old to be dated with carbon-14 dating (more than 40,000 years ago).

third party relationships   relationships among other individuals. For example, monkeys and apes are believed to understand something about the nature of kinship relationships among other group members.

thoracic vertebrae   the vertebrae connected to ribs.

thymine   one of the four bases of the DNA molecule. The complementary base of thymine is adenine.

tibia   the larger of the two long bones in the lower leg.

torque   a twisting force that generates rotary motion.

toxins   chemical compounds that are poisonous or toxic.

traits   characteristics of organisms.

transfer RNA (tRNA)   a form of RNA that facilitates protein synthesis by first binding to amino acids in the cytoplasm and then binding to the appropriate site on the mRNA molecule. There is at least one distinct form of tRNA for each amino acid.

transitive   a property of triadic (three-way) relationships in which the relationships between the first and second elements and the second and third elements automatically determine the relationship between the first and third elements. For example, if A is greater than B and B is greater than C, then A is greater than C. In many primate species, dominance relationships are transitive.

tRNA   *see* transfer RNA.

trophoblast   the outer layer of cells of an embryo.

type specimen   the individual specimen that served as the basis for the original naming of a new species.

unhaft   *see* haft.

unlinked   refers to genes on different chromosomes. *See also* linkage.

Upper Paleolithic   the period from about 40 kya to about 10 kya in Europe, North Africa, and parts of Asia. The tool kits from this period are dominated by blades. Later Stone Age tool kits in Africa are very similar to Upper Paleolithic tools.

Upper Pleistocene   the period from 137 kya to 10 kya. The Pleistocene is divided into three parts: the Lower, Middle, and Upper Pleistocenes. The beginning of the Upper Pleistocene coincides with the temination of the next-to-last glacial period, and the end is marked by the conclusion of the last glacial period.

uracil   one of the four bases of the RNA molecule. It corresponds to the base thymine in DNA; as with DNA, its complementary base is adenine.

variant   the particular form of a trait. For example, blue eyes, brown eyes, and gray eyes are variants of the trait eye color.

variation among groups   differences in phenotype or genotype between individuals in a group.

variation within groups   differences in the average phenotype or genotype between groups.

vertebrae   the bones of the spinal column.

vertebral canal   a tube or canal that carries the spinal cord and is formed by the vertebrae.

vertical clinging and leaping   a form of locomotion in which the animal clings to vertical supports and moves by leaping from one vertical support to another.

vestibular system   a system of tubes that are embedded in the inner ear (inside the cranium) and part of the system animals use to maintain balance.

vitamins   organic substances necessary in small quantities for normal metabolism.

viviparity   giving birth to live young.

zygomatic arches   the cheek bones.

zygote   the cell formed by the union of an egg and sperm.

Credits

Bark cloth designs from *MBUTI DESIGN: Paintings by Pygmy Women of the Ituri Forest* by Georges Meurant and Robert Farris Thompson. Photos by Heini Schneebeli and Brian Forrest. Reproduced courtesy Hansjörg Mayer, Editions Hansjörg Mayer.

Title page and contents page detail from Zadok Ben-David. *The Mystical Experience of the Wildcat.* 1986. Resin and pigment on metal armature. Private collection, England. Photo courtesy of Jason & Rhodes Limited.

Prologue

1 Robert Harding Picture Library, London. 2 Corbis/Bettmann. 3 By courtesy of the National Portrait Gallery, London. 4 Marcelo Brodsky/Science Photo Library/Photo Researchers, Inc. 5 Robert Boyd. 6 Irven DeVore/Anthro-Photo. 7 (c) Murray Alcosser, Louis K. Meisel Gallery, New York. 8 (c) A. Bannister/NHPA.

Chapter 1

1.3 (c) Wolfgang Kaehler. 1.4 Robert Harding Picture Library, London. 1.5a Courtesy of John Byram. 1.5b New York Public Library (from *Zoology of the HMS Beagle under Captain Fitzroy*, edited by Charles Darwin, vol. 3, London: Smith, 1841). 1.7a Peter T. Boag. 1.7b Peter T. Boag. 1.8 Adapted from Figure 56C in P. R. Grant, 1986, *The Ecology and Evolution of Darwin's Finches*, Princeton University Press, Princeton, N.J. 1.10 Adapted from Figure 55 in P. R. Grant, 1986, *The Ecology and Evolution of Darwin's Finches*, Princeton University Press, Princeton, N.J. 1.11 Adapted from Figure 56D in P. R. Grant, 1986, *The Ecology and Evolution of Darwin's Finches*, Princeton University Press, Princeton, N.J. 1.14 Adapted from Figure 11 in L. V. Salvin-Plawin and E. Mayr, 1977, On the evolution of photoreceptors and eyes, *Evolutionary Biology* 10:207–263. 1.15 Drawing by Robert Boyd. 1.16a (c) Jeffrey L. Rotman. 1.16b (c) Jeffrey L. Rotman. 1.17 Redrawn from Figures 2 and 72 in P. R. Grant, 1986, *The Ecology and Evolution of Darwin's Finches*, Princeton University Press, Princeton, N.J. 1.19 By permission of the Syndics of Cambridge University Library (from Charles Darwin, *The Variations of Animals and Plants Under Domestication*, vol. 1, London: John Murray, 1868). 1.20 Adapted from Figure 2 in D. Nilsson and S. Pleger, 1994, A pessimistic estimate of the time required for an eye to evolve, *Proceedings of the Royal Society*, London, Series B, 256:53–58. 1.21a (c) Art Wolfe. 1.21b (c) Gerard Lacz/Peter Arnold, Inc. 1.21b2 (c) BIOS/Peter Arnold, Inc.

Chapter 2

2.1 Mendelianum, Museum Moraviae, Brno, Czechoslovakia. 2.2 Location unknown. 2.4 Courtesy M. W. Shaw, University of Michigan, Ann Arbor. 2.5 Dr. Jeremy Burgess/Science Photo Library/Photo Researchers, Inc. 2.16. Adapted from Figure 2.5 in P. Berg and M. Singer, 1992, *Dealing with Genes: The Language of Heredity*, University Science Books, Mill Valley, Calif. 2.17 Adapted from Figure 2.10 in P. Berg and M. Singer, 1992, *Dealing with Genes: The Language of Heredity*, University Science Books, Mill Valley, Calif. 2.22 Adapted from Figure 2.15 in P. Berg and M. Singer, 1992, *Dealing with Genes: The Language of Heredity*, University Science Books, Mill Valley, Calif. 2.23 Adapted from Figure 2.17 in P. Berg and M. Singer, 1992, *Dealing with Genes: The Language of Heredity*, University Science Books, Mill Valley, Calif.

Chapter 3

3.1 (c) G. Büttner/OKAPIA/Photo Researchers, Inc. 3.4a A. Barrington Brown/Science Photo Library/Photo Researchers, Inc. 3.4b AP/Wide World Photos. 3.4c AP/Wide World Photos. 3.11 Adapted from J. F. Crow, 1986, *Basic Concepts in Population, Quantitative, and Evolutionary Genetics*, W. H. Freeman, New York. 3.12 (c) Jean-Michel Labat/AUSCAPE International. 3.23 From P. Hedrick, 1983, *Genetics of Populations*, Science Books International, Boston. 3.24 Adapted from D. Nilsson, 1989, Vision optics and evolution, *BioScience*, 39:298–307. 3.26 Robert Boyd.

Chapter 4

4.1a Jane Goodall, courtesy of The Jane Goodall Institute. 4.1b (c) Andrew Plumptre/Oxford Scientific Films. 4.4a Carrol Henderson, Blaine, Minn. 4.4b Peter T. Boag. 4.4c Peter T. Boag. 4.5 Camille Parmesan/Michael C. Singer. 4.7 Redrawn from Figure 1 in D. Schulter, T. D. Price, and P. R. Grant, 1986, Ecological character displacement in Darwin's finches, *Science*, 227:1056–1059. 4.8 (c) Thalia Grant. 4.9a–d Joan Silk. 4.11a Carrol Henderson, Blaine, Minn. 4.11b–c Peter T. Boag. 4.12 Painting by H. Douglas Pratt. 4.14 (c) Ben Cropp/AUSCAPE International. 4.17a (c) T. Whittaker/Premaphotos Wildlife. 4.17b (c) Rod Williams/Bruce Coleman Collection. 4.17c (c) Tim Laman/The Wildlife Collection. 4.17d (c) Richard Wrangham/Anthro-Photo. 4.18a Robert Boyd and Joan Silk. 4.18b Erwin & Peggy Bauer/Bruce Coleman Inc. 4.23 (c) Warren Garst/Tom Stack & Associates. 4.28 (c) John Giannicchi/Science Source/Photo Researchers, Inc. 4.30 From Figure 8 in Sibley and Alquist, 1987, *Journal of Molecular Evolution*, 26:99–121.

Chapter 5

5.1a (c) Liz Bonford/Ardea London. 5.1b (c) Grospas/Nature. 5.1c (c) I. Polunin/NHPA. 5.1d (c) Jan Lindblad/Photo Researchers, Inc. 5.1e (c) Tom Vezo/The Wildlife Collection. 5.2a (c) Dr. U. Nebelsiek/Peter Arnold, Inc. 5.2b (c) Karl & Kay Ammann/Bruce Coleman Inc. 5.2(c) Alan Root/Survival Anglia/Oxford Scientific Films. 5.3 (c) Doug Cheeseman/Peter Arnold, Inc. 5.4 (c) Norbert Wu/www.norbertwu.com. 5.5 Redrawn from Figure 3.1 in J. G. Fleagle, 1988, Primate Adaptation and Evolution, Academic Press, San Diego, Calif. 5.6 Robert Boyd. 5.7 (c) Mantis Wildlife Films/Oxford Scientific Films. 5.8 (c) K. G. Preston-Mafham/Premaphotos Wildlife. 5.9b Robert Boyd and Joan Silk. 5.10a (c) Art Wolfe. 5.10b (c) K. G. Preston-Mafham/Premaphotos Wildlife. 5.11b Robert Boyd and Joan Silk. 5.12 From Figures 6.21 and 6.23 in R. D. Martin, 1990, *Primate Origins and Evolution*, Princeton University Press, Princeton, N.J. 5.14a (c) Karl & Kay Ammann/Bruce Coleman Inc. 5.14b (c) Michael J. Doolittle/Peter Arnold, Inc. 5.16a Joan Silk. 5.18 (c) Lee Lyon/Bruce Coleman Inc. 5.19a–b Redrawn from Figures 1 and 2 in G. Teleki, 1989, Population status of wild chimpanzees (*Pan troglodytes*) and threats to survival, pp. 312–353 in *Understanding chimpanzees*, ed. by P. G. Heltne and L. A. Marquardt, Harvard University Press, Cambridge.

Chapter 6

6.1 Joan Silk. 6.3 Joan Silk. 6.5 From Figure 8.9 in J. G. Fleagle, 1988, *Primate Adaptation and Evolution*, Academic Press, San Diego, Calif. 6.6 Redrawn from Figure 5.13 in A. Richard, 1985, *Primates in Nature*, W. H. Freeman, New York. 6.7 (c) Norbert Wu/www.norbertwu.com. 6.8d Joan Silk. 6.9 Adapted from Figure 4.1 in J. Terborgh, 1983, *Five New World Primates*, Princeton University Press, Princeton, N.J. 6.12 (c) Andrew Plumptre/Oxford Scientific Films. 6.13 Joan Silk. 6.15 Dieter & Mary Plage/Bruce Coleman Collection. 6.17a–b Joan Silk. 6.18a–c Robert Boyd. 6.19a–b, d–e Robert Boyd. 6.19c (c) Wolfgang Kaehler. 6.20 b Robert Boyd.

Chapter 7

7.1 Joan Silk. 7.2 (c) Henry Holdsworth/The Wildlife Collection. 7.4 (c) Rudie H. Kuiter, OSF/Animals Animals. 7.6 (c) Ardella Reed Stock Photography. 7.7 Adapted from data in A. Mori, 1979, Analysis of population changes by measurement of body weight in the Koshima troop of Japanese monkeys, *Primates* 20:371–397. 7.8 (c) Sarah Blaffer Hrdy/Anthro-Photo. 7.10a–b From Figures 1 and 2 in A. Pusey, J. Williams, and J. Goodall, 1997, The influence of dominance rank on the reproductive success of female chimpanzees, *Science* 277:828–831. 7.10c Adapted from Table 15.1 in R. I. M. Dunbar, 1984, The social ecology of gelada baboons, pp. 332–351 in *Ecological Aspects of Social Relationships*, ed. by D. I. Rubenstein and R. W. Wrangham, Princeton University Press, Princeton, N.J. 7.11 Joan Silk. 7.12 Robert Boyd and Joan Silk. 7.13 Adapted from Appendix 1 in P. C. Lee, P. Majluf, and I. J. Gordon, 1991, Growth, weaning, and maternal investment from a comparative perspective, *Journal of Zoology*, London 225:99–114. 7.15 Joan Silk. 7.16 Joan Silk. 7.17 Joan Silk. 7.18 Joan Silk. 7.19 From Figure 16.1 in A. H. Harcourt, 1992, Coalitions and alliances: Are primates more complex than nonprimates? pp. 445–471 in *Coalitions and Alliances in Humans and Other Animals*, ed. by A. H. Harcourt and F. B. M. de Wall, Oxford University Press, Oxford. 7.21 Adapted from Figure 6 in C. H. Janson, 1988, Food competition in brown capuchin monkeys (*Cebus apella*): quantitative effects of group size and tree productivity, *Behavior* 105:53–76. 7.22 (c) BIOS/Peter Arnold, Inc. 7.23 From Figure 3 in J. R. Ruiter, 1985, The influence of group size on predator scanning and foraging behavior of wedgecapped capuchin monkeys (*Cebus olivaceous*), *Behavior* 98:240–258. 7.24 (c) Gunter Ziesler/Peter Arnold, Inc. 7.25 (c) Norman Myers/Bruce Coleman, Inc. 7.27 Robert Boyd and Joan Silk. 7.28 From Figure 10.1 in S. J. Andelman, 1986, Ecological and social determinants of cercopithecine mating patterns, pp. 201–216 in *Ecological Aspects of Social Evolution: Birds and Mammals*, ed. by D. I. Rubenstein and R. W. Wrangham, Princeton University Press, Princeton, N.J. 7.29 (c) K. G. Preston-Mafham/Premaphotos Wildlife. 7.30a (c) Art Wolfe. 7.30b (c) BIOS/Peter Arnold, Inc. 7.31a–b Robert Boyd. 7.31c From Figure 23.2 in C. Packer et al., Reproductive success of lions, pp. 363–383 in *Reproductive Success*, ed. by T. H. Clutton-Brock, University of Chicago Press, Chicago. 7.32 Joan Silk. 7.33 From Figure 2 in P. H. Harvey and A. H. Harcourt, 1984, Sperm competition, testes size, and breeding systems in primates, pp. 589–599 in *Sperm Competition and the Evolution of Animal Mating Systems*, ed. by R. L. Smith, Academic Press, New York. 7.34 From Figure 2 in P. H. Harvey and A. H. Harcourt, 1984, Sperm competition, testes size, and breeding systems in primates, pp. 589–599 in *Sperm Competition and the Evolution of Animal Mating Systems*, ed. by R. L. Smith, Academic Press, New York. 7.36 (c) Schafer & Hill/Peter

Arnold, Inc. 7.37 (c) Jean-Paul Ferrero/AUSCAPE International. 7.38 Jersey Wildlife Preservation Trust/Phillip Coffey. 7.39 Data from Figure 2 in P. A. Garber, 1997, One for all and breeding for one: Cooperation and competition as a tamarin reproductive strategy, *Evolutionary Anthropology* 5:187–199. 7.40 From Figure 2 in R. Palombit, in press, Infanticide and the evolution of pair bonds in nonhuman primates, *Evolutionary Anthropology*. 7.41a (c) Sarah Blaffer Hrdy/Anthro-Photo. 7.41b (c) K. G. Preston-Mafham/Premaphotos Wildlife. 7.42 Adapted from Figure 2 in C. M. Crockett and R. Sekulic, 1984, Infanticide in red howling monkeys, pp. 173–191, in *Infanticide: Comparative and Evolutionary Perspectives*, ed. by G. Hausfater and S. B. Hrdy, Aldine, New York. 7.43 Joan Silk. 7.44a Robert Boyd and Joan Silk. 7.44b Robert Boyd. 7.45a From Figure 7 in J. R. de Ruiter, et al., 1992, Male social rank and reproductive success in wild long-tailed macaques, pp. 175–191 in *Paternity in Primates: Genetic Tests and Theories*, ed. by R. D. Martin, A F. Dixon, and E. J. Wickings, Karger, Basel. 7.45b From Table I in D. A. Gust, T. McCaster, T. P. Gordon, W. F. Gergits, N. J. Casna, and H. M. McClure, 1998, Paternity in sooty mangabeys, *International Journal of Primatology* 19:83–94. 7.46 Joan Silk.

Chapter 8

8.1 (c) K. G. Preston-Mafham/Premaphotos Wildlife. 8.7b Joan Silk. 8.8a–b Joan Silk. 8.9 Joan Silk. 8.10 Joan Silk. 8.11c Joan Silk. 8.12 Adapted from Figure 4 in E. Kapsalis and C. M. Berman, 1996, Models of affiliative relationships among free-ranging rhesus monkeys (*Macaca mulatta*) I. Criteria for kinship, *Behaviour* 133:1209–1234. 8.13 Adapted from Figure 2 in F. Aureli, C. P. van Schaik, and J.A.R.A.M. van Hooff, 1989, Functional aspects of reconciliation among captive long-tailed macaques (*Macaca fasicularis*), *American Journal of Primatology* 19:39–51. 8.14 Joan Silk. 18.5a From Figure 8.4 in J. B. Silk, 1992, Patterns of intervention in agonistic contests among male bonnet macaques, pp. 215–232 in *Coalitions and Alliances in Humans and other Animals*, ed. by A. H. Harcourt and F. B. M. De Waal, Oxford University Press, Oxford. 8.16a From Table 1 in J. A. Johnson, 1987, Dominance rank in juvenile olive baboons, *Papio anubis*: the influence of gender, size, maternal rank and orphaning, *Animal Behaviour* 35:1694–1708. 8.16b Joan Silk. 8.17 Adapted from Figure 1 in G. Hausfater, J. Altmann, and S. Altmann, 1982, Long-term consistency of dominance relationships among female baboons (*Papio cyncocephalus*), *Science* 217:752–755. 8.18 Joan Silk. (Rdg 3) Joan Silk. 8.19 From Figure 1 in R. M. Seyfarth and D. L. Cheney, 1984, Grooming, alliances, and reciprocal altruism in vervet monkeys, *Nature* 308:541–543. 8.20 From Figure 2 in F. B. M. de Waal, 1997, The chimpanzee's service economy: Food for grooming, *Evolution & Human Behavior* 18:375–386.

Chapter 9

9.1 From Figure 8.7 in R. D. Martin, 1990, *Primate Origins and Evolution*, Princeton University Press, Princeton, N.J.

9.2a (c) Norbert Wu/www.norbertwu.com. 9.2b (c) J. M. Pearson/Biofotos. 9.4a Joan Silk. 9.5 Joan Silk. 9.6a–b Redrawn from p. 109 in S. J. Jones, R. D. Martin, and D. A. Pilbeam, 1992, *The Cambridge Encyclopedia of Human Evolution*, Cambridge University Press, Cambridge. 9.7a–c From R. I. M. Dunbar, 1992, Neocortex size as a constraint on group size in primates, *Journal of Human Evolution* 20:469–493. 9.8 Robert Boyd and Joan Silk. 9.9a Joan Silk. 9.11a–b Joan Silk. 9.13 Courtesy of Richard Byrne.

Chapter 10

10.1 Figure from pp. 38–39 in R. J. G. Savage and M. R. Long, 1986, *Mammal Evolution*, Facts on File and the British Museum (Natural History), Oxford, England. 10.2 From Figure 5.3 in R. D. Martin, 1990, *Primate Origins and Evolution: A Phylogenetic Reconstruction*, Princeton University Press, Princeton, N.J. 10.3 (c) Art Wolfe. 10.5 Adapted from p. 21 in R. Lewin, 1989, *Human Evolution*, Blackwell Scientific Publishers, Cambridge. 10.6 From Figure 1 in R. D. Martin, 1993, Primate origins: Plugging the gaps, *Nature* 363:223–234. 10.7 From p. 200 in S. J. Jones, R. D. Martin, and D. A. Pilbeam, 1992, *The Cambridge Encyclopedia of Human Evolution*, Cambridge University Press, Cambridge. 10.8a–b From Figure 3 in K. D. Rose, 1994/5, The earliest primates, *Evolutionary Anthropology* 3:159–173. 10.9 From Figure 11.9 in J. G. Fleagle, 1988, *Primate Adaptation and Evolution*, Academic Press, San Diego. 10.10 (c) Art Wolfe. 10.11 From Figure 12.5 in J. G. Fleagle, 1988, *Primate Adaptation and Evolution*, Academic Press, San Diego. 10.12 From Figure 12.6 in J. G. Fleagle, 1988, *Primate Adaptation and Evolution*, Academic Press, San Diego. 10.13 From Figure 12.13 in J. G. Fleagle, 1988, *Primate Adaptation and Evolution*, Academic Press, San Diego. 10.14a–c From Figure 1.10 in G. Conroy, 1990, *Primate Evolution*, W. W. Norton, New York. 10.15 From Figure 4.14 in G. Conroy, 1990, *Primate Evolution*, W. W. Norton, New York. 10.16 Courtesy Robert Boyd. 10.18a–c From Figure 2 in R. D. Martin, 1993, Primate origins: Plugging the gaps, *Nature* 363:223–234. 10.19 From Figure 13.7 in J. G. Fleagle, 1988, *Primate Adaptation and Evolution*, Academic Press, San Diego. 10.20 From Figure 2.23 in R. D. Martin, 1990, *Primate Origins and Evolution: A Phylogenetic Reconstruction*, Princeton University Press, Princeton, N.J. (Rdg. 1–2) Redrawn from pp. 50 and 51 in R. Potts, 1996, *Humanity's Descent*, William Morrow, New York.

Chapter 11

11.1 (c) Institute of Human Origins/Nanci Kahn. 11.2 John Reader/Science Photo Library/Photo Researchers, Inc. 11.3 Redrawn from p. 237 in S. Jones, R. Martin, and D. Pilbeam, 1992, *The Cambridge Encyclopedia of Human Evolution*, Cambridge University Press, Cambridge. 11.6a–c From Figures 20.15 and 20.20 in L. Aeillo and C. Dean, 1990, *An Introduction of Evolutionary Human Anatomy*, Academic Press, New York. 11.8 From p. 67 in R. Lewin, 1989, *Human Evolution*, Blackwell Scientific, Boston. 11.9 (c) 1999 John

Reader. 11.10 Joan Silk. 11.12 Robert Boyd. 11.14 Robert Boyd. 11.15 (c) Kenya National Museum. 11.16 Redrawn from Figure 1 in M. G. Leakey, C. S. Feibel, I. McDougall, and A. Walker, 1995, New four-million-year-old hominid species from Kanapoi and Allia Bay, Kenya, *Nature* 376:565–571. 11.17 Robert Boyd. 11.18 Adapted from Figure 1 in T. D. White, D. C. Johanson, and W. H. Kimbel, 1981, *Australopithecus africanus*: Its phyletic position reconsidered. *South African Journal of Science* 77:445–470. 11.21 Joan Silk. 11.24a–c From p. 80 in R. Lewin, 1989, *Human Evolution*, Blackwell Scientific, Boston. 11.26 Adapted from Figure 7.1 in R. Foley, 1987, *Another Unique Species*, Longman Scientific and Technical, Harlow, U. K. 11.27 Adapted from Figure 1 in A. Sillen, 1992, Strontium-calcium ratios (Sr/Ca) of *Australopithecus robustus* and associated fauna from Swartkrans, *Journal of Human Evolution* 23:495–516. 11.28 (c) Wolfgang Kaehler. (Rdg. 1–3) Courtesy of the Getty Conservation Institute, Los Angeles, California, (c) 1999 the J. Paul Getty Trust. 11.29 (c) Kenya National Museum. 11.30 (c) Kenya National Museum. 11.35 From p. 113 in K. Schick and N. Toth, 1993, *Making Silent Stones Speak*, Simon & Schuster, New York. 11.36 Redrawn from p. 141 in K. Schick and N. Toth, 1993, *Making Silent Stones Speak*, Simon & Schuster, New York. 11.40 Illustration (c) Christa Hook, courtesy of Boxtree/Granada Television.

Chapter 12

12.1 (c) Karl & Kay Ammann/Bruce Coleman Inc. 12.3b Joan Silk. 12.3c Robert Boyd. 12.3d (c) Lee Lyon/Bruce Coleman Inc. 12.6 Adapted from Figure 1 in C. Stanford, J. Wallis, H. Matama, and J. Goodall, 1994, Patterns of predation by chimpanzees on red colobus monkeys at Gombe National Park, 1982–1991, *American Journal of Physical Anthropology* 94:213–228. 12.7 Adapted from Table 9 in C. Boesch and H. Boesch, 1989, Hunting behavior of wild chimpanzees in the Tai National Park, *American Journal of Physical Anthropology* 78:547–573. 12.8a–b Joan Silk. 12.9a–b Joan Silk. 12.10 Joan Silk. 12.11 From Figure 3.1 in R. I. M Dunbar, 1988, *Primate Social Systems*, Cornell University Press, Ithaca, N. Y. 12.12 From Figure 6.1 in R. Potts, 1988, Early Hominid Activities at Olduvai, Aldine de Gruyter, New York. 12.13 Adapted from Figure 2 in R. Potts, 1984, Home bases and early hominids, *American Scientist* 72:338–347. 12.14 Adapted from Figure 3 in R. Potts, 1984, Home bases and early hominids, *American Scientist* 72:338–347. 12.15 (c) Art Wolfe. 12.16b–c National Museum of Natural History, Smithsonian Institution, Washington, D.C. (Rdg. 1–2) Courtesy of Kathy D. Schick and Nicholas Toth, CRAFT Research Center, Indiana University. 12.17a–b (c) Y. Arthus-Bertrand/Peter Arnold, Inc. 12.18 (c) Art Wolfe. 12.19 Joan Silk. 12.20 Robert Boyd. 12.21 Robert Boyd and Joan Silk. 12.22 Robert Boyd. 12..23 Drawing by Christian Gurney from Figure 2 in Henry T. Bunn and Ellen M. Kroll, 1986, Systematic butchery by Plio/Pleistocene hominids at Olduvai Gorge, Tanzania, *Current Anthropology*, 27:432–52. 12.25 From Table 5.2 in W. C. McGrew, 1992, *Chimpanzee Material Culture*, Cambridge University Press, Cambridge. 12.28 Adapted from Table 13 in C. Boesch and H. Boesch, 1989, Hunting behavior of wild chimpanzees in the Tai National Park, *American Journal of Physical Anthropol-*

ogy 78:547–573. 12.30 From Figure 9.18 in B. Campbell, 1985, *Human Evolution*, Aldine de Gruyter, Hawthorne, N.Y. 12.32 Adapted from Figure 4 in R. Potts, 1984, Home bases and early hominids, *American Scientist* 72:338–347.

Chapter 13

13.3 (c) Alan Walker. 13.4 Photograph courtesy of Alan Walker, reprinted with permission of the National Museums of Kenya. 13.5 From p. 232 in K. D. Schick and N. Toth, 1993, *Making Silent Stones Speak*, Simon & Schuster, New York. 13.7 From p. 244 in S. Jones, R. Martin, and D. Pilbeam, 1992, *The Cambridge Encyclopedia of Human Evolution*, Cambridge University Press, Cambridge. 13.8 Redrawn from Figure 17 in C. Stringer and C. Gamble, 1993, *In Search of the Neanderthals*, Thames & Hudson, New York. 13.10 Illustration (c) Christa Hook, courtesy of Boxtree/Granada Television. 13.12 National Museums of Kenya. 13.13 National Museums of Kenya. 13.14 Robert Boyd. 13.17 Courtesy of Richard Klein. 13.18 Javier Trueba/Madrid Scientific Films. 13.19 Courtesy of Richard Klein. 13.21 Adapted from Figure 1, Greenland Ice-Core Project, 1993, Climate instability during the last interglacial period recorded in the GRIP ice core, *Nature* 364:203–207. 13.24 From Figure 34 in C. Stringer and C. Gamble, 1993, *In Search of the Neanderthals*, Thames & Hudson, New York. 13.25 Adapted from Figure 3.2 in C. Stringer, 1984, Adaptation in the Pleistocene, pp. 55–83 in *Hominid Evolution and Community Ecology*, ed. by R. Foley, Academic Press, New York. 13.29a (c) BIOS/Peter Arnold, Inc. 13.29b (c) D. Robert Franz/The Wildlife Collection. 13.30 (c) Martin Harvey/The Wildlife Collection. 13.31 (c) Erik Trinkaus. 13.32 Joan Silk.

Chapter 14

14.3 (c) Jean-Michel Labat/AUSCAPE International. 14.4 From Figure 56 in F. Bordes, 1968, *La Paleolithique dans le monde*, Hachette, Paris. 14.7 From Figure 13.4 in P. Mellars, 1996, *The Neanderthal Legacy*, Princeton University Press, Princeton, N.J. 14.8 (c) Jean Clottes/Sygma, New York. 14.9 Ulmer Museum, Ulm, Germany. (Rdg. 1) Musée national de Préhistoire, Les Ezyies, France. 14.10 Adapted from Figure 7.9 in R. Klein, 1989, *The Human Career*, University of Chicago Press, Chicago. 14.11 From Figure 34.8 in O. Soffer, 1989, The Middle to Upper Paleolithic transition on the Russian Plain, pp. 714–742 in *The Human Revolution*, ed. by P. Mellars and C. Stringer, Princeton University Press, Princeton, N.J. 14.13 Reprinted by permission from *Nature*, in P. Bahn, Neanderthals emancipated, 394:719–721, (c) 1998 Macmillan Magazines Ltd. 14.14 Redrawn from Figure 15 in P. Mellars, 1989, Major issues in the emergence of modern humans, *Current Anthropology* 30:349–385. 14.18 Adapted from Figure 4 in H. C. Harpending, S. T. Sherry, A. R. Rogers, and M. Stoneking, 1993, The genetic structure of ancient human populations, *Current Anthropology* 34:483–496. 14.19 From p. 25 in M. Krings, A. Stone, R. W. Schmitz, H. Krainitzki, M. Stoneking, and S. Pääbo,

1997, Neanderthal DNA sequences and the origin of modern humans, *Cell* 90:19–30.

Chapter 15

15.1a–b Redrawn from p. 135 in P. Lieberman, 1992, Human speech and language, pp. 134–137 in *The Cambridge Encyclopedia of Human Evolution*, ed. by J. Jones, R. Martin, and D. Pilbeam, Cambridge University Press, Cambridge. 15.2 (c) K. G. Preston-Mafham/Premaphotos Wildlife. 15.3 Dr. Ronald H. Cohn/The Gorilla Foundation. 15.4 Steve Winters/courtesy of the Language Research Center. 15.5 (c) Anthony Bannister/ABPL Photo Library. 15.6 (c) A. Bannister/NHPA. 15.9 Redrawn from p. 121 in T. Deacon, 1992, The human brain, pp. 115–123 in *The Cambridge Encyclopedia of Human Evolution*, ed. by J. Jones, R. Martin, and D. Pilbeam, Cambridge University Press, Cambridge. 15.10 From Figure 22.5 in P. Lieberman, 1989, The origins of some aspects of human language and cognition, pp. 371–414 in *The Human Revolution*, ed. by P. Mellars and C. Stringer, Princeton University Press, Princeton, N.J. 15.11 Adapted from Figures 12.8, 12.10, and 12.11 in P. Lieberman, 1984, *The Biology and Evolution of Language*, Harvard University Press, Cambridge, Mass.

Chapter 16

16.1 From Figure 20.12 in W. Bodmer and L. L. Cavalli-Sforza, 1976, *Genetics, Evolution, and Man*, W. H. Freeman, New York. 16.2 Tim O'Dell/*Sports Illustrated*. 16.3 (left) George Long/*Sports Illustrated*, (right) Richard Mackson/*Sports Illustrated*. 16.4 Adapted from p. 4 of M. Gopnik and M. B. Crago, 1991, Familial aggregation of a developmental language disorder, *Cognition* 39:1–50. 16.5a–b Courtesy of Marion I. Barnhart, Wayne State University Medical School, Detroit, Michigan. 16.7 Adapted from Figures 2.14.1B and 2.14.9 in L. L. Cavalli-Sforza, P. Menozzi, and A. Piazza, 1994, *The History and Geography of Human Genes*, Princeton University Press, Princeton, N.J. 16.8 Sarah Errington/Panos Pictures. 16.9 Redrawn from Figure 100 in G. Flatz, 1987, The genetics of lactose digestion in humans, *Advances in Human Genetics* 16:60. 16.10 (c) Jean Higgins/Unicorn Stock. 16.11 Adapted from Figure 2.3.3 in L. L. Cavalli-Sforza, P. Menozzi, and A. Piazza, 1994, *The History and Geography of Human Genes*, Princeton University Press, Princeton, N.J. 16.13 Redrawn from Figure 2.4.5 in L. L. Cavalli-Sforza, P. Menozzi, and A. Piazza, 1994, *The History and Geography of Human Genes*, Princeton University Press, Princeton, N.J. 16.15 From Figure 1 (p. 241) in D. Hatl, 1980, *Principles, of Population Genetics*, Sinauer, Sunderland, Mass. 16.16 (c) John W. McDonough. 16.17 From Figure 19.12 in W. F. Bodmer and L. L. Cavalli-Sforza, 1976, *Genetics, Evolution, and Man*, Freeman, New York. 16.19 From Figure 16.6 in G. A. Harrison, J. M. Tanner, D. R. Pilbeam, and P. T. Baker, 1988, *Human Biology: An Introduction to Human Evolution, Variation, Growth, and Adaptability*, 3rd ed., Oxford Science Publications, Oxford. 16.20 From Figure 2.11.1 in L. L. Cavalli-Sforza, P. Menozzi, and A. Piazza, 1994, *The History and Geography of Human Genes*, Princeton University Press, Princeton, N.J. 16.21 From Figure 2.13.4 in L. L. Cavalli-Sforza, P. Menozzi, and A. Piazza, 1994, *The History and Geography of Human Genes*, Princeton University Press, Princeton, N.J. 16.22a–b From Figures 2.2.3 and 2.3.3 in L. L. Cavalli-Sforza, P. Menozzi, and A. Piazza, 1994, *The History and Geography of Human Genes*, Princeton University Press, Princeton, N.J.

Chapter 17

17.1 (c) SIU/Photo Researchers, Inc. 17.2 Isidor Jeklin/Cornell Laboratory of Ornithology. 17.3 From Figure 6.10 in J. Wood, 1994, *Dynamics of Human Reproduction*, Aldine de Gruyter, New York. 17.4 Photograph by Lennart Nilsson/Albert Bonniers Forlag AB, *Behold Man*, Little, Brown and Company. 17.5 (c) Keith/Custom Medical Stock Photo. 17.6 Joan Silk. 17.7 From Table 4.6 in N. Howell, 1979, *Demography of the Dobe !Kung*, Academic Press, New York. 17.9 From Tables 7.1 and 13.3 in N. Howell, 1979, *Demography of the Dobe !Kung*, Academic Press, New York. 17.11 From Figure 4.1 in M. Rose, 1991, *Evolutionary Biology of Aging*, Oxford University Press, Oxford. 17.12 By permission of the Syndics of Cambridge University Library, Darwin Collection. 17.14 Data from Tables 5.1 and 5.2 in R. G. Godsen, 1985, *The Biology of Menopause*, Academic Press, London. 17.15 Adapted from Figure 4.2 in J. Wood, 1994, *Dynamics of Human Reproduction*, Aldine de Gruyter, New York. 17.16 Adapted from Figure 4.4 in J. Wood, 1994, *Dynamics of Human Reproduction*, Aldine de Gruyter, New York. 17.17 From Figure 4.9 in J. Wood, 1994, *Dynamics of Human Reproduction*, Aldine de Gruyter, New York. 17.18 From Figure 9.16 in J. Wood, 1994, *Dynamics of Human Reproduction*, Aldine de Gruyter, New York. 17.19 From Figure 1 in S. N. Austad, 1994, Menopause: An evolutionary perspective, *Experimental Gerontology* 29:255–263. 17.20a Joan Silk. 17.21 Adapted from Figure 4 in T. M. Caro, D. W. Sellen, A. Parish, R. Frank, D. M. Brown, E. Voland, and M. Borgerhoff Mulder, 1995, Termination of reproduction in nonhuman and human primates, *International Journal of Primatology* 16:205–220. 17.23 Data from Table I in K. Hawkes, J. F. O'Connell, and N. G. Blurton Jones, 1989, Hardworking Hadza grandmothers, pp. 341-365 in *Comparative Socioecology: The Behavioural Ecology of Humans and Other Mammals*, ed. by V. Standen and R. A. Foley, Blackwell Scientific, Oxford.

Chapter 18

18.1 (c) Joe Cavanaugh/DDB Stock Photo. 18.2 (c) Stephen Dalton/Photo Researchers, Inc. 18.3, 18.4a–b, 18.5a–b Courtesy of the University of California Press and of Munsell Color McBeth Division. 18.6 (c) Art Wolfe. 18.7 Modified from Figure 6.6 in W. H. Durham, 1992, *Coevolution*, Stanford University Press, Stanford, Calif. 18.8 (c) Jean-Paul Ferrero/AUSCAPE International. 18.9a Adapted from Table 12.6 in A. P. Wolf, 1995, *Sexual Attraction and Childhood Association*, Stanford University Press, Stanford, Calif. 18.9b

Adapted from Tables 11.1, 11.2, and 11.3 in A. P. Wolf, 1995, *Sexual Attraction and Childhood Association*, Stanford University Press, Stanford, Calif. 18.11 Adapted from Table 71 in J Needham, 1988, *Science and Civilization in China,* vol. 4, pt. 3, Cambridge University Press, Cambridge, UK. 18.12a–b From Figures 929 and 943 in J. Needham, 1988, *Science and Civilization in China,* vol. 4, pt. 3, Cambridge University Press, Cambridge, UK. 18.15 (c) Kennosuke Tsuda/Nature Production, Tokyo. 18.16 Robert Boyd. 18.17 (c) Kennan Harvey. 18.20a–b From Figures 6.1 and 6.2 in H. Kaplan and K. Hill, 1992, The evolutionary ecology of food acquisition, pp. 167–202 in *Evolutionary Ecology and Human Behavior,* ed. by E. A. Smith and B. Winterhalder, Aldine de Gruyter, Hawthorne, N.Y.

Chapter 19

19.1 English Heritage Photo Library/G. P. Darwin on behalf of Darwin Heirlooms Trust. 19.3 From Tables 2 and 6 in D. Buss, 1989, Sex differences in human mate preferences: Evolutionary hypotheses tested in 37 cultures, *Behavioral and Brain Sciences* 12:1–49. 19.4 From Figure 3.2 in M. Borgerhoff Mulder, 1988, Kipsigis bridewealth payments, pp. 65–82 in *Human Reproductive Behavior,* ed. by L. Betzig, M. Borgerhoff Mulder, and P. Turke, Cambridge University Press, Cambridge. 19.7 From N. L. Levine and J. B. Silk, 1997, When polyandry fails: Sources of instability in polyandrous marriages, *Current Anthropology* 38:375–398. 19.8a–b From

Figure 4.8 in M. Daly and M. Wilson, 1988, *Homicide,* Aldine de Gruyter, New York. 19.9 Stephen Dalton/Photo Researchers, Inc. 19.10 Adapted from Table III in J. B. Silk, 1980, Adoption in Oceania, *American Anthropologist* 82:799–820. 19.11 (c) Irven DeVore/Anthro-Photo. 19.14 From Figure 14.1 in N. Blurton Jones, 1989, The costs of children and the adaptive scheduling of births: Towards a sociobiological perspective on demography, pp. 265–282 in *The Sociobiology of Sexual and Reproductive Strategies,* ed. by A. E. Rasa, C. Vogel, and E. Voland, Chapman & Hall, London. 19.15 From Figure 14.2 in N. Blurton Jones, 1989, The costs of children and the adaptive scheduling of births: Towards a sociobiological perspective on demography, pp. 265–282 in *The Sociobiology of Sexual and Reproductive Strategies,* ed. by A. E. Rasa, C. Vogel, and E. Voland, Chapman & Hall, London.

Epilogue

1 By courtesy of the National Portrait Gallery, London.

Appendix

1 Adapted from Figure B.1.b–c in G. Conroy, 1990, *Primate Evolution,* W. W. Norton, New York.

Index